Methods in Molecular Biology™

Series Editor
John M. Walker
School of Life Sciences
University of Hertfordshire
Hatfield, Hertfordshire, AL10 9AB, UK

For further volumes:
http://www.springer.com/series/7651

Human Cell Culture Protocols

Third Edition

Edited by

Ragai R. Mitry

Institute of Liver Studies, King's College Hospital, London, UK

Robin D. Hughes

Institute of Liver Studies, King's College London School of Medicine, London, UK

Humana Press

Editors
Ragai R. Mitry, Ph.D
Institute of Liver Studies
King's College Hospital
London, UK
ragai.mitry@kcl.ac.uk

Robin D. Hughes, Ph.D
Institute of Liver Studies
King's College London
School of Medicine
London, UK
robin.hughes@kcl.ac.uk

ISSN 1064-3745 e-ISSN 1940-6029
ISBN 978-1-4939-6208-2 ISBN 978-1-61779-367-7 (eBook)
DOI 10.1007/978-1-61779-367-7
Springer New York Dordrecht Heidelberg London

Printed on acid-free paper

Humana Press is part of Springer Science+Business Media (www.springer.com)

Dedication

My sincere thanks to my mother Elizabeth, sister Antoinette, and brother Nabil for their love, encouragement, and support throughout my studies and career.

Ragai R. Mitry

Preface

Welcome to the 3rd edition of *Human Cell Culture Protocols*. The 1st edition was published in 1996 with Gareth Jones as editor and the 2nd edition in 2004 with Joanna Picot as editor. We hope our edition is a worthy addition to this series. Some might say, "Why do we need another book on human cell culture?" Human cell culture is not a new topic, but development of new molecular techniques and reagents which can be used to investigate cell function and the intracellular mechanisms responsible make it a continuing requirement. Human stem cells have become an important area for cell culture but are not described here as they are dealt with by other volumes. We describe protocols for isolation and culture of a range of primary cells from human tissues. It is impossible to include all cell types, and, for instance, we have not included neural cells or cancer cells. Of increasing importance, the techniques can be adapted to obtain cell cultures from diseased tissues to help understand the pathogenetic mechanisms involved.

A number of the chapters are new versions of protocols from authors involved in the previous editions of *Human Cell Culture Protocols*, who, with this, are bringing years of experience of working on particular cell types. There are chapters from authors new to the series on different cell types. For example, we have included pancreatic cells needed for basic studies on the pathogenesis of diabetes and also to their application for islet transplantation. There are three chapters on hepatocytes, which is the primary interest of the editors who work on hepatocyte transplantation. There are also protocols for skin cells, lung cells, parathyroid cells, gastric cells, renal cells, adipocytes, ovarian cells, bone cells, vascular smooth muscle cells, vascular endothelial cells, regulatory T cells, and blood mononuclear cells. There are chapters on new techniques being applied to human cell culture, particularly the use of biocompatible scaffolds to grow cells, the in vitro use of laser microdissection to isolate cells from culture, and automated cell culture.

The detailed protocols described in the different chapters make it possible for a worker with basic cell culture training to prepare cell cultures of the specific cell type. It is hoped that this volume will be of use to scientists in the fields of cell biology, gene therapy, and cell transplantation.

We would like to thank all the contributors for making this book possible.

London

Ragai R. Mitry
Robin D. Hughes

Contents

Contributors

ANKITA AGRAWAL • *The Mellanby Centre for Bone Research, Academic Unit of Bone Biology, The University of Sheffield, Sheffield, UK*

JEAN-FRANÇOIS BEAULIEU • *CIHR Team on Digestive Epithelium, Département d'anatomie et de biologie cellulaire, Faculté de médecine et des sciences de la santé, Université de Sherbrooke, Sherbrooke, QC, Canada*

MICHAEL BITZER • *Department of Gastroenterology & Hepatology, Medical University Hospital, Tuebingen, Germany*

PEYMAN BJÖRKLUND • *Department of Surgical Sciences, Uppsala University, Uppsala, Sweden*

DIMITRIOS P. BOGDANOS • *Institute of Liver Studies, King's College London School of Medicine, London, UK*

PETER BÜCHLER • *Department of General Surgery, TU Munich, MRI, Munich, Germany*

KIRSTIN A. CARSWELL • *Department of Surgery, King's College Hospital and King's College London, London, UK*

JOSÉ V. CASTELL • *Unidad de Hepatología Experimental, Centro de Investigación, Hospital La Fe, Valencia, Spain; CIBERehd: Centro de Investigación Biomédica en Red en el Área Temática de Enfermedades Hepáticas y Digestivas, Fondo de Investigación Sanitaria, Barcelona, Spain; Departamento de Bioquímica y Biología Molecular, Facultad de Medicina, Universidad de Valencia, Spain*

CATHY CENAYKO • *Institute of Liver Studies, King's College London School of Medicine, London, UK*

PIERRE CHAILLER • *CIHR Team on Digestive Epithelium, Département d'anatomie et de biologie cellulaire, Faculté de médecine et des sciences de la santé, Université de Sherbrooke, Sherbrooke, QC, Canada*

RICHARD R. H. COOMBS • *Department of Musculoskeletal Surgery, Imperial College London, Charing Cross Hospital, London, UK*

NICOLE DAUM • *Drug Delivery, Helmholtz Institute for Pharmaceutical Research Saarland, Helmholtz Center for Infection Research, Saarbruecken, Germany*

ANIL DHAWAN • *Institute of Liver Studies, King's College Hospital, London, UK*

JANE P. DILLON • *Bone and Joint Research Group, Department of Musculoskeletal Biology, Institute of Ageing and Chronic Disease, The University of Liverpool, Liverpool, UK*

MARK E. C. DOCKRELL • *SWT Institute for Renal Research, St. Helier Hospital, Carshalton, Surrey, UK*

MARÍA TERESA DONATO • *Unidad de Hepatología Experimental, Centro de Investigación, Hospital La Fe, Valencia, Spain; CIBERehd: Centro de Investigación Biomédica en Red en el Área Temática de Enfermedades Hepáticas y Digestivas, Fondo de Investigación Sanitaria, Barcelona, Spain; Departamento de Bioquímica y Biología Molecular, Facultad de Medicina, Universidad de Valencia, Spain*

CECILIA DRAGOMIR • *Laboratory for Cartilage Biology, The Hospital for Special Surgery, Weill Cornell Medical College, New York, NY, USA*
SABRINA EHNERT • *Department of Traumatology, Eberherd karls Universität Tübingen, Tübingen, Germany*
KARIM EL HACHEM • *Laboratory for Cartilage Biology, The Hospital for Special Surgery, Weill Cornell Medical College, New York, NY, USA*
MARTA FAVERO • *Laboratory for Cartilage Biology, The Hospital for Special Surgery, Weill Cornell Medical College, New York, NY, USA*
SUSAN K. FRIED • *Section of Endocrinology, Diabetes, and Nutrition, Department of Medicine, Boston University, Boston, MA, USA; Boston Obesity Nutrition Research Center, School of Medicine, Boston University, Boston, MA, USA*
MIZUHO FUKUNAGA-KALABIS • *Molecular and Cellular Oncogenesis Program, Melanoma Research Center, The Wistar Institute, Philadelphia, PA, USA*
JERZY F. GALAS • *Department of Endocrinology and Tissue Culture, Institute of Zoology, Jagiellonian University, Kraków, Poland*
JAMES A. GALLAGHER • *Bone and Joint Research Group, Department of Musculoskeletal Biology, Institute of Ageing and Chronic Disease, The University of Liverpool, Liverpool, UK*
ALISON GARTLAND • *The Mellanby Centre for Bone Research, The University of Sheffield, Sheffield, UK*
PATRICIO GODOY • *IFADO, Leibniz Research Centre for Working Environment and Human Factors, Dortmund University, Dortmund, Germany*
MARY B. GOLDRING • *Laboratory for Cartilage Biology, The Hospital for Special Surgery, Weill Cornell Medical College, New York, NY, USA*
MARÍA JOSÉ GÓMEZ-LECHÓN • *Unidad de Hepatología Experimental, Centro de Investigación, Hospital La Fe, Valencia, Spain; CIBERehd: Centro de Investigación Biomédica en Red en el Área Temática de Enfermedades Hepáticas y Digestivas, Fondo de Investigación Sanitaria, Barcelona, Spain*
AL-SHYMAA M. HAFEZ • *Department of Musculoskeletal Surgery, Imperial College London, Charing Cross Hospital, London, UK*
MOUSTAFA I. HAFEZ • *Department of Musculoskeletal Surgery, Imperial College London, Charing Cross Hospital, London, UK*
HANS HAUNER • *Else Kröner-Fresenius-Zentrum für Ernährungsmedizin, Technische Universität München, Freising-Weihenstephan, Germany*
STEPHANIE HEIN • *Biopharmaceutics and Pharmaceutical Technology, Saarbruecken, Germany*
PER HELLMAN • *Department of Surgical Sciences, Uppsala University, Uppsala, Sweden*
JAN HENGSTLER • *IFADO, Leibniz Research Centre for Working Environment and Human Factors, Dortmund University, Dortmund, Germany*
MEENHARD HERLYN • *Molecular and Cellular Oncogenesis Program, Melanoma Research Center, The Wistar Institute, Philadelphia, PA, USA*
GUO CAI HUANG • *Department of Diabetes and Endocrinology, King's College London School of Medicine, London, UK*
ROBIN D. HUGHES • *Institute of Liver Studies, King's College of Medicine, London, UK*
HANNO HUWER • *Cardiothoracic Surgery, Heart Center Voelklingen, Saarbruecken, Germany*

PETER M. JONES • *Diabetes Research Group, King's College London, London, UK*
JÖRG KLEEFF • *Department of General Surgery, TU Munich, MRI, Munich, Germany*
DANIEL KNOBELOCH • *Cytonet Gmbh, Weinheim, Germany*
ANNA KUEHN • *Drug Delivery, Helmholtz Institute for Pharmaceutical Research Saarland, Helmholtz Center for Infection Research, Saarbruecken, Germany*
AGUSTÍN LAHOZ • *Unidad de Hepatología Experimental, Centro de Investigación, Hospital La Fe, Valencia, Spain*
ULRICH M. LAUER • *Department of Gastroenterology & Hepatology, Medical University Hospital, Tuebingen, Germany*
CLAUS-MICHAEL LEHR • *Drug Delivery, Helmholtz Institute for Pharmaceutical Research Saarland, Helmholtz Center for Infection Research, Saarbruecken, Germany*
MI-JEONG LEE • *Division of Endocrinology, Diabetes, and Nutrition, Boston University, Boston, MA, USA; Boston Obesity Nutrition Research Center, School of Medicine, Boston University, Boston, MA, USA*
LING LI • *Molecular and Cellular Oncogenesis Program, Melanoma Research Center, The Wistar Institute, Philadelphia, PA, USA*
BO LIU • *Diabetes Research Group, King's College London, London, UK*
MARIA SERENA LONGHI • *Institute of Liver Studies, King's College Hospital, London, UK*
YUN MA • *Institute of Liver Studies, King's College Hospital, London, UK*
ATHANASIOS MAVROPOULOS • *Institute of Liver Studies, King's College London School of Medicine, London, UK*
IAN D. MCCARTHY • *Department of Musculoskeletal Surgery, Imperial College London, Charing Cross Hospital, London, UK*
DANIEL MÉNARD • *CIHR Team on Digestive Epithelium, Département d'anatomie et de biologie cellulaire, Faculté de médecine et des sciences de la santé, Université de Sherbrooke, Sherbrooke, QC, Canada*
RAGAI R. MITRY • *Institute of Liver Studies, King's College London School of Medicine, London, UK*
AHLAM MUSTAFA • *Institute of Liver Studies, King's College London School of Medicine, London, UK*
ANDREAS K. NUSSLER • *Department of Traumatology, Eberherd karls Universität Tübingen, Tübingen, Germany*
NATASCHA C. NUSSLER • *Department of General and Visceral Surgery, Klinikum Neuperlach, Munich, Germany*
MIGUEL OTERO • *Laboratory for Cartilage Biology, The Hospital for Special Surgery, Weill Cornell Medical College, New York, NY, USA*
WILLIAM R. OTTO • *Histopathology Laboratory, Cancer Research UK, London, UK*
SHANTA J. PERSAUD • *King's College London, London, UK*
CHRISTINA PHILIPPEOS • *Institute of Liver Studies, King's College Hospital, London, UK*
DARREN A. PLUMB • *Laboratory for Cartilage Biology, The Hospital for Special Surgery, Weill Cornell Medical College, New York, NY, USA*
DIANE PROUDFOOT • *Department of Medicine, University of Cambridge, Cambridge, UK*
ALBERTO QUAGLIA • *Institute of Liver Studies, King's College London School of Medicine, London, UK*
ELIZABETH RATCLIFFE • *Centre for Biological Engineering, Wolfson School of Mechanical and Manufacturing Engineering, Loughborough University, Leicestershire, UK*

Eirini I. Rigopoulou • *Department of Medicine, University of Thessaly Medical School, Mezourlo, Larissa, Greece; Institute of Biomedical Research and Technology (BIOMED), Center for Research and Technology – Thessaly (CE.RE.TE.TH), Larissa, Greece*
Robin M. H. Rumney • *The Mellanby Centre for Bone Research, The University of Sheffield, Sheffield, UK*
Anne Sandison • *Department of Musculoskeletal Surgery, Imperial College London, Charing Cross Hospital, London, UK*
Catherine E. Sarraf • *Department of Biomedical Science, University of Westminster, London, UK*
Ulrich F. Schaefer • *Biopharmaceutics and Pharmaceutical Technology, Saarbruecken, Germany*
Michael Schoenberg • *Department of Surgery, Rotkrentklinikum Münchern, Munich, Germany*
Lilianna Schyschka • *Department of Traumatology, MRI, Technische Universität München, Munich, Germany*
Catherine Shanahan • *Cardiovascular Division, King's College London, London, UK*
Claire C. Sharpe • *Renal Medicine, King's College London, London, UK*
Richard C. M. Siow • *Cardiovascular Division, King's College London School of Medicine, London, UK*
Thomas Skurk • *Else Kröner-Fresenius-Zentrum für Ernährungsmedizin, Technische Universität München, Freising-Weihenstephan, Germany*
Daniel Smyk • *Institute of Liver Studies, King's College London School of Medicine, London, UK*
Keith N. Stewart • *School of Medicine & Dentistry, University of Aberdeen, Aberdeen, Scotland, UK*
Haibin Su • *Department of Liver Transplantation, Beijing 302 Hospital, Beijing, China*
Wolfgang E. Thasler • *Department of Surgery, Ludwigs-Maximillian Universität Münchern, Munich, Germany*
Robert Thomas • *Centre for Biological Engineering, Wolfson School of Mechanical and Manufacturing Engineering, Loughborough University, Leicestershire, UK*
Diego Vergani • *Institute of Liver Studies, King's College Hospital, London, UK*
Pengyun Wang • *Institute of Liver Studies, King's College Hospital, London, UK*
Timo Weiland • *Department of Gastroenterology & Hepatology, Medical University Hospital, Tuebingen, Germany*
Heather M. Wilson • *Division of Applied Medicine, School of Medicine & Dentistry, University of Aberdeen, Aberdeen, Scotland, UK*
Min Zhao • *Department of Diabetes and Endocrinology, King's College London – School of Medicine, London, UK*
Martina Zimmermann • *Department of Gastroenterology & Hepatology, Medical University Hospital, Tuebingen, Germany*
Ko Hashimoto • *Laboratory for Cartilage Biology, The Hospital for Special Surgery, Weill Cornell Medical College, New York, NY, USA*

Chapter 1

Introduction to Cell Culture

Christina Philippeos, Robin D. Hughes, Anil Dhawan, and Ragai R. Mitry

Abstract

The basics of cell culture as applied to human cells are discussed. Biosafety when working with human tissue, which is often pathogenic, is important. The requirements for a tissue culture laboratory are described, particularly the range of equipment needed to carry out cell isolation, purification, and culture. Steps must be taken to maintain aseptic conditions to prevent contamination of cultures with micro-organisms. Basic cell-handling techniques are discussed, including choice of media, primary culture, and cryopreservation of cells so they can be stored for future use. Common assays which are used to determine cell viability and activity are considered.

Key words: Biosafety, Tissue culture laboratory, Aseptic technique, Cell viability

1. Background

The culture of human cells is a widely used technique that allows the study of human metabolism and physiology that is not readily possible in vivo. The alternatives are to use tissue slices and biopsies, which, although retaining the in vivo structure, usually have to be used immediately because of their short-term cell viability. With cells, if they can be isolated from the tissue, there is the possibility of establishing cell cultures for a period of days to weeks. Cells can be obtained from normal tissue (e.g. skin biopsy) if clinical procedures and ethical considerations allow and also from diseased tissue (e.g. liver tumour biopsy) removed during surgery as part of the patients' treatment for their disease. With cell culture, the behaviour of individual cell types without the influence of systemic variations that might arise during normal homeostasis in vivo can be investigated. Cell culture is usually in the form of a suspension of

Ragai R. Mitry and Robin D. Hughes (eds.), *Human Cell Culture Protocols*, Methods in Molecular Biology, vol. 806,
DOI 10.1007/978-1-61779-367-7_1, © Springer Science+Business Media, LLC 2012

dispersed cells taken from original tissue (by enzymatic, mechanical, or chemical dissociation), a primary culture, or a cell line and is carried out under strict laboratory conditions of sterility and a controlled environment involving temperature, gases, and pressure. It should mimic the in vivo environment successfully such that the cells are capable of survival and proliferation in a controlled manner. Although human cell culture has been performed for many years and despite the rapid advancement in the techniques of molecular biology and highly sensitive analytical techniques and assays which now can be used, gives an on-going need for culture of cells from all organs/tissues of the human body.

1.1. Advantages of Cell Culture

Cell culture allows for the control of the physiochemical environment (pH, temperature, osmotic pressure, and O_2 and CO_2 tension). This provides many advantages as cytology and immunostaining can be easily performed, quantitation is straightforward, and experiments can be performed with reduced volumes which lowers the costs. Cultures may be exposed to a reagent at a low and defined concentration with direct access to the cell (1).

Tissue samples are regularly heterogeneous; however, after one or two passages, cultured cell lines assume a uniform constitution, as the cells are randomly mixed at each transfer and the selective pressure of the culture conditions tends to produce a homogeneous culture of the most vigorous cell type. This, therefore, allows for experimental replicates to be very similar, reducing statistical variance.

1.2. Limitations

Culture techniques need to be carried out under strict aseptic conditions, as common contaminants (e.g. bacteria, moulds, and yeast) grow more rapidly than mammalian cells. There needs to be strict environmental control as cells from multicellular animals do not normally exist in isolation and consequently are not able to sustain an independent existence without the necessary support. Other limitations include the high cost of consumables and media, cross-contamination, and dedifferentiation and selection. Dedifferentiation is the overgrowth of undifferentiated cells that produces a loss of the phenotypic characteristics typical of the tissue from which the cells have been isolated. This can be overcome under the right conditions with selective media.

1.3. Biosafety

When working with human tissue samples, understanding possible risks and a clear concept of biological safety are essential to prevent occupational acquired infections as well as the release of pathogens into the environment. There are four categories of biosafety ranging from low to high risk (Table 1).

Biohazard materials include (1) human pathogens (bacteria, fungi, viruses, parasites, prions); (2) all human blood products, tissues, and body fluids; (3) cultured cells; (4) toxins; and (5) infected tissues.

Table 1
Categories of biosafety and risk levels

Safety level	Risk	Description	Safety requirements
Biosafety Level 1	Low	Laboratories appropriate for training and teaching Performed with defined and characterised strains of viable micro-organisms not known to cause any disease in healthy adult humans	Sink for hand washing
Biosafety Level 2	Moderate	Laboratories appropriate for diagnostics and teaching Performed with agents that are associated with human diseases (micro-organisms like Hep. B, HIV, most bacteria) as well as human body fluids, tissues, and primary human cell lines	Primary barriers: Face protection, gowns, gloves, Biosafety Class II cabinet Secondary barriers: Sinks for hand washing, waste decontamination facilities
Biosafety Level 3	Moderate–High	Laboratories appropriate for diagnostics, teaching, research, or production facilities Performed with exotic agents with a potential of respiratory transmission which may cause serious and potentially lethal infections (Myc. Tub. Cox. Burnetti)	Primary barriers: Aerosol-tight chamber for work Secondary barriers: Controlled access to the laboratory
Biosafety Level 4	High	Laboratory appropriate for research Performed with dangerous and exotic agents that pose a high individual risk of life-threatening disease, which is transmitted via the aerosol route and for which there is no vaccine or therapy available (*Marburg virus*, *Ebola virus*)	Primary barriers: complete full-body air-supplied, positive pressure personal suit (Biosafety Class III cabinets) Secondary barriers: Complete isolated zone in a separate building

Recommended work practices, such as the use of appropriate labware, aseptic culture techniques, maintaining a clean working environment, as well as general hygiene (e.g. hand washing, tying back long hair, not eating, drinking, or smoking in the laboratory), are necessary to ensure safety in the lab.

A number of chemicals used in the laboratory are hazardous. All manufacturers of hazardous materials are required to supply the user with pertinent information on any hazards associated with their chemicals. This information is supplied in the form of Material Safety Data Sheets (MSDSs). The MSDS contains the chemical name, CAS#, health hazard data including first aid treatment, physical data, fire and explosion hazard data, reactivity data, spill or leak procedures, and any special precautions needed when handling this chemical.

2. Tissue Culture Laboratory: Design and Layout

The most important factor that distinguishes a tissue culture laboratory/room from a general laboratory is the need to maintain sterility. Although it is not economically viable to create large aseptic areas, it is important that it is dust free and has no through traffic. The room should be designed to allow for easy cleaning and supplied with filtered air to usual industrial or office standards. Six main functions are performed in a tissue culture laboratory and need to be accommodated for sterile handling, incubation, preparation, wash-up, sterilisation, and storage. If only one room is available, then it should be designed in such a way that the clean area for sterile handling is located at one end of the room, furthest from the door, and the wash-up and sterilisation facilities placed at the other end, with preparation, storage, and incubation in between (Fig. 1).

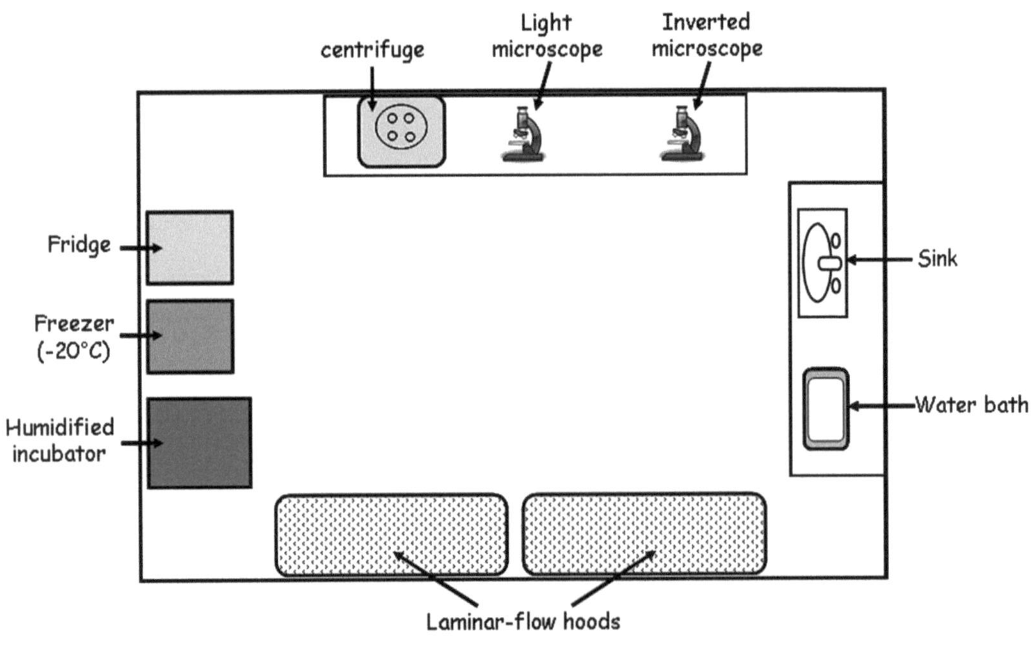

Fig. 1. A typical layout of a tissue culture laboratory.

Table 2
A list of main equipment required for a tissue culture laboratory

Basic equipment requirements	Non-essential, but beneficial
Biohazard laminar flow hood	Peristaltic pump
Humid CO_2 incubator with CO_2 cylinders	pH meter
Water bath	Cell counter
Inverted microscope	
Bench centrifuge	
Refrigerator	
Freezer (for −20°C storage)	
Autoclave	
Drying oven	

2.1. Equipment

Every cell culture laboratory has essential basic equipment requirements as well as non-essential, but beneficial, equipment. These are summarised in Table 2.

2.1.1. Laminar Flow Hood

A laminar flow hood contains a high-efficiency, bacteria-retentive filter and is widely used in cell culture applications to avoid contamination. Room air is taken into the unit and passed through a pre-filter to remove gross contaminants (lint, dust, etc.). The air is then compressed and channelled up behind and through the high-efficiency particulate air (HEPA) filter so that the purified air flows out over the entire work surface in parallel lines at a uniform velocity (laminar flow). The HEPA filter removes nearly all of the bacteria from the air. Such hoods exist in both horizontal and vertical configurations (Fig. 2), and there are many different types of cabinets with a variety of airflow patterns and acceptable uses. There should be sufficient working space as well as clearance space for ventilation and servicing the hood. A Class II or Class III biohazard cabinet should be used for potentially hazardous materials; these include any primate (including human) cell lines, virus-producing cultures, radioisotopes, and carcinogenic or toxic drugs. Laminar flow cabinets may have a UV-C germicidal lamp to sterilize the shell and contents when not in use. This light should be turned off during use to ensure personnel safety. Prevention of cross-contamination is ensured more by good aseptic technique than by any action of the cabinet itself. The actions of the operator must always complement the operation of the cabinet.

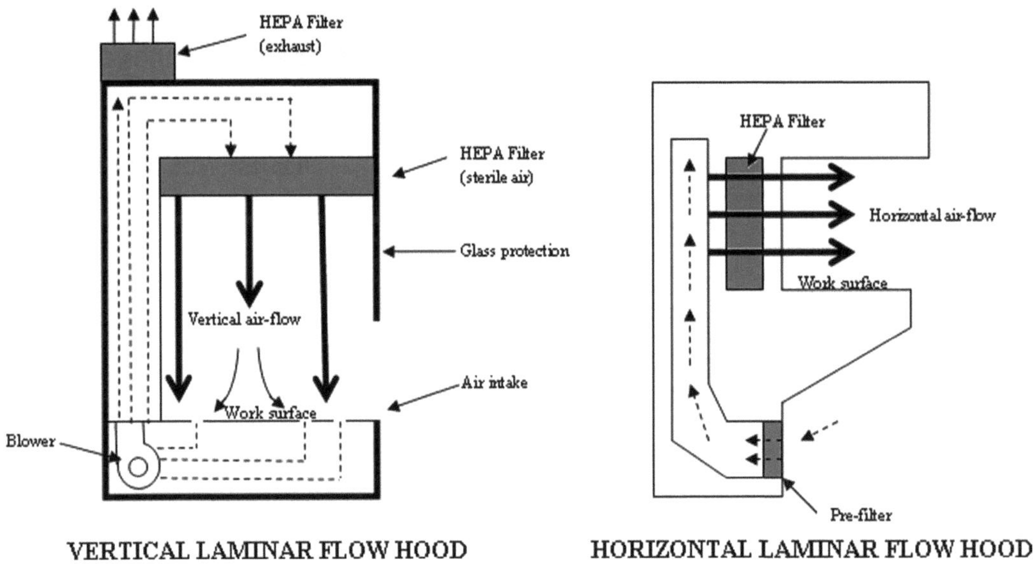

Fig. 2. A schematic diagram showing laminar flow hood configurations (*left*: vertical; *right*: horizontal;).

2.1.2. Incubator

A carbon dioxide (CO_2)-humidified incubator operates on fairly simple parameters based on three elements which are CO_2, temperature, and relative humidity (RH). This allows the mammalian environment (in vivo) outside of its natural state (in vitro) to be replicated. The incubator combines three elements that create an environment needed for cells to thrive by establishing a balanced and controlled pH at 7.2–7.4: stable temperature at 37°C, high RH at 95%, and controlled CO_2 level at 5%. CO_2 incubators have become acceptable, reliable equipment given the growth in cell culture research and the ability of incubators to grow cells in vitro. But on the flip side of this benefit is the constant threat of contamination to the cell culture environment. The incubator should be resistant to corrosion and easily cleaned. Humidified CO_2 incubators, by their very nature, are a haven for contamination since they create an environment for the support and growth of bacteria, mould, and spores. CO_2 incubators incorporate copper in their cabinet design to deter contamination. As the copper breaks down, it releases copper oxide, which destroys microbes present in the chamber. Frequent cleaning of incubators, especially humidified ones, is essential to prevent contamination. One main source of contamination in humidified incubators is the humidity pan itself. Unless this is cleaned regularly, mould, bacteria, and yeast can settle in comfortably.

2.1.3. Centrifuge

A centrifuge is an essential piece of equipment for the laboratory, and often a significant investment. With so many recent advances in both science and technology, there is a wide range of centrifuges to choose from that should be tailored to your laboratory needs. Key factors that need to be taken into account when purchasing a centrifuge are:

- Applications and protocols the centrifuge is used to support; does the centrifuge need to be versatile to a range of applications?
- The maximum and minimum *g*-force and volume requirements
- The expected activity level of the machine, i.e. number of users, number of samples required to be processed in a day, etc.
- The types of sample formats used, i.e. microplates, blood collection tubes, disposable conical tubes
- Laboratory space (which model to be chosen, i.e. floor model or benchtop)
- Types of rotors required, i.e. swinging bucket or fixed angle
- Budget

A small refrigerated bench-top centrifuge, preferably with proportionally controlled braking, is sufficient for most purposes. Cells sediment reasonably at 80–100 g, and higher gravity may cause damage and promote agglutination of the pellet.

2.1.4. Inverted Microscope

A simple inverted microscope is essential for any tissue culture laboratory as looking at cultures regularly ensures that early detection of the deterioration of a culture is recognised and/or microbiological infection is identified. Culture flasks fit with ease onto the stage as its light source and condenser are on the top, above the stage pointing down, while the objectives and turret are below the stage pointing up. This allows for cultures to be viewed under more natural conditions rather than as a sample on a glass slide, and allows the researcher to observe the culture over a period of time. If the photographing of cultures is required, microscopes with high-quality optics, and a long-distance, phase-contrast condenser and objectives, with provisions for a camera are available.

2.1.5. Autoclave

Sterilisation can be achieved using a number of techniques: irradiation (UV and other), filtration, chemical, etc.; however, the most widely used sterilisers are autoclaves as these eradicate microorganisms by the use of pressurized steam causing thermal stress. To ensure that items have been adequately sterilised, it is necessary to raise the temperature such that even the most thermo-tolerant are inactivated, hence the most typical programme being 121°C for 15–20 min.

Modern autoclaves have evolved considerably from the days when large "pressure cookers" were widely used due to the increased requirement to meet various laboratory safety and quality standards. This has resulted in a whole range of autoclave alternatives, and choosing one depends on finances as well as the primary purpose for which the autoclave is used. Ensure that your autoclave can cope with the volume and type of material you need to put through it, as placing too much material in an autoclave does not allow sufficient room for thorough steam circulation and complete sterilisation may not be achieved.

There are physical, chemical, and biological indicators that can be used to ensure that an autoclave reaches the correct temperature for the correct amount of time. Medical packaging and autoclave tape contain a chemical that changes colour once the correct conditions have been met. Biological indicators contain spores of a heat-resistant bacterium, *Geobacillus stearothermophilus*. If the autoclave does not reach the right temperature, when incubated the spores will germinate, and their metabolism will change the colour of a pH-sensitive chemical. Some physical indicators consist of an alloy designed to melt only after being subjected to the relevant holding time. If the alloy melts, the change will be visible. These changes indicate that the object inside the package or under the tape has been processed.

3. Aseptic Technique

Contamination by micro-organisms is a major problem in cell culture, and proper aseptic technique is required to eliminate this from occurring. Bacteria, yeast, fungal spores, and mycoplasma may be introduced through several sources; however, the biggest contributors are usually from the operator, atmosphere, work surfaces, and solutions. All materials that come into direct contact with the culture must be sterile to prevent micro-organisms from entering the flask or dish.

Tips for maintaining aseptic conditions:

- Always maintain good personal hygiene. Hand washing removes any excess dry skin that could otherwise be blown onto the culture. Regularly spray your gloves with 70% alcohol.
- Frequently swab the work surface with 70% alcohol, and wipe up any spills immediately.
- Always ensure that bottles and tubes are swabbed with 70% alcohol after being placed in the water bath and before entering the laminar flow hood.
- Flasks, plates, and dishes from humid incubators should always be swabbed with 70% alcohol before being placed in the hood, as they are particularly prone to outside contamination.
- Only bring items you require for the particular procedure into the laminar flow hood.

- Arrange the work area so that there is easy access to all items without having to reach over one to get to another. Also ensure that there is a large clear open space in the centre to work.
- Do not stick pipettes into solution bottles directly, rather decant them into a disposable sterile tube or bottle for use during the procedure, and discard the remainder after use.
- Do not share bottles of media among cell lines.
- Never leave bottles uncapped in the flow hood.
- When pouring from one bottle to another, ensure that bottles do not touch and that solution is not spilled as a bridge of liquid between the outside of the bottle and the inside may permit contamination to enter the bottle.

4. Cell-Handling Basic Techniques

4.1. Media

The choice of medium is very important and depends on the type and use of cells. This is usually made by what was previously being used by others in the literature for the same cells. Currently, there are two major paths: cells that are isolated and propagated for a specific lineage or cells that are used as substrates for the formation of products, as a host for viral propagation or for non-cell-specific molecular studies. The choice of media for continuous cell lines is not critical as long as conditions are constant, but for specialised cell types, primary cultures, and growth in the absence of serum, this choice is important (2). Production techniques use more selective media, usually serum-free, while viral and molecular work relies mainly on Eagle's minimal essential medium (EMEM) (3), Dulbecco's modified Eagle's medium (DMEM) (4), or Roswell Park Memorial Institute (RPMI) medium 1640 (5). EMEM is one of the most widely used of all synthetic cell culture media as a wide variety of cells grown in monolayers can be cultivated in it. The formulation has been further modified by optional elimination of calcium to permit the growth of cells in suspension. EMEM's amino acid concentrations mimic the protein composition of mammalian cells, and optional supplementation of non-essential amino acids broadens the medium's usefulness. Many modifications of Eagle's medium have been developed since the original formulation appeared in the literature. Among the most widely used of these modifications is Dulbecco's modification. DMEM contains a higher concentration of vitamins, amino acids, as well as additional supplementary components. Commonly used media are available commercially and do not require sterilisation; however, special formulations or any additions that require preparation need to be sterilised. Although organisms need to be eradicated, the growth characteristics of the media should not be compromised by the sterilisation cycle. Some components are thermolabile and suffer from a degree of heat degradation during the autoclaving/sterilisation process, possibly leading to reduced performance. In general, stable solutions (water, salts, and media supplements, such as tryptone or peptone) may be autoclaved at 121°C (100 kPa or 1 atm above ambient) for 20 min while labile solutions (media, trypsin, and serum) must be filter sterilised through a 0.2 μm porosity membrane filter (Millipore, Sartorius, Gelman, Pall) (2). Additional supplements that are usually added to media to improve cell growth and functionality are hormones, growth factors, glucose, serum, and antibiotics (to reduce contamination). The most common additive is serum as it contains growth factors, which promote cell proliferation, minerals, lipids, and hormones, and adhesion factors and antitrypsin activity, which promote cell attachment. The most used sera in tissue culture are calf, foetal bovine, horse, and human sera, although these need to be screened for viruses, such as HIV and hepatitis B.

4.2. Cells

4.2.1. Primary Culture

Cells that are cultured directly from a tissue are known as primary cells. A primary culture may be produced either by allowing cells to migrate out from the tissue after sterile dissection, which is adhering to a substrate, or by disaggregating the tissue mechanically or enzymatically to produce a suspension of cells. These cells are, therefore, more representative of the cell types in the tissue from which they were isolated from, although many cells are unable to attach and survive in vitro. Most primary cultures are usually heterogeneous, have a low growth fraction, and have a limited lifespan, with the exception of some derived from tumours. Cells that have attached are trypsinised and reseeded in a fresh flask and become a secondary culture.

4.2.2. Subculture

Once cells become confluent, it is necessary to transfer and dilute them into a second vessel by dissociating the cells of the monolayer in trypsin or diluting cell suspensions. This is done by rinsing the monolayer with PBS, removing the rinse and adding trypsin–EDTA onto the monolayer in the flask, and incubating at 37°C for 2–5 min. Cells are then centrifuged and re-suspended in medium, counted, and reseeded. Usually, cells are subcultured at 1:3 or higher ratio (dependent on the doubling time and growth rate of the cell line used) for up to several passages (sometimes up to 40) before discarding cell line. These lines are known as finite cell lines, as they have a limited number of passages determined by the number of doublings that the cell population can go through before it stops growing due to senescence.

4.2.3. Growth Curve

The growth characteristics of a cell line can be determined by the generation of a growth curve. This is constructed from samples taken at intervals throughout the growth cycle. When cells are initially seeded, they enter a latent period of no growth, called the *lag period*. This period usually lasts between 2 and 24 h and allows the cells to recover from trypsinisation, reconstruct their cytoskeleton, and secrete attachment molecules, enabling them to re-enter the cell cycle. Once the cells have readjusted, they enter a period of exponential growth, known as the *log phase*, during which the cell population doubles over a definable period, known as the *doubling time*. When the cells have reached confluency, they then enter the *plateau* or *stationary phase*, where the growth fraction drops to close to zero. Cells may be subcultured from plateau, but it is preferable to subculture before plateau is reached, as the growth fraction will be higher and the recovery time (lag period) will be shorter if the cells are harvested from the top end of the log phase (6). These phases are characteristic for each cell line and are important when designing routine subculture and experimental protocols, as cell biochemistry changes significantly during each phase.

4.2.4. Contamination

Microbial contamination is one of the biggest drawbacks when culturing cells, although it has been greatly reduced by the use of aseptic technique, laminar flow hoods, and antibiotics. Contamination is easily detected by rapid changes in pH (i.e. changes of colour of the medium), cloudiness in the medium, extracellular granularity under the microscope, or any unidentified particles growing in medium. One of the most serious contaminations is mycoplasma, which can affect cellular biochemistry, antigenicity, and cell growth, and is not detected by routine microscopy. It can be introduced through media, sera, trypsin, or the operator. Any cell culture laboratory should have a mycoplasma screening programme in operation (usually tested every 1–3 months), especially those collecting tissue for primary culture. If a culture is contaminated, it should be discarded immediately before it can spread the contamination to other cultures. The incubator and laminar flow hoods should be cleaned and decontaminated to maintain sterility.

Another major risk in a tissue culture laboratory is cross-contamination and its severity is often underestimated (7). To ensure that cross-contamination does not occur, do not share bottles of media or reagents among cell lines or operators and do not return a pipette that has been in a vessel with cells.

4.2.5. Cryopreservation

Cells can be preserved for many years by freezing if the cell line can be expanded sufficiently. This allows secure stocks to be maintained without aging and protects them from problems of contamination, incubator failure, or medium and serum crises. The survival of the cells after freezing and thawing varies depending on the type of cell and its ability to withstand a variety of stresses imposed by the physical and physico-chemical changes occurring in the freezing medium during cooling to and warming from the storage temperature (8).

There are many factors that determine the success of cell survival; however, the most important aspects are:

1. Type and concentration of cryoprotectants (an additive, such as glycerol or dimethyl sulphoxide, that can protect cells against freezing injury)
2. Cell density in cryopreservation solution at the time of freezing
3. Cooling and thawing rates of cell suspension
4. Dilution rate of thawed cell suspension

Cryoprotectants can have detrimental toxic and osmotic effects on cells, and therefore critical points that determine cell survival occur when cryoprotectants are added to the bathing medium before freezing and, in particular, the procedure used to return the cells to a medium without cryoprotectants after thawing.

There are differences of opinion regarding some of the conditions for freezing and thawing; however, the general cryopreservation protocol is as follows. Cultures to be cryopreserved should be healthy, free from contamination, and should be maintained in log-phase growth for several days before freezing. Ideally, cells and freezing medium should be chilled to diminish the toxic effects of the preservative and 1×10^6 to 1×10^7 cells/ml should be frozen in aliquots, with slow cooling, 1°C/min, down to −70°C, and then rapid transfer to a liquid nitrogen freezer. Thawing should occur rapidly followed by slow dilution of tenfold to dilute out the cryoprotectant (rapid dilution results in osmotic shock). If cells are sensitive, then the preservative must be removed by centrifugation following the dilution and resuspended in fresh medium. However, this also results in cell loss as centrifugation itself may be damaging to freshly thawed cells. If cells are more tolerant, it is best to allow them to attach overnight and change the medium the next day to remove the preservative. Follow protocols stated in the literature for the particular cell line used to ensure the best cryopreservation method for your cells.

4.2.6. Cell Viability and Activity Assays

There are several "easy"-to-carry-out assays/tests to check cell viability and activity. The following list gives some examples of such assays/tests, and the chemicals used could be purchased from many suppliers, e.g. Sigma–Aldrich.

- Cell attachment: Sulforhodamine B (SRB) dye. It is a 96-well plate culture-based assay, which requires a simple staining procedure and a plate reader.
- Overall activity: MTT using 96-well plate culture. The assay is based on mitochondrial dehydrogenases activity, and requires a plate reader.
- Cell viability: Fluorescein diacetate/ethidium bromide (FDA/EtBr) staining; viable cells convert FDA to fluorescein (bright

green), and dead cells have bright red nuclei only due to EtBr staining. Requires a fluorescent microscope.

- Apoptosis: Propidium iodide staining of cell nuclei (whole or fragmented). Analysis requires the use of flow cytometry in order to obtain the percentage of cells in each stage of the cell cycle, including sub-G1 (reflects apoptosis). However, there more recently developed sensitive assays, such as the M30 CytoDeath™ ELISA (PEVIVA AB, Bromma, Sweden), which offer a unique possibility to quantify apoptosis in, e.g. epithelially derived cells.
- Protein synthesis: [^{14}C]leucine (radioactive isotope; could be purchased from, e.g. PerkinElmer) incorporation assay using 96-well culture. Requires a cell harvester and β-counter.

References

1. Freshney, R.I., *Culture of Animal Cells: A Manual of Basic Technique.* Fourth ed. 2000, New York: Wiley-Liss, Inc.
2. Freshney, R.I., *Introduction to basic principles.* Second ed. Animal Cell Culture: A Practical Approach, ed. R.I. Freshney. 1994, Oxford: Oxford University Press. 1–14.
3. Eagle, H., *Amino Acid Metabolism in Mammalian Cell Cultures.* Science, 1959. **130**: p. 432–437.
4. Dulbecco, R. and G. Freeman, *Plaque production by the polyoma virus.* Virology, 1959. **8**: p. 396–397.
5. Moore, G.E., R.E. Gerner, and H.A. Franklin, *Culture of Normal Human Leukocytes.* Journal of the American Medical Association, 1967. **199**: p. 519–524.
6. Freshney, R.I., *Basic Principles of Cell Culture.* Culture of Cells for Tissue Engineering, ed. G. Vunjak-Novakovic and R.I. Freshney. 2006, New York: John Wiley & Sons, Inc. 3–22.
7. Marcovic, O. and N. Marcovic, *Cell cross-contamination in cell cultures: the silent and neglected danger.* In Vitro Cellular & Developmental Biology, 1998. **34**: p. 108.
8. Armitage, W.J., *Cryopreservation of animal cells.* Symposia of the Society for Experimental Biology, 1987. **41**: p. 379–393.

Chapter 2

Isolation and Cultivation of Dermal Stem Cells that Differentiate into Functional Epidermal Melanocytes

Ling Li, Mizuho Fukunaga-Kalabis, and Meenhard Herlyn

Abstract

Human melanocytes have been extensively studied, but a melanocyte stem cell reservoir in glabrous skin has not yet been found. Human dermis contains cells that are nonpigmented but can differentiate to several different cell types. We have recently shown that multipotent dermal stem cells isolated from human neonatal foreskins are able to differentiate to multiple cell lineages, including pigmented melanocytes. The dermal stem cells grow as three-dimensional spheres in human embryonic stem cell medium and express some neural crest stem cell and embryonic stem cell markers. Melanocytes derived from dermal stem cells express melanocytic markers and act the same way as mature epidermal melanocytes. Dermal spheres, embedded in the reconstructed dermis consisting of collagen with fibroblasts, can migrate to the basement membrane, where they become pigmented in the same way as epidermal melanocytes suggesting that dermal stem cells can give rise to epidermal melanocytes.

Key words: Melanocyte, Stem cells, Dermal reservoir, 3D skin reconstruct

1. Introduction

Melanocytes originate from the neural crest in vertebrate development, and undergo a complex process of fate specification, proliferation, survival, and differentiation, before finally homing in the epidermis, where they contact with surrounding keratinocytes (1). Different techniques of the isolation and subsequent culture of human epidermal melanocytes had been attempted (2–5) until Eisinger and Marko established the method to grow homogeneous human melanocytes to yield cells in sufficient quantity for biological, biochemical, and molecular analyses (6). Since then, epidermal melanocytes have been extensively studied. Recently, the presence of a stem cell niche for melanocytes has been shown in hair follicles

Ragai R. Mitry and Robin D. Hughes (eds.), *Human Cell Culture Protocols*, Methods in Molecular Biology, vol. 806,
DOI 10.1007/978-1-61779-367-7_2, © Springer Science+Business Media, LLC 2012

of mouse and human skin (7, 8). Glabrous skins are also abundant in melanocytes; however, no obvious, spatially restricted stem cell niche for melanocytes has been found in such areas. We have recently shown that multipotent dermal stem cells isolated from human foreskins lacking hair follicles are able to home to the epidermis to differentiate into melanocytes (9). These dermal stem cells grow as three-dimensional spheres in human embryonic stem cell medium, self-renew, and express a neural crest marker NGFRp75 and an embryonic stem cell marker OCT4, but not melanocytic markers, such as HMB45 or S100. In addition, these cells are able to differentiate into multiple neural crest-derived cell types, including melanocytes. In a three-dimensional skin reconstruct model, the dermal stem cells are differentiated into HMB45/E-cadherin-positive melanocytes, which migrated from the dermis to the epidermis and reside singly among the basal-layer keratinocytes. In this chapter, we outline our procedures for the isolation of human dermal stem cells and differentiation toward mature functional melanocytes that home to the epidermis in skin reconstructs, produce pigment, and express the appropriate melanocytic markers MITF, DCT, TYRP1, S100, HMB45, and E-cadherin.

2. Materials

2.1. Reagent Setup

1. Normal skin-transporting medium: The medium for collecting normal skin is composed of Hanks' balanced salt solution without Ca^{++} and Mg^{++} (HBSS; Gibco-BRL Grand Island, NY) supplemented with penicillin (100 U/mL; USB Cleveland, OH), streptomycin (100 μg/mL; USB), gentamicin (100 μg/mL; Bio Wittaker Walkersville, MD), and Fungizone (0.25 μg/mL; JRH Biosciences, Lenexa, KS). After sterilization through a 0.2-μm filter, the skin-transporting medium is transferred into sterile containers in 20-mL aliquots and stored at 4°C for up to 1 month.
2. Epidermal isolation solution: Dissolve 0.48 g of dispase (grade II, 0.5 U/mg; Boehringer Mannheim, Indianapolis, NJ) in 100 mL of phosphate-buffered saline (PBS) without Ca^{++} and Mg^{++} (Cellgro by Mediatech, Herndon, VA) containing 0.1% bovine serum albumin (BSA) (fraction V; Sigma, St. Louis, MO) to yield a final dispase activity of 2.4 U/mL. Sterilize the enzyme solution through a 0.2-μm filter, aliquot into a 5-mL tube, and store at –20°C for up to 3 months.
3. Dermal isolation solution: Collagenase type IV (Invitrogen, Grand Island, NY) 100 mg is dissolved in 100 mL of Dulbecco's modified Eagle's medium (DMEM) (Cellgro) to yield a final concentration 1 mg/mL. Sterilize the enzyme

solution through a 0.2-μm filter, aliquot into a 5-mL tube, and store at −20°C for up to 3 months.

4. Mouse embryonic fibroblast (MEF) derivation medium: Combine 870 mL DMEM (Invitrogen), 100 mL defined FBS (Invitrogen; heat inactivate for 30 min at 57°C), 10 mL 200 mM L-glutamine (Invitrogen – keep frozen, thaw immediately before use), 10 mL nonessential amino acids 100× (Invitrogen), and10 mL penicillin–streptomycin 100×.
5. MEF growth medium: Combine 880 mL DMEM (Invitrogen), 100 mL defined FBS (Invitrogen; heat inactive for 30 min at 57°C), 10 mL 200 mM L-glutamine (Invitrogen – keep frozen, thaw immediately before use), and 10 mL nonessential amino acids 100× (Invitrogen).
6. Human embryonic stem cell medium (HES): Combine 800 mL DMEM/F-12 (Invitrogen), 200 mL Knockout-Serum Replacer (Invitrogen), 10 mL 100 mM L-glutamine + β-mercaptoethanol (Invitrogen – add 7 μL β-mercaptoethanol to 10 mL), 10 mL nonessential amino acids 100× (Invitrogen), and 1 mL basic fibroblast growth factor [bFGF; Fitzgerald Industries – dissolve in 0.1% BSA/1× Dulbecco's PBS (DPBS; Cellgro, Manassas, VA) at a concentration of 4 μg/mL and store at −70°C].
7. Human embryonic stem cell medium 4 (HESCM4): Mix 700 mL MEF-conditioned HES medium and 300 mL HES medium, sterilize through a 0.2-μm filter.
8. L-Wnt3a cell medium: Add 100 mL FBS and 8 μL of 50 mg/mL G418 (Sigma) to 900 mL DMEM (Cellgro).
9. L-Wnt3a conditioning medium: Add 10 mL FBS to 990 mL DMEM (Cellgro).
10. Melanocyte differentiation medium (Mel-1): Combine 30 mL DMEM-Low Glucose (Invitrogen), 20 mL MCDB201, and 50 mL Wnt3a conditioned medium, add 20 μL Dexamethasone (Sigma – dissolve in ddH$_2$O at a concentration of 0.25 M and store at −20°C), 1 mL ITS Liquid Medium Supplement (Sigma), 1 mL Linoleic Acid-BSA (Sigma), 1 mL L-ascorbic acid (Sigma), 1-mL stem cell factor (SCF; Fitzgerald Industries, Concord, MA – dissolve in 0.1% BSA/1× DPBS at a concentration of 10 μg/mL and store at −70°C), 100 μL 4 μg/mL bFGF (Fitzgerald Industries), 100 μL endothelin-3 (ET-3; American Peptide Co., Sunnyvale, CA – dissolve in 0.1% BSA/1× DPBS at a concentration of 264 μg/mL and store at −70°C), 150 μL Cholera toxin (Sigma – dissolve in 0.1% BSA/1× DPBS at a concentration of 3.32 μg/mL and store at 4°C), and 12.5 μL TPA [Sigma – dissolve in dimethyl sulfoxide (DMSO); Fisher Scientific, Fair Lawn, NJ] at a concentration of 250 μg/mL and store at −20°C.

11. Skin reconstruct medium preparation: Basic medium (500 mL): Add the following reagents to 490 mL keratinocyte serum-free medium (Invitrogen), 1.8 mL bovine pituitary extract (BPE; Invitrogen 1×), 10 mL dialyzed fetal bovine serum (dFBS) (Hyclone, Logan, UT), 500 μL of 10 μg/mL SCF (Fitzgerald Industries), 125 μL 4 μg/mL bFGF (Fitzgerald Industries), and 500 μL of 264 μg/mL ET-3 (American Peptide Co.). Medium I: To 100 mL basic medium, add 10 μL Epidermal Growth Factor (EGF; Invitrogen, Camarillo, CA – dilute to 100 μg/mL in 0.1% BSA in 1× DPBS and store at −70°C). Medium II: To 100 mL basic medium, add 2 μL EGF. Medium III: To 300 mL basic medium, add 720 μL CaCl2 (Sigma – dissolve in ddH_2O at 1 M and store in room temperature).
12. 0.1% gelatin (500 mL): Dissolve 500 mg gelatin powder (Sigma), 500 mL endotoxin-free, reagent-grade water. Autoclave gelatin solution for 45 min on a liquid cycle and store at room temperature.
13. Fibronectin (Advanced Biomatrix, Sandiego, CA): Dissolve 1 mg in 2 mL 1× DPBS, store in −20°C. For coating flasks, dilute to 10 ng/mL in DPBS.
14. Soy bean trypsin inhibitor (Invitrogen): Dissolve 250 mg in 1,000 mL of 1× DPBS.

2.2. Additional Reagents

1. 0.25% Trypsin/EDTA (Cellgro, Manassas, VA).
2. Minimal essential medium with Earle's salts (10× EMEM) (Lonza, Walkersville, MD).
3. FBS (Hyclone, Logan, UT).
4. Bovine tendon acid-extracted collagen I (Organogenesis, Canton, MA).
5. Optimal cutting temperature (OCT) freezing media (Sakura, Torrance, CA).

2.3. Equipment

1. Forceps (Roboz, Gaithersburg, MD).
2. Scissors (Roboz, Gaithersburg, MD).
3. Iris scissors (Roboz, Gaithersburg, MD).
4. Surgical blades (Feather, Japan).
5. 100-mm culture dish (Corning Incorporated, Corning, NY).
6. 100, 70, and 40 μm cell strainers (Becton Dickinson, Franklin Lakes, NJ).
7. 50-mL sterile centrifuge tubes (Becton Dickinson, Franklin Lakes, NJ).
8. 15-mL sterile centrifuge tubes (Becton Dickinson, Franklin Lakes, NJ).
9. T25 and T75 flasks (Becton Dickinson, Franklin Lakes, NJ).

10. Shaker (Taitec, microincubator M-36).
11. CO_2 incubator (CO_2 at 5% (vol/vol); humidified, $T = 37°C$).
12. Centrifuge (Eppendorf Centrifuge 5810).
13. Inverted light microscope.
14. 4-Well chamber slides (Fisher, Pittsburgh, PA).
15. Tissue culture 6-well trays with inserts (Organogenesis, Canton, MA).
16. Multi cassettes (Surgipath, Richmond, IL).
17. TBS biopsy papers (Triangle Biomedical Sciences, Durham, NC).
18. Plastic boats (Fisher, Pittsburgh, PA).

2.4. Cells and Sources

1. CF-1 female mice (days 13–14 gestation) for MEF derivation (Jackson Laboratory).
2. L-Wnt-3A cells (ATCC #CRL-2647).
3. Neonatal human foreskin.
4. Human fibroblasts.
5. Human keratinocytes.

3. Methods

3.1. MEF-Conditioned HES Medium

1. Sacrifice a female mouse (CF-1, WiCell) at day 13 or 14 of pregnancy.
2. Soak the abdomen with 70% ethanol.
3. Using a forceps, pull up the skin separating the hide from the peritoneum and cut a nick in the skin with scissors.
4. Using a new set of forceps and scissors, cut the peritoneum to expose the abdominal cavity.
5. Grab hold of the uterine horns with a blunt-point forceps and using a scissors cut them from the abdominal cavity. Place the uterine horns in a 100-mm cell culture dish that contains 10 mL of DPBS without Ca^{++} and Mg^{++}.
6. Wash the uterine horns three times with 10 mL of DPBS without Ca^{++} and Mg^{++}.
7. Using two fine-pointed forceps, tease open the uterine walls or cut using a scissor to release the embryos into the culture dish.
8. Separate the embryos from the placenta and fetal membranes.
9. Transfer the embryos to a new culture dish and wash them three times using 10 mL of DPBS without Ca^{++} and Mg^{++}.

10. Using a fine-tipped forceps, individually, dissect out and discard the viscera, liver, and heart, which appear as red spots, from each embryo.
11. Wash the embryos three times with 10 mL of DPBS without Ca^{++} and Mg^{++}.
12. Remove the DPBS and add 2 mL of trypsin/EDTA solution to the washed embryos.
13. Using curved Iris scissors, finely mince each embryo. Add 5 mL trypsin/EDTA solution.
14. Incubate at 37°C for 20–30 min on shaker until individual cells are visible (using an inverted culture microscope).
15. Add 20 mL of MEF derivation media to the plate after the incubation. Transfer the individualized cells to a 50-mL tube.
16. Rinse remaining tissue in the plate with a few milliliters of MEF derivation medium. Transfer to the 50-mL tube. Mix by pipetting a few times. Allow the debris in the suspension to settle to the bottom of the tube for 1 min.
17. Remove the top 12 mL of the suspension containing the individualized cells and spin down to remove all the trypsin, and then resuspend the pellet and divide into T75 flask (three embryos/flask).
18. Mix the remaining 8 mL with the debris, spin, resuspend, and add into one flask.
19. Add additional MEF derivation medium to each flask; bring the final volume to 20 mL/flask.
20. Incubate flasks in a 37°C, humidified incubator at 5% CO_2 for 2–3 days, until 80–90% confluent. At this time, the MEFs are ready to be harvested and passaged.
21. Split MEF cells at 1:4 into gelatin-coated T75 flasks with MEF growth medium and grow to reach 80% confluence.
22. Remove MEF growth medium. Add 20 mL HES medium to each flask and incubate for 24 h.
23. Collect conditioned medium in a bottle and store at −70°C. This is the HES conditional medium.

3.2. L-Wnt3a-Conditioned Medium for Melanocyte Differentiation Medium (Mel-1)

1. The L-wnt3a cells are cultured in L-Wnt3a medium (DMEM containing 10% FBS and 0.4 mg/mL G418).
2. When cells reach confluence, split 1:10 into 100-mm culture dishes and add 10 mL L-Wnt3a conditioning medium (1% FBS without G418) in each dish.
3. Incubate cells for 4 days at 37°C 5% CO_2.
4. Collect the medium and filter sterilize. This is Batch 1.

5. Add 10 mL fresh L-Wnt3a conditioning medium and culture for another 3 days.
6. Collect the medium and sterile filter. This is Batch 2. Discard the cells.
7. Mix Batch 1 and Batch 2 medium at 1:1 ratio. This is the L-Wnt3a-conditioned medium.

3.3. Dermal Stem Cell Culture

Day 1

1. Prepare the following in a laminar flow hood: One pair each of sterile forceps, curved scissors, and surgical scalpel blade; 5 mL of epidermal isolation solution (see Subheading 2.1) in a sterile centrifuge tube; 10 mL of Ca^{++}- and Mg^{++}-free HBSS in a 100-mm culture dish; and 10 mL of 70% ethanol in a separate sterile100-mm culture dish.
2. Soak the neonatal human foreskin specimens in 70% ethanol for 1 min. Transfer skin to the dish containing HBSS to rinse off ethanol (see Notes 1 and 2).
3. Cut skin ring open, and trim off fat and subcutaneous tissue with scissors (see Note 3).
4. Cut skin into pieces (approximately 5×5 mm^2) using the surgical scalpel blade with one-motion cuts (see Note 4).
5. Transfer the skin pieces into the tube containing epidermal isolation solution. Cap, invert, and incubate the tube in the refrigerator at 4°C for 18–24 h (see Note 5).

Day 2

1. Remove the tube containing the sample from the refrigerator and incubate at 37°C for 5 min.
2. Prepare the following in a laminar flow hood: Two pairs of sterile forceps and a surgical scalpel blade; two empty sterile Petri dishes; 2 mL of dermal isolation solution.
3. Pour tissue in epidermal isolation solution into one of the empty 100-mm culture dishes. Separate the epidermis (thin, brownish, translucent layer) from the dermis (thick, white, opaque layer) using the forceps. Hold the dermal part of the skin piece with one pair of forceps, and the epidermal side with another. Gently tease them apart. Discard the epidermis immediately (see Note 5). Transfer the harvested dermis to an empty culture dish. Repeat the above-described procedure for each piece of tissue and then mince them as small as possible with a surgical scalpel blade (see Note 5). Transfer the minced tissue to a 50-mL tube with 2 mL of dermal isolation solution and leave the tube for 24 h at room temperature.

Day 3

1. Add 25 mL of HBSS without Ca^{++} and Mg^{++} in the tube of dermal tissue with dermal isolation solution. Mix well and serially filter through 100-, 70-, and 40-μm cell strainers (see Note 6).
2. Spin down at 200 × *g* for 5 min.
3. Resuspend pellets in 5 mL of HESCM4 medium at 4×10^6 per T25 flask.
4. Incubate in 37°C and 5% CO_2.
5. Change ½ medium with fresh HESCM4 medium twice a week.

3.4. Melanocyte Differentiation

3.4.1. Monolayer Differentiation Culture

1. Coat T25 flask or tissue culture-grade 4-well chamber slides with 10 ng/mL fibronectin (3 mL for T25 flask, 0.5 mL/well for chamber slides) overnight at 37°C.
2. Collect dermal spheres into a 50-mL tube and let spheres settle to the bottom (see Note 7).
3. Remove as much medium without disturbing settled spheres.
4. Add 5-mL melanocyte differentiation medium (Mel-1) to spheres. Pipette up and down five times. Aspirate fibronectin from chamber slide wells or T25 flask.
5. Gently remove as much medium from spheres as possible without disturbing.
6. Add 5 mL Mel-1 and transfer spheres to fibronectin-coated chamber slide well or T25 flask.
7. Incubate at 37°C and in a 5% CO_2 tissue culture incubator for 3 weeks, changing ½ of Mel-1 medium twice a week (see Note 7).
8. After a 3-week culture, cells can be passaged in T25 flask at a ratio of 1:3 and grown in melanocyte differentiation medium without TPA. Cells in chamber slide wells are ready for staining.

3.4.2. Three-Dimensional Skin Reconstruct Culture

1. Acellular layer: Mix the following reagents in a 50-mL tube: 0.59 mL 10× minimal essential medium (EMEM), 50 μL 200 mM L-glutamine (keep frozen; thaw immediately before use), 0.6 mL FBS, 120 μL 7.5% sodium bicarbonate, 4.6 mL bovine collagen I. Add 1 mL of mixture into one insert of tissue culture trays. Incubate for 30 min at room temperature.
2. Detach human fibroblasts from culture flasks with 0.25% trypsin/EDTA, and add DMEM containing 10% FBS to neutralize. Collect cells by centrifugation and resuspend 0.45×10^6 cells in 0.75-mL skin-reconstruct medium.
3. Collect 6600 dermal spheres in 0.75-mL HESCM4 medium (see Note 8).

4. Cellular layer: Mix the following in a 50-mL tube: 1.65 mL 10× MEM, 150 μL 200 mM L-glutamine, 1.85 mL FBS, 350 μL 7.5% sodium bicarbonate, 14 mL bovine collagen I, 0.75 mL fibroblasts suspension from step 2, and 0.75 mL dermal spheres from step 3 and mix well. Add 3 mL to each acellular layer-coated insert. Incubate for 45 min at 37°C in a 5% CO_2 tissue culture incubator. Add skin-reconstruct medium I (2 mL inside and 10 mL outside of insert). Incubate for 4 days.
5. Detach human keratinocytes from culture flasks with 0.05% trypsin/EDTA, neutralize trypsin with soy bean trypsin inhibitor, spin down, and resuspend 3×10^6 cells in 600 μL skin-reconstruct medium I.
6. Remove skin-reconstruct tray from incubator, and aspirate medium from both inside and outside of insert.
7. Add skin-reconstruct medium I (1.5 mL inside and 10 mL outside of insert).
8. Drop 100 μL keratinocyte suspension to inside insert. Incubate for 2 days.
9. Aspirate skin-reconstruct medium I from both inside and outside of insert. Add skin-reconstruct medium II (2 mL inside and 10 mL) outside. Incubate for another 2 days.
10. Aspirate skin-reconstruct medium II both inside and outside, and add 7.5 mL skin-reconstruct medium III to only outside of insert (see Note 9). Change medium III every other day until day 18.
11. Harvest skin reconstruct at day 18.
    (a) Aspirate media from insert both inside and outside.
    (b) Remove insert from tray with forceps.
    (c) Cut out the reconstruct (including the polycarbonate filter) by tracing a circle close to the edge with a scalpel blade.
    (d) Cut the reconstruct in half on a hard surface.
12. For paraffin sections: Place a half of the reconstruct in a histology cassette between two black TBS biopsy papers and soak the whole cassette in 10% formalin for 4–6 h. Then, place the cassette in 70% ethanol and store it at 4°C until you are ready to have it processed for paraffin embedding.
13. For frozen sections:
    (a) Place the other half of reconstruct in 50% sucrose at 4°C. After 1–2 h, change the sucrose to 2 M and store it at 4°C for another 1–2 h.
    (b) Dispense OCT freezing media into a plastic boat so that it is about ½ full. Let the OCT polymerize at room temperature for about 15 min. Avoid any bubbles in the OCT.

(c) Remove the reconstruct from the sucrose using forceps to grab the edge and a spatula underneath the polycarbonate filter, and place it on 4 Kimwipes. Dispense enough OCT on top to cover the surface and let it sit for about 10 min (the Kimwipes absorb the sucrose and the OCT starts to bind to the reconstruct).

(d) Transfer the reconstruct into the boat on top of the OCT using forceps and a spatula. Dispense more OCT over the top of the reconstruct until the boat is completely full (make sure that you do not have any bubbles in the OCT which make cutting difficult).

(e) Place the boat evenly on crushed dry ice, pit a lid on the bucket for faster freezing, and allow the OCT to freeze completely (see Note 10).

(f) Wrap the boat in tin foil and store it at −70°C until you are ready to cut it using a cryostat.

3.4.3. Immunostaining

For monolayer differentiated cells:

1. Fix differentiate cells in chamber slides with 4% paraformaldehyde for 20 min.
2. Rinse slides three times with 1× PBS.
3. Unless extracellular antigens are being targeted, permeabilize cells using 0.5% Triton X-100 (Sigma #T9284) for 5 min.
4. Rinse slides twice with 1× PBS.
5. Incubate slides in blocking solution (3% BSA) for 30 min.
6. Removing blocking solution and add primary antibodies diluted in 1× PBS without Ca^{++} and Mg^{++}. Incubate overnight at 4°C.
7. Wash off primary antibody solution three times with 1× PBS without Ca^{++} and Mg^{++}.
8. Incubate slides in fluorescence-conjugated secondary antibodies for 45 min at room temperature in the dark.
9. Rinse twice with 1× PBS without Ca^{++} and Mg^{++}.
10. Dispense one drop of VECTASHIELD mounting medium onto the slide. Use VECTASHIELD mounting medium with DAPI if nuclear counterstaining is desired. Then coverslip and allow the mounting media to disperse over the entire slide.

For skin reconstructs:

1. Deparaffinization in xylene twice for 10 min.
2. Redehydrate in 100, 100, 95, 70, and 50% ethanol for 2 min.
3. Rinse three times with 1× PBS without Ca^{++} and Mg^{++}.
4. Retrieve antigen with trypsin.
5. Repeat steps 4–10 for monolayer cells.

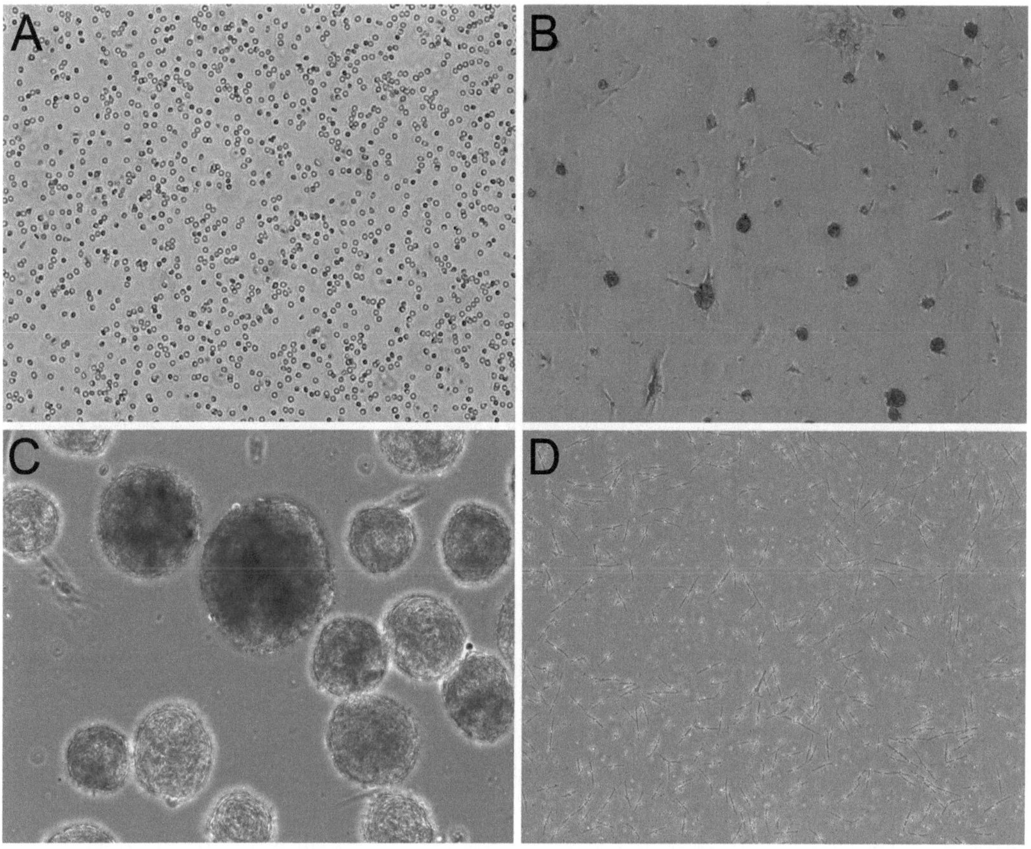

Fig. 1. Isolating and culturing dermal spheres in HESCM4 medium. (**a**) Day 0: Single cells were isolated from neonatal foreskin dermis and grown in HESCM4 medium. (**b**) Day 5: Small cell clusters were formed. (**c**) Day 14: Cell clusters became large 3-dimensional spheres which are loosely adherent or nonadherent to culture flasks. (**d**). Mature melanocytes isolated from epidermis died in HESCM4 medium (day 7 picture).

3.5. Results

3.5.1. Dermal Sphere Formation and Characterization

HESCM4 medium is sufficient to maintain human embryonic stem cells in an undifferentiated state in the absence of feeder cells (10). After enzymatic digestion, dermis-derived single cells were grown in HESCM4 (Fig. 1a). Although most of the single cells gradually died, some cells started forming small cell clusters within 5–7 days of culture (Fig. 1b). After 10–14 days, a majority of theses cell clusters became loosely adherent to culture plates and formed characteristic 3D (Fig. 1c). In contrast to dermal spheres, mature melanocytes isolated from foreskins did not survive in HESCM4 medium more than 1 week (Fig. 1d). The dermal spheres expressed a neural crest cell marker NGFRp75 and an embryonic stem cell marker OCT4, and were capable of self-renewal and differentiated into multiple neural crest-derived cell lineages, including melanocytes (9).

3.5.2. Differentiated Melanocytes Express Melanocytic Markers

When the dermal spheres were cultured for 2–3 weeks in melanocyte differentiation medium (11), most of the cells died and floated, whereas some of the attached cells developed dendritic processes

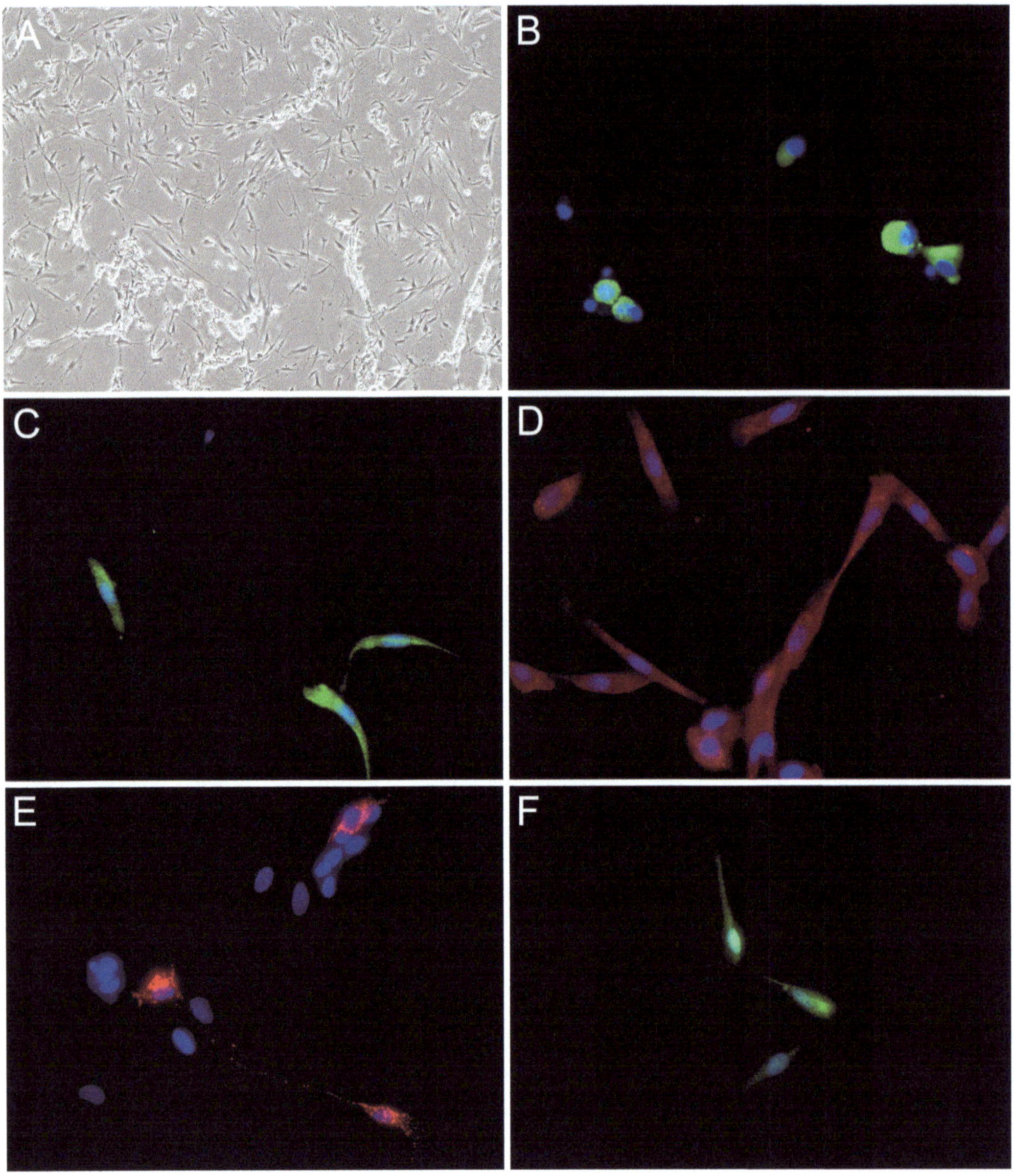

Fig. 2. Dermal stem cell-derived melanocytes express different melanocytic markers. (**a**) When the dermal spheres were cultured for 2–3 weeks in melanocyte differentiation medium, cells migrated from dermal spheres and became adherent single cells. Some of the attached cells developed dendritic processes. (**b**–**f**) After 3 weeks cultured in melanocyte differentiation medium, immunofluoresent staining was performed. The differentiated cells were positive for melanocytic markers MITF (**b**), DCT (**c**), TYRP1 (**d**), HMB45 (**e**), and S100 (**f**). Nuclei were counterstained with DAPI (*blue*).

(Fig. 2a). These adherent cells expressed the melanocyte markers MITF, DCT, TYRP1, HMB45, and S100 (Fig. 2b–f). Undifferentiated dermal spheres were embedded into the dermis of skin reconstructs. At day 14, after seeding keratinocytes, single cells which migrated out from dermal spheres were observed at the

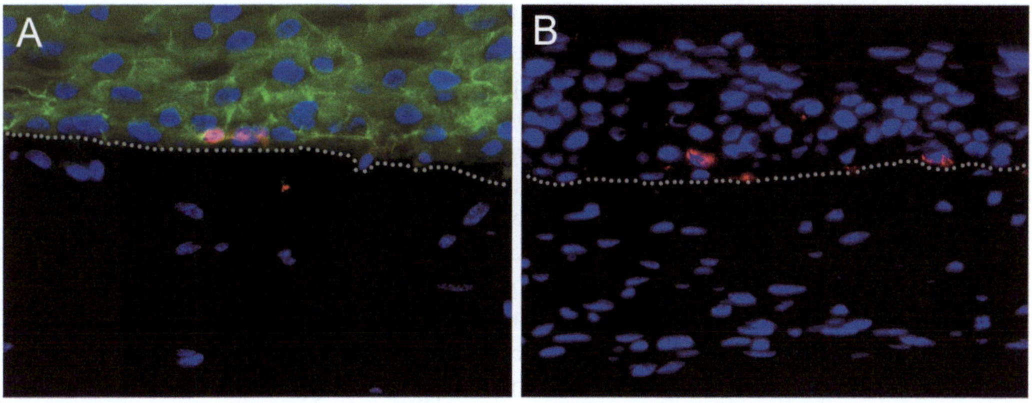

Fig. 3. Melanocyte differentiation in three-dimensional skin-reconstruct model. Undifferentiated dermal spheres were embedded in the reconstructed dermis consisting of collagen with fibroblasts, and after 4 days keratinocytes were added on top of the dermis. At day 14, after seeding keratinocytes, single cells which migrated out from dermal spheres were observed at the basement membrane at the epidermis–dermis junction. (**a**) These cells derived from dermal spheres expressed a melanocytic marker S100 (*red*). They expressed E-cadherin (*green*) when they resided within the basal layer of epidermis. Surrounding keratinocytes are also positive for E-cadherin. (**b**) These single cells were positive for another melanocytic marker HMB45 (*red*). White dotted lines indicate the epidermis–dermis junction. Nuclei were counterstained with DAPI (*blue*).

basement membrane at the epidermis–dermis junction. These cells expressed only the melanocytic markers HMB45, S100, and E-cadherin (Fig. 3a, b) and not neural crest marker NGFRp75 (9).

4. Notes

1. Tissue source and collection: The source of tissue for dermal stem cell cultures is human neonatal foreskins obtained from routine circumcision. At the time of excision, the skin is placed into a sterile container with 20 mL of normal skin-transporting medium supplied in advance and kept near the surgical area at 4°C. Specimens are delivered immediately to the tissue culture laboratory or stored at 4°C. Neonatal foreskins can be kept for up to 72 h. However, the fresher the specimens, the higher the yield of live cells upon isolation.
2. Sterilization of skin specimens: Reduce contamination by a short treatment (1 min) of intact skin with 70% ethanol in a laminar flow hood. After sterilization, rinse samples thoroughly with Ca^{++}- and Mg^{++}-free HBSS.
3. Preliminary tissue preparation: Place tissue on a 100-mm dish, and remove most of the subcutaneous fat and membranous material with curved scissors.
4. Adjustment of tissue size for enzymatic digestion: To improve reagent penetration, cut the skin samples into small pieces (approximately 5 × 5 mm²) rinsed in Ca^{++}- and Mg^{++}-free HBSS.

5. Separate epidermis from dermis: Pieces of skin are incubated in epidermal isolation solution for up to 24 h at 4°C to allow detachment of epidermis from dermis. Each piece of skin is secured with two pairs of forceps; one holds the epidermis and the other the dermis. The epidermal sheet is then peeled apart from the dermis and discarded immediately. Since melanocytes are located just above the basement membrane in the epidermis, scrape the basement membrane with a scalpel blade to try to get rid of as many epidermal melanocytes as possible even though epidermal melanocytes cannot survive in stem cell medium. Dermis is transferred to a 100-mm dish and minced with a scalpel blade as small as possible (approximately 1×1 mm^2).
6. Cell dispersal techniques: A single-cell suspension is generated from clumps of dermal tissue by enzymatic treatment with dermal isolation solution for 24 h at room temperature. The suspensions are diluted in HBSS and are serially filtered through 100-, 70-, and 40-μm cell strainers. Cells are collected by centrifugation, resuspended with stem cell medium, and seeded in a T25 culture flask.
7. Change ½ medium: Remove half volume of medium from flask or chamber slide wells, and add same amount of fresh medium.
8. Harvest dermal spheres: The majority of derma spheres adhere to plastic, and only a small number of spheres floated. The dermal spheres are easily detached by tapping the flasks as they separate from the monolayer cells, which are strongly attached to the culture flasks.
9. Air-lift epidermis: Add skin-reconstruct medium III only outside of insert to expose epidermis in air. This step induces keratinocyte differentiation to form thick epidermis.
10. OCT embedding: Make sure that you do not have any bubbles in the OCT, which makes cutting difficult. The OCT becomes white when frozen.

References

1. Le Douarin, N.M. (1999). The Neural Crest. Cambridge, UK: Cambridge University Press.
2. Hu, F., Staricco, R.J., Pinkus, H., and Fosnaugh, R. (1957) Human melanocytes in tissue culture. J. Invest. Dermatol. 28, 15–32.
3. Kitano, Y. (1976) Stimulation by melanocyte stimulating hormone and dibutyryl adenosine 3′, 5′-cyclic monophosphate of DNA synthesis in human melanocytes in vitro. Arch. Derm. Res. 257, 47–52.
4. Mayer, T.C. (1982) The control of embryonic pigment cell proliferation in culture by cyclic AMP. Dev. Biol. 94, 509–614.
5. Wilkins, L.M. and Szabo, G.C. (1981) Use of mycostatin-supplemented media to establish pure epidermal melanocyte culture (abstract). J Invest. Dermatol. 76, 332.
6. Eisinger, M. and Marko, O. (1982) Selective proliferation of normal human melanocytes in vitro in the presence of phorbol ester and cholera toxin. Proc. Natl. Acad. Scil USA 79, 2018–2022.
7. Nishimura, E.K., Jordan, S.A., Oshima, H., Yoshida, H., Osawa, M., Moriyama, M., Jackson, I.J., Barrandon,Y., Miyachi, Y. and Nishikawa, S. (2002) Dominant role of the niche in melanocyte stem-cell fate determination. Nature 416, 854–860.

8. Yu, H., Fang, D., Kumar, S.M., Li, L., Nguyen, T.K., Acs, G., Herlyn, M. and Xu, X. (2006) Isolation of a novel population of multipotent adult stem cells from human hair follicles. Am. J. Pathol. 168, 1879–1888.
9. Li, L., Fukunaga-Kalabis, M., Yu, H., Xu, X., Kong, J. Lee, J. and Herlyn, M. (2010) Human dermal stem cells differentiate into functional epidermal melanocytes. J Cell Sci. 123, 853–860.
10. Xu C, Inokuma MS, Denham J, Golds K, Kundu P, Gold JD *et al.* (2001). Feeder-free growth of undifferentiated human embryonic stem cells. *Nat Biotechnol*, **19**, 971–974.
11. Fang D, Leishear K, Nguyen TK, Finko R, Cai K, Fukunaga M *et al.* (2006). Defining the conditions for the generation of melanocytes from human embryonic stem cells. *Stem Cells*, **24**, 1668–1677.

Chapter 3

Isolation, Cultivation, and Application of Human Alveolar Epithelial Cells

Nicole Daum, Anna Kuehn, Stephanie Hein, Ulrich F. Schaefer, Hanno Huwer, and Claus-Michael Lehr

Abstract

The blood–air barrier formed by the alveolar epithelium of the peripheral lung is crucial for the pulmonary delivery of drugs. Most existing in vitro models mimicking the blood–air barrier are represented by tumor cells or immortalized cells and lack biological relevance due to their genetic alterations and underexpressed essential physiological functions. However, the increasing interest of aerosol administration of medicines to the respiratory system requires the development and use of representative in vitro models. Thereby, human alveolar epithelial cells (hAEpC) are a suitable test system allowing standardized toxicity and transport studies for newly developed compounds and delivery systems. The isolation, purification, and cultivation of hAEpC are described as well as their possible application in the so-called Pharmaceutical Aerosol Deposition Device On Cell Cultures (PADDOCC) mimicking the complete inhalation process of a powder aerosol in vitro.

Key words: hAEpC, Alveolar epithelial cells, ATI, ATII, Primary culture, Transwell®, Snapwell®, Transepithelial electrical resistance, Pulmonary drug delivery, PADDOCC

1. Introduction

The so-called blood–air barrier is formed by the alveolar epithelium of the peripheral lung consisting mainly of two cell types. The very thin alveolar type I (ATI) cells are large in size and covering 90–95% of the alveolar surface while the smaller, cuboidal alveolar type II (ATII) cells display approximately 5% of the surface (1). Whereas ATI provide a large area for gas exchange and are involved in ion and protein transport, the physiological function of ATII includes

Ragai R. Mitry and Robin D. Hughes (eds.), *Human Cell Culture Protocols*, Methods in Molecular Biology, vol. 806, DOI 10.1007/978-1-61779-367-7_3,

surfactant production and secretion as well as to serve as progenitor cells for the regeneration of ATI cells (2, 3).

In a pharmaceutical and medical context, the blood–air barrier is crucial for the pulmonary delivery of drugs (either systemically or locally). Several in vitro models of the blood–air barrier based on pulmonary epithelial cells have already been described and widely used, including their culture and drug transport studies at an *a*ir–*l*iquid *i*nterface (ALI), to mimic the physiological situation best (4). However, most existing models represented by tumor cells or immortalized cells lack biological relevance due to their genetic alterations and underexpressed essential physiological functions (e.g., tight intercellular junctions) (5, 6). Up to date, alveolar epithelial cells in primary culture are the only reliable model for studying alveolar transport processes (7). Although several animal species (e.g., rat, rabbit, swine) have been used as a source for primary cells to mimic the in vivo situation (8, 9), the extrapolation of results within species may in general be questionable. Therefore, the use of human tissue is essential, although access is very limited.

Pulmonary drug delivery offers promising advantages in comparison to "classical" administration routes as oral or transcutaneous delivery, which is strongly reflected by an increasing interest and number of marketed products for inhalation therapy (10). Aerosol administration of medicines to the respiratory system represents a significant opportunity for the delivery of both small molecules and macromolecules. Advantages are a more rapid absorption into the systemic circulation and avoidance of the first-pass effect, among others. However, developments in pulmonary drug delivery again require useful in vitro models for the evaluation of inhaled therapeutics. Thereby, *h*uman *a*lveolar *ep*ithelial *c*ells (hAEpC) are a suitable test system allowing standardized toxicity and transport studies for newly developed compounds and delivery systems. Additionally, this in vitro model can help to elucidate infection pathways across the respiratory tract and to facilitate the discovery of drug compounds capable of treating these diseases.

Recently, a new in vitro method for testing of pharmaceutical aerosol formulations on cell cultures was developed to aerosolize and deposit a dry powder aerosol realistically onto ALI-grown lung epithelial cells. The so-called *P*harmaceutical *A*erosol *D*eposition *D*evice *O*n *C*ell *C*ultures (PADDOCC) is thereby able to mimic the complete inhalation process of a powder aerosol in vitro combining aerosol generation, aerosol deposition onto pulmonary epithelial cells, and subsequent drug transport across the monolayer (11). This system in combination with air-interface cultured lung epithelial cells helps to investigate new aerosol formulations at an early stage of development and allows their reproducible application onto cell cultures in vitro.

2. Materials

2.1. Isolation and Purification

1. Samples of healthy lung tissue (5–30 g) from the distal part of the lung from patients undergoing lung resection (see Notes 1 and 2).
2. Antibiotics: Penicillin (Pen) (10,000 U/ml) and streptomycin (Strep) (10 mg/ml) in 0.9% NaCl.
3. Balanced salt solution buffer (BSSB): 137 mM NaCl (8.0 g), 5.0 mM KCl (0.4 g), 0.7 mM $Na_2HPO_4{\cdot}7H_2O$ (0.14 g), 10 mM HEPES (2.38 g), 5.5 mM glucose (1.0 g), 1.2 mM $MgSO_4{\cdot}7H_2O$ (0.29 g), 1.8 mM $CaCl_2{\cdot}2H_2O$ (0.27 g), 5 ml Pen/Strep solution. Adjust pH 7.4 at 37°C and filter it by a 0.2-μm bottle-top filter into an autoclaved flask. BSSB can be stored for 3 weeks in a refrigerator at 4°C (the amounts in brackets are calculated for a final volume of 1.0 l).
4. 10× PBS: 130 mM NaCl (7.59 g), 5.4 mM KCl (0.40 g), 11 mM glucose (1.98 g), 10.6 mM HEPES (2.53 g), 2.6 mM $Na_2HPO_4{\cdot}7H_2O$ (0.70 g), solve in 100-ml distilled water, filtrate by a 0.2-μm bottle-top filter into an autoclaved flask. The 10× PBS is stable for 6 weeks at 4°C (the amounts in brackets are given for a final volume of 100 ml). The molarities are of the final 1× PBS.
5. Washing buffer for the magnetic beads column (for a final volume of 100 ml): 10 ml 10× PBS, 0.5 g bovine serum albumin (BSA), 54.5 mg ethylenediaminetetraacetic acid (EDTA), and filter via a 0.2-μm bottle-top filter into an autoclaved flask. The washing buffer can be stored for 4 weeks in a refrigerator at 4°C.
6. DMEM/F12: Dulbecco's modified Eagle's medium (DMEM)/ Ham's nutrient mixture F12 in equal amounts supplemented with 1% Pen/Strep (Gibco, Darmstadt, Germany).
7. SAGM™ BulletKit®: Small Airway Epithelial Growth Medium (Lonza, Verviers, Belgium) with 1% Pen/Strep and 1% fetal calf serum (FCS).
8. Enzyme solutions: DNase (Sigma): Deoxyribonuclease I Type IV from bovine pancreas, 150,000 U, resolve in 10 ml BSSB, prepare aliquots of 1 ml. Trypsin (Sigma-Aldrich, Deisenhofen, Germany): Trypsin type I from bovine pancreas, ~10,000 BAEE U/mg, resolve in 10 ml BSSB, prepare aliquots of 1.5 ml. Elastase (Worthington, Lakewood, USA): Elastase from porcine pancreas, resolve in 10 ml BSSB, prepare aliquots of 300 μl.
9. Inhibition solution: Mix 30 ml of DMEM/F12, 10 ml FCS, and 1 ml DNase. Use the solution at room temperature.

10. Adhesion medium: Mix 22.5 ml DMEM/F12, 22.5 ml SAGM™, and 1 ml DNase. Let the solution warm up to 37°C in the water bath before use.
11. Coating solution: Fibronectin (Becton Dickinson, Heidelberg, Germany): Resolve 1 mg of fibronectin from human serum in 1 ml aqua dest. Prepare aliquots of 100 µl. Collagen (Sigma): Collagen solution 3 mg/ml ultrapure from bovine. Add 100 µl fibronectin solution and 100 µl collagen solution to 10 ml aqua dest. 200 µl of coating solution for one Transwell® filter insert (0.33-cm^2 growth area) as well as one well of a 16-well chamber slide or 700 µl for one Snapwell® filter insert (1.12-cm^2 growth area).
12. Percoll® solutions (Sigma): Autoclave the whole 500 ml before use. Light Percoll® solution: Mix 4 ml 10× PBS, 10.9 ml Percoll®, and 25.1 ml aqua dest. Heavy Percoll® solution: Mix 4 ml 10× PBS, 26 ml Percoll®, and 10 ml aqua dest (see Note 3).
13. Anti-CD326 (EpCAM) conjugated MicroBeads® (Miltenyi Biotec, Bergisch Gladbach, Germany) and a matching separation column system (MiniMACS Starting Kit, Miltenyi Biotec) for positive selection of cells of epithelial origin.
14. 0.4% Trypan Blue in PBS for cell counting.
15. Clean bench (laminar flow bench).
16. Sterile Petri dishes, sterile beakers (100 ml), sterile Erlenmeyer flask (250 ml), sterile tubes for centrifugation (15 and 50 ml), sterile 2.0-ml reaction tubes, sterile-graded pipettes (10 and 25 ml), and Pasteur pipettes.
17. Autoclaved dissection kit with curved scissors and forceps.
18. McIlwain® tissue chopper.
19. Sterile cell strainers of nylon (40- and 100-µm pores, Becton Dickinson) and gauze mesh (see Note 4).
20. Centrifuge with swing-out rotor.
21. Water bath with shaking function.
22. Neubauer hemocytometer.
23. Incubator (5% CO_2, 95% relative humidity, 37°C).

2.2. Cultivation and Differentiation

1. Cell culture medium: SAGM™ BulletKit®: Small Airway Epithelial Growth Medium (Lonza) with 1% Pen/Strep and 1% fetal calf serum (FCS).
2. Cell culture-optimized plastic or glassware, depending on the experimental setup [e.g., Transwell® or Snapwell® permeable filters (Corning Costar, Bodenheim, Germany) or Lab-Tek chamber slides (Nunc, Langenselbold, Germany)].
3. Epithelial Voltohmmeter (EVOM, World Precision Instruments, Berlin, Germany) with chopstick electrodes.
4. Incubator (5% CO_2, 95% relative humidity, 37°C).

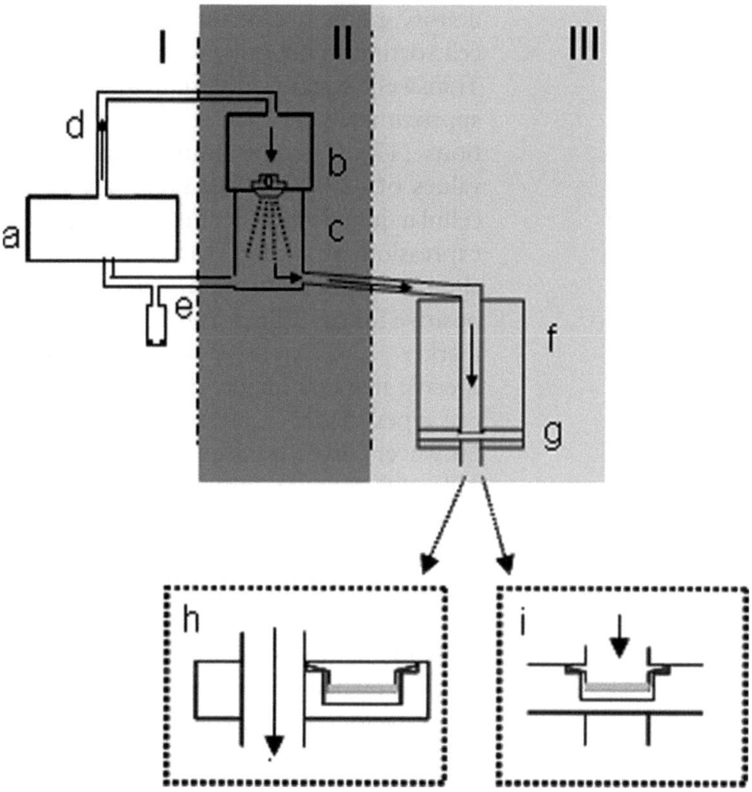

Fig. 1. Schematic view of the PADDOCC system to aerosolize and deposit a dry powder aerosol realistically onto air–liquid interface grown lung epithelial cells (for detailed description, see ref. 11). I: air flow control unit, II: aerosolization unit, III: deposition unit, a: Akita® device, b: HandiHaler® chamber with capsule, c: aerosolization chamber, d: aerosolization tube, e: *y*-shaped tube with syringe, f: sedimentation chamber, g: sampling unit, h: sampling unit (with Snapwell®) in ventilation mode, i: sampling unit (with Snapwell®) in deposition mode.

2.3. Application

1. KRB buffer containing 1.41 mM $CaCl_2$, 3.00 mM KCl, 2.56 mM $MgCl_2$, 142.03 mM NaCl, 0.44 mM K_2HPO_4, 4.00 mM D-glucose, and 10.0 mM HEPES.
2. PADDOCC system (11) consisting of an aerosolization unit fitting a commercial dry powder inhaler (HandiHaler®, Boehringer Ingelheim, Germany), an air flow control unit (Akita®, Activaero, Germany), and a custom-made sedimentation chamber (Fig. 1).
3. Coni-Snap® capsule, size 3, volume 0.3 ml (Capsugel, Bornem, Belgium) containing drug powder.

3. Methods

In summary, the tissue is chopped followed by a digestion step using a combination of trypsin and elastase. The ATII cell population is then purified by a combination of cell attachment procedure, density gradient centrifugation via Percoll®, and positive magnetic cell sorting. ATII cells, when seeded on permeable supports (e.g., Transwell® system), differentiate into an ATI-like phenotype within approximately 1 week when applying appropriate culture conditions (12). Thereby, transelectrical epithelial resistance (TEER) values of ~2,000–2,500 Ωcm^2 are reached displaying tight intercellular junctions reflecting intact barrier properties. In addition, expression of typical marker proteins for ATI and ATII cells is altered during cultivation, detectable by an increasing cell number positive for caveolin-1 as ATI marker and negative for Sp-C as ATII marker (13). Extensive research has led to the discovery of other specific markers further allowing the differentiation between both cell types besides a pure morphological characterization (14–17). However, also a transient phenotype may be observed when ATII cells differentiate into ATI-like cells. The developed epithelial monolayers consist of coexisting ATI-like and ATII-like cells and correspond to the normal composition of the alveolar epithelium of the donor tissue.

The differentiated epithelial monolayers can be used for transport and uptake studies or cytotoxicity assays. Furthermore, when grown at ALI, they can be applied to the PADDOCC which was developed to aerosolize and deposit a dry powder aerosol realistically onto ALI-grown lung epithelial cells mimicking the complete inhalation process of a powder aerosol in vitro. The device is separated into three compartments: the air flow control unit, aerosolization unit, and deposition unit, connected by silicon tubes. The system is schematically depicted in Fig. 1.

3.1. Isolation and Purification

1. Success of the whole procedure strongly depends on careful and sterile handling of all materials. Any step should be performed in a laminar flow bench. Preheat BSSB and SAGM™ medium to 37°C before use.
2. Weigh the tissue to calculate the number of cells which can be obtained (see Notes 1 and 2). Eventually, take pictures of the tissue to document the quality.
3. Rinse the lung tissue in a sterile Petri dish containing BSSB for several times to reduce the number of blood cells. Cut the tissue in small pieces of approximately 1 cm^3 using forceps and scissors (see Note 5).
4. Mince the tissue piece by piece with the tissue chopper (cut size 5 μm). It is highly recommended to repeat this step for each piece. Use forceps to immediately transfer the minced lung into a centrifuge tube containing about 25 ml preheated (37°C) BSSB. Collect all the tissue and mix it gently with BSSB in the tube.
5. Put a 100-μm cell strainer on top of a 50-ml centrifuge tube to separate the tissue from the liquid by gently stirring.

Discard the flow through. Transfer the tissue from the filter into a new 50-ml sterile centrifuge tube filled with 25 ml preheated (37°C) BSSB. To remove blood cells and mucus, it is crucial to repeat this washing step at least three times or until the BSSB remains clear.

6. Prepare the solution for the enzymatic digestion of the tissue in a sterile Erlenmeyer flask. Mix up 30 ml warm BSSB, 1.5 ml trypsin solution, and 300 μl elastase solution. Put the minced tissue directly from the 100-μm cell strainer into the Erlenmeyer flask and disperse it carefully in the solution. Incubate the solution for 40 min at 37°C in a shaking water bath (see Note 6).
7. After incubation, the enzymatic reaction is stopped by applying the inhibition solution to the tissue solution. Use a 25-ml pipette to disperse the digested tissue for at least 5 min to obtain a suspension (see Note 7).
8. Filter the tissue solution through the self-made gauze mesh filter (see Note 4) into a sterile beaker to remove the solid portions of the mixture. Discard the filter together with the tissue remains.
9. Filter the resulting flow through once more through a 40-μm cell strainer to discard small tissue pieces. Collect and distribute equally the resulting crude alveolar cell extract in two 50-ml centrifuge tubes.
10. Centrifuge for 5 min at 300 × *g* at room temperature. Resuspend each pellet in 15 ml of the adhesion medium and transfer the resulting liquid into two cell culture-treated Petri dishes. To let the alveolar macrophages adhere at the surface of the plastic dishes, it is necessary to incubate the plates for 90 min at 37°C in an incubator without movement.
11. In the meantime, it is recommended to prepare the coating solution and incubate all surfaces which are seeded with cells with sufficient amounts of the coating solution. The coating should remain for at least 2 h on the plastic or glassware or Transwell®/Snapwell® filter in an incubator.
12. Proceed with the Petri dishes. Collect the solution with a pipette, rinse the Petri dishes gently once with 10 ml BSSB, and transfer everything into two centrifuge tubes. Discard the Petri dishes.
13. Centrifuge the cell suspension for 5 min at 300 × *g* at room temperature. Meanwhile, the Percoll® solution gradient should be prepared (see Note 8).
14. Discard the supernatant and resuspend the cell pellets together in 2 ml BSSB. Layer the cell suspension carefully on top of the Percoll® gradients. Centrifuge the preparation for 20 min at 300 × *g* at room temperature without using the brake (see Note 9).

15. After centrifugation, the cells build up a slightly visible layer, where both Percoll® solutions meet. Harvest the cells carefully using a Pasteur pipette and transfer them to two previously prepared 50-ml centrifuge tubes each containing 30 ml BSSB. Centrifuge the cells for 5 min at $300\times g$ and room temperature. Discard the supernatant and resuspend the pellets in each 40 ml BSSB. Repeat the centrifugation and discard the supernatant.
16. Both cell pellets have to be solved and reunited in only 1.5 ml BSSB. Put the cell suspension into a 2-ml reaction tube and add 200 μl of Anti-CD326 conjugated MicroBeads® to select cells of epithelial origin. Incubate the mixture for 30 min at 4°C in a refrigerator using an orbital shaker to maintain circulation (see Note 10).
17. The coating solution has to be removed from the cell culture plastic or glassware by pipetting and the coated surfaces should be allowed to dry in the sterile bench.
18. Place the MiniMACS holder with magnet in the sterile bench and attach the separation column to the magnet. To collect the liquid waste, put a beaker under the column. Equilibrate the column with 3 ml of the washing buffer. Pipette the hAEpC suspension on the column and let it flow through. After four washing steps with each 3 ml BSSB, eluate the cells from the column with 5 ml BSSB (see Note 11). Prepare a new column and repeat the whole procedure with the obtained eluate, but this time wash the cells off the column finally with 5 ml preheated SAGM™ (see Note 12).
19. Estimate the total number of viable cells and percentage of dead cells using a Neubauer hemocytometer (see Note 13). Calculate the number of cells obtained per gram of tissue (see Note 14).

3.2. Cultivation and Differentiation

1. Dilute the hAEpC suspension in SAGM™. Per apical compartment of a Transwell® filter or well of a 16-well chamber slide, respectively, 200 μl of the hAEpC suspension with 0.2×10^6 cells is applied to a surface of 0.33 cm^2 while the 1.12 cm^2 of a Snapwell® is filled with 500 μl of cell suspension, including 0.7×10^6 cells (see Note 15). Additional 800 or 1,500 μl SAGM™ are needed for the basolateral compartments of the Transwell® or Snapwell® plate, respectively.
2. Cultivation takes place in an appropriate incubator (5% CO_2, 95% relative humidity, 37°C). Control the attachment and morphology of the cells regularly in a light microscope with a high magnification.
3. First replacement of SAGM™ is necessary 72 h after seeding. The following medium exchanges should be performed every 48 h (see Note 16).

4. To culture the cells as an air-interface culture, proceed after 24 h by carefully removing the medium. Replenish 500 μl to the basolateral compartment. The cultured monolayer is now fed through the Transwell® or Snapwell® filter from below. This approach can be useful, if one wants to study the secretion ability of the hAEpC culture or use it in the PADDOCC system.
5. After 5–10 days, the cells differentiate into a confluent monolayer of ATII cells and ATI-like cells. To investigate the development of the cell layer on the filter insert, the TEER is measured using an EVOM. To sterilize the chopstick electrodes, put them in 70% ethanol for 15 min and let the ethanol, which remains on the electrodes, evaporate inside the laminar flow bench. Place the Transwell® or Snapwell® plate in the sterile bench; optionally, put it onto a heating plate (37°C) to avoid the cells cooling to room temperature. Put the longer one of the electrodes into the basolateral compartment and the shorter one into the apical compartment and measure the TEER value (see Note 17). The unit of TEER value is Ω cm^2; therefore, the measured TEER value has to be multiplied with the cell culture-growing surface area. The highest TEER is normally observed around day 8 of culture with values about 2,000–2,500 Ωcm^2 (see Notes 14, and 18).

3.3. Application

1. Preparation of Snapwells® with ALI-grown hAEpC: Remove medium from the basolateral compartment of the Snapwells® and wash this basolateral compartment twice with 1 ml of fresh, prewarmed KRB buffer (see Note 19).
2. Equilibrate the Snapwells® for at least 30 min with 1 ml of KRB buffer in the basolateral compartment in an incubator (37°C, 5% CO_2, >90% humidity) (see Note 20).
3. Put the Snapwells® into the sampling wells of the PADDOCC system which contain 500 μl KRB buffer.
4. Put a capsule containing the drug powder in the HandiHaler®, tap the capsule by pressing the green button at the HandiHaler®, and assemble the PADDOCC system.
5. For the deposition process: Turn the Akita® device on and put the sampling unit into the ventilation mode.
6. Pull and push the plunger of the syringe to generate the dispersion impulse to release the powder from the capsule into the aerosolization chamber. After the dispersion impulse, the transport flow occurs transporting the aerosol to the sedimentation chamber.
7. Akita® device is turned off and the sampling unit is put into deposition mode by rotating the sampling unit through 60°. The aerosol is now sedimenting for 10 min onto the hAEpC.

8. Steps 5–7 are repeated twice to empty the capsule completely.
9. After the three aerosolization–deposition steps, the Snapwells® are transferred into a 6-well culture plate and subsequent studies can be performed (see Note 21).

4. Notes

1. The use of human tissue for isolation of primary cells was reviewed and approved by the respective local ethics committee (State Medical Board of Registration, Saarland, Germany).
2. All materials being in contact with human tissue or blood should be treated as potentially infectious in accordance to the local safety rules. Do not forget the operator's protection.
3. For a more comfortable handling and distinguishing of both Percoll® solutions, beginners are recommended to dye one of the solutions with one droplet of phenol red (Sigma).
4. The gauze mesh filters can easily be self-made from a 50-ml centrifuge tube and gauze bandage. Cut the bottom of the tube and tape some gauze (at least two layers) firmly to it. Sterilize the self-made filter before use.
5. Sometimes, it is advantageous to remove the bigger blood vessels and bronchi from the lung parenchyma. It depends on which part of the lung is used.
6. Use the waiting time to tidy up the clean bench, to clean up the tissue chopper from blood and tissue remains, and to prepare the inhibition solution and the adhesion medium for the further procedure.
7. It is crucial to release as many cells as possible from the tissue. The longer the triturating is performed, the higher the yield is. By trying to eliminate all bigger pieces, it should be easier to pipette the solution afterward.
8. Both Percoll® solutions have to be cooled to 4°C before pouring the gradient. Put 10 ml light Percoll® solution into two 50-ml centrifuge tubes each. Take a 25-ml pipette filled up with 12 ml heavy Percoll® solution, cross the light solution, and place the tip on the bottom of the tube. Let 10 ml of the solution gently stream out of the pipette under the light solution and prevent shaking, disturbing, or bubbling. Dyeing with phenol red aids identification of the solutions interface. Furthermore, it is helpful to cool the gradients on ice before applying the cells.
9. The acceleration of the swing-out rotor should be diminished after centrifugation by turning off the brake.
10. The separation method is based on positive selection with an *ep*ithelial *c*ell *a*dhesion *m*olecule (EpCAM) antibody.

EpCAM (CD326) is a cell marker which is known to be highly expressed on epithelial cells.

11. To eluate the hAEpC from the column, remove it from the magnet, place it onto a 15-ml centrifugation tube, put the medium on top, and press the column plug rapidly.
12. Every single step should be performed with reasonable care to obtain an appropriate amount of cells. The quality of the tissue is a very important factor in respect to the yield. Nevertheless, 1×10^6 hAEpC per gram of tissue can be obtained.
13. Mix up 25 μl of the hAEpC suspension and 25 μl of the Trypan Blue solution. Use the Neubauer hemocytometer to calculate the number of viable and dead cells per ml. Dead cells appear blue under the light microscope; viable cells do not take up the dye. Since the uptake of Trypan Blue is time dependent, viable cells can also become blue after a while. Time is of the essence.
14. Results may be donor dependent and influenced by anamnesis and medical pretreatment of the patient.
15. The seeding density should be 0.6×10^6 cells/cm^2. If the seeding density is too low or too high, the cells will not grow or not differentiate well. Too less space to spread or insufficient cell–cell contact hinders the development of a tight monolayer and affects the TEER value in a negative manner.
16. It is important to exchange the medium of the Transwell® and Snapwell® plate in a certain manner. When removing the medium, start with the basolateral compartment, and then empty the apical compartment. When refilling the compartments, start with apical followed by basolateral one. If only the basolateral compartment remains filled, it is possible that your adherent cells detach off the filter insert due to hydrostatic pressure from below.
17. Sometimes, the TEER values vary while measuring between two numbers, hence a second round of measurement could be helpful.
18. Do not touch or even scratch the monolayer or filter insert while measuring TEER values and keep the culture sterile. Pay attention to correct orientation of the chopstick electrodes. Do not forget measuring the TEER value of one Transwell® filter with medium but without cells to subtract this background value from the culture TEER values.
19. This protocol was first adapted for other cell types (e.g., Calu-3) that are able to grow under ALI conditions.
20. It is very important to maintain 37°C during the experiments.
21. Besides transport studies, uptake studies or cytotoxicity assays can be performed.

References

1. Crapo, J. D., Barry, B. E., Gehr, P., Bachofen, M., and Weibel, E. R. (1982) Cell number and cell characteristics of the normal human lung, *American Review of Respiratory Disease* ***125***, 740–745.
2. Williams, M. C. (2003) Alveolar Type I Cells: Molecular Phenotype and Development, in *Annual Review of Physiology*, pp 669–695.
3. Fehrenbach, H. (2001) Alveolar epithelial type II cell: Defender of the alveolus revisited, *Respiratory Research* ***2***, 33–46.
4. Grainger, C. I., Greenwell, L. L., Lockley, D. J., Martin, G. P., and Forbes, B. (2006) Culture of Calu-3 cells at the air interface provides a representative model of the airway epithelial barrier, *Pharmaceutical Research* ***23***, 1482–1490.
5. Forbes, B., and Ehrhardt, C. (2005) Human respiratory epithelial cell culture for drug delivery applications, *European Journal of Pharmaceutics and Biopharmaceutics* ***60***, 193–205.
6. Van Den Bogaard, E. H. J., Dailey, L. A., Thorley, A. J., Tetley, T. D., and Forbes, B. (2009) Inflammatory response and barrier properties of a new alveolar type 1-like cell line (TT1), *Pharmaceutical Research* ***26***, 1172–1180.
7. Kim, K. J., Borok, Z., and Crandall, E. D. (2001) A useful in vitro model for transport studies of alveolar epithelial barrier, *Pharmaceutical Research* ***18***, 253–255.
8. Steimer, A., Laue, M., Franke, H., Haltner-Ukomado, E., and Lehr, C. M. (2006) Porcine alveolar epithelial cells in primary culture: morphological, bioelectrical and immunocytochemical characterization, *Pharmaceutical Research* ***23***, 2078–2093.
9. Corti, M., Brody, A. R., and Harrison, J. H. (1996) Isolation and Primary Culture of Murine Alveolar Type II Cells, *American Journal of Respiratory Cell and Molecular Biology* ***14***, 309–315.
10. Henning, A., Hein, S., Schneider, M., Bur, M., and Lehr, C. M. (2010) Pulmonary drug delivery: Medicines for inhalation, *Handbook of Experimental Pharmacology* ***197***, 171–192.
11. Hein, S., Bur, M., Kolb, T., Muellinger, B., Schaefer, U. F., and Lehr, C. M. (2010) Pharmaceutical Aerosol Deposition Device On Cell Cultures (PADDOCC) *in vitro*: design and experimental protocol, *ATLA Alternatives to Laboratory Animals* ***38***, 1–11.
12. Elbert, K. J., Schaefer, U. F., Schaefers, H. J., Kim, K. J., Lee, V. H., and Lehr, C. M. (1999) Monolayers of human alveolar epithelial cells in primary culture for pulmonary absorption and transport studies, *Pharm Res* ***16***, 601–608.
13. Fuchs, S., Hollins, A. J., Laue, M., Schaefer, U. F., Roemer, K., Gumbleton, M., and Lehr, C. M. (2003) Differentiation of human alveolar epithelial cells in primary culture: Morphological characterization and synthesis of caveolin-1 and surfactant protein-C, *Cell and Tissue Research* ***311***, 31–45.
14. Cheek, J. M., Evans, M. J., and Crandall, E. D. (1989) Type I cell-like morphology in tight alveolar epithelial monolayers, *Experimental Cell Research* ***184***, 375–387.
15. Kasper, M., and Singh, G. (1995) Epithelial lung cell marker: Current tools for cell typing, *Histology and Histopathology* ***10***, 155–169.
16. Newman, G. R., Campbell, L., Von Ruhland, C., Jasani, B., and Gumbleton, M. (1999) Caveolin and its cellular and subcellular immunolocalisation in lung alveolar epithelium: Implications for alveolar epithelial type I cell function, *Cell and Tissue Research* ***295***, 111–120.
17. Shirasawa, M., Fujiwara, N., Hirabayashi, S., Ohno, H., Lida, J., Makita, K., and Hata, Y. (2004) Receptor for advanced glycation end-products is a marker of type I lung alveolar cells, *Genes to Cells* ***9***, 165–174.

Chapter 4

Culture of Parathyroid Cells

Peyman Björklund and Per Hellman

Abstract

The parathyroid cells are highly differentiated with more or less their only function to secrete parathyroid hormone in response to the extracellular calcium level. Tumours from the parathyroid glands are >99% benign, and have a slow proliferation rate. Culture of parathyroid cells is known to be very difficult most likely due to the high differentiation level. This chapter reveals some details in order how to get parathyroid cells to survive in culture after dispersion of normal bovine parathyroid glands or pathological human parathyroid tumours. Detailed protocols describing cell dispersion with collagenase, short-term cultures, and establishment of long-term cultures are presented.

Key words: Parathyroid, Calcium, Vitamin D, Lithium, Bovine, Human, Collagenase, Chief cell

1. Introduction

The parathyroid glands are confined by the parathyroid-specific chief and oxyphil cells as well as cells from the supporting tissue: endothelial cells, fat-storing cells, fibroblasts, etc. The parathyroid hormone (PTH)-secreting cells are the most abundant chief cells. The function of the acidophilic and mitochondria-rich oxyphil cells is somewhat obscure (1). The amount of fat in the gland, mostly confined to the fat-storing cells, correlates to the activity of the gland. Thus, cells in proliferation or highly secreting PTH presumably use more energy, which depletes the fat storages (2). The amount of fat – correlating to a degree of pathology in the gland – may be used as a diagnostic tool by using the rather easy oil red O fat staining (1).

A normal parathyroid gland is small, with a fat-free dry weight of approximately 40 mg. The different variants of pathology in the glands may be divided into diffuse hyperplasia, nodular hyperplasia,

Ragai R. Mitry and Robin D. Hughes (eds.), *Human Cell Culture Protocols*, Methods in Molecular Biology, vol. 806,
DOI 10.1007/978-1-61779-367-7_4, © Springer Science+Business Media, LLC 2012

and adenoma. Parathyroid carcinoma is extremely rare, and not further commented in this text. The diffuse hyperplasia may be induced by several different stimuli, where low calcium levels and low vitamin D3 levels (1,25-dihydroxy-cholecalciferol) are typical most commonly due to renal insufficiency, but may also occur in case of intestinal malabsorption with uptake deficiency of calcium and vitamin D3. This situation usually confines to secondary hyperparathyroidism (sHPT), and is usually present in early stages of renal insufficiency. Diffuse hyperplasia may also be present as hereditary primary HPT due to mutations in the *menin* gene, causing the MEN1 syndrome, or the *ret* gene, causing MEN-2. Diffusely hyperplastic glands demonstrate low fat content and general signs of diffusely proliferating cells throughout the gland. This condition may turn into nodular hyperplasia presumably due to clonal expansion of cells in which additional genetic disturbances ensue. Nodules of various sizes occur, which may be more autonomous and less reactive to the external calcium stimulus normally regulating the release of PTH and the proliferative activity of the cells. Nodular hyperplasia is, thus, present in later stages of renal insufficiency or in MEN1 (and MEN2) patients. Parathyroid hyperplasia may also occur in preferably older individuals without having overt renal insufficiency or possessing deranged *menin* or *ret* genes. The reason for this situation is somewhat obscure, but reasons as a minor renal insufficiency with age leading to lower vitamin D3 levels have been suggested, but also contradicted. Another stimulus associated with hyperplasia is lithium. Patients with lithium treatment have been known to get a mild form of HPT, with a presentation mimicking nodular hyperplasia. A primary genetic disturbance within a parathyroid cell may lead to an adenoma, leading to sporadic pHPT. Similar to the nodules in nodular hyperplasia, the adenomas are less reactive to external calcium, secrete higher amounts of PTH, and have a higher proliferative activity.

Until recently, there have not been any parathyroid cell lines available. The cells are extremely difficult to culture for long term, but some reports are present (6–8). They have a very low proliferative activity. Therefore, physiological studies of parathyroid cells have mostly been performed in short-term cultures (3, 4). It is clear that normal parathyroid cells are more or less impossible to culture and induce division within. Human as well as bovine cells have been used. Hyperplastic or adenomatous human cells have offered more success in this matter. On the other hand, parathyroid cells – normal or pathological – may, even though they fail to proliferate, attach to a plastic surface and function in terms of secreting PTH in a calcium-dependent way.

In this chapter, a method for dispersion of glands for short-term cultures is described, but also ways of establishing a parathyroid cell line for long-term culture.

2. Materials

2.1. Reagents

1. Phosphate-buffered saline (PBS).
2. Transport buffer: HEPES-buffered Ham's F10 with 10% fetal calf serum (FCS).
3. Collagenase (Sigma Chemical Co., St Louis, MO, USA).
4. DNAse (Sigma).
5. Percoll (Pharmacia, Stockholm, Sweden).
6. Secretion buffer: 125 mM NaCl, 5.9 mM KCl, 5 mM $MgCl_2$, 0.625 mM HEPES.
7. Growth medium: DMEM containing 10% FCS, glutamine, streptomycin, penicillin.
8. Epidermal growth factor (EGF; Gibco, Uxbridge, Great Britain).
9. Wash buffer: HEPES-buffered Ham's F10 with 0.136 M NaCl, 4.7 mM KCl, 0.65 mM $MgSO_4 \cdot 7H_2O$, 1.22 mM $CaCl_2 \cdot 2H_2O$, 25 mM HEPES.
10. Bovine pituitary extract (Gibco).
11. Dulbecco's minimal essential medium (DMEM; Gibco).
12. RPMI (Gibco).
13. Keratinocyte culturing medium (Gibco).
14. Cytobuster Protein Extract Reagent (Novagen Inc., Madison, Wisconsin, USA) supplemented with complete protease inhibitor cocktail (Roche Diagnostics GmbH, Mannheim, Germany).
15. Immunoradiometric assay/radioimmunoassay for measurements of PTH (e.g. Nichol's Institute, San Capistrano, CA, USA).
16. Anti human parathyroid hormone goat polyclonal antibody (Santa Cruz Biotechnology Inc., Santa Cruz, USA, # sc-9678).
17. Normal goat serum (# X0907) and FITC-labelled secondary antibody (#F0250, Dakocytomation Denmark A/S, Glostrup, Denmark).
18. Vectashield mounting medium for fluorescence with DPI (Vector Laboratories, Inc., Burlingame, CA, USA).
19. RNeasy Minikit (# 74104 Qiagen Sciences, Germantown, MD, USA).
20. First-Strand cDNA Synthesis kit (GE Healthcare Europe GmbH, Uppsala, Sweden).
21. Platinum Taq polymerase (Invitrogen Corp., Carlsbad, CA, USA).

2.2. Equipment

1. Shaking incubator allowing 300 rpm.
2. Sharp scissors.
3. Conical-shaped container.
4. 24-Well plates (Nunclon, Lincoln Park, NJ, USA).
5. Chemiluminescence system (GE Healthcare Europe GmbH, Uppsala, Sweden).
6. Fluorescence microscope (Leica DMRB, Leica Mikroskopie und Systeme GmbH, Wetzlar, Germany).
7. NucleoCounter (ChemoMetec A/S, Allerod, Denmark).

3. Methods

3.1. Parathyroid Cell Dispersion

Human parathyroid glands are obtained during surgery for primary or secondary HPT, and normal bovine glands from the local slaughterhouse. Biopsies of the excised human glands are immediately taken care of after excision by being placed in ice-cold saline solution. Informed consent of the patients is obtained, and approval of the Local Ethics Committee to store human tissue in a biobank and use for physiological studies is ensured.

1. Place biopsies of the excised human glands in an ice-cold saline and transfer to the laboratory (see Note 1).
2. On arriving in the laboratory, human glands are minced with small and sharp scissors after removal of visible fat and connective tissue surrounding the glands. It is essential to remove all periparathyroid tissue and to leave the naked capsule before mincing of the actual parathyroid gland (see Note 2).
3. Digest the minced preparation in 10 ml (per ≈100 mg minced tissue) of 1 mg/ml collagenase, 0.05 mg/ml DNase, 1.5% bovine serum albumin, and 1.25 mM Ca^{2+} at 37°C (5). Transfer to a shaking incubator and digest at 300 rpm for 20–30 min (see Note 3).
4. After 20–30 min, wash the cell suspension two times with wash buffer to inactivate and dilute the collagenase (see Note 4).
5. Eventually, one may expose the cell suspension to 1 mM EGTA in 25 mM HEPES buffer (pH 7.4) containing 142 mM NaCl and 6.7 mM KCl during 10–20 s, which allows further dispersion to single-cell suspensions, if needed (see Note 5).
6. Remove dead cells and debris by centrifugation (5 min; 300 × *g*) through 25% standard isotonic Percoll diluted in PBS. If needed, remove blood cells by centrifugation (10 min; 600 × *g*) through 75% standard isotonic Percoll diluted in PBS. Wash the suspensions thoroughly with wash buffer (times 2 or 3) (see Notes 6 and 7).

7. Determine cell viability by Trypan blue exclusion. Add a small amount, e.g. 5 μl Trypan blue and 5 μl cell suspension, and count cell in which the dye is not excluded (=dead cells). In our hands, this routinely exceeded 95%.
8. The entire procedure generally yields small clusters of 2–20 cells, which may be used for short-term cultures for measurements of, e.g. PTH release, or for long-term cultures. Needless to mention are the demands of performing the whole procedure in a sterile fashion.

3.2. Short-Term Cultures for PTH Measurements

1. Suspend the cells in a balanced buffer solution ("secretion buffer") or in culture medium (either RPMI with low calcium concentration allowing additions of calcium to be made or DMEM with an approximate physiological ionized calcium concentration). Cells are incubated at 37°C in duplicate or triplicate preferably using a 24-well plate (see Note 8). One may include incubations at different calcium concentrations (0.5, 1.25, and 3.0 mM) in order to characterize the actual cell batch and its responsiveness to calcium (see Note 9). For physiological studies of other compounds, additions should be made at an external calcium concentration of 1.25 mM, resembling the physiological level (reached by using RPMI with calcium additions or balanced salt solution – "secretion buffer") (see Note 10).
2. After the incubation period, aspirate the contents of the wells and centrifuge quickly to separate non-attached cells and debris from the supernatant.

3.2.1. Measurements of PTH

1. Determine the PTH concentrations of the supernatants from above. Always use at least duplicates, preferably triplicates (see Note 11).
2. Analyses of PTH may be correlated to the amount of cells either by using at least duplicates and a thorough addition of similar number of cells in each well (500,000) or by determining the total protein amount in each well after separation of the cell pellet.
3. Interpretation and presentation of the assay results usually are done by setting the PTH release at 0.5 mmol/l to 100%, and maximal inhibition of PTH release set to the amount at an external Ca^{2+} concentration of 3.0 mmol/l. Typically, the PTH release at external Ca^{2+} is about 60% of that at 0.5 mM Ca^{2+} (see Note 12).

3.3. Long-Term Cultures

1. Suspend the cells in the above-mentioned keratinocyte medium in a 25-cm^2 cell culture flask. An alternative is overnight culture in DMEM with 10% FCS to allow the cells to attach to the culture vial plastics somewhat more efficiently than in the keratinocyte medium, despite the negative effect on the growth of fibroblasts (see Note 13).

2. After the eventual overnight culture in DMEM/10% FCS, cells are fed with the keratinocytes medium as above. In this environment, parathyroid cells may be cultured up to 60 days at our laboratory. The most success has been when using pathological parathyroid cells, in terms of achieving dividing, proliferating parathyroid cells (see Note 14). Normal human parathyroid cells attach to the plastic surfaces but are unwilling to divide in this environment. However, they still secrete PTH in a calcium-dependent manner.
3. When the cells fill the bottom layer of the culture flask, contact inhibition ensues. At this stage, detach the cells with or without trypsin and culture them further in new flasks. Regardless of the procedure at this stage, the cells usually cease to proliferate in the next passage, although they still function in terms of secreting PTH and responding to different external calcium concentrations (6).

3.4. Cell Line Establishment

A recent successful method of establishing a parathyroid cell line derived from pathological parathyroid glands of HPT secondary to uremia is described (7).

1. Prepare parathyroid cells as above – we used cells from a hyperplasitc gland from a patient with secondary HPT (see Note 15).
2. Suspend the cells in 96-well plates at a concentration of less than one cell/100 μl of growth medium.
3. Culture for 45 days, with occasional medium exchange.
4. Replate colonies in 35-mm plates. Proliferating colonies may be stimulated by addition of growth medium supplemented with 10 mM lithium chloride for a couple of passages (we used four) (see Note 15).
5. Harvest cells and investigate parathyroid origin by:
   (a) Western blotting:
      - Description of method with PTH antibody. Western blotting analysis can be done on extracts prepared in CytoBuster Protein Extract Reagent supplemented with complete protease inhibitor cocktail. Anti-human parathyroid hormone goat polyclonal antibody was used. After incubation with the appropriate secondary antibodies, bands were visualised using the enhanced chemiluminescence system.

   (b) Fluorescent immunostaining analysis using an antibody specific for PTH:
      - Cells were grown on glass coverslip and fixed in 4% paraformaldehyde for 10 min, followed by incubation in ice-cold 70% ethanol for 20 min.

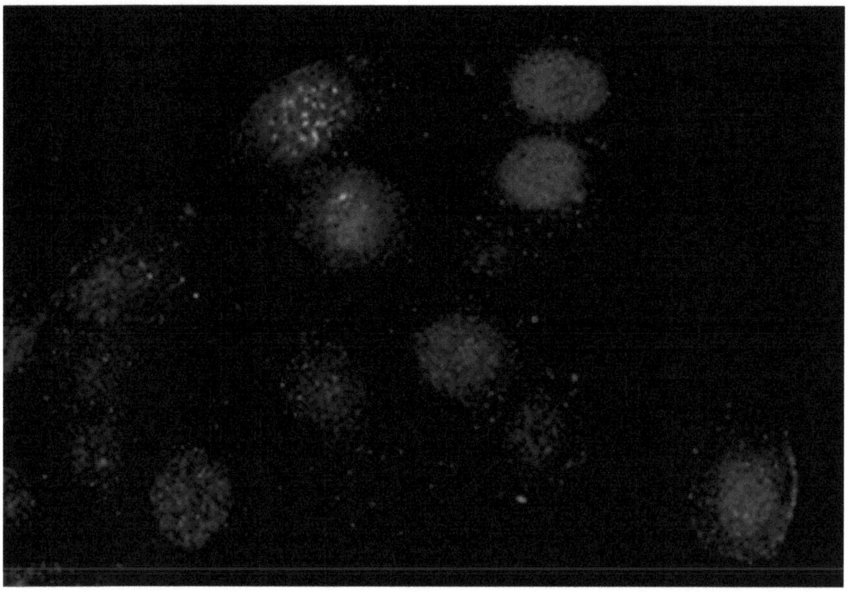

Fig. 1. Fluorescent immunostaining utilising antibody against PTH in the established parathyroid cell line (7).

Cells were washed in PBS, blocked in normal goat serum (1:10), and then were incubated with the PTH antibody (1:100) for 60 min.

After three times wash in PBS follows, cells were incubated with FITC-labelled secondary antibody (1:500).

- Coverslips were mounted with a Vectashield mounting medium for fluorescence with DPI.
- Cells can be examined by a fluorescence microscope (Fig. 1).

(c) Expression of PTH mRNA detected by RT-PCR: Total RNA was extracted using an RNeasy Minikit according to the manufacturer's instructions. Reverse transcription of total DNA-free RNA was performed with hexamer random primers using the First-Strand cDNA Synthesis kit according to the manufacturer's instructions. Primers used for detection of PTH mRNA were: Forward: gac atg gct aaa gtt atg at and reverse: cag ctt ctt acg cag cca ttc tac. The PCR reaction parameters for PTH were 95°C for 5 min followed by 95°C for 20 s, 58°C for 20 s, and 72°C for 20 s repeated in 30 cycles followed by an extension at 72°C for 7 min. The PCR reaction contained, in a total of 50 μl, 2 μl of cDNA, 5 μl of 10X PCR buffer, 0.2 mM dNTP, 1.5 mM $MgCl_2$, and 1.0 U of Platinum Taq polymerase.

(d) Cell growth determination:

- Cells were harvested at different time points.
- Number of viable cells determined using the NucleoCounter.

3.5. Long-Term Cultures in the Literature

Reports of long-term cultures in the literature are scarce. Parathyroid glands from patients with uremia seem to be most successful (8, 9). An impressive 140 doublings were achieved with cells isolated from bovine parathyroid glands cultured in Coon's modified Ham's F-12 medium containing low (0.3 mM) concentrations of calcium and supplements of bovine hypothalamic extract, bovine pituitary extract, EGF, insulin, transferrin, selenous acid, hydrocortisone, triiodothyronine, retinoic acid, and galactose (10). We have made own attempts in trying to repeat such cultures without obvious success. Attempts of co-culture with irradiated 3T3 fibroblast cell lines and even irradiated isolated parathyroid fibroblasts have also been fruitless in order to achieve long-term cultures of proliferating and functioning parathyroid cells.

4. Notes

1. Reduce the time of warm ischemia as much as possible. Place the parathyroid glands in an ice-cold buffer after an initial and at this stage rather rough mincing with a sharp pair of small scissors. Transport of the glands from the operating theatre to the laboratory should be done in a buffered isotonic solution, for instance HEPES-buffered Ham's F10 with 10% FCS on ice, which keeps pH stable and metabolism low.
2. We have used a conical-shaped small container for this procedure, enabling the glands to be kept together in the narrow bottom of the container and facilitating the mincing.
3. We have used incubators which use circulating shaking movements, as well as transversely shaking, with similar results. The goal may differ depending on the purpose of dispersion – we have mostly tried to achieve clusters of approximately 2–8 cells and not a fully dispersed suspension. Single-cell suspensions seem to have been more harmed during the dispersion, expressing less number of cell surface receptors, etc. compared to small clusters of cells in which the presumably important cell contact still is present.
4. The parathyroid gland dispersion is the most critical step in obtaining cells for further culture. The collagenase treatment is harmful to the cells, and should be kept at as short time as possible. To reduce the exposure time to collagenase, we also use manual pipetting every 5–15 min.

5. This procedure further disrupts bounds between the cells but should be limited in time since the somewhat sensitive cells otherwise may be disrupted and a high rate of dead cells follows. The EGTA treatment has been performed if, for instance, measurements of intracellular calcium concentration are planned.
6. Since the collagenase treatment is harmful to the cells, a significant number of dead cells are present after the initial steps. Debris has lesser density than viable cells, and therefore is collected on top of the Percoll gradient while the viable cells remain in the bottom pellet. In case of a cell suspension rich in erythrocytes, a possible additive step is to include a 50% standard isotonic Percoll gradient as well. The comparably high-density erythrocytes are collected in the bottom pellet, debris on the top while the viable cells are to be collected in the interphase between the 25 and 50% Percoll cushions. It is essential to add these Percoll cushions thoroughly, starting either with the 50% in the bottom of a tube, followed by the 25% on top, and lastly the cell suspension to be separated – or to first add the cell suspension followed by 25% Percoll using a long syringe adding the Percoll underneath the cell suspension. The 50% Percoll may be added similarly.
7. For studies of PTH release, incubation times of between 30 min and 6 h are generally sufficient. For studies of immediate responses, we have used 30-min incubations (occasionally shorter; 5–15 min) while impacts on PTH release after genomic interactions (e.g. vitamin D3) take longer time (3, 6, 11). Similar short-term cultures have been made after incubation of cells in a serum-free medium containing 5 ng/ml EGF and 50 μg/ml bovine pituitary extract as well as in regular DMEM or RPMI containing 5–10% FCS (6).
8. A minimum of 200,000 cells in each well in a 24-well plate are needed, although 500,000 give much more reliable results, incubated in 500 μl buffer.
9. The ability of high external calcium concentrations to reduce PTH release has been used as a reflection of the viability of the cells. Thus, healthy cells express functioning cell surface calcium receptors and a functioning internal pathway while damaged cells seem to merely leak PTH in an uncontrolled way. Incubation of cells with 3.0 mM calcium should reduce PTH release with approximately 20–40% after 30 min compared to 0.5 mM external calcium concentration (3). It is wise to check the concentration of ionised calcium in the buffer by, e.g. an ion-selective electrode, in order to make sure that the BSA, serum, and phosphate in the buffer solution or medium do not dramatically interfere with the free calcium concentration. The used serum-free medium originally intended for keratinocytes

has an extremely low calcium concentration of 0.09 mM while DMEM has 1.8 mM and RPMI approximately 0.4 mM.

10. For proliferation studies, a low calcium concentration should be ideal for parathyroid cells, since this signal causes a diffuse parathyroid hyperplasia and presumably would initiate proliferation in the culture flask. Further, fibroblasts do not grow in extremely low calcium concentration, which is an advantage with the "keratinocyte" medium.
11. PTH may be analysed in different ways, depending on which fragments are wished to be evaluated. Generally, intact PTH is most common to investigate; therefore, an assay using two antisera should be used.
12. The set point – the external calcium concentration at which half-maximal inhibition of PTH release is achieved – is shifted to the right in pathological cells. The right-shifted set point is also seen after measuring the intracellular calcium concentration ($[Ca^{2+}]_i$) changes to external calcium variations. Thus, the relation between external calcium and PTH release is inverse and sigmoidal.
13. Cells may be cultured in different flasks, but regarding the usual limited number of cells not larger than 25-cm^2 cell culture flasks are recommended. The low calcium concentration of this medium triggers parathyroid chief cell division and proliferation and also reduces the proliferation and survival of occasional fibroblasts. The usual problem by performing cultures in higher calcium concentration has been a soon overgrowth of fibroblasts. Culture of parathyroid cell dispersions in DMEM/10% FCS leads to overgrowth of fibroblasts within 2–3 days. Nevertheless, this medium may be used to allow cells to attach to the culture flask plastics if used overnight. The use of surface coating with polylysine or serum, etc. has not been proven superior to culturing without.
14. Parathyroid cells derived from hyperplastic glands seem to be the easiest to obtain cell cultures from. In particular, cells from patients suffering from multiple endocrine neoplasia type 1 (MEN-1) are especially prone to proliferate in this environment. The reason may be their inborn tendency for proliferation, which is further stimulated in the low calcium concentration (6). The parathyroid pathophysiology of MEN-1 is characterised by a proliferative disorder rather than a set-point shift.
15. Selection of glands for establishing parathyroid long-term cultures is most likely extremely important. We have been most successful with hyperplastic glands, either from patients with uremic HPT or with MEN-1. Since lithium was successful in the medium, one would assume that a lithium-induced pathological parathyroid gland would be a good source as well.

References

1. **Grimelius L, Åkerström G, Johansson H, Juhlin C, Rastad J** 1991 The parathyroid glands. In: Kovacs K, Asa L eds. Functional endocrine pathology. Boston: Blackwell Scientific Publishers; 375–395.
2. **Liu W, Hellman P, Li Q, Yu W-R, Juhlin C, Nordlinder H, Rollman O, Åkerström G, Törmä H, Melhus H** 1996 Biosynthesis and function of all-trans- and 9-cis-retinoic acid in parathyroid cells. Biochem Biophys Res Commun 229:922–929.
3. **Ridefelt P, Nygren P, Hellman P, Larsson R, Rastad J, Åkerström G, Gylfe E** 1992 Regulation of parathyroid hormone release in normal and pathological parathyroid cells exposed to modulators of protein kinase C. Acta Endocrinol 126:505–509.
4. **Brown E** 1991 Extracellular Ca2+ sensing, regulation of parathyroid cell function, and role of Ca2+ and other ions as extracellular (first) messengers. Physiol Rev 71:371–411.
5. **Rudberg C, Grimelius L, Johansson H, Ljunghall S, Odselius R, Pertoft H, Wide L, Åkerström G** 1986 Alterations in density, morphology and parathyroid hormone release of dispersed parathyroid cells from patients with hyperparathyroidism. Acta Path Microbiol Immunol Scand Sect A 94: 253–261.
6. **Liu W, Ridefelt P, Akerstrom G, Hellman P** 2001 Differentiation of human parathyroid cells in culture. J Endocrinol 168:417–425.
7. **Bjorklund P, Akerstrom G, Westin G** 2007 Activated beta-catenin in the novel human parathyroid tumor cell line sHPT-1. Biochem Biophys Res Commun 352:532–536.
8. **Roussanne M, Gogusev J, Hory B, Duchambon P, Souberbielle J, Nabarra B, Pierrat D, Sarfati E, Drueke T, Bourdeau A** 1998 Persistance of Ca2+-sensing receptor expression in functionally active, long-term human parathyroid cultures. J Bone Min Res 13:354–362.
9. **Roussanne MC, Lieberherr M, Souberbielle JC, Sarfati E, Drueke T, Bourdeau A** 2001 Human parathyroid cell proliferation in response to calcium, NPS R-467, calcitriol and phosphate. Eur J Clin Invest 31:610–616.
10. **Brandi ML, Fitzpatrick LA, Coon HG, Aurbach GD** 1986 Bovine parathyroid cells: Cultures maintained for more than 140 populations doublings. Proc Natl Acad Sci USA 83:1709–1713.
11. **Ridefelt P, Hellman P, Wallfelt C, Åkerström G, Rastad J, Gylfe E** 1992 Neomycin interacts with Ca2+ sensing of normal and adenomatous parathyroid cells. Mol Cell Endocrinol 83: 211–218.

Chapter 5

Functional Analysis of Human Islets of Langerhans Maintained in Culture

Shanta J. Persaud, Bo Liu, and Peter M. Jones

Abstract

Islets of Langerhans isolated from experimental animals, such as mice and rats, have contributed much to our understanding of the mechanisms by which pancreatic β-cells secrete insulin in a regulated manner, and this knowledge is important in identifying potential novel therapies for Type 2 diabetes. However, although many of the signal transduction pathways identified in rodent islets are common to humans, some critical differences have been demonstrated experimentally. It is, therefore, essential that experiments are performed using islets isolated from human pancreas to provide robust data defining whether the key observations made in rodents are also applicable to the human situation. The rate-limiting factor in this area of research is the supply of high-quality human islets isolated from pancreases retrieved from organ donors, and the protocols described in this chapter aim to provide methods of maximising information that can be obtained when human islets are maintained in culture.

Key words: Human islets, α-Cell, β-Cell, δ-Cell, Insulin, Glucagon, Somatostatin, Viability, Hormone secretion, Gene knock-down

1. Introduction

Islets of Langerhans are small, approximately spherical endocrine cell clusters that are scattered throughout the pancreas. They show a fairly consistent size across species, with islets isolated from both mouse and human pancreas having 2,000–3,000 cells and an average diameter of approximately 200 μm. The majority islet cell type is the insulin-secreting β-cell, which is vital for maintaining glucose homeostasis, and the β-cells constitute approximately 80–90% of the total islet mass in rodents but only 50–60% of the total cell number in human islets (1, 2). Islets also contain glucagon-secreting α-cells and somatostatin-secreting δ-cells, which account for

Ragai R. Mitry and Robin D. Hughes (eds.), *Human Cell Culture Protocols*, Methods in Molecular Biology, vol. 806, DOI 10.1007/978-1-61779-367-7_5, © Springer Science+Business Media, LLC 2012

around 15 and 5% of the cell population in rodents and around 35 and 5% in humans, respectively (1).

There has been a concerted effort over the past four decades to define the intracellular signalling mechanisms by which β-cells respond to changes in nutrient levels, and it is clear that improved understanding of these pathways is a prerequisite for developing novel insulin secretagogue therapies for Type 2 diabetes and for generating β-cell replacements for Type 1 diabetes. The majority of experimental studies in this area have been performed using readily available islets isolated from animal species (mainly rats and mice) and insulin-secreting cell lines, since human pancreatic tissue for experimentation is scarce. Many of the studies in which human islets have been used have indicated that human and rodent β-cells have much in common, suggesting that experiments using rodent β-cells can provide a solid foundation for our understanding of the events occurring in human β-cells. Thus, in the context of Ca^{2+} handling, there is good evidence that human β-cells express voltage-dependent Ca^{2+} channels (VDCCs; (3)), that increased glucose concentrations cause Ca^{2+} elevations in human islets (4, 5), and that human islets express plasma membrane Ca^{2+}/ATPases that have been identified in rat islets (6).

However, one potentially confounding aspect of much of the past and current research is the realisation that, although many parallels do exist between rodent and human β-cell-signalling pathways, there are also several significant interspecies differences. For example, the expression and functions of some β-cell receptors varies between species. Thus, the primary action of adrenergic agonists on rat β-cells is to inhibit insulin secretion through α_2-adrenoreceptors, whereas human β-cells also express β-adrenoreceptors through which adrenergic agonists can stimulate insulin secretion (7); different classes of purinergic receptor are coupled to the regulation of intracellular Ca^{2+} ($[Ca^{2+}]_i$) in rodent (8) and human (9) β-cells; and melatonin inhibits glucose-induced insulin secretion from mouse β-cells, but not from human islets (10). Findings such as these imply that we cannot assume that mechanisms identified using rodent β-cell lines and islets from experimental animals will automatically apply to human β-cells, and reinforce the absolute requirement for confirming and extending the key observations to human tissue. Furthermore, the mechanisms regulating glucagon and somatostatin secretion from islets have been studied less extensively than β-cell secretory pathways, and it has not been established whether pathways identified in rodent α- and δ-cells also operate in humans.

The protocols in this chapter describe methods for the maintenance of isolated human islets in culture, assessment of islet purity and viability, quantification of islet hormone content, measurements of dynamic hormone secretion, and siRNA-mediated knock-down of human islet genes. An obvious prerequisite for all of these studies are islets of Langerhans isolated from human pancreases retrieved from organ donors with appropriate ethical permission for their use for research: we are grateful to the families of donors for their contributions to our understanding of stimulus–response coupling cascades in human islet cell populations.

2. Materials (See Note 1)

2.1. Maintenance of Islets in Culture

1. Connaught Memorial Research Laboratories (CMRL)-1066 medium (Invitrogen) supplemented with 10% fetal bovine serum (FBS; Invitrogen) and stored at 4°C. The medium may also be supplemented with penicillin (100 U/ml) and streptomycin (100 μg/ml; Sigma–Aldrich), if required.
2. Petri dishes (Sterilin).
3. Tissue culture incubator (Jencons).
4. Tissue culture hood (Jencons).
5. Microscope (Nikon).

2.2. Assessment of Islet Purity

2.2.1. Dithizone Staining of Islet β-Cells

1. Dithizone (diphenylthiocarbazone; Sigma–Aldrich); diluted to 10% in DMSO. Store at room temperature.

2.2.2. Electron Microscopy

1. EM fixative composition: 2.5% Glutaraldehyde (VWR International) in 0.1 M sodium cacodylate (Sigma–Aldrich). Store at 4°C.
2. Standard EM preparation and visualisation reagents (see Note 2).

2.3. Viability Assays

2.3.1. Trypan Blue Exclusion Test

1. Trypan blue (Sigma–Aldrich), diluted to 0.2% in phosphate-buffered saline (PBS; Sigma–Aldrich). Store at room temperature.
2. 96-Well plates (NUNC).
3. Incubator (Jencons).

2.3.2. ATP Synthesis

1. White-walled 96-well plates (Greiner Bio-One).
2. Cell-titer glo reagents (Promega).
3. Plate luminometer (Turner Biosystems).

2.3.3. MTT Assay

1. 96-Well plates (NUNC).
2. CMRL-1066 medium (Invitrogen) supplemented with 10% FBS (Invitrogen).
3. 1 mg/ml dimethylthiazol-diphenyltetrazolium bromide (MTT; Sigma–Aldrich) in CMRL; make up fresh as required.

4. Incubator (Jencons).
5. Acidified isopropanol: 48 ml isopropanol (Sigma–Aldrich) plus 250 μl concentrated HCl. Store at room temperature.
6. Absorbance microplate reader (Thermo Labsystems).

2.3.4. Caspase Assay

1. White-walled 96-well plates (Greiner Bio-One).
2. Caspase-glo 3/7 reagents (Promega).
3. Plate luminometer (Turner Biosystems).

2.3.5. Quantification of Islet Hormone Content

1. 10 mM HCl, diluted as appropriate from concentrated HCl stock. Store at room temperature.
2. 0.5 M NaH_2PO_4; pH to 7.4 with 5 M NaOH and add 0.5 mg/ml BSA.
3. Sonicator (Fisons Scientific).
4. ^{125}I-insulin; iodinated in-house. Also available from PerkinElmer.
5. Anti-insulin antibody generated in-house in guinea pigs. Also available from Dako.
6. Human insulin standard from Tocris Bioscience.
7. ^{125}I-glucagon from PerkinElmer.
8. Anti-glucagon antibody generated in-house in guinea pigs. Also available from Sigma–Aldrich.
9. Human glucagon standard from Bachem.
10. Somatostatin radioimmunoassay kit from Euro-Diagnostica, Malmo, Sweden.
11. Borate buffer composition: 133.4 mM boric acid, 10 mM EDTA, 67.5 mM NaOH in distilled water, pH to 8.0 with 10 mM HCl and then add 1 mg/ml BSA. Store at 4°C.
12. PBS (Sigma–Aldrich).
13. 30% (w/v) polyethylene glycol (PEG; VWR International): Dissolve 600 g PEG in 1 l distilled water and then make up to 2 l final volume with distilled water.
14. Precipitant: Dissolve 0.1% w/v γ-globulins (Sigma–Aldrich) in PBS, add an equal volume of 30% PEG, and supplement with 0.05% v/v polyoxyethylene-sorbitan monolaurate (Tween 20; Sigma–Aldrich).
15. Wizard Automatic Gamma counter, PerkinElmer.

2.4. Measurements of Dynamic Hormone Secretion

1. Perifusion chambers (Swinnex® 13 Filter Holders from Millipore).
2. 1-μm nylon mesh (Plastok Associates Ltd.).
3. Peristaltic pumps and tubing (Gilson).

4. Physiological salt solution composition: 111 mM NaCl, 27 mM $NaHCO_3$, 5 mM KCl, 1 mM $MgCl_2 \cdot 6H_2O$, 0.28 mM $MgSO_4 \cdot 7H_2O$, 0.12 mM KH_2PO_4 in distilled water. Store at 4°C. Adjust pH to 7.4 with 5% CO_2 and then add 0.5 mg/ml BSA and 2 mM glucose on the day of use.
5. Temperature-controlled chamber or room.
6. 2-ml 96-well plates (Thermo Scientific).
7. Radioimmunoassay reagents as for Subheading 2.4.

2.5. siRNA-Mediated Knock-Down of Human Islet Genes

1. Research Park Memorial Institute (RPMI)-1640 medium (Invitrogen).
2. 1% w/v Gelatin (Type B, from bovine skin; Sigma–Aldrich) in distilled water.
3. 24-Well plates (NUNC).
4. ON-TARGETplus SMARTpool siRNAs (Dharmacon).
5. Control (non-silencing) siRNAs (QIAGEN).
6. Metafectene Pro transfection reagent (Biontex Laboratories GmbH).
7. 0.02% w/v ethylenediaminetetraacetic acid (EDTA) solution (Sigma–Aldrich).
8. FBS (Invitrogen).

3. Methods

The methods described below assume availability of high-purity (>80%) isolated human islets ((5); see Note 1), with minimal contamination by pancreatic exocrine cells. Islets may be used for experimentation immediately after isolation, but it is more usual to maintain them in culture at least overnight to allow cellular recovery from the isolation procedure and permit re-synthesis of membrane-associated proteins that may have been digested during the islet isolation process. This is especially important when studies are directed at investigating the expression and/or function of cell surface receptors, transporters, or ion channels. Subheading 3.1 describes the appropriate procedures for islet maintenance in culture.

Researchers who have experience of using isolated rodent islets may find that human islets are more difficult to work with: although they are similar in size to rat and mouse islets, islets retrieved by density gradient purification from human pancreas (5) do not appear as such discrete structures and may be difficult to distinguish from contaminating exocrine cells that form small clusters in culture. The protocols described in Subheading 3.2 should allow confidence in identification of human islets and provide information on the purity of isolated islet populations.

Islets are metabolically active micro-organs and are susceptible to compromised function if they are not maintained under appropriate conditions. For example, glucose metabolism by β-cells is temperature dependent and β-cell function declines at reduced temperatures. Assessment of islet viability, after isolation or after exposure to experimental procedures, can be carried out using the protocols described in Subheading 3.3.

The essential function of islet cells is to secrete hormones that are involved in maintenance of glucose homeostasis, so quantification of islet hormone content or dynamic release in response to potential modulators of secretion is required. Appropriate protocols are described in Subheadings 3.4 and 3.5.

For some experiments, it may be necessary to functionally inhibit or decrease expression of a particular protein to identify the role that it plays in regulating hormone secretion. Receptor antagonists and inhibitors of enzymes, ion channel function, etc. may be commercially available and their effects can be investigated by measuring hormone secretion using the protocols described in Subheading 3.5. It may be necessary to knock down mRNAs coding for the protein of interest if an appropriate selective inhibitor or antagonist is not available or to provide additional supporting data for the inhibitor studies. This may be achieved using small interfering RNAs (siRNAs) as described in Subheading 3.6.

3.1. Maintenance of Islets in Culture

1. After isolation by density gradient centrifugation (10), transfer islets to 100-mm Petri dishes (see Note 3) at a density of approximately 500–1,000 islets/ml in 10 ml CMRL-1066 medium supplemented with 10% FBS.
2. Replace the culture medium under aseptic conditions (see Note 4) every 2–3 days until the islets are used for determination of purity (Subheading 3.2), viability (Subheading 3.3), or for experimental purposes.
3. The general status of the islets can be assessed by viewing them using a standard tissue culture microscope (see Note 5; Fig. 1).

3.2. Assessment of Islet Purity

3.2.1. Dithizone Staining of Islet β-Cells (See Note 6)

1. Dilute 10% dithizone stock 1/100 in PBS to give a final concentration of 0.1%.
2. Re-suspend islets maintained in culture and transfer approximately 1% of the islet material from the Petri dish (100 μl from 10 ml) to a 1.5-ml Eppendorf tube.
3. Centrifuge at room temperature for 30 s using a bench Eppendorf centrifuge at 1,000 × *g*.
4. Remove the supernatant fluid using a 200-μl Gilson pipette.
5. Add 1 ml of 0.1% dithizone to the islet pellet.

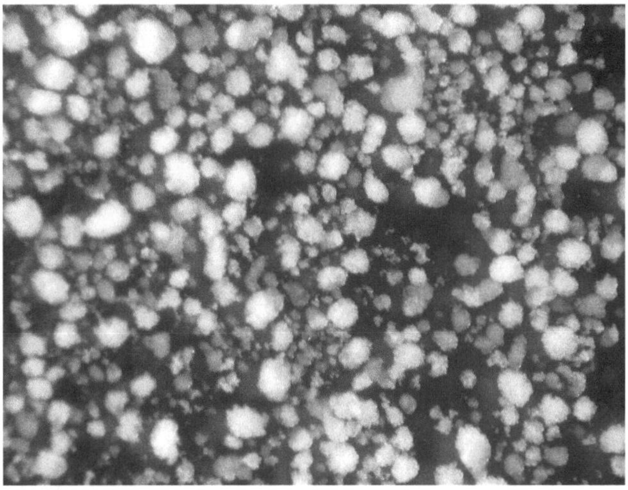

Fig. 1. Isolated human islets of Langerhans. Image obtained under a phase-contrast microscope showing human islets of Langerhans after density gradient purification.

6. Mix the dithizone with the islets by pipetting up and down for 1–2 min. The red stain can be seen to develop by eye over this time.
7. Plate out islets in PBS in a Petri dish and view staining under a standard binocular microscope (see Note 7).
8. Count the proportion of red-stained islet clusters to determine the purity of the source islet population.

3.2.2. Electron Microscopy

1. Remove EM fixative from the fridge and allow it to reach room temperature.
2. Re-suspend islets maintained in culture and transfer material equivalent to approximately 1,000–2,000 islets from the Petri dish to a 1.5-ml Eppendorf tube.
3. Centrifuge at room temperature for 30 s using a bench Eppendorf centrifuge at 1,000 × *g*.
4. Remove supernatant fluid using a 200-μl Gilson pipette so that only approximately 100 μl remains.
5. Add an equivalent volume of EM fixative at room temperature.
6. Leave at room temperature for approximately 5 min.
7. Centrifuge islet material at room temperature for 30 s using a bench Eppendorf centrifuge at 1,000 × *g*.
8. Remove all supernatant fluid using a 200-μl Gilson pipette or a fine glass capillary.
9. Add 400 μl EM fixative to the islet pellet and store at 4°C until standard transmission EM processing (see Note 2; Fig. 2).

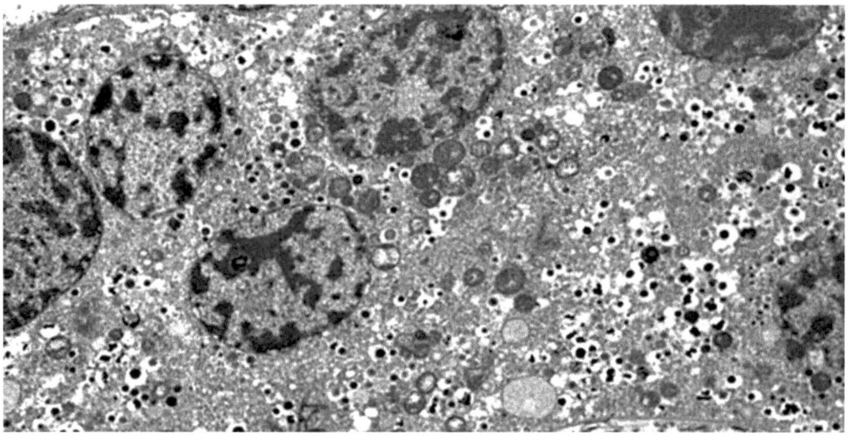

Fig. 2. Transmission electron micrograph of cells in isolated human islets. Transmission electron micrograph showing β-cells with typical dense core secretory granules within isolated human islets.

3.3. Viability Assays (See Note 8)

3.3.1. Trypan Blue Exclusion Test

1. Under aseptic conditions, transfer groups of ten islets in CMRL medium to 4–5 wells of a 96-well plate in a volume of 100 μl using a 100-μl Gilson pipette under a standard binocular microscope (see Notes 7 and 9).
2. Add 100 μl 0.2% (w/v) trypan blue to the wells containing islets.
3. Incubate at 37°C, 5% CO_2 for 15 min.
4. Remove trypan blue, ensuring that islets remain in the well (see Note 10).
5. Wash islets once with 200 μl PBS, again making sure that islets are not removed.
6. Add 200 μl PBS to each well containing islets.
7. View islets in wells (see Note 5) and take photographs as required (see Note 11).

3.3.2. ATP Synthesis

1. Under aseptic conditions, transfer groups of ten islets in CMRL medium to 6–8 wells of a white-walled 96-well plate in a volume of 50 μl using a 100-μl Gilson pipette (see Notes 7, 9, and 12).
2. Add 50 μl of Promega CellTiter-Glo reagents, reconstituted according to the manufacturer's recommendations.
3. Cover the plate and briefly mix the contents on a shaking platform.
4. Incubate for approximately 10 min at room temperature.
5. Determine ATP production by quantifying luminescence with a plate luminometer.

3.3.3. MTT Assay

1. Under aseptic conditions, transfer groups of five islets in CMRL medium to 6–8 wells of a 96-well plate in a volume of 50 μl using a 100-μl Gilson pipette (see Notes 7, 9, and 12).
2. Add 50 μl of 1 mg/ml MTT to each of the wells containing cells.
3. Incubate for 4 h at 37°C, after which time purple formazan crystals should be visible.
4. Add 100 μl acidified isopropanol to each well containing islets.
5. Solubilise the purple/brown crystals of formazan by pipetting up and down several times.
6. Determine formazan production by quantifying absorbance at 570 nm using a plate reader.

3.3.4. Caspase Assay

1. Under aseptic conditions, transfer groups of ten islets in CMRL medium to 6–8 wells of a white-walled 96-well plate in a volume of 100 μl using a 100-μl Gilson pipette (see Notes 7, 9, and 12).
2. Add 100 μl of Promega Caspase-Glo 3/7 reagents, reconstituted according to the manufacturer's recommendations.
3. Cover the plate and briefly mix the contents on a shaking platform.
4. Incubate for approximately 1 h at room temperature.
5. Determine caspase 3/7 activities by quantifying luminescence with a plate luminometer.

3.4. Quantification of Islet Hormone Content (See Note 13)

1. Select groups of ten isolated human islets using a 200-μl Gilson pipette or a fine glass capillary under a standard binocular microscope (see Notes 9 and 12).
2. Transfer the islets into a 0.5-ml Eppendorf tube.
3. Centrifuge at room temperature for 30 s using a bench Eppendorf centrifuge at 1,000 × *g*.
4. Remove all supernatant fluid using a 200-μl Gilson pipette or a fine glass capillary under a standard binocular microscope (see Note 14).
5. Re-suspend the islet pellet in 50 μl 10 mM HCl and leave at room temperature for approximately 15 min (see Notes 15 and 16).
6. Neutralise the HCl by adding 200 μl 0.5 M NaH_2PO_4 buffer supplemented with 0.5 mg/ml BSA.
7. Sonicate the islet samples on ice (see Note 17).
8. Serially dilute the islet extracts with borate buffer for immunoassay of islet hormones (see Note 18).

9. For the hormone radioimmunoassays, prepare standard curves in triplicate over 0.08–10 ng/ml (insulin and glucagon) or 3.9–250 pM (somatostatin) by carrying out doubling dilutions of standard stocks (10 ng/ml insulin and glucagon; 250 pM somatostatin; see Notes 19 and 20). Borate buffer is used as a diluent for insulin and glucagon standards while phosphate buffer is used for the somatostatin assay (provided with the immunoassay kit). Each standard should be in a final volume of 100 µl.
10. For each hormone, assay dilutions of the islet extracts as duplicate samples of 100 µl.
11. Add 100 µl hormone-specific antibody to all standards and samples (see Note 21).
12. Add ^{125}I-labelled hormone, diluted to approximately 5,000 cpm/100 µl, to all standards and samples (see Note 22).
13. Prepare control samples to determine the total amount of radioactivity added (^{125}I-labelled hormone alone), non-specific binding (assay buffer plus ^{125}I-labelled hormone), and maximum binding (assay buffer, antibody, and ^{125}I-labelled hormone).
14. Leave all samples and standards at 4°C for 48–72 h for antibody–hormone binding to reach equilibrium (see Note 23).
15. Precipitate hormone–antibody complexes by adding 1 ml of precipitant to each of the assay tubes and centrifuging at $2,000 \times g$ for 15 min at 4°C.
16. Aspirate the supernatant fluid to leave a firm pellet (see Note 24).
17. Detect the γ emissions from all pellets using a gamma counter and interpolate the hormone concentrations in the diluted islet extracts from the appropriate standard curves.

3.5. Measurements of Dynamic Hormone Secretion (See Note 25)

1. Select groups of 100 isolated human islets using a 200-µl Gilson pipette or a fine glass capillary under a standard binocular microscope (see Notes 7, 12, 26, and 27).
2. Transfer the islets onto 1-µm filters in individual perifusion chambers connected to buffer reservoirs and peristaltic pumps by 2-mm diameter tubing (see Fig. 3).
3. Fill the chambers with physiological salt solution and remove bubbles when connecting tubing to the chambers (see Note 28).
4. Use peristaltic pumps to perifuse the islets with the salt solution at 0.5 ml/min for at least 1 h at 37°C to allow hormones released from damaged cells to be washed through the system (see Note 29).

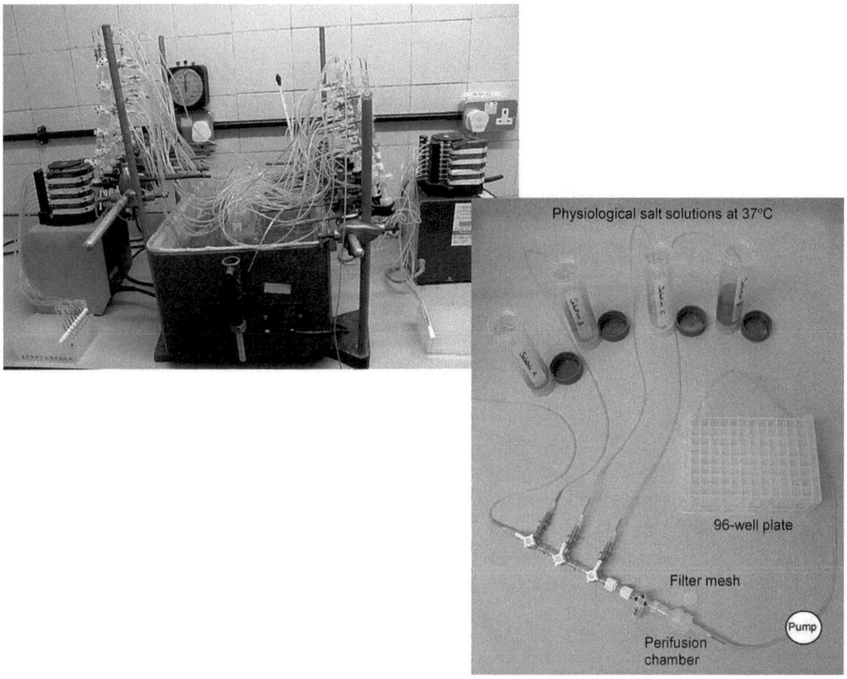

Fig. 3. Islet perifusion apparatus. Physiological salt solutions are pumped over islets in perifusion chambers using peristaltic pumps. Sixteen groups of islets can be perifused simultaneously using two 8-channel pumps. Buffers are maintained at 37°C in a water bath (not shown in the simplified image) and changed using three-way taps. Perifusate samples are collected into 96-well blocks and assayed for islet hormones.

5. Select experimental buffers in a heated reservoir using a three-way tap system and transfer buffers to the perifusion chambers using peristaltic pumps at a flow rate of 0.5 ml/min (see Note 30).
6. Every 2 min, collect 1-ml outflow fractions containing secreted islet hormones into 2-ml 96-well plates.
7. Quantify hormone concentrations of perifusate fractions according to the hormone immunoassay protocols described in Subheading 3.4.
8. Hormone output can be expressed as rate of secretion with time (see Fig. 4).

3.6. siRNA-Mediated Knock-Down of Human Islet Genes

3.6.1. Intact Islets

1. Add 350 μl/well, pre-warmed, serum- and antibiotic-free RPMI medium into 6 wells of a gelatin-coated 24-well plate (see Notes 31 and 32).
2. Under aseptic conditions, transfer groups of 50 islets in RPMI to 6 wells of the above plate in a volume of 50 μl using a 100-μl Gilson pipette (see Notes 7, 12, and 33).
3. Prepare the siRNAs in 150 μl serum- and antibiotic-free RPMI (see Notes 31, 32, 34–36).

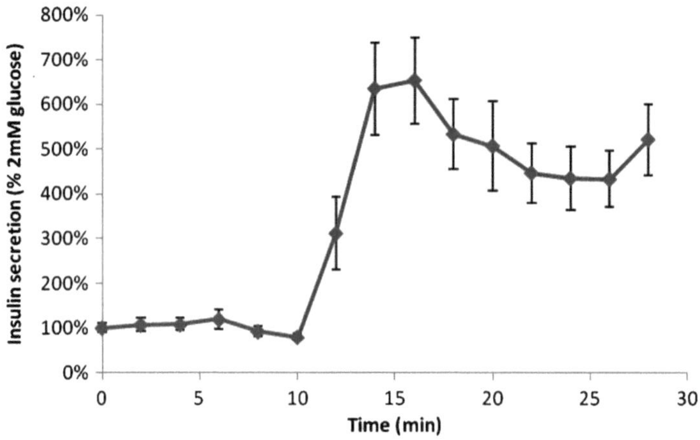

Fig. 4. Dynamic glucose-stimulated insulin secretion from isolated human islets. Isolated human islets were perifused with a buffer containing 2 mM glucose for the first 10 min and with 20 mM glucose thereafter. Insulin secreted from the islets at 2-min intervals was measured by radioimmunoassay. 20 mM glucose induced a rapid, sustained increase in insulin secretion. Data show mean ± SEM of islets in eight separate perifusion chambers.

4. Prepare the transfection reagent solution by addition of 3 μl Metafectene Pro to 147 μl serum- and antibiotic-free RPMI (see Notes 31, 32, 37, and 38).
5. Add the siRNA solution to the transfection reagent solution and incubate at room temperature for 20 min (see Note 39).
6. Prepare 300 μl control transfection solution as described in steps 3–5 above, replacing target siRNAs with non-targeting/scrambled siRNAs (see Note 40).
7. Immediately following incubation, add 100 μl of the target/control siRNA–lipid complexes dropwise to the wells containing islets (see Note 41).
8. Gently mix by swirling to ensure overall distribution of siRNA–lipid complexes and incubate the islets at 37°C, 5% CO_2 for 8 h (see Note 42).
9. Add 500μl RPMI containing 20% v/v FBS 8 h after transfection, and incubate islets for a further 16 h at 37°C, 5% CO_2 (see Note 31).
10. After the 16-h incubation, remove all medium, wash the islets once with serum- and antibiotic-free RPMI, and re-suspend them in 400 μl of this medium (see Note 14).
11. Carry out a second round of transfection by following steps 3–9 above (see Note 43).
12. Gently swirl the plate so that the islets accumulate at the centre of each well. Islets can then be used for determination of target gene silencing or further functional analysis (see Notes 44 and 45).

3.6.2. Dissociated Islets (See Note 46)

1. Under aseptic conditions, transfer 400 islets into a 15-ml centrifuge tube (see Notes 7 and 12).
2. Pellet by brief centrifugation at room temperature (1,000 × *g*, 1 min) and discard the supernatant fluid.
3. Re-suspend the islets in 1 ml 0.02% EDTA solution (see Note 47).
4. Incubate the islets at 37°C for 7–8 min, pipetting up and down every 2 min to facilitate islet dissociation.
5. Pellet by brief centrifugation at room temperature (1,000 × *g*, 5 min) and discard the supernatant fluid.
6. Re-suspend islet cell clusters/dissociated islet cells in 400 μl serum- and antibiotic-free RPMI medium (see Notes 31 and 32).
7. Transfer 50-μl aliquots (equivalent to 50 islets) to 6 wells of a gelatin-coated 24-well plate using a 200-μl Gilson pipette.
8. Prepare siRNA–lipid complexes and perform transfections as described in Subheading 3.6.1 (steps 1–10).

4. Notes

1. Human islets are isolated from cadaver organ donors with the appropriate ethical approvals.
2. Electron microscopy protocols require dedicated experimental equipment and should be carried out by experienced personnel with access to transmission electron microscopy facilities.
3. Islets are maintained in culture in uncharged Petri dishes to minimise cell adhesion.
4. Islets should be maintained in culture under aseptic conditions by handling them in a category two laminar flow hood and maintaining them in a standard tissue culture incubator in an atmosphere of 5% CO_2.
5. Islets in culture are viewed using a Nikon inverted microscope (final magnification ×10 to ×40). Other suitable microscopes may be used.
6. Dithizone (diphenylthiocarbazone) binds to zinc ions in β-cell secretory granules, and this results in red staining of β-cells (and islets). Exocrine and other tissues, which do not contain zinc, do not stain. This is especially useful for detecting human islets which are difficult to distinguish from clusters of exocrine tissue.
7. Islets are viewed and selected using an Olympus SZ40 microscope (final magnification ×10 to ×20). Other suitable microscopes may be used.

8. There are a range of methodologies available to assess islet viability. The ones described here depend on the principles of dye uptake by compromised plasma membranes (trypan blue exclusion test), synthesis of ATP by metabolically active cells (ATP assay), NAD-dependent dehydrogenase cleavage of a tetrazolium salt (MTT assay), and loss of cell viability by the induction of apoptosis (caspase assay). Depending on the circumstances, one or more of these protocols may be used to determine whether the islets are functioning normally.
9. When retrieving islets for viability assays, it may be preferable to transfer an aliquot of islets from the culture dish to a separate container so that they can be used without the possibility of contaminating the stock population.
10. It may be easiest to remove supernatant fluid with the aid of a microscope to make sure that islets are not lost during medium removal and islet washing steps.
11. Unless islets have been exposed to membrane perturbing agents or are of very low viability, it is unlikely that trypan blue will gain access to cells beyond the islet periphery.
12. Islets of variable size should be picked to distribute islets of different sizes randomly between the groups.
13. The protocols described in Subheadings 3.4 and 3.5 describe quantification of the major islet hormones, but the samples may also be assayed for other islet hormones, such as pancreatic polypeptide and peptide YY, although more islets may be required per group as these hormones are less abundant.
14. Care should be taken that islets are not removed when withdrawing medium.
15. Samples may be frozen at this stage and samples assayed for hormone content at a later stage.
16. If only insulin content is to be quantified, acidified ethanol (ethanol:water:HCl, 52:17:1) rather than HCl/NaH_2PO_4 can be used to solubilise the islet cells, and extracts must be diluted at least 1/500 for immunoassay.
17. Any suitable method may be used to disrupt the islet cells and release the hormones from secretory granules.
18. As a general guideline, the required dilutions of islet extracts can be calculated by assuming that each human islet contains 40 ng insulin, 15 ng glucagon, and 3 ng somatostatin. In this way, the diluted samples should fall on the appropriate hormone assay standard curve.
19. There is no necessity to assay hormone levels by radioimmunoassay. There is a range of commercially available enzyme immunoassays for human islet hormones, or homogeneous time-resolved fluorescence assays may be used.

20. Different immunoassays may be sensitive over different standard curve ranges.
21. The antibody dilutions used for the assays depend on the titre of the antibodies being used.
22. Care should be taken when handling radioactivity and ionising radiation local rules should be followed.
23. The binding reaction can be accelerated by incubation at room temperature rather than 4°C, but this may reduce sensitivity of the assays.
24. Ensure that the radioactive waste is handled appropriately and disposed of in an appropriate manner.
25. Islet hormone secretion may also be measured in "static incubation" experiments, but dynamic perifusion experiments provide detailed information on the time course and reversibility of secretory events.
26. More islets may be required if quantification of secretion of a low-abundance hormone is required.
27. If large numbers of islets are required, fixed-volume aliquots of islets can be dispensed into perifusion chambers and secretion expressed as % basal rather than as pg/islet/min, etc.
28. The presence of bubbles within the closed perifusion system may lead to unregulated hormone release through shear stress on the islets.
29. Islet hormone secretion is temperature sensitive, so perifusion experiments should take place in a temperature-controlled cabinet or in a room maintained at 37°C.
30. The perifusion flow rate can be reduced if a low-abundance hormone is being investigated and/or if islets are rate limiting.
31. Serum may inhibit formation of siRNA/lipid complexes, so serum-free medium should be used for complex formation and serum-free conditions should be maintained for the first 8 h after transfection.
32. The presence of antibiotics during transfection may reduce the efficiency of transfection.
33. Based on experimental requirements, the number of islets used can be varied.
34. A set of several siRNAs targeting different sites for the gene of interest may be used to increase the probability of high-efficiency silencing.
35. The lowest optimal concentration of siRNAs should be identified for each siRNA and experimental condition.
36. RNA oligos are susceptible to RNase-mediated degradation and should therefore be handled in RNase-free conditions.

37. Metafectene Pro transfection reagent contains cationic lipids with co-lipids in water. Other suitable transfection reagents may be used, and these should be optimised accordingly.
38. A 1:1 siRNA:Metafectene Pro ratio is often effective, but optimal siRNA:Metafectene Pro ratios should be established for each siRNA and experimental condition.
39. Mixing procedures should be avoided while combining the siRNA solution and transfection reagent solution, as the shear stress generated might be detrimental to the siRNA–lipid complex.
40. Non-targeting or scrambled siRNAs should be used as negative controls in all transfection experiments. Comparison of the results between non-targeting negative-control cells and siRNA-transfected cells should reveal a clear pattern of siRNA-mediated target reduction.
41. Three replicates for control and target siRNAs should be sufficient, but further replicates may be used if islets are not rate limiting.
42. Gently shake the plate before transferring to the incubator to facilitate the uptake of siRNAs by islets during the 8-h incubation period.
43. Repeated administration of siRNAs may have a cumulative effect on target reduction. We have observed that a repeated round of transfection using identical experimental conditions significantly increases transfection efficiency without increased cell death (11).
44. Time-course studies may be required to identify the optimal transfection period for reduction of target gene expression. Typically, time points for detection of target reduction at the mRNA level are 24–48 h, with protein and other functional analysis requiring longer transfection period of between 48 and 72 h.
45. siRNA-treated human islets may be used for studies investigating regulation of apoptosis or hormone secretion using protocols described in Subheadings 3.3.4 and 3.5, and also for other functional analyses, such as regulation or preproinsulin mRNA expression (12).
46. Higher transfection efficiencies may be obtained by using dissociated islet cells rather than intact islets. Islets can be dissociated prior to transfection when the integrity of the islet architecture is not essential for functional analysis.
47. Other suitable dissociation solutions may be used.

Acknowledgements

The authors are grateful to the relatives of organ donors for human pancreases and to Professor S. Amiel and Dr. G.C. Huang for provision of isolated human islets.

References

1. Cabrera, O., Berman, D.M., Kenyon, N.S., Ricordi, C., Berggren, P.O. and Caicedo, A. (2006) The unique cytoarchitecture of human pancreatic islets has implications for islet cell function. *Proc Natl Acad Sci USA.* **103**, 2334–9.
2. Kim, A., Miller, K., Jo, J., Kilimnik, G., Wojcik, P. and Hara, M. (2009) Islet architecture. A comparative study. *Islets* **1**, 129–136.
3. Davalli, A.M.,Biancardi, E., Pollo, A., Socci, C., Pontiroli, A.E., Pozza, G., Clementi, F., Sher, E. and Carbone, E. (1996) Dihydropyridine-sensitive and -insensitive voltage-operated calcium channels participate in the control of glucose-induced secretion from human pancreatic beta cells. *J.Endocrinol.* **150**, 195–203.
4. Martin, F. and Soria, B. (1996) Glucose-induced (Ca^{2+}) oscillations in single human pancreatic islets. *Cell Calcium* **20**, 409–414.
5. Huang, G.C., Zhao, M., Jones, P.M., Persaud, S., Ramracheya, R., Lobner, K., et al. (2004) The development of new density gradient media for the purification of human islets and islet quality assessments. *Transplantation* 77, 143–145.
6. Varadi, A., Molnar, E. and Ashcroft, S.J.H. (1996) A unique combination of plasma membrane Ca^{2+}/ATPase isoforms is expressed in islets of Langerhans and pancreatic β-cell lines. *Biochem. J.* **314**, 663–669.
7. Lacey, R.J., Berrow, N.S., London, N.J.M., Lake, S.P., James, R.F.L., Scarpello, J.H.B. and Morgan, N.G. (1990) Differential effects of β-adrenergic agonists on insulin secretion from pancreatic islets isolated from rat and man. *J. Mol. Endocrinol.* **5**, 49–54.
8. Li, G., Milani, D., Dunne, M.J., Pralong, W., Theler, J., Petersen, O.H. and Wollheim, C.B. (1991) Extracellular ATP causes Ca^{2+}-dependent and -independent insulin secretion in RINm5F cells. *J. Biol. Chem.* **266**, 3449–3452.
9. Squires, P.E., James, R.F.L., London, N.J.M. and Dunne, M.J. (1994) ATP-induced intracellular Ca^{2+} signals in isolated human insulin-secreting cells. *Pflugers Arch.* **427**, 181–183.
10. Ramracheya, R.D., Muller, D., Squires, P.E., Brereton, H., Sugden, D., Huang, G.C. Amiel, S.A., Jones, P.M. and Persaud, S.J. (2008) Function and expression of melatonin receptors on human pancreatic islets. *J. Pineal Res.* **44**, 273–279.
11. Muller, D., Jones, P.M. and Persaud, S.J. (2008) Expression of a potent anti-apoptotic protein in MIN6 cells using METAFECTENE PRO, Application note. http://www.biontex.com/con_4_6_4/cms/upload/pdf/Muller_MP_en.pdf.
12. Muller, D., Huang, G.C., Amiel, S., Jones, P.M. and Persaud, S.J. (2006) Identification of insulin signalling elements in human β-cells: autocrine regulation of insulin gene expression. *Diabetes* **55**, 2835–2842.

Chapter 6

Conversion of Non-endocrine Human Pancreatic Cells to Insulin-Producing Cells for Treatment of Diabetes

Min Zhao and Guo Cai Huang

Abstract

Type I diabetes results from the autoimmune destruction of the insulin-secreting pancreatic β-cells, affecting many millions of people worldwide. The optimal treatment is to restore the endogenous supply of insulin either through the transplantation of pancreas or the transplantation of islets of langerhans or simply the β-cells. However, the donated pancreas organs are limited and the available organs are only able to treat a small portion of the diabetes patients. Thus, glucose-responsive, insulin-producing cells from human origin are urgently needed. The aim of this chapter is to give some insight views to how to turn the potential human pancreatic non-endocrine cells into cells that are capable of secreting insulin in response to glucose and ameliorating insulin-deficient diabetes conditions after transplantation.

Key words: Diabetes, Human islet transplantation, Pancreatic beta cells, Dedifferentiation, Transdifferentiation

1. Introduction

There are substantial evidences suggesting that the adult pancreas can generate new β-cells in response to pancreatic damage or increased demand for insulin (1–3). The sources of these new β-cells are now known to be pre-existing β-cells (4). However, the capacity of mature β-cells to proliferation and then re-differentiate back into as β-cells has not been demonstrated in vitro. An alternative source for new β-cells may be pancreatic non-endocrine cells (5, 6), as the endocrine and exocrine cells share the same origin in development – the epithelia of the foregut, under the complex control of transcription factors (7–9). These cells could hold some chromatin structures similar to the endocrine cells and could be diverted into endocrine cells. In culture, both endocrine and

Ragai R. Mitry and Robin D. Hughes (eds.), *Human Cell Culture Protocols*, Methods in Molecular Biology, vol. 806, DOI 10.1007/978-1-61779-367-7_6, © Springer Science+Business Media, LLC 2012

exocrine cells dedifferentiate rapidly and express the same ductal cell marker, CK-19 (human) or CK-20 (rodent) (10–12), indicating that they are dedifferentiating in a similar pattern. Furthermore, exocrine and ductal cells of both rodent and man have been made to express Pdx1/Ipf1 gene (13, 14), an important transcription factor in pancreatic development and maintenance of β-cell phenotype (15), suggesting that these non-endocrine pancreatic cells can differentiate into endocrine cells in a suitable environment and that cells with the ductal phenotype could be an intermediate product between undifferentiated and terminally differentiated cells. Indeed, human pancreatic duct cells have been implicated as the origin of new β-cells generated without (16, 17) or with the help of ectopic gene expression (18). Rodent exocrine cells can be induced to express insulin under several protocols (19, 20), supporting the concept that the exocrine cell may be a candidate for new β-cell surrogates. However, the goal of generating human glucose-responsive β-cell surrogates sufficient to reverse diabetes in clinical setting remains elusive. In an attempt to enhance efficacy of β-cell generation from human non-endocrine pancreatic cells, we developed a protocol by culturing the non-endocrine cells to achieve a phase with partial dedifferentiation. Secondly; the cells are differentially induced by culturing the cells in a three-dimensional structure together with a cocktail of chemicals which favored the differentiation of endocrine cells in vitro with or without prior ectopic expression of Pdx1/Ipf1, NuroD1, and Ngn3 genes. Finally, the partially differentiated cells are allowed to mature in a diabetic environment in vivo in a SCID mouse. This leads to the reduction in hyperglycemia in the diabetes mouse.

2. Materials

2.1. Human Islet Isolation and Purification

1. Collagenases NB1, 2000 U and Neutral Protease NB, and 50 DMC-U are from SERVA electrophoresis GmbH Germany. The enzymes are stored at 4°C. The stock solution is prepared as follows: Collagenases NB1 is dissolved in 50 ml Hank's balanced salt solution (HBSS) buffer solution with gentle agitating and incubated for 15–30 min at 4°C. Neutral Protease NB is dissolved separately in 5 ml of pyrogen-free water and incubated at 4°C for 15–30 min. The working solution is prepared by mixing the two freshly prepared enzyme stock solutions with HBSS buffer (containing ≥3.1 mM $CaCl_2$) to obtain the desired volume for the tissue dissociation. After mixing, working solution must be used immediately (see Note 1).

2. Serine Protease inhibitor – Pefabloc SC PLUS and PSC-protector solution (Roche Diagnostics GmbH Germany) are stored at 4°C. Perfabloc SC is dissolved in pyrogen-free water to give a final concentration of 30 mg/ml and this stock solution is aliquoted as 1 ml per microfuge tube, wrapped in aluminum foil, and then stored at −20°C. The working solution contains 0.8 mM of Perfabloc SC and 1 ml of PSC-protector solution (see Note 2).
3. University of Wisconsin solution (UW), ViaSpan™ (Du Pont Critical Care, Waukegan, USA). Store at 4°C.
4. Buffer solutions pack: Custom-made pack produced by Lonza Wokingham Ltd, Wokingham, UK. The pack contains HBSS (350 ml) for dissolving enzymes, minimum essential medium (MEM) for dilution (8 L), HBSS (150 ml) for pancreas decontamination, and HBSS (500 ml) for final rinse. Pack stored at room temperature.
5. Biocoll cell separating solution with density at 1.10 g/ml (Biochrom AG, Berlin, Germany). Store at 4°C.
6. Dithizone solution: Freshly prepared by dissolving 10 mg of dithizone (Diphenyl Thiocarbazine) in 1 ml of dimethyl sulfoxide (DMSO) solution and then further diluted in 200 ml HBSS and mixed using a magnetic stirrer. To avoid precipitation of dithizone, the solution must be added dropwise to HBSS. In order to stain islets, islet suspension is added to dithizone/HBSS solution at 1:1 (v/v) dilution. This gives dithizone final concentration of around 25 μg/ml.

2.2. Isolated Cell Viability Analyses

1. Fluorescein diacetate (FDA; Sigma, UK) stock solution: Dissolve FDA in acetone (9.9 mg/ml) to give a concentration of 24 μM. Working solution should be 0.46 μM.
2. Propidium iodide (PI, Sigma, UK) stock solution: Dissolve PI in PBS (0.5 mg/ml) to give a concentration of 748 μM. Working solution should be 14.34 μM.

2.3. Culture of Human Pancreatic Non-endocrine Cells

1. Washing solution: HBSS containing 1% human serum albumin (HSA; Bio Products Laboratory UK).
2. Cell culture medium: CMRL1066 medium supplemented with 10% FCS, 2 mM L-glutamine, 500 U/ml of penicillin, 500 μg/ml of streptomycin, and 0.25 μg/ml of amphotericin B (Invitrogen, UK) is used as the culture medium for both human islets and non-endocrine pancreatic cells.
3. Streptozotocin (STZ) solution (Sigma UK): 50 mg of STZ (stored at −80°C) is thawed at room temperature and dissolved in 2 ml citron buffer, pH4.5, and use immediately (for the

Table 1
Primary antibodies list

Antibody	Supplier	Cat. No.	Source
Anti-glucagon mAb	SIGMA	G2654	Mouse IgG1
Anti-human alpha amylase	SIGMA	A8273	Rabbit polyclone
Anti-insulin mAb	SIGMA	12018	Mouse IgG-1
Anti-human somatostatin	AbD serotec	AHP533	Rabbit polyclone
Anti-human pancreatic polypeptide Ab-1	Thermo scientific	RB-1310-A	Rabbit polyclone
Anti-human nestin	Chemicon International	MAB5326	Mouse monoclone IgG1
Anti-human Ki-67	DAKO	A0047	Rabbit
Anti-human, mouse, rat PDX-1 (A-17)	Santa Cruz Biotech	Sc-14664	Goat polyclonal IgG
Anti-human neuro-D (A-10)	Santa Cruz Biotech	Sc-46684	Mouse monoclone IgG1
Anti-human neurogenin-3(H-80)	Goat Polyclonal Ab	Sc-25654	Rabbit polyclonal IgG
Anti-human cytokeratin-19 (A53B/A2)	Santa Cruz Biotech	Sc-6278	Mouse monoclone IgG1
Anti-human cytokeratin-19	Sigma	C6930	Mouse monoclone IgG2a
Anti-human vimentin clone V9 mAb	Cell sciences Inc.	MON3005-1	Murine monoclonal IgG-1
Anti-human fibroblast clone 5B5	DAKO	M0877	Mouse monoclonal IgG1
Anti-human C-peptide	LINCO	4020-01	Rabbit polyclonal
Anti-human insulin	CHEMICON	AB3440	Guinea pig polyclonal
Anti-human endoglin (CD105)	Santa Cruz Biotech	Sc-71043	Mouse monoclonal IgG1

treatment of pancreatic non-endocrine cells; final concentration is 250 mg/L).

4. 0.05% of Trypsin with EDTA is purchased from Invitrogen, UK.
5. Nunc® Tissue Culture dishes are from Cole-Parmer Instrument Company Ltd UK.

2.4. Analyzing the Non-endocrine Cells in Their Capacity for Proliferation and Dedifferentiation

1. 5-bromo-2′-deoxyuridine (BrdU): Dissolve in methanol and a final concentration of 20 μg/ml.
2. SV Total RNA Isolation System (Promega, UK) is used to prepare total RNA from cells.
3. The SuperScript® II Reverse Transcriptase from Invitrogen, UK, is used to reverse transcribe mRNA into single-strand cDNA.
4. Hot-start Taq polymerase from Qiagen, UK, is used to amplify genes of interest.
5. SYBR Green I fluorescence Kit (Roche Diagnostics, Germany) is used to amplify genes of interest using real-time PCR.
6. Primary antibodies used are listed in Table 1. FITC antibodies: Anti-mouse IgG kit (FI-2100), anti-rabbit IgG kit (FI-1200), anti-goat IgG (FI-5000); Texas anti-goat IgG (TI-5000); and Vectastain universal quick kit (PK8800) were all purchased from Vector Laboratories, UK.

2.5. Re-Differentiation Induction of Non-endocrine Cells into Insulin-Producing Cells In Vitro and In Vivo

1. Differentiation induction cocktail medium: CMRL1066 medium from Invitrogen, UK, supplemented with 10% FCS plus 10 ng/ml of hepatocyte growth factor (HGF), 25 pM activin A, 200 pM betacellulin (R&D Systems, Oxford, UK), 10 mM nicotinamide, and 16.5 mM glucose (Sigma, Dorset, UK) (termed HABNG) are used as the differentiation medium.
2. Lipofectamine™ 2000 Transfection Reagent from Invitrogen, UK, is used as a mediator to deliver DNA into cells.
3. Sterilin petri dishes are purchased from Lennox Laboratory Supplies Ltd, Ireland, and used to culture cells at suspension and for differentiation induction.
4. Anesthetize reagent: Hypnorm, Hypnovel, and pyrogen-free water were as ratio of 1:1:2, prepared freshly just before use.

3. Methods

The average adult human pancreas weighs 75 g. Although there are about a million of islets, they constitute only ~2% of the total tissue mass of the pancreas. The rest 98% are majority of non-endocrine tissues (exocrine, duct tissues, etc.). The pancreatic exocrine tissues are the by-products of the preparation of human islets.

3.1. Human Islet Isolation from Donor Pancreas

The isolations of human islets are carried out in the King's Cell Isolation Unit, a purpose-built clean laboratory dedicated to the isolation of cells for human use. Consent for using human pancreases for research was obtained from donor relatives in all cases and the proposal of usage of such cells is approved by King's College Hospital Ethical Committee.

1. The pancreata were retrieved after hypothermic perfusion with UW solution in situ.
2. The pancreas organ is transported to the Cell Isolation Laboratories in UW solution at 4°C. The cold ischemia times are between 3 and 8 h.
3. Under sterile environment, the pancreas is quickly dissected away from the duodenum and other organs (i.e., spleen) and fat tissues without making any cuts in the pancreas.
4. Transfer the pancreas into a sterile jar with lid on and weigh the pancreas (rough weight).
5. The pancreas goes through three steps decontamination solutions in three sterile containers, one containing ~150 ml 10% beta-dine, one with disinfections (150 ml antibiotics containing solution), and other for final rinse solutions (500 ml) to remove the remaining beta-dine solution. Briefly, the pancreas is submerged in the 10% beta-dine solution for ~20 s and then transferred to the disinfection solution (HBSS BESP250L150 containing antibiotics) for ~30 s and transferred to final rinse solution (HBSS BESP251F).
6. Insert a drawing needle into the pancreatic common duct at the head or cut the pancreas at its neck and insert two drawing needles, one into the body and one into the head through pancreatic common duct. The needle(s) is (are) secured by clamp(s).
7. Prepare the collagenases working solution by top-up of the Collagenase NB1 to 150 ml with perfusion solution, followed by 1 ml of Pefabloc and 1 ml SC Plus. 2.5 ml of neutral protease NB solution (25 U) are added to the collagease/Pefabloc/SC Plus mixture at the end of pancreas distension.
8. Distend the pancreas with pre-made collagenase solution using a 50-ml syringe.
9. Following distension of the pancreas, both fat and fibrous tissues are quickly removed. Chop the pancreas into several pieces with relatively equal sizes, hook them onto a triple porter, and secure the tissues onto it.
10. Tissue digestion: In a microbiological safety cabinet, the digestion chamber (Ricordi Chamber) is connected to a blood bag warmer through three tubings (Fig. 1). Transfer the tissue (see step 9) into Ricordi's chamber inside and digestion is started at 37°C. Top up the digestion fluid with perfusion HBSS and continue the process until the pancreas is completely digested.

Fig. 1. Ricordi Chamber before assembly (*left*) and after assembly (*right*).

11. Monitor the digestion from 12 min onward by staining a small amount of digested fluid (around 2 ml) with dithizone solution (around 2 ml). Repeat this step until the majority of islets are free from acinar tissues.
12. Once a desired degree of digestion is achieved, dilute the digestion fluid with dilution MEM and start collecting the outflow-digested tissues.
13. Continue the collection into bottles containing a final concentration of ~1% human albumin until almost no tissues come out through the system.
14. Centrifuge the digested tissues at 240 × *g* for 1 min at 4°C as the collection continued. The tissues are pooled and re-suspended in UW solution, and incubate at 4°C for 1 h before the purification processing.
15. Turn on the air conditioner-controlled Cobe 2991 machine and set up the centrifuge bag. Connect the Cobe 2991 centrifuge bag with gradient makers (Fig. 2) loading and unloading tubes. The loading tube is passed through pump that provides the force for liquid movement.
16. Make up the base medium by mixing 68 ml of Biocoll cell-separating medium (at 1.10 g/ml) with 62 ml UW solution (Dupont) and then feed the 130 ml solution into the Cobe 2991 centrifuge bag. Start the centrifuge at 157 × *g* and increase to 626 × *g* and press the supernatant-out button to push all the air out of the centrifuge bag. Repeat this for three times.
17. Make up the continuous density gradient medium in gradient solution maker with 68 ml of Biocoll-separating solution and 62 ml of UW solution (this mixture ratio could be changed based on the judgment of digested islets) and the lighter of

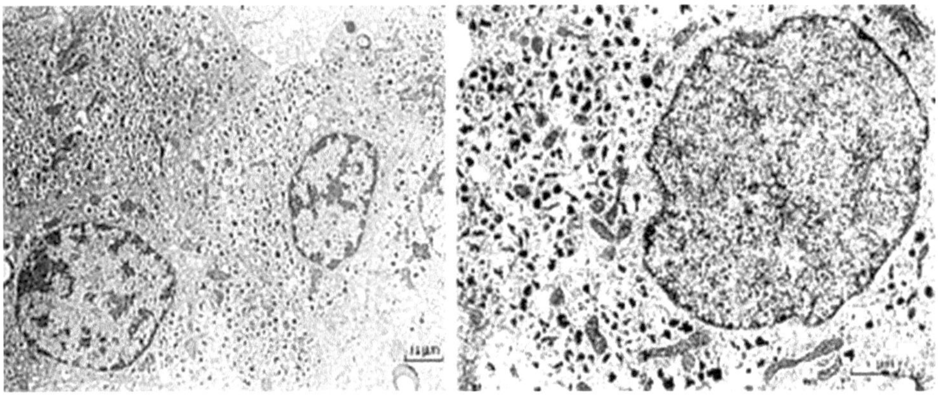

Fig. 2. Transmission electron microscopy images. *Left*: Presence of insulin secretory granules in the non-endocrine/PNN cells (*black dots*). *Right*: Presence of insulin protein in non-endocrine/PNN cells (*black spots*) assessed by immunogold labeling.

39 ml of Biocoll-separating solution and 91 ml UW solution. The density media are loaded at a speed of ~25 ml/min while increasing the centrifugation speed to 1409 × *g* followed by loading the digested tissues (in 150 ml UW solution).

18. Set Cobe 2991 to spin continually at 3,000 rpm for additional 4 min and start the harvest of the islet cells by pressing the supernatant-out button. Discard the first 100 ml or until you can see the tissue come out (around 130 ml).
19. Collect the tissues into 12 tubes with 30 ml each (each containing 20 ml of pre-made and pre-cooled washing medium containing 1% human albumin).
20. Remove a small portion from each of the 12 tubes, mix them with dithizone, and determine which tubes contain islet tissues with desired islet purity under an inverted microscope. Pool the islet containing fractions (purer and less pure fractions separate or together) and remove 100 μl from the final 50 ml of tissue for counting (number, sizes, and purity) to determine the final islet IEQ. 100 μl are used for islet viability analyzed by FDA/PI staining.
21. Non-endocrine tissues were collected from the centrifuge bag into the pre-cooled washing medium.
22. Wash both islet and non-endocrine tissues three times with the pre-cooling washing medium with centrifugation at 1,000 rpm for 1 min each. Islet cells were incubated in CMRL1066 culture medium and assessed whether they meet the criteria for clinical use while the non-endocrine tissue fraction was brought to research laboratories for research use.

3.2. Dedifferentiate and Expand the Human Non-endocrine Cells (Most Are Exocrine Cells) In Vitro

1. Removal of the existing beta cells from the exocrine cell fraction by the STZ treatment: The gradient-enriched exocrine cells were collected, washed twice with HBSS, and cultured in suspension as clusters in complete CMRL1066 medium. Freshly prepared STZ solution (250 mg/L final concentration) is adding the culture twice, one at the beginning of the culture and the other on the third day of culture. The cells are cultured at 37°C in a humid 5% CO_2 incubator for a period of 5 days, with medium changed daily. A small portion of the cell culture is collected daily for a): to analyse genomic DNA quantity for the determination of cell number; b) to assess the presence of insulin containing cells by immunocytochemical staining and c) to extract total RNA for RT-PCR analysis. Culture medium is collected for the assessment of presence of insulin protein by radioimmunoassay (RIA) (22).
2. Expansion of cultured non-endocrine cells: After the STZ treatment, the cells are cultured to 10-cm tissue culture plates in the CMRL1066 medium with 50 mg/L of G418 to inhibit the growth of fibroblast cells (11), feeding cells every 2 days. These cluster cells are adherent onto plates and spread out gradually. After 1 week of culture, the cluster cells are spread out to form a monolayer. The cells are continually passaged when the cells reach to ~90% confluence and the cells are split at a ratio of 1:3 for expansion up to 7 weeks. The cell-doubling time is determined by the cell counting. Cell samples need to be analyzed at intervals of 5–7 days by immunocytochemistry and RT-PCR for the phenotype changes.
3. Analyses for cell proliferation: Cells from the monolayer cultures are dissociated by using trypsin/EDTA solution and counted at intervals. Cell proliferation is assessed for the incorporation of BrdU. For this, cultured cells are dissociated and seeded for 2 h before the addition of BrdU (20 μg/ml). The BrdU is in the culture medium for 2 h and then replace with fresh culture medium for another 16 h. The cells are then dissociated using trypsin/EDTA solution and centrifuged down into a pellet. Pellets are then cryosectioned and immunostained for the proliferation markers Ki67 and anti-BrdU.

3.3. The Differentiation of These Cells into Insulin-Producing Cells In Vitro

1. To induce the partially dedifferentiated non-endocrine cells into low-degree differentiation in vitro, 2–3 weeks expanded non-endocrine cells are converted into cluster culture and cultured in normal CMRL1066 medium without G418 antibiotics overnight to allow the formation of clusters at a cell concentration of 3×10^6 cells per 5-cm diameter Sterilin petri dish. Next morning, replace with fresh serum-free CMRL1066 (plain CMRL1066) supplemented with 50 nM GLP-1 (Sigma–Aldrich, UK) and incubated for 3 h. At the end of the incubation, a cocktail of differentiation induction reagents are added

Table 2
Primers used in RT-PCR

Genes	Primer set	Product length
Pdx-1	F 5′CGGGCCGCAGCCATGAACG3′ R 5′CTCCTGCCTCTCATCGTGGTTCCTG3′	874 bp
NeuroD1	F 5′AAATGAATTCATGACCAAATCGTACAGCGAGA3′ R 5′TCGCGTCGACCTAATCATGAAATATGGCATT GAGCTG3′	1,093 bp
Ngn3	F 5′GCGCCGGTAGAAAGGATGACGCC3′ R 5′CCCGGCTCCCTCCCTCTCCCTTAC3′	677 bp

at final concentrations of 25 pM activin A, 200 pM betacellulin, 10 mM nicotinamide, and 16.5 mM glucose (ABNG), supplemented with 10% FCS. The cells are cultured for additional 6–8 days, with fresh medium containing ABNG replaced every 2 days.

2. The transfection of human Pdx1/Ipf1, NuroD1, and Ngn3 expression plasmids: The coding regions of the human PDX-1, NeuroD1, and Ngn3 genes are amplified, respectively, by RT-PCR using specific primer sets (Table 2). The genes are then, respectively, cloned into the expression plasmid pIRES-N1 (21) under the control of CMV promoter and the three genes were partially sequenced to confirm the authenticity. For gene transfection, the 3×10^6 cells per 10-cm tissue culture plate were seeded at 90% confluence for 3 h before transfection with lipofectamine 2000 (Invitrogen, Paisley, UK) according to manufacturer's instructions as described previously (25). Briefly, plasmid DNA of Pdx-1, NeuroD1, and Ngn3 was mixed in a ratio of 1:3:1 in weight prior to the transfection. Twenty-four-microgram plasmids DNA (per 10-cm petri dish) are diluted in 1.5 ml of OPTI-MEM medium (Invitrogen, UK) in tube A. Meanwhile, 60 μl of lipofectamine 2000 (per 10-cm petri dish) are diluted with 1.5 ml of Opti-MEM medium in tube B. Following an incubation at room temperature for 5 min, DNA in tube A was mixed with the lipofectamine 2000 in tube B and incubated at room temperature for another 20 min. At the end of this incubation, the cells for DNA transfection are rinsed once with Opti-MEM medium and the 3 ml of DNA/lipofectamine 2000 mixture are added dropwise onto the cells. The cells are incubated at 37°C in a humid 5% CO_2 incubator for 45 min following by addition of 7 ml of CMRL1066 culture medium, and the cells are incubated

for another 4 h. Thereafter, the transfection mixture is removed and replaced with culture medium, and cultured for another 48 h. The cells are dissociated using trypsin/EDTA solution and following by the transdifferentiation induction mentioned above. Empty plasmid pIRES-N1 is used as a negative transfection control. Cell samples are taken and assessed for expression of β-cell markers by the following assays: semi-quantitative RT-PCR, quantitative PCR, immunochemical staining, insulin secretion assay and human C-peptide immunoassay, and electron microscopy (EM) analysis.

3.4. Further Maturation of the Partial Differentiated Non-endocrine Cells In Vivo

1. Animals: Male SCID mice (20–25 g, C.B-17/Icr), purchased from Charles River UK Ltd., Margate, UK, are selected as recipients for the partially differentiated non-endocrine cells. The mice are maintained in filter cages in the Comparative Biology Center at King's College London, according to the guidelines of Home Office (UK) for Animal Scientific Procedures. Diabetes is induced in the mice by injection of a single dose (180 mg/kg i.p) of STZ. Diabetic mice with blood glucose levels >15 mM for 1 week are used as recipients. The diabetic mice are randomly classed into three transplantation groups. The mice in group 1 receive ~1,500 clustered, differentiation-induced non-endocrine cells with gene transfection; group 2 mice transplant same amount of 1,500 clustered, differentiation-induced non-endocrine cells without gene transfection and the group 3 mice are used as sham control with saline injections.
2. Surgical procedures: Recipient mice are anesthetized by an intraperitoneal injection of the mixture of Hypnorm and Hypnovel. The side for transplant is shaved and decontaminated with 70% alcohol. Under anesthesia, the left lateral laparotomy is performed. A small incision is made on the mouse and the kidney is exposed. Keep the kidney moist by applying normal saline with cotton-tipped swab. The ~1,500 cell clusters are loaded into a blunt-end, 50-μl Hamilton syringe. A hole in subcapsule membrane at the low pole of the kidney is made via a pin of a 23-gauge needle. The syringe is inserted into the kidney through the hole till it reaches the subcapsule membrane on the other end of the kidney. The cells are discharged into the parenchyma as the syringe is slowly withdrawn. In the sham control group, a cell-free saline injection is made in a precise fashion. Close the peritoneum with a running stitch using 5–0 silk sutures. Use 4–0 silk sutures to put the skin together with 3–4 stitches. Clean the skin of the mouse of any blood using cotton-tipped swab and saline. Plasma glucose is measured using a G2 blood glucose sensor from MediSense (Abbott Laboratories Ireland Ltd, Ireland).

3. Intraperitoneal glucose tolerance test (IPGTT): The mice are fasted overnight and plasma glucose level is determined. Glucose solution at 1.5 g/kg body weight is injected intraperitoneally and plasma glucose levels are determined at 30, 60, 90, 120, and 180 min post injection.
4. At the end of the experiments, blood, kidneys, and pancreases are retrieved. Immunohistochemistry is performed on the tissues for the proteins of insulin, glucagon, somatostatin, and pancreatic polypeptide Y. Cell nuclei are counterstained with propidium iodide (Sigma–Aldrich, UK) and analyzed by fluorescent or confocal microscopy. Hormone-positive cells and the total transplanted cells (total cluster cells) are counted under light fluorescence microscopy and the percentage of positive cells derived from counting 1,500 total graft cells per staining. The level of human-specific C-peptide is determined by immunoradioassay (DSL, USA). Organs from sham mice are studied in parallel as negative control.
5. Data analysis: All data should be expressed as mean ± SD unless specified and statistical differences between groups determined by ANOVA.

4. Notes

1. For calculation of the enzyme concentration of Collagenase NB1, please consider the total enzymatic activity per vial as stated on the Certificate of Analysis. The solutions should be prepared freshly, just before usage. If sterile filtration is desired, a sterile filter with low protein binding properties (e.g., PES, PVDF) should be used with the filter being rinsed with an additional small volume of buffer solution before and after filtration.
2. In the PSC-protector solution, a precipitate may occur during storage. This can be solved by warming the tube up to 15–25°C.

References

1. Wang RN, Kloppel G, Bouwens L (1995) Duct- to islet-cell differentiation and islet growth in the pancreas of duct-ligated adult rats. Diabetologia. 38:1405–11.
2. Bonner-Weir S, Baxter LA, Schuppin GT, Smith FE (1993) A second pathway for regeneration of adult exocrine and endocrine pancreas. A possible recapitulation of embryonic development. Diabetes. 42:1715–20.
3. Guz Y, Nasir I, Teitelman G (2001) Regeneration of pancreatic beta cells from intra-islet precursor cells in an experimental model of diabetes. Endocrinology 142: 4956–68.
4. Dor Y, Brown J, Martinez OI, Melton DA (2004) Adult pancreatic beta-cells are formed by self-duplication rather than stem-cell differentiation. Nature. 429:41–6.

5. Slack JM (1995) Developmental biology of the pancreas. Development. 121:1569–80.
6. Edlund H (1999) Pancreas: how to get there from the gut?. Curr Opin Cell Biol. 11:663–8.
7. Edlund H (1998). Transcribing pancreas. Diabetes. 47:1817–23.
8. St-Onge L, Wehr R and Gruss P (1999) Pancreas development and diabetes. Curr Opin Genet 9: 295–300.
9. von Schonfeld J, Goebell H, Muller MK (1994) The islet-acinar axis of the pancreas. Int J Pancreatol. 16:131–40.
10. Hall PA, Lemoine NR (1992) Rapid acinar to ductal transdifferentiation in cultured human exocrine pancreas. J Pathol 166:97–103.
11. Githens S, Schexnayder JA, Moses RL, Denning GM, Smith JJ, Frazier ML (1994). Mouse pancreatic acinar/ductular tissue gives rise to epithelial cultures that are morphologically, biochemically, and functionally indistinguishable from interlobular duct cell cultures. In Vitro Cell Dev Biol Anim 30A: 622–35.
12. Kerr-Conte J, Pattou F, Lecomte-Houcke M et al., (1996) Ductal cyst formation in collagen-embedded adult human islet preparations. A means to the reproduction of nesidioblastosis in vitro. Diabetes 45:1108–14.
13. Rooman I, Heremans Y, Heimberg H, Bouwens L (2000) Modulation of rat pancreatic acinoductal transdifferentiation and expression of PDX-1 in vitro. Diabetologia 43:907–14.
14. Gmyr V, Kerr-Conte J, Belaich S et al., (2000) Adult human cytokeratin 19-positive cells reexpress insulin promoter factor 1 in vitro: further evidence for pluripotent pancreatic stem cells in humans. Diabetes 49: 1671–80.
15. Miller CP, McGehee RE Jr, Habener JF (1994) IDX-1: a new homeodomain transcription factor expressed in rat pancreatic islets and duodenum that transactivates the somatostatin gene. EMBO J. 13:1145–56.
16. Bonner-Weir S, Taneja M, Weir GC et al., (2000) In vitro cultivation of human islets from expanded ductal tissue. Proc Natl Acad Sci USA. 97:7999–8004.
17. Gao R, Ustinov J, Pulkkinen MA, Lundin K, Korsgren O, Otonkoski T (2003) Characterization of endocrine progenitor cells and critical factors for their differentiation in human adult pancreatic cell culture. Diabetes. 52:2007–15.
18. Heremans Y, Van De Casteele M, in't Veld P et al., (2002) Recapitulation of embryonic neuroendocrine differentiation in adult human pancreatic duct cells expressing neurogenin 3. J Cell Biol. 159:303–12.
19. Lardon J, Huyens N, Rooman I, Bouwens L (2004) Exocrine cell transdifferentiation in dexamethasone-treated rat pancreas. Virchows Arch. 444:61–5.
20. Baeyens L, De Breuck S, Lardon J, Mfopou J, K,Rooman I, Bouwens L. (2005) In vitro generation of insulin-producing cells from adult exocrine pancreatic cells. Diabetologia 48: 49–57.
21. Hobbs S.M (1997) Artificial DNA; bicistronic eukaryotic expression vector pIRES-N1 submitted to EMBL/GenBank/DDBJ databases. AC no.: Y11035.
22. Zhao M, Amiel SA, Christie MR, Rela M, Heaton N, Huang GC. Insulin-producing cells derived from human pancreatic non-endocrine cell cultures reverse streptozotocin-induced hyperglycaemia in mice. Diabetologia. 2005 Oct; 48(10):2051–61.

Chapter 7

Evaluation of Cytochrome P450 Activities in Human Hepatocytes In Vitro

María José Gómez-Lechón, Agustín Lahoz, José V. Castell, and María Teresa Donato

Abstract

Major hepatic cytochrome P450 activities (CYP1A2, CYP2A6, CYP2B6, CYP2C9, CYP2C19, CYP2D6, CYP2E1, and CYP3A4) can be simultaneously examined in human hepatocytes by incubation with a cocktail of multiple specific probes. Cocktail strategy in combination with mass spectrometry is shown to be a robust, fast, and sensitive procedure for P450 activity assessment. This procedure allows a drastic reduction of the number of cells required in the assay and sample analysis time and increases throughput and reproducibility. Major applications of the probe cocktail strategy are P450 phenotyping of hepatocytes and induction studies.

Key words: Human hepatocytes, Cytochrome P450, HPLC-MS/MS, Mass spectrometry, ADME, Substrates cocktail

1. Introduction

Nowadays, a major challenge facing the pharmaceutical industry is to find new ways to increase productivity and decrease costs in the development of new therapies that enhance human health. At the early stage of drug discovery, thousands of new chemical entities (NCEs) may be screened before a single candidate can be identified for development. A leading candidate needs adequate bioavailability, the appropriate physicochemical properties to enable formulation development, the ability to cross crucial membranes, a reasonable metabolic stability, and the appropriate safety and efficacy in humans. Drug metabolism is not only one of the major determinants of drug clearance, but also the factor that is most frequently responsible

Ragai R. Mitry and Robin D. Hughes (eds.), *Human Cell Culture Protocols*, Methods in Molecular Biology, vol. 806, DOI 10.1007/978-1-61779-367-7_7,

for inter-individual differences in drug pharmacokinetics and toxic effects. Cytochrome P450 enzymes are mainly concentrated in the liver and are responsible for metabolic clearance of many oral marketed drugs or/and xenobiotics. The main in vitro systems for studying NCE metabolism in a discovery setting have been using recombinant liver enzymes, microsomes, and hepatocytes. These models can help predict in vivo metabolism, clearance, and potential drug–drug interactions of these new compounds by P450 (1, 2). To help address these challenges, analytical technologies and high-throughput automated platforms should be employed for the performance of more experiments in a shorter time frame with increased data quality. Therefore, reliable methods for determining enzyme activities are needed to characterize individual P450 enzymes and to obtain a tool for the evaluation of their role in drug metabolism in humans (3). Different liquid chromatography tandem mass spectrometry (LC/MS/MS) methodologies have been developed for the fast and routine analysis of major in vivo and in vitro P450 activities simultaneously (4–6). The high sensitivity and selectivity of MS allow traditional assays to be minimized, with a marked reduction in the number of hepatocytes needed for each P450 activity measurement, a fact that is particularly critical concerning human hepatocytes, thus saving time, efforts, and money. Therefore, this technology has become the method of choice for the fast assessment of P450 enzyme activities in early drug discovery development (5, 7).

2. Materials

2.1. Hepatocyte Culture and Treatment

1. Seeding culture medium: Ham's F-12/Williams (1:1) medium (Gibco BRL, Paisley, Scotland) supplemented with 2% newborn calf serum (NBCS; Gibco BRL, Paisley, Scotland), 0.1% bovine serum albumin (BSA) fraction V (Sigma, Madrid, Spain), 10 nM insulin (Novo Nordisk, A/S Bagsvaerd, Denmark), 25 μg/ml transferrin, 0.1 μM sodium selenite, 65.5 μM ethanolamine, 7.2 μM linoleic acid, 7 mM glucose, 6.14 mM ascorbic acid, 0.64 mM *N*-omega-nitro-L-arginine methyl ester (Sigma, Madrid, Spain), and 50 mU/ml penicillin and 50 μg/ml streptomycin (Gibco BRL, Paisley, Scotland).
2. 24-h chemically defined medium: The same culture medium composition as above but NBCS-free and supplemented with 10 nM dexamethasone (Sigma, Madrid, Spain).
3. Coating mixture for culture plates: Dissolve 1 mg of human fibronectin (Sigma, Madrid, Spain) in 97 ml of DMEM supplemented with 0.1% BSA. Add to the former solution 3 ml of 0.1% collagen type I from calf skin solution in 0.1 M acetic acid (Sigma, Madrid, Spain).

4. Viability assessment: Trypan blue 0.4% solution, sterile filtered, cell culture tested (Sigma, Madrid, Spain).
5. P450 inducer solutions: 1 mM 3-Methylcholanthrene (MC), 10 mM omeprazole (OMP), 400 mM phenobarbital (PB), and 20 mM rifampicin (RF) (Sigma, Madrid, Spain) stock solutions are prepared in dimethylsulfoxide (DMSO; Merck Chemicals, Barcelona, Spain) and stored in aliquots at -20°C.

2.2. P450 Substrate Cocktail

1. Substrate stock solutions: 50 mM Coumarin; 20 mM bupropion; 120 mM chlorzoxazone; 30 mM diclofenac (Sigma–Aldrich, Madrid, Spain). 3 mM midazolam; 30 mM phenacetin; 30 mM bufuralol; 100 mM mephenytoin (Ultrafine, Manchester, UK). All are prepared separately in organic solvent (methanol, DMSO, or acetonitrile; Merck Chemicals, Barcelona, Spain) and stored in aliquots at −80°C.
2. Substrate mixture stock solution: Appropriate volumes of each individual substrate stock solution and solvent (DMSO) are mixed to obtain a cocktail of probe substrates at concentrations indicated in Table 1. This substrate mixture is stored in aliquots at −80°C.
3. β-Glucuronidase/arylsulfatase (Roche, Barcelona, Spain).
4. Incubation medium: Na_2HPO_4 1 mM, 137 mM NaCl, 5 mM KCl, 0.5 mM $MgCl_2$, 2 mM $CaCl_2$, 10 mM glucose, and 10 mM HEPES. Adjust pH to 7.4. Store at -20°C. Just prior to use, add $CaCl_2$ (1 mM final concentration).
5. Incubation cocktail: Dilute 5 μl of substrate mixture stock solution in 1 ml of incubation medium plus $CaCl_2$ (see Note 1). Final incubation concentrations of probe substrates are shown in Table 1.

Table 1
Substrate cocktail composition for multiple P450 assessment

P450	Substrate	Stock mixture (mM)	Concentration assay (μM)	Metabolite
CYP1A2	Phenacetin	2	10	Acetominophen
CYP2A6	Coumarin	1	5	7-Hydroxycoumarin
CYP2B6	Bupropion	2	10	Hydroxybupropion
CYP2C9	Diclofenac	2	10	4′-Hydroxydiclofenac
CYP2C19	Mephenytoin	10	50	4′-Hydroxymephenytoin
CYP2D6	Bufuralol	2	10	1-Hydroxybufuralol
CYP2E1	Chlorzoxazone	10	50	6-Hydroxychlorzoxazone
CYP3A4	Midazolam	1	5	1′-Hydroxymidazolam

2.3. Sample Analysis

1. Instrumentation: Micromass Quattro Micro (Waters, Milford MA, USA) triple quadrupole mass spectrometer in electrospray ionization mode, interfaced with an Alliance 2795 HPLC (Waters Chromatography).
2. Liquid chromatography: Teknokroma C18 column (100 × 2.1 mm, 3-μm particle size).
3. Mobile-phase solutions: 0.1% formic acid in water (A) and 0.1% formic acid in acetonitrile (B).
4. Metabolite stock solutions: 14 mM Acetaminophen; 1 mM 7-hydroxycoumarin; 5 mM 4′-hydroxydiclofenac; 20 mM 1-hydroxybufuralol maleate; 5 mM 6-hydroxychlorzoxazone; 1 mM 1′-hydroxymidazolam (Sigma–Aldrich, Madrid, Spain). 12.5 mM hydroxybupropion (BD Gentest, Madrid, Spain). 20 mM 4-hydroxymephenytoin (Toronto Research Chemicals, Canada). All are separately dissolved in organic solvent (methanol, DMSO, or acetonitrile) and stored in aliquots at –80°C.
5. Metabolite cocktail stock solution: Mix adequate volumes of each individual metabolite stock solution to reach the desired concentrations in the metabolite cocktail (250 μM acetaminophen, 125 μM 7-hydroxycoumarin, 250 μM hydroxybupropion, 250 μM 4′-hydroxydiclofenac, 1,250 μM 4-hydroxymepheny toin, 250 μM 1-hydroxybufuralol, 1,250 μM 6-hydroxychlorzoxazone, and 125 μM 1 mM 1′-hydroxymidazolam). Stored in aliquots at –80°C.

3. Methods

Different types of hepatic tissue, including whole or split livers from organ donors or from cadavers, waste liver from therapeutic hepatectomies or small-sized surgical biopsies, are used to prepare human hepatocyte cultures (8). The two-step collagenase perfusion to isolate hepatocytes from human liver allows obtaining high yields of viable human hepatocytes (9). Hepatocytes in chemically defined culture conditions express most typical hepatic biochemical functions, among which is the ability to metabolize drugs (8–11). Primary human hepatocytes are recognized to be the closest model to human liver and are currently considered a valuable in vitro tool for determining drug metabolism and for assessing the risk of drug hepatotoxicity in man (12, 13). The restricted and irregular accessibility to suitable liver samples has led to optimize and miniaturize the protocols so that available tissue can be used more efficiently for drug metabolism studies.

3.1. Coating of Culture Plates Procedure

1. Culture plates are coated with 10 μl/cm^2 of fibronectin/collagen coating mixture above described and left stand at 37°C for 1 h before hepatocyte seeding.
2. Excess of coating mixture is removed and hepatocytes are seeded on the plates at the appropriate density.

3.2. Cell Counting and Viability Assessment

1. After resuspension of hepatocytes in seeding culture medium, viability is determined in a cell aliquot by cell counting using the Trypan blue exclusion method.
2. An aliquot of a mixture of hepatocyte suspension and 0.4% Trypan blue solution in saline (1:1 v/v) was immediately loaded in a counter chamber.
3. Viable cells were counted in five different fields under the optical microscope as non-blue cells. Cell preparations with viability below 60% are generally not adequate for further cultivation.

3.3. Culture of Human Hepatocytes

1. A hepatocyte suspension is prepared at a density of 5×10^5 alive cells/ml in seeding culture medium.
2. Hepatocytes are then plated on fibronectin/collagen-coated plates at a density of 8×10^4 cells/cm^2 suspended in an appropriate volume of culture medium.
3. One hour after cell seeding, the medium is aspirated to remove unattached cells and debris and the same volume of fresh culture medium is added. The attachment efficiency of hepatocyte suspension to fibronectin/collagen-coated plates 1–2 h after cell seeding is usually an 80% of viable cells.
4. Twenty-four hours after cell plating, cells are shifted to serum-free, chemically defined medium.
5. Chemically defined medium is renewed every following days.
6. Under these culture conditions, hepatocytes survive up to 7–10 days.

3.4. Treatment of Cultured Hepatocytes with Inducers

1. Stock solutions of chemical inducers are thawed at room temperature and conveniently diluted in serum-free, chemically defined medium to reach desired concentrations: 2 μM MC, 50 μM OMP, 1 mM PB, and 50 μM RF (see Note 2). Concentration of the solvent (DMSO) in the cultured medium does not exceed 0.5% (v/v).
2. Hepatocytes are incubated in the presence of inducers for 48 h. Treatments start 24 h after cell plating by removing cultured medium and adding fresh medium containing the inducer or solvent (control cells). Inducers are re-added 24 h later with medium change.

3.5. Incubation of Hepatocytes with the P450 Substrate Cocktail

P450 enzymes are heme-thiolate proteins that are responsible for the oxidative metabolism of a wide variety of xenobiotics. They comprise a superfamily of related enzymes that are grouped into families and subfamilies based on similarities in amino acid sequences. Currently, the human CYP superfamily contains 57 functional genes with members of the 1–3 families playing an important role in the metabolism of therapeutic drugs, other xenobiotics, and some endogenous compounds (2). The five major human P450 enzymes responsible for the metabolism of xenobiotics are CYP1A2, CYP2C9, CYP2C19, CYP2D6, and CYP3A4 and account for 70% of all drug clearance and it is estimated that these five enzymes are responsible for approximately 99% of P450-mediated drug metabolism (2). P450 function can be easily assayed in human hepatocytes using cocktail mixtures containing selective probes (4–6). In particular, a cocktail for the assessment of eight individual P450 enzymes (CYP1A2, 2A6, 2B6, 2C9, 2C19, 2D6, 2E1, and 3A4) in cultured human hepatocytes was recently reported (4) (see Note 3). Cell monolayers are directly incubated with the substrate cocktail and the corresponding metabolites formed and released into the incubation medium are quantified by using the high sensitivity and selectivity of MS (4). The selection of probe substrates is a critical point, as not all substrates are suited for cocktail approaches. The composition of substrate mixture used in the present study was carefully designed considering a number of factors, including substrate specificity (for a particular P450) (see Note 4), substrate solubility and concentration (to minimize P450 inhibition and avoid potential cytotoxicity), or the lack of probe-to-probe interactions at analytical level (4, 7). This fast and sensitive method allows a rapid evaluation of P450 activities in hepatocytes cultured in miniaturized formats (i.e. 96-well plates) (7, 14). This procedure can be used for phenotyping P450 activities in human hepatocytes and for the assessment of metabolic capacity of hepatocyte preparations. Moreover, a major application of the assay is high-throughput screening of potential inducers of P450 enzymes involved in drug metabolism.

1. Remove culture medium from culture plates and add an appropriate volume of incubation cocktail (recommended volumes: 150, 300, or 750 μl for 96-well, 24-well, or 3.5-cm-diameter plate formats, respectively) (see Note 5).
2. After 60 min into a cell culture incubator (37°C 5% CO_2), incubation medium is recovered into clean tubes and centrifuged for 5 min at 500 × *g* (see Note 6). Thereafter, the pellet containing detached cells is discarded and supernatants are transferred to clean tubes and kept frozen at –80°C until used.

3. At this stage, cell monolayers are washed out with PBS, frozen in liquid N_2, and kept at −20°C for further cell protein quantification.

3.6. Sample Preparation

1. Prior to metabolite analysis, incubation samples are thawed at room temperature and metabolite conjugates are hydrolysed with β-glucuronidase/arylsulfatase in 0.1 M acetate, pH 4.5 (15).
2. After 120 min in a shaking incubator (37°C), hydrolysis reactions are stopped by diluting (1:2) in cold acetonitrile (quenching solution, protein precipitation).
3. Centrifuge for 10 min at 10,000 × *g* at 4°C and transfer 100 μl to a clean HPLC vial.

3.7. Preparation of Standard Calibration Curves

1. Dilute (1:40) an aliquot of thawed metabolite cocktail stock solution in incubation medium (saline solution plus Ca^{2+}).
2. Make serial dilutions (1:2) in incubation medium (saline solution plus Ca^{2+}) to prepare a calibration curve (up to ten different metabolite mixture concentrations).
3. Transfer to clean HPLC vials.

3.8. HPLC/MS/MS Analysis

1. A volume of 20 μl of sample (or standard) is injected for chromatographic analysis.
2. Liquid chromatography is performed at 35°C and a flow rate of 0.4 ml/min.
3. The proportion of acetonitrile (see Subheading 2.3 for details in mobile-phase composition) increases linearly from 0 to 90% in 6 min and then the column was re-equilibrated at the initial conditions for 4 min.
4. The column eluent is directed to an electrospray ionization interface without splitting, operating at 320°C, and using nitrogen as cone gas (50 l/h).
5. MS/MS analysis is carried out with a triple quadrupole analyzer operating in the multiple reaction monitoring mode. MS conditions for quantification of each metabolite are shown in Table 2.
6. For each P450 enzyme, activity is expressed as pmol of the corresponding metabolite formed per minute and per mg of cell protein (see Note 7). In induction studies, activity values in treated hepatocytes are compared with those in control (untreated) hepatocytes (Fig. 1)

Table 2
MS conditions for the eight metabolites formed during P450 cocktail assay

	Acetaminophen (CYP1A2)	7-OH-coumarin (CYP2A6)[a]	OH-bupropion (CYP2B6)	4′-OH-mephenytoin (CYP2C19)
Mode	Positive	Negative	Positive	Positive
Capillary voltage	3	−3	3.2	3
Cone voltage	30	35	30	15
Desolvation temperature	320	320	320	320
Source temperature	130	130	130	130
Collision energy	32	25	25	20
Analyte *m/z* transition	152 → 110	161 → 133	256 → 139	235 → 186
Retention time	3.35	4.4	4.63	4.29
Standard curve (nM)	6,000-10	3,000-5	6,000-10	15,000-30
LOQ (nM)	<4	<5	<5	<15

	4′-OH-diclofenac (CYP2C9)	1-OH-bufuralol (CYP2D6)	6-OH-chlorzoxazone (CYP2E1)	1′-OH-midazolam (CYP3A4)
Mode	Positive	Positive	Negative	Positive
Capillary voltage	3	3	−3.3	3.05
Cone voltage	20	35	40	40
Desolvation temperature	320	320	320	320
Source temperature	130	130	130	130
Collision energy	35	20	30	25
Analyte *m/z* transition	312 → 230	278 → 187	184 → 120	342 → 324
Retention time	5.1	3.6	4.2	4.41
Standard curve (nM)	6,000-10	6,000-10	15,000-30	3,000-5
LOQ (nM)	<1	<4	<10	<10

[a]OH: hydroxyl

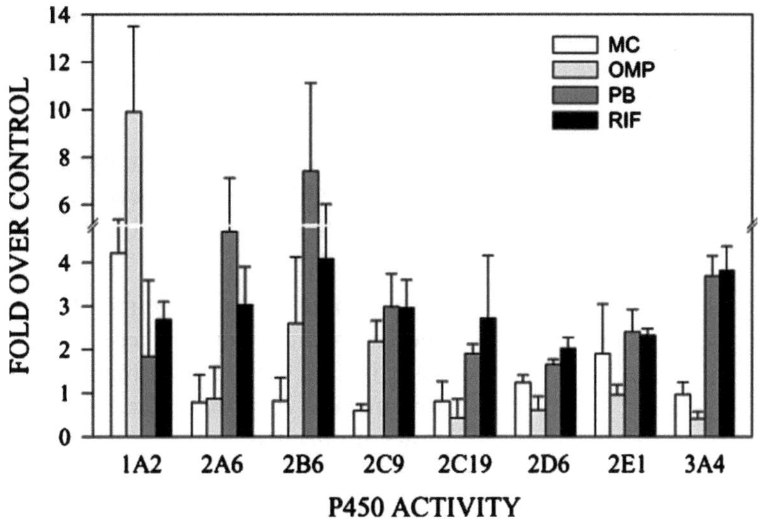

Fig. 1. Effects of model inducers on the P450 activities in primary cultured human hepatocytes. After 24 h in culture, hepatocytes were exposed to 2 μM 3-methylcholanthrene (MC), 50 μM omeprazole (OMP), 25 μM rifampicin (RIF), and 1 mM phenobarbital (PB). After 48 h of treatment, P450 activities were determined by incubation of cell monolayers with a cocktail of selective probes: phenacetin (CYP1A2), coumarin (CYP2A6), bupropion (CYP2B6), diclofenac (CYP2C9), S-mephenytoin (CYP2C19), bufuralol (CYP2D6), chlorzoxazone (CYP2E1), and midazolam (CYP3A4). Data are mean ± SD of three independent cell cultures and are expressed as fold increased over activity measured in control cells.

4. Notes

1. Substrate cocktail can be prepared in other pH 7.4-buffered saline solutions or in culture media.
2. MC, OMP, PB, and RF are widely recognized as model inducers of P450 enzymes. Other compounds can be used for P450 induction studies. For a general induction screening, at least three different concentrations of the chemical are recommended. Cytotoxicity of the compound to hepatocytes should be previously tested and only sub-cytotoxic concentrations are used for P450 induction assays.
3. A cocktail approach has also been developed for the simultaneous determination of five P450 enzymes (CYP1A2, CYP2A6, CYP2C9, CYP2E1, and CYP3A4) and two conjugating (phase II) activities (UDP-glucuronyltransferase and sulfotransferase) in human hepatocyte preparations. The assay consists of the co-incubation of cells with multiple-probe drugs (7-ethoxyresorufin, coumarin, diclofenac, chlorzoxazone, and midazolam) followed by HPLC/MS/MS analysis and quantification of metabolites formed and released into incubation medium (16).

4. Although the composition of this substrate cocktail has been specifically defined for human liver-derived cells, it can also be used for cells from other species (i.e. rodent hepatocytes) (17).
5. This cocktail approach can also be used for the assessment of P450 activities in hepatocyte suspensions. A cell density of 10^6 cells/ml and an incubation time of 30 min are recommended (18).
6. Sixty minutes is the recommended incubation time for activity measurement in human hepatocyte cultures. However, if the assay is used for the assessment of P450 activities in cells with low metabolic capacity (i.e. HepG2 cells), longer incubation times are needed.
7. Activity values can also be referred as pmol of metabolite formed per minute and per million of cells (i.e. incubations in cell suspension) or per pmol of P450 enzyme (i.e. assays in recombinant P450 preparations).

References

1. Ingelman-Sundberg M. (1998) Functional consequences of polymorphism of xenobiotic metabolising enzymes. *Toxicol Lett.* **102–103**, 155–160.
2. Zhou S.F, Liu J.P, and Chowbay B. (2009) Polymorphism of human cytochrome P450 enzymes and its clinical impact. *Drug Metab Rev.* **41**, 89–295.
3. Gómez-Lechón M.J., Castell J.V., and Donato M.T. (2007) Hepatocytes–the choice to investigate drug metabolism and toxicity in man: *in vitro* variability as a reflection of in vivo. *Chem Biol Interact.* **168**, 30–50.
4. Lahoz A., Donato M.T., Castell J.V., and Gómez-Lechón M.J. (2008) Strategies to in vitro assessment of major human CYP enzyme activities by using liquid chromatography tandem mass spectrometry. *Curr Drug Metab.* **9**, 12–19.
5. Petsalo A., Turpeinen M., Pelkonen O., and Tolonen A. (2008) Analysis of nine drugs and their cytochrome P450-specific probe metabolites from urine by liquid chromatography-tandem mass spectrometry utilizing sub 2 microm particle size column. *J Chromatogr A.* **1215**, 107–115.
6. Youdim K.A., Saunders K.C. (2010) A review of LC-MS techniques and high-throughput approaches used to investigate drug metabolism by cytochrome P450s. *Chromatogr B Analyt Technol Biomed Life Sci.* [Epub ahead of print].
7. Lahoz A., Donato M.T., Picazo L., Castell J.V., and Gómez-Lechón M.J. (2008) Assessment of cytochrome P450 induction in human hepatocytes using the cocktail strategy plus liquid chromatography tandem mass spectrometry. *Drug Metab Lett.* **2**, 205–209.
8. Gómez-Lechón M.J., Donato M.T., Castell J.V., and Jover R. (2004) Human hepatocytes in primary culture: the choice to invsetigate drug metabolism in man. *Curr Drug Metab.* **5**, 443–462.
9. Serralta A., Donato M.T., Martinez A., Pareja E., Orbis F., Castell J.V., et al. (2005) Influence of preservation solution on the isolation and culture of human hepatocytes from liver grafts. *Cell Transplant.* **14**, 837–843.
10. Gómez-Lechón M.J., Castell J.V., and Donato M.T. (2008) An update on metabolism studies using human hepatocytes in primary culture. *Expert Opin. Drug Metab Toxicol.* **4**, 837–854.
11. Soars M.G., Mcginnity D.F., Grime K., and Riley R.J. (2007) The pivotal role of hepatocytes in drug discovery. *Chem Biol Interact.* **168**, 2–15.
12. Farkas D., Tannenbaum S.R. (2005) In vitro methods to study chemically-induced hepatotoxicity: a literature review. *Curr Drug Metab.* **6**, 111–25.
13. Gómez-Lechón M.J., Lahoz A., Gombau L., Castell J.V., and Donato M.T.(2010) In vitro evaluation of potential hepatotoxicity induced by new drugs. *Curr. Pharm. Design.* [Epub ahead of print].
14. Lahoz A., Donato M.T., Picazo L., Gómez-Lechón M.J., and Castell J.V. (2007) Determination of major human cytochrome P450s

activities in 96-well plates using liquid chromatography tandem mass spectrometry. *Toxicol In Vitro* **21**, 1247–1252.

15. Donato M.T., Gómez-Lechón M.J., and Castell J.V. (1993) A microassay for measuring cytochrome P450IA1 and P450IIB1 activities in intact human and rat hepatocytes cultured on 96-well plates. *Anal Biochem.* **213**, 29–33.
16. Lahoz A., Donato M.T., Montero S., Castell J.V., and Gómez-Lechón M.J. (2008) A new in vitro approach for the simultaneous determination of phase I and phase II enzymatic activities of human hepatocyte preparations. *Rapid Commun Mass Spectrom.* **22**, 240–244.
17. Mathijs K., Kienhuis A.S., Brauers K.J., Jennen D.G., Lahoz A., Kleinjans J.C., et al. (2009) Assessing the metabolic competence of sandwich-cultured mouse primary hepatocytes. *Drug Metab Dispos.* **37**, 1305–1311.
18. Donato M.T., Lahoz A., Montero S., Bonora A., Pareja E., Mir J., et al. (2008) Functional assessment of the quality of human hepatocyte preparations for cell transplantation. *Cell Transplant.* **17**, 1211–1219.

Chapter 8

Human Hepatocytes: Isolation, Culture, and Quality Procedures

Daniel Knobeloch, Sabrina Ehnert, Lilianna Schyschka, Peter Büchler, Michael Schoenberg, Jörg Kleeff, Wolfgang E. Thasler, Natascha C. Nussler, Patricio Godoy, Jan Hengstler, and Andreas K. Nussler

Abstract

The use of isolated human liver cells in research and development has gained increasing interest during the past years. The possible application may vary between elucidation of new biochemical pathways in liver diseases, drug development, safety issues, and new therapeutic strategies up to direct clinical translation for liver support. However, the isolation of human liver cells requires a well-developed logistic network among surgeons, biologists, and technicians to obtain a high quality of cells. Our laboratories have been involved in various applications of human liver cells and we have long-lasting experiences in human liver cell isolation and their application in R&D. We here summarize the present protocol of our laboratories for cell isolation from normal resected liver tissue, the most common tissue available. In addition, we discuss the necessary network in the clinic and quality controls to maintain human liver cells in culture and the effect of 3D extracellular matrix in cultured cells which results in preservation of hepatocyte epithelial polarity in the form of bile canaliculi and repression of epithelial to mesenchymal transitions occurring in 2D cultures.

Key words: Human hepatocytes, Cell isolation, Hepatocyte culture, Quality control, Metabolic competence

1. Introduction

1.1. Hepatocyte Isolation

Hepatocyte isolation was first applied in the late fifties of the last century to isolate cells for cell physiological research and drug metabolism studies (1). With a continuous progress in cell isolation techniques and defined logistics, transplantation of cells into patients with pathological liver diseases was also performed in the past 20 years (2). The increased demand for hepatocytes in the field of research, clinical use, and toxicological prediction is accompanied by their limited availability. In the past decades, isolated primary human hepatocytes

Ragai R. Mitry and Robin D. Hughes (eds.), *Human Cell Culture Protocols*, Methods in Molecular Biology, vol. 806, DOI 10.1007/978-1-61779-367-7_8, © Springer Science+Business Media, LLC 2012

in 2D culture have been proven to be a valuable tool to study drug toxicity, interaction, and metabolism in the liver. Thus, the need for cells for pharmaceutical research has got more and more in the focus, as metabolism of xenobiotics can hardly be predicted and transferability from animals to humans is restricted. Furthermore, to provide increased safety information on substances in the European Union (EU), the EU has established the European Community Regulation on chemicals and their safe use called Registration, Evaluation and Authorisation of Chemicals (REACH) (2). For this aim, a huge amount of primary human hepatocytes would be theoretically needed to determine, in animal replacement methods, the toxicity of all these substances. Over the past decades, optimization of human liver cell isolation has always been in the center of interest. However, it is a costly and time-consuming process and requires a well-organized network. More and more clinical centers have expanded their surgical repertory and handle with primary benign and malign tumors, e.g., hemangioma, focal nodular hyperplasia, cysts, adenoma, liver metastasis (colorectal and noncolorectal), as well as cholangio- and hepatocellular carcinoma. Upon that, only a few clinics have a well-associated research and clinical network to isolate human liver cells from tissue for the clinical and research purposes.

In the process of tumor resection, surrounding healthy liver tissue is also resected. This tissue can be used, assuming that the patients have given their consent to isolate parenchymal and non-parenchymal cells. Between surgical resection of pathological liver tissue and isolation of parenchymal liver cells, tissue must be transported from the operation room to the research facility. During the phase of transport, it is mandatory to keep tissue in a transport medium on ice similar as to what is used when organs are transported for liver transplantation. This is important to preserve cells from increased warm ischemia time. In general, it is important to maintain the time between liver resection and organ harvest as short as possible. The majority of all used liver tissue from human donors comes from liver resections, including partial hepatectomy. The size and weight of the liver tissue for parenchymal cell isolation are limited. Liver pieces smaller than 5 g are sometimes difficult to perfuse since they have no complete vessel architecture to be entirely perfused. In contrast, pieces of more than 100 g are likewise not easy to perfuse since an equal perfusion pressure in the whole tissue requires a pump with a sufficiently strong flow rate. In summary, the transport logistics are extremely important; thus, the cell yield decreases with increased tissue transportation time, increased warm or cold ischemia time, and sometimes when the liver piece is too large and the right equipment is not available. However, beside all these technical matters, it is mandatory to have a trustful relation between the surgeon and laboratory members to avoid additional logistics problems.

Beside transport and logistics, there are additional reasons why cell isolation may be restricted or the cell isolation is simply not possible. Livers with severe infections, e.g., human immunodeficiency virus (HIV), hepatitis B (HBV) and C virus (HCV), and methicillin-resistant *Staphylococcus aureus* (MRSA), are usually excluded from cell isolation because of a high risk of infection of the laboratory members. Further exclusion criteria or yield-limiting reasons are advanced cirrhosis, fibrosis, or bile duct stasis. In the latter, cells are chronically stressed and further harmed during the isolation procedure. Moreover, intoxication via drugs and steatohepatitis due to alcohol or in patients with a very high body mass index limits the successful outcome. Some surgical preoperation methods and neoadjuvant therapies can affect the liver parenchyma, e.g., downstaging/-sizing via chemotherapy or an embolization of the portal vein. Furthermore, tumor recurrence is evidence of mostly effected liver parenchyma. The diagnosis and patient-related data cannot be influenced, but can give a good approximation of the expected cell yield and possible viability. No difference can be observed regarding the gender of the liver donors or moderate tobacco and/or alcohol consumption (3, 4), except the liver is severely harmed, such as through fibrosis or cirrhosis.

The ideal tissue for liver cell isolation is a capsule tissue from a young patient (below 50 years) with a local pathological finding (e.g., single metastasis, hemangioma) or trauma. The best capsule is achieved by taking it from a left hepatectomy, as the capsule runs out flat. In this case, the ratio of open resection area to liver tissue weight is the best. The smaller the open resection area is, the better the perfusion can be performed.

The method of human parenchymal liver cell isolation is based on the principles of the two-step liver cell isolation using a collagenase perfusion of the rat liver, published by Seglen in 1976 (5). Several modified in situ perfusion techniques have been described in the past (6).

1.2. Functional Analysis

The definition of a hepatocyte found in the Merriam-Webster dictionary is as follows: any of the polygonal epithelial parenchymatous cells of the liver that secrete bile (entry for: hepatocyte, http://medical.merriamwebster.com/medical/hepatocyte, Merriam-Webster's Medial Dictionary). Unfortunately, this characterization is not sufficient to answer the question if any cell resembles a primary hepatocyte. Therefore, an expression profile (markers) for the desired cell type has to be defined further. This is supported by studies showing that expression of albumin and an epithelial morphology alone does not guarantee all functions that make up a hepatocyte (7–9). Hepatocytes play a central role in the processing of endogenous and exogenous substances. By biotransformation, they detoxify and inactivate compounds, such as steroids, bilirubin and fatty acids, and drugs and chemicals. Formation and following secretion of bile salts, phospholipids, and cholesterol into bile allows the hepatocytes to excrete metabolized biogenics and xenobiotics from the body. Another major hepatic function is the conversion of ammonia into urea for renal excretion. Furthermore, hepatocytes play a central role in glycolysis and gluconeogenesis. Therefore, we (9) suggest to perform not only qualitative assays (e.g., RT-PCR, IHC), but also a quantitative analysis of phase I and II drug-metabolizing enzyme activities, e.g., CYPs 1A1, 2D6, 3A4, UDP--glucuronosyltransferase, and glutathione-S-transferase, as well as measurement of synthesis products, e.g., aspartate aminotransferase (AST), alanine aminotransferase (ALT), lactate dehydrogenase (LDH), urea, glucose, and fibrinogen.

1.3. Drug Metabolism

Drug metabolism (biotransformation) in the liver can be divided into two phases (10). Phase I metabolism adds a functional group (e.g., OH, SH, or NH_2) to the substrate by oxidative, reductive, and hydrolytic pathways. Phase II metabolic enzymes modify the newly introduced functional group to *O*- and *N*-glucuronides, sulfate esters, various α-carboxyamides, and glutathionyl adducts in order to increase their polarity (11) making secretion from the cells more convenient. Thus, hepatocytes mediate detoxification by activation of the phase I and II enzymatic pathways. An overview of how to measure the main phase I and phase II is given below.

All members of the cytochrome P450 (CYP) superfamily, belonging to phase I drug-metabolizing enzymes, can be identified by a highly conserved heme-thiolate functionality, responsible for their catalytic mechanism (12). Amino acid variations in their substrate-binding sites confer compound, region, and stereo selectivity of the enzymes (13) often being involved in the rate-limiting step of drug biotransformation processes. Experimental data suggest that most biotransformation of xenobiotics is done by enzymes of the first three families (CYP1, 2, and 3) while other CYPs are involved in "housekeeping" metabolism of endogenous molecules (11, 14).

Drug metabolism is one of the major determinants of drug clearance and thus the factor that is most frequently responsible for the interindividual differences in drug pharmacokinetics (15). Adverse pharmacokinetics can result in an inadequate or altered response of the drug affecting its use as therapeutic (16). Therefore, in vitro screenings became an invaluable tool to identify the metabolic profile of drug candidates, potential drug interactions, or the role of polymorphic enzymes before starting clinical trials.

To date, primary human hepatocyte cultures appear to be the most powerful tool for in vitro studies (17), having major limitations due to donor organ scarcity and rapid cellular changes during culture (18). Along with the morphologic changes, the expression and activity of phase I (Fig. 1) and II drug-metabolizing enzymes are quickly reduced. Strategies to preserve enzyme activities include culture in the so-called "sandwiches" consisting of proteins of the extracellular matrix (ECM).

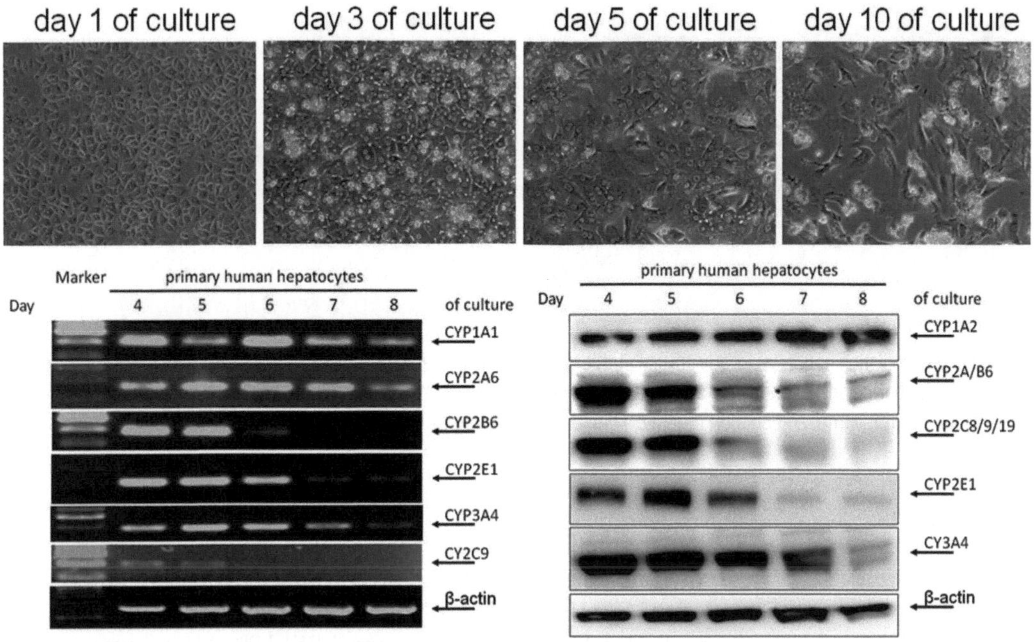

Fig. 1. Expression of Cytochrome P450 isoforms decreases during culture of primary human hepatocytes along with morphologic changes. Hepatocytes lose their characteristic octagonal shape during culture. Already after 3 days, bile canaliculi are less pronounced and after 5 days first changes toward a mesenchymal phenotype can be observed. This is accompanied by loss of Cytochrome P450 expression as measured by RT-PCR and Western blot.

1.4. Methods for Quality Control

Quality of isolated hepatocytes should be checked routinely. For doing this, a panel of different characteristics can be chosen, depending on the intended use of the cells. If the hepatocytes shall be used for drug metabolism studies, expression of phase I and II drug-metabolizing enzymes is very important. Especially, certain cytochromes (CYP1A1/2 and CYP3A4) can be regulated in vitro; therefore, we recommend stimulation of the hepatocytes with known inducers (e.g., rifampicin, 3-methylcholantrene, or phenobarbital) accepted by pharmaceutical companies. If hepatocytes shall be used for transplantation, an efficient detoxification is needed; therefore, urea and glucose metabolism as well as albumin production should be analyzed. Methods to measure selected hepatic functions are described below. The methods are downscaled to 96-well-plate format in order to ensure a fast screening without wasting cells.

2. Materials

2.1. Preparation of the Isolation Setup

1. Before starting, all necessary instruments and solutions have to be prepared sterile and ready to use (Fig. 2a, b).
2. When starting the isolation, all working steps must be done at 4°C without interruption.

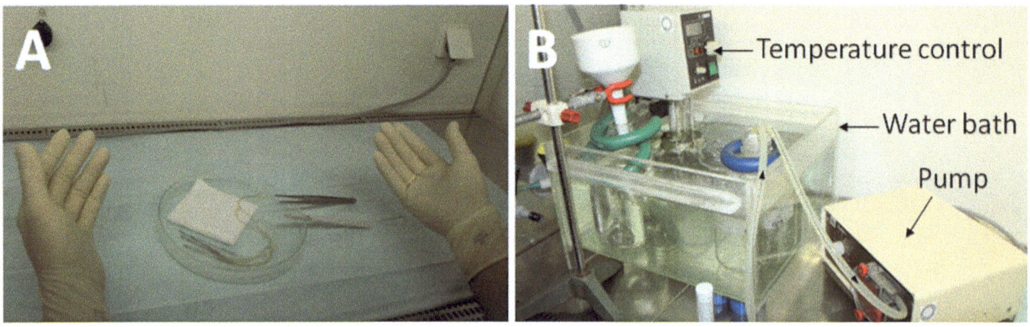

Fig. 2. Liver cell isolation setup. (**a**) Ready-to-use sterile work area with sterile drape sheet, glass dish, scalpel, forceps, gauze, and cannula for perfusion. (**b**) Under the same sterile bench as the work area is located the perfusion system with water bath, temperature control, pump, and suction (Buchner) filter. *Dotted arrows* indicate the flow path of the perfusion solutions.

3. Prepare medium for hepatocytes with Williams medium E that contains penicillin 100 U, streptomycin 100 μM, fortecortin 1.6 μM, HEPES 15 mM, sodium pyruvate 1 mM, human insulin 1 mM, fetal bovine serum (FBS) 10%, MEM (nonessential amino acids) 1%.
4. Furthermore, perfusion solution I with and without ethylene glycol tetra acetic acid (EGTA) and perfusion solution II has to be prepared before starting the cell isolation.

Perfusion solution I (with EGTA):

Additives	Concentration
NaCl	142 mM
KCl	6.7 mM
HEPES	100 mM
EGTA	2.4 M

Perfusion solution I (without EGTA):

Additives	Concentration
NaCl	142 mM
KCl	6.7 mM
HEPES	100 mM
FBS	2%

Perfusion solution II:

Prepare solutions 1 and 2 separately.

Solution 1:

Additives	Concentration
NaCl	67 mM
KCl	6.7 mM
HEPES	100 mM
Albumin	0.5%

Adjust to 1,300 ml with distilled water.

Solution 2:

Additives	Concentration
$CaCl_2 \cdot 2H_2O$	4.8 mM

Adjust to 150 ml with distilled water.

Mix solution 1 and solution 2 and top up distilled water up to 1,500 ml.

5. In general, the pH must be adjusted to 7.4 for all solutions.
6. Medium and solutions have to be filtered sterilely and stored at 4°C for up to 4 weeks. If the solution is not prepared freshly, make sure to check pH immediately before the isolation procedure and if necessary adjust the pH.
7. Set the temperature for the water bath for the isolation at 39°C. Keep the temperature at the outflow between 36 and 38°C to obtain optimal collagenase activity (see Note 1).
8. Place a bottle of perfusion solution I with EGTA into the water bath and fix it with a lead ring to prevent it from floating. Secure the Buchner funnel onto the tripod, and place an empty sterile bottle (500 ml) underneath the funnel for the waste. Place both ends of the sterile silicon tube into the bottle of perfusion solution I and let the solution circulate until no air bubbles remain in the hose (Fig. 2b). A sterile work area is prepared when a surgical drape sheet is unwrapped. Place a sterile glass Petri dish in the middle and put a scalpel, forceps, gauze, and cannula inside the glass dish (Fig. 2a).
9. Furthermore, prepare a beaker with ice and 50-ml tubes for collecting cell suspension.
10. Place a sterile gauze on top of a plastic filter and place it on the 50-ml tube to filter the cells.

3. Methods

3.1. Isolation Procedure

1. Take the liver piece out of the transport medium and place it onto the Petri dish. The resection area dissects the vessels and bile duct. By using the biological architecture of vessels, the

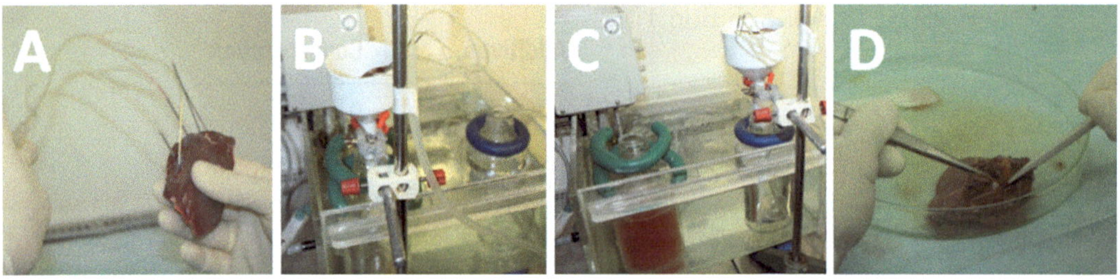

Fig. 3. Two-step perfusion of liver tissue. (**a**) Liver capsule cannulated and sealed with tissue adhesive. (**b**) First perfusion step to wash blood out of tissue into waste. (**c**) Recirculation of collagenase P perfusion solution to singularize parenchymal cells in the tissue. (**d**) By cutting the capsule with a scalpel orthogonal to the resection area, the cells are released and washed out with a stop solution [perfusion solution I (without EGTA) and 5–10% FBS).

whole liver tissue can be perfused with a set of at least three cannulas.

2. Fix cannulas into the biggest vessels and close other vessels with tissue glue (Fig. 3a), until an adequate perfusion pressure has been reached. To prevent coagulation of remaining residual blood in the tissue, the additive EGTA is present in the first perfusion solution.
3. Prepare the instrument setup as shown in Fig. 3b and use at least 500 ml of perfusion solution I (with EGTA) to wash out all blood from the tissue and warm up the tissue. If blood is left in the tissue, repeat perfusion until blood is completely removed; otherwise, digesting quality and cell outcome in further steps are decreased.
4. In the following step, the liver is perfused by the standard digesting enzyme collagenase P solution. Therefore, while washing the liver tissue with perfusion solution I (with EGTA), prepare approximately 100 ml perfusion solution II containing the amount of collagenase needed to dissolve the tissue collagen. Subsequently, tubing has to be modified to enable the recirculation of perfusion solution II (Fig. 3c). The recirculation reduces the amount of expensive collagenase solution. Terminate the process after approximately 20 min or as soon as the liver stays irreversibly deformed through slight pressure. Collagenase is a calcium-dependent enzyme and must therefore be present in $CaCl_2$ solution during the second stage of perfusion.
5. Now, place the liver in the Petri dish with perfusion solution I (without EGTA) to inactivate collagenase activity. In addition, a serum-containing solution (5–10%) is added to inactivate trypsin residues in the collagenase to prevent overdigestion of the cells. Cut the liver tissue in two halves with the scalpel and open the liver capsule (Fig. 3d). Use scalpel and forceps to "comb" the cells free. Rinse the remaining liver tissue in the Petri dish with ice-cold perfusion solution I (without EGTA)

and 5–10% serum. Pour the cell suspension through the prepared funnel with gauze into 50-ml tubes on ice in order to eliminate tissue debris.

6. Cells and collagenase suspension are separated by a short centrifugation for 5 min with 50–100 g at 4°C (for fatty liver, it is recommended to use higher gravitational acceleration). Discard the supernatant and resuspend the cell pellet in hepatocyte incubation medium and store at 4°C.
7. In the following, the vitality and cell number, using trypan blue exclusion, can be estimated using a hemocytometer.
8. For counting, mix together 100 μl cell suspension with 900 μl trypan blue (diluted with PBS 1:4) to reach a suspension of 1:10.

   Calculation: Total cells = counted cells × 10^4 × 10 (dilution) × ml cell suspension:

$$\text{Viability}\ [\%] = \frac{\text{vital cells}}{\text{total cells}} \times 100.$$

9. If the viability is low (less than 75%), a separation of dead and living cells is recommended by density gradient centrifugation using a 25% Percoll gradient.

3.2. Percoll Purification

1. Prepare for each tube 20 ml of a 25% Percoll solution (15 ml PBS and 5 ml Percoll solution in a 50-ml Falcon tube). The calculated number of necessary tubes depends on the initially isolated cell number. The best separation between dead and nonparenchymal cells from viable hepatocytes is reached when no more than 50×10^6 viable cells are overlaid onto the Percoll solution (Fig. 4).
2. Centrifuge the 50-ml tubes with the two layers with 1,278 × g at 4°C for 20 min. It is important to turn off the brake to have a better layer separation. After centrifugation, a further third layer is formed: the upper layer contains dead cells and cell debris (Fig. 4). The upper and a part of the middle layer have to be aspirated. Leave approximately 5 ml of the Percoll on top of the pellet. Wash the cell pellet with 35 ml PBS and centrifuge at 50 × g at 4°C for 5 min. The cells form a compact pellet. The supernatant can be discarded. Resuspend the cell pellet with hepatocyte incubation medium and determine the cell number and cell viability again by using the trypan blue exclusion technique. At this point, human hepatocytes can be stored before seeding or shipping for a few hours on ice in a high-density cell suspension.
3. During the isolation procedure, plates for culturing the cells can be prepared. When culturing the cells in 2D monolayer, the cells need a cell–cell contact and a prepared surface for attachment.

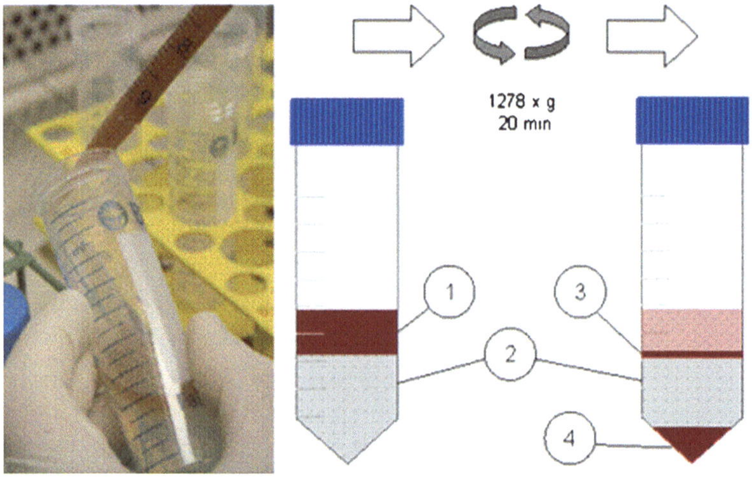

Fig. 4. Separation of live hepatocytes from dead cells and cell debris by density gradient centrifugation. The picture on the left shows careful covering of Percoll solution with cell suspension. The schematic picture shows cell suspension (1) that covers in the beginning the Percoll solution (2). After centrifugation, the cell debris and dead cells left over on top of Percoll (3) and all live cells form a cell pellet on the bottom (4).

The cell density has to be selected carefully to ensure cell confluence. A suitable cell density for cell seeding on different dishes is $0.1–0.15 \times 10^6$ cells/cm^2.

4. Different coating strategies were tried to preserve cell integrity and function. To maintain close in vivo environmental conditions, the plates have to be coated with a protein matrix. In a subset, matrices were tested. Hepatocytes synthesize actin filaments when being attached to the different matrices by integrins. Thus, the integrin-independent coating poly-L-Lysine caused only very few stress fibers. On rat tail collagen-coated surfaces, the cells attached fastest; consequently, most actin filament was formed. Regarding the shape of the actin fibers at day 7 of culture, rat tail collagen and laminin coating showed best integrity of the hepatocytes, whereas fibronectin, hydromatrix, and matrigel (growth factor reduced) showed strong expression of stress fibers. In poly-L-lysine-coated dishes, hepatocytes could not be kept in culture for long times. Cells quickly started to detach and die. Although laminin showed promising results, the attachment phase with rat tail collagen was significantly shorter (Fig. 5). Thus, the use of rat tail collagen is the most accepted coating in hepatocyte research. Rat tail collagen mainly consists of collagen type I (19).
5. Rat tail collagen can be bought or prepared by the researcher. Collagen fibers are isolated from rat tail tendons by breaking the skinned tails between every vertebrae and pulling the tendons out (start at the top of tail). Fibers are air dried and sterilized using UV light. Afterward, collagen fibers are stirred

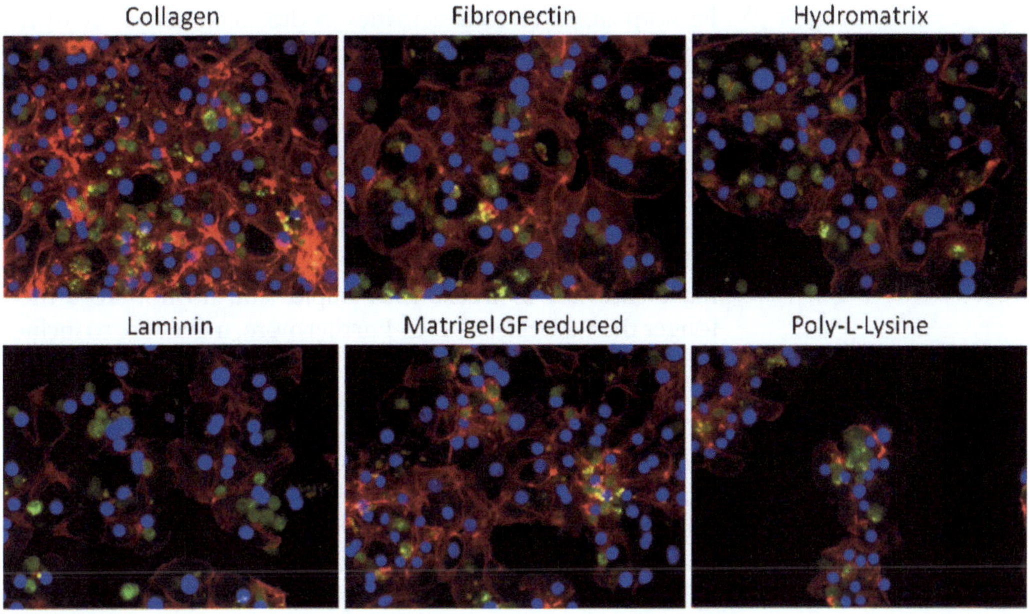

Fig. 5. Attachment characteristics of primary human hepatocytes visualized by fluorescence on different coated surfaces after 7 days of culture. Red labels phalloidin/actin filaments, blue Hoechst/nuclei, and green shows dying or dead cells (×100 magnification).

for 48 h in a 0.02 N acetic acid solution to solubilize the collagen (4 g of collagen in 1 l of acetic acid). This stock solution is then diluted (1:50 in PBS) to cover cell culture plastics and incubated for at least 30 min at room temperature (RT). After incubation, the solution must be discarded and the dishes have to be air dried. When using 2D culture systems, the hepatocytes flatten over long culture times and lose their characteristics. To overcome this problem, cells can also be cultured in 3D culture systems (bioreactors) or as a sandwich culture surrounded by a gel composed of ECM proteins, e.g., collagen (see Note 2).

3.3. Transaminase and Lactate-Dehydrogenase Measurement

1. Depending on the experimental setup, AST, ALT, and LDH measurement can be used both for characterizing hepatocytes' function and damage. For assessing the basal expression levels of AST, ALT, and LDH, cells are lysed with low concentrations of Triton X-100 (0.1%). Total AST, ALT, and LDH activities are measured. Transaminase and LDH activities are very sensitive toward freezing. Avoid freezing and thawing of samples before analysis (samples can be stored at +4°C for up to 1 week).
2. To assess cellular damage, transaminase and LDH activities are measured in the culture supernatants. The so-called AST, ALT, and LDH leakage into the culture supernatant is measured when determining toxic concentrations of various substances and can

be normalized to total activities as described above. With this approach, additional care has to be taken. Medium without cells has to be included into the measurement to determine effects of the chemicals investigated on the activities measured.

3. Test kits for AST, ALT, and LDH activity measurement can be obtained, e.g., from HITADO, Clonetech, or Roche. We measured the enzyme activities as indicated by the manufacturer (HITADO). For measuring AST, ALT, and LDH activity, make sure not to freeze the samples and do not store them longer than 1 week at +4°C. Furthermore, make sure to include all necessary controls, especially when measuring toxic effects of various substances as they might interfere with the Test kit.

3.4. Urea Formation and NH_4Cl Metabolism

1. Ammonia is produced as toxic side product during various cellular reactions. One major function of hepatocytes is the conversion of ammonia into urea for renal excretion to protect the body from intoxication. Thus, we can measure urea production by its release into the culture supernatant. Supplementation of reaction buffer with ammonium chloride (NH_4Cl) increases the basal urea production. Further addition of ornithine, an essential cofactor for urea formation, enhances this effect (Fig. 6).
2. Briefly, wash cells three times with Dulbecco's phosphate-buffered saline (DPBS) to remove residual medium.
3. Incubate cells with DPBS (supplemented with 29 mM KCl; 1 mM $MgCl_2$; 1 mM sodium pyruvate; ±5 mM NH_4Cl; ±1 mM ornithine) at 5% CO_2, 37 ± 0.1°C.
4. After 24 h, transfer 80 µl of supernatant to a transparent 96-well plate.

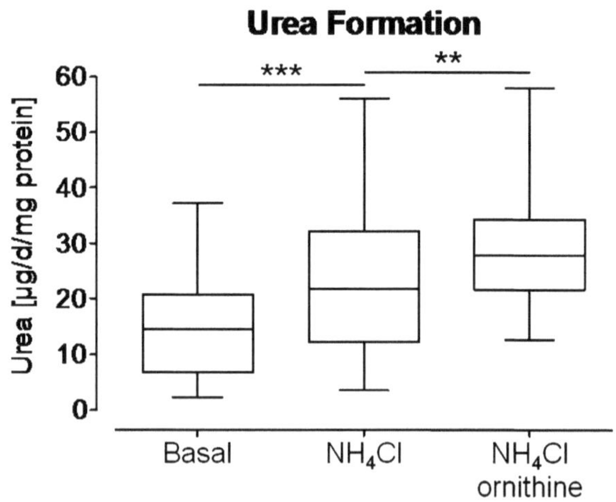

Fig. 6. Hepatocytes convert ammonia into urea. Increased activation of urea conversion by NH_4Cl, NH_4Cl and ornithine compared to basal activity by hepatocytes. Results are normalized against total protein amount.

5. Onto the same plate, transfer 80 µl of a dilution series of urea (0, 0.78, 1.56, 3.13, 6.25, 12.5, 25, and 50 µg/ml).
6. Add to each sample 60 µl solution A (7.4% H_2SO_4; 0.0002% O-phthalaldehyde; 0.03% Brij 35) and 60 µl solution B (22.2% H_2SO_4; 0.5% boric acid; 0.0006% N-(1-naphthyl)ethylenediamine dihydrochloride; 0.03% Brij 35).
7. Incubate reaction mixture at 37.0 ± 0.1°C for 1–2 h.
8. Measure optical density at λ = 505 nm.
9. Fix cells and stain with sulforhodamine B (SRB) for normalization.

3.5. Glycolysis and Gluconeogenesis

1. One major function of hepatocytes is the regulation of blood glucose levels by glycolysis and gluconeogenesis. An important enzyme to produce glucose is glucose-6-phosphatase (G6Pase). G6Pase can be visualized by an enzymatic-dependent staining. During this staining, phosphate from glucose-6-phosphate is transferred to $Pb(NO_3)_2$ to form $Pb(PO_4)_2$, which in the presence of ammonium sulfide form a brownish black precipitate Pb(S) (Fig. 7).
2. Hepatocytes store energy in form of glycogen or as lipids. When stimulated with insulin, hepatocytes convert glucose into neutral lipids, e.g., cholesteryl ester and triglyceride. Neutral lipids can be visualized by Oil Red O staining, during which the hydrophobic Oil Red O diffuses into the lipid droplets and marks them red (Fig. 8).
3. Briefly, wash cells three times with DPBS to remove residual medium.
4. Incubate cells with DPBS (supplemented with 29 mM KCl; 1 mM $MgCl_2$; 1 mM sodium pyruvate; ±10 mM sodium L-lactate) at 5% CO_2, 37 ± 0.1°C.
5. After 24 h, transfer 100 µl of the culture supernatant to a transparent 96-well plate.
6. Onto the same plate, transfer 100 µl of a dilution series of glucose (0, 10, 20, 30, 60, 90, 120, and 150 mM).

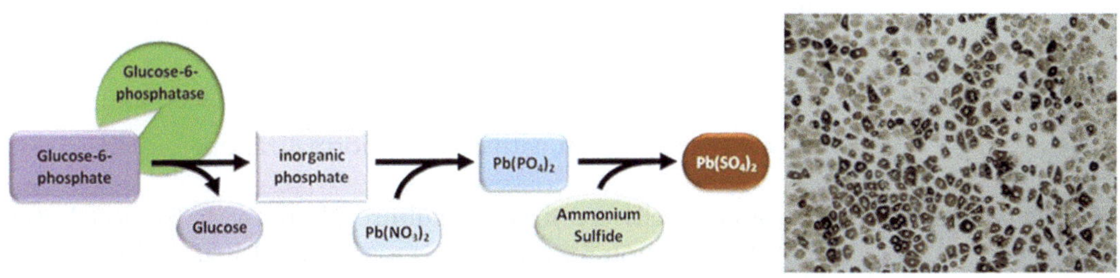

Fig. 7. Mechanism of glucose-6-phosphate (G6PAse) staining and representative staining. G6PAse is transferred to $Pb(NO_3)_2$ to form $Pb(PO_4)_2$, which in the presence of ammonium sulfide form a brownish black precipitate $Pb(SO_4)_2$.

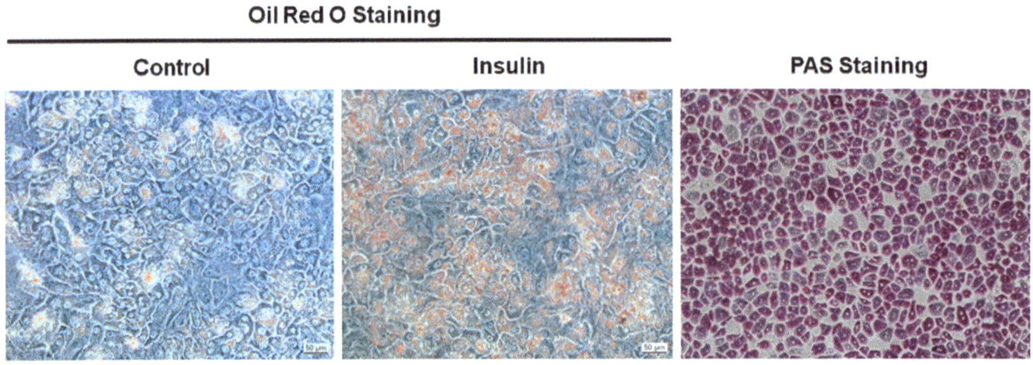

Fig. 8. Hepatocytes stained for neutral lipid accumulation and glycogen. Hepatocytes incubated for 48 h with 170 nM insulin strongly accumulate neutral lipids as visualized by Oil Red (*left panel*). Untreated hepatocytes store energy in form of glycogen, as visualized by PAS staining (*right*).

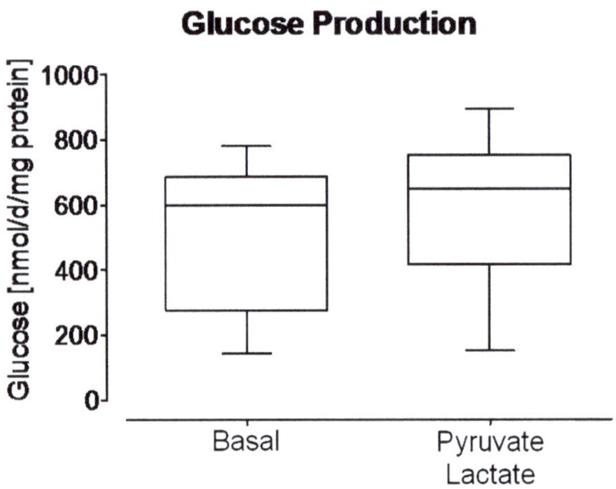

Fig .9. Hepatocytes perform glycolysis and gluconeogenesis. Basal glucose production is due to degradation of glycogen. Additional supplementation with pyruvate and lactate is used to show the rate of gluconeogenesis.

7. To each sample, add 100 µl GLOX solution (250 mM Tris–HCl, pH = 8.0; 0.04% glucose-oxidase; 0.007% peroxidase; 0.2 mM EDTA; 0.01% o-dianisidine).
8. Incubate reaction mixture at 37.0 ± 0.1°C, 5.0 ± 0.1% CO_2 for 1–2 h.
9. Assess optical density at λ = 420 nm (Fig. 9) (see Note 3).
10. Fix cells and stain with SRB for normalization.

3.6. Sulforhodamine B Staining

1. Briefly, cells are fixed to culture plastic with fixation buffer (95% ethanol; 5% acetic acid) for at least 1 h at −20°C.
2. After fixation, wash cells five times with H_2O before staining with 0.4% SRB dissolved in 1% acetic acid.

3. After 30 min (dark; RT) of incubation, wash cells five times with 1% acetic acid to remove unbound dye.
4. Bound SRB is solubilized with 10 mM unbuffered Tris ($pH \approx 10.5$). Optical density is measured at $\lambda = 565$ nm (SRB) and $\lambda = 690$ nm (impurities).

3.7. Phase I and II Enzyme Activity

1. Fluorescence-based CYP assays are performed by direct incubation of cells, cultured in 96-well plates, with selected substrates as published previously (1).
2. Briefly, add 100 μl of incubation medium (1 mM Na_2HPO_4, 137 mM NaCl, 5 mM KCl, 0.5 mM $MgCl_2$, 2 mM $CaCl_2$, 10 mM O-(+)-glucose, and 10 mM HEPES; pH = 7.4) containing fluorogenic substrates to each well. Substrates were added in acetonitrile solution in concentrations for which the final solvent concentration did not exceed 0.3% (v/v). Experimental conditions are summarized in the Table 1 below.
3. After a 60 to 120-min incubation at 37.0 ± 0.1°C, 5.0 ± 0.1% CO_2, assays are stopped by transfer of incubation medium to a white/black 96-well plate.
4. Fix cells and stain with SRB for normalization.
5. Potential metabolite conjugates formed during the activity assays are hydrolyzed by incubation of supernatants with β-glucuronidase/arylsulfatase (150 Fishman units/ml and 1,200 Roy units/ml, respectively) for 2 h at 37.0 ± 0.1°C, 5.0 ± 0.1% CO_2 (1).
6. Finally, samples were diluted (1:4) in the corresponding quenching solution and fluorescent metabolite formation was quantified fluorometrically.
7. Methanol fixed cells are used as negative control. Results are expressed as picomoles of metabolite formed per minute and per SRB signal.

3.8. Periodic Acid Schiff Staining

1. Before staining, wash cells twice with DPBS.
2. For fixation, cover cells with 4% PFA in DPBS for 5 min at RT.
3. Incubate cells for 5 min with 1% periodic acid solution (in ddH_2O).
4. Remove excessive periodic acid by washing cells twice with DPBS.
5. After washing, incubate cells for 15 min at RT (shaker) with Schiff's reagent, which stains glycogen pink.
6. Residual Schiff's reagent is washed off the cells with tap water (Fig. 8).

Table 1
Experimental conditions for phase I and II enzyme activity measurement

	Substrate	Assay conc.	Incubation time	Metabolite	96-well plate	Quenching solution	Ex/Em (nm)
CYP1A1/2	CEC	30 µM	90 min	CHC	Black	DPBS	390/460
CYP2A6	Coumarin	50 µM	60 min	HC	Black	0.1 M Tris (pH 9.0)	355/460
CYP2B6	EFC	30 µM	120 min	HFC	White	0.1 M Tris (pH 9.0)	390/520
CYP2C8/9	DBF	10 µM	90 min	Fluoresceine	Black	10 mM NaOH	485/538
CYP2D6	AMMC	10 µM	120 min	AHMC	White	0.1 M Tris (pH 9.0)	390/460
CYP2E1	MFC	10 µM	120 min	HFC	White	0.1 M Tris (pH 9.0)	390/520
CYP3A4	BFC	100 µM	120 min	HFC	White	0.1 M Tris (pH 9.0)	390/520
Phase II	CHC	1 µM	–	Conjugated	Black	DPBS	390/460
Phase II	HC	0.25 µM	–	Conjugated	Black	0.1 M Tris (pH 9.0)	355/460
Phase II	HFC	1 µM	–	Conjugated	White	0.1 M Tris (pH 9.0)	390/520
Phase II	Fluoresceine	1 µM	–	Conjugated	Black	10 mM NaOH	485/538
Phase II	AHMC	1 µM	–	Conjugated	White	0.1 M Tris (pH 9.0)	390/460
Phase II	4-MU	1 µM	–	Conjugated	Black	DPBS	355/460
Phase II	Resorufin	2 µM	–	Conjugated	Black	DPBS	544/590
Phase II	MCB	1 µM	–	Conjugated	White	DPBS	355/460

3.9. Glucose-6-Phosphatase Staining

1. Before staining, wash cells twice with DPBS.
2. For fixation, cover cells with 4% PFA in DPBS for 5 min at RT.
3. Cover cells with G6Pase staining solution (100 mM Tris, pH = 6.5, 0.06% glucose-6-phosphate, and 0.08% lead-(II)-nitrate) for 45 min at 37.0 ± 0.1°C, 5.0 ± 0.1% CO_2.
4. Excessive staining solution is washed off the cells twice with DPBS.
5. After washing, cells are briefly covered (approximately 30 s) with 1% ammonium sulfide solution (fume hood!). During this short incubation, the G6Pase containing cells turns brownish black.
6. Excessive ammonium sulfide solution is washed off the cells with tap water (Fig. 7).

3.10. Oil Red O Staining

1. Before staining, wash cells twice with DPBS.
2. For fixation, cover cells with 4% PFA in DPBS for 5 min at RT.
3. Wash cells once with 60% isopropanol.
4. Cover cells with Oil Red O staining solution (0.2% Oil Red O; 60% isopropanol; prepare freshly before use!) and incubate for 10–15 min at RT. During this time, Oil Red O diffuses into neutral lipid droplets and excessive Oil Red O starts to precipitate.
5. Remove unbound stain by washing four to five times with tap water. It is possible to quantify this staining by resolving the bound Oil Red O in 100% isopropanol.
6. After incubation for 10 min at RT (shaking), the optical density is measured at $\lambda = 500$ nm (Fig. 7).

3.11. Preservation of Hepatocytes Polarity by 3D Extracellular Matrix

1. The addition of a collagen gel overlay on cells plated onto collagen-I-coated dishes (CM-Gel) strongly represses hepatocyte dedifferentiation which includes the loss of the typical hepatocytes honeycomb morphology and formation of actin stress fibers, typical for 2D-cultured hepatocytes (CM) (Fig. 10).
2. A collagen gel is obtained by neutralizing the pH of a collagen-I solution (1 mg/ml) by adding NaOH 1 M dropwise while keeping the collagen solution on ice. This solution is composed of nine parts of collagen-I and one part of DMEM 10× (Biozol, Eching, Germany).
3. 500 μl of the neutralized gel solution is added on top of the hepatocytes plated onto six-well plates, which can be done 6–24 h after plating when the cells are well-attached.

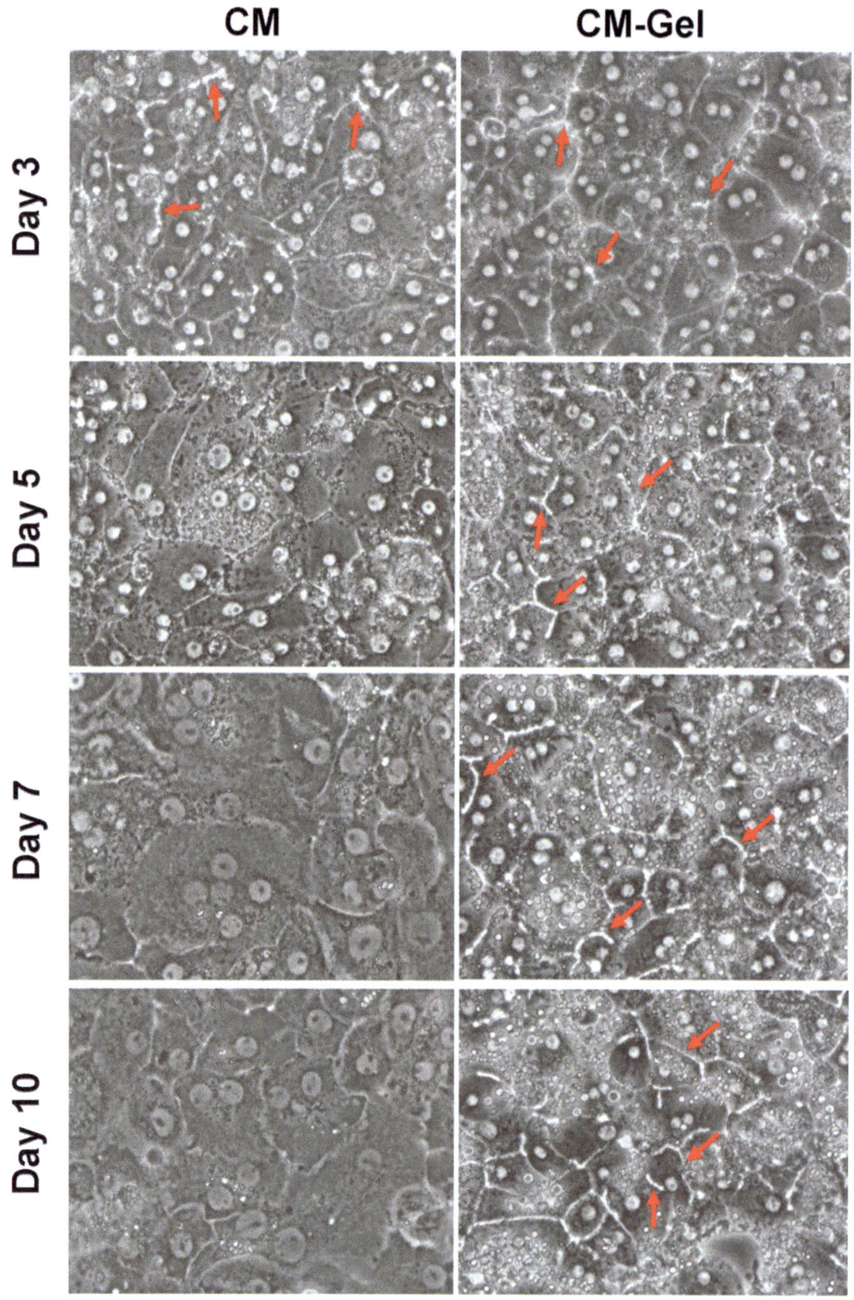

Fig. 10. Addition of collagen gel overlay preserves hepatocytes' morphology and bile canaliculi formation. Phase-contrast photographs of human hepatocytes cultured in conventional collagen-coated dishes alone (CM) or with an additional collagen gel overlay (CM-Gel) for the indicated time. Bile canaliculi are indicated by *red arrows*.

4. Ensure that the collagen solution is well-spread over the cells by tiling the well in all directions, and allow the solution to gel for 45 min in the incubator at 37°C.
5. Then, add culture medium softly to avoid disrupting the gel.

3.12. Staining for DPP-IV

1. Besides the loss of honeycomb morphology, hepatocytes in CM also fail to form extensive bile canaliculi network and to preserve these structures in a long culture period. Conversely, cells in CM-Gel cultures show progressive formation of bile canaliculi, which are clearly seen as bright refringent structures in phase-contrast microscopy (Fig. 10). To visualize molecular details of dedifferentiation and polarity, the cells are fixed and stained for actin fibers and dipeptidyl peptidase-IV (DPP-IV), a typical marker of bile canaliculi in vivo (20).
2. Briefly, cells are fixed with 4% PFA for 15 min at room temperature, permeabilized with 0.5% Triton X-100 in PBS for 15 min, and washed three times with PBS.
3. The cells are blocked with 5% BSA in PBS for 2 h at room temperature.
4. Then, cells are incubated with 10 μg/ml of polyclonal goat antihuman DPP-IV (RnD Systems, AF1180) in 3% BSA/PBS over night at 4°C, washed three times in PBS, and incubated with rabbit anti-goat-Cy5-labeled secondary antibody overnight at 4°C.
5. The cells are then incubated with 2 U/ml rhodamine-labeled phalloidin (Invitrogen, Karlsruhe, Germany) and DAPI for 1 h at room temperature.
6. After washing three times in PBS, cells are mounted on glass slides with Slowfade Glod (Invitrogen, Karlsruhe, Germany).
7. Epifluorescent pictures clearly show a typical actin accumulation in stress fibers on CM-cultured hepatocytes (Fig. 11a), whereas in CM-Gel cultures actin fibers mainly localize at the cell borders and in bile canaliculi (Fig. 11b). Confocal microscopy gives further details of these structures, and clearly demonstrates a strong canalicular localization for DPP-IV which colocalizes well with actin fibers in CM-Gel-cultured cells (Fig. 11d) while in CM-cultured hepatocytes DPP-IV is found mainly at the cytoplasm, and only a weak staining is observed at the cell borders (Fig. 11c).

4. Notes

1. Collagenase (EC 3.424.3) is used to digest intercellular collagen. There are many collagenases on the market to isolate liver cells. All of them have in common that it is mandatory to test the used collagenase before using it routinously, since its composition varies from batch to batch and subsequently its proteolytic activity strongly varies among different types. In our laboratories, we use Collagenase P from *Clostridium histolyticum* which has

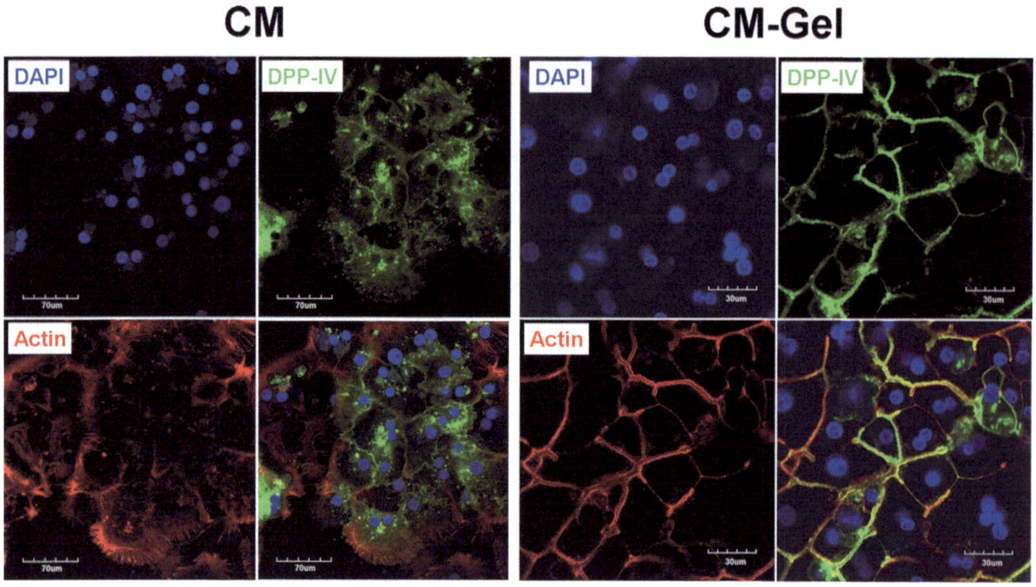

Fig. 11. Actin stress fibers and DPP-IV localization in CM and CM-Gel-cultured hepatocytes. Epifluorescence of hepatocytes stained by rhodamine-labeled phalloidin for actin fibers and DAPI for nuclei. Cells in CM display typical actin stress fibers (**a**) while cells cultured in CM-Gel exhibit actin fibers which are mainly distributed over the cell borders and preferentially at the bile canaliculi, indicated by arrows (**b**). Confocal scanning microscopy of hepatocytes in CM or CM-Gel stained with rhodamine-labeled phalloidin, anti-DPP-IV antibody, and DAPI. DPP-IV localizes primarily in the cytoplasm of CM-cultured cells, which show a weak staining at the cell contacts (**c**). Conversely, the localization of DPP-IV is clearly restricted to the cell borders in CM-Gel-cultured cells, accumulating preferentially at bile canaliculi, similarly to actin fibers (**d**).

been initially developed to isolate pancreatic islets cells, but it is very convenient for human liver tissue when it is mildly fibrotic. However, other collagenases can be used equally. It is important to filter the final solution and warm it up to 37°C in a water bath. It is important to warm up the solution, since collagenase can lose its activity below <37°C. Potent inhibitors of Collagenase P are oleic and palmitic acid which are the main components seen in fatty livers (21). Sometimes, incubation with collagenase P has to be extended when a liver is overloaded with fat. An overload of fat, however, can be easily seen macroscopically by a slightly yellow tissue color.

2. It is possible to send the cells to other laboratories. Plated cells can be sent after an attachment time of at least 4 h by using warm packs and closely sealed dishes. The cells survive the sending procedure for at least 12 h without changes compared to classical cultivation. Sometimes, it is necessary to send cells in suspension. Classical organ storage solution (e.g., University of Wisconsin cold storage solution) is insufficient for cell suspension storage (nonpublished data). However, cold storage solution for human hepatocyte preservation can be bought (*hepa*cult GmbH, Regensburg, Germany). Following the manufacturer's instructions, transport with a loss of approximately 20% of cell viability is possible up to 24 h.

3. To measure the degradation from glycogen to glucose, cells are incubated with plain buffer for 24 h and glucose is measured in the culture supernatant. In order to determine the rate of gluconeogenesis, the buffer is additionally supplemented with pyruvate and lactate. Measuring glucose in the culture supernatant, one has to take special care with washing the cells so that any residual medium (containing glucose) is removed from the cells.

Acknowledgments

This work was supported by the German Ministry of Research FKZ-BMBF 0315741 (Virtual Liver) and FZK-BMBF 01 GG 0732 (3-D TOX).

References

1. Branster, M. V., and Morton, R. K. (1957) Isolation of intact liver cells. *Nature.* **180**, 1283–1284.
2. de Boer, J., and van Bavel, B. (2009) European "REACH" (Registration, Evaluation, Authorisation and Restriction of Chemicals) program. *J Chromatogr A.* **1216**, 301.
3. Alexandre, E., Cahn, M., Abadie-Viollon, C., Meyer, N., Heyd, B., Mantion, G., et al. (2002) Influence of pre-, intra- and post-operative parameters of donor liver on the outcome of isolated human hepatocytes. *Cell Tissue Bank.* **3**, 223–233.
4. Lloyd, T. D., Orr, S., Patel, R., Crees, G., Chavda, S., Vadyar, H., et al. (2004) Effect of patient, operative and isolation factors on subsequent yield and viability of human hepatocytes for research use. *Cell Tissue Bank.* **5**, 81–87.
5. Seglen, P. O. (1976) Preparation of isolated rat liver cells. *Methods Cell Biol.* **13**, 29–83.
6. Dorko, K., Freeswick, P. D., Bartoli, F., Cicalese, L., Bardsley, B. A., Tzakis, A., et al. (1994) A new technique for isolating and culturing human hepatocytes from whole or split livers not used for transplantation. *Cell Transplant.* **3**, 387–395.
7. Sharma, A. D., Cantz, T., Richter, R., Eckert, K., Henschler, R., Wilkens, L., et al. (2005) Human cord blood stem cells generate human cytokeratin 18-negative hepatocyte-like cells in injured mouse liver. *Am J Pathol.* **167**, 555–564.
8. Beerheide, W., von Mach, M. A., Ringel, M., Fleckenstein, C., Schumann, S., Renzing, N., et al. (2002) Downregulation of beta2-microglobulin in human cord blood somatic stem cells after transplantation into livers of SCID-mice: an escape mechanism of stem cells? *Biochem Biophys Res Commun.* **294**, 1052–1063.
9. Hengstler, J. G., Brulport, M., Schormann, W., Bauer, A., Hermes, M., Nussler, A. K., et al. (2005) Generation of human hepatocytes by stem cell technology: definition of the hepatocyte. *Expert Opin Drug Metab Toxicol.* **1**, 61–74.
10. Williams, R. T. (1959) *Detoxification Mechanisms.* New York: John Wiley & Sons.
11. Parkinson, A. (1996) Biotransformation of xenobiotics. *In Casarett and Doull's Toxicology: the basic Science of Poinsons,* ed. Klaassen, C. D. New York: McGraw-Hill. 113–186.
12. Nelson, D. R., Zeldin, D. C., Hoffman, S. M., Maltais, L. J., Wain, H. M., and Nebert, D. W. (2004) Comparison of cytochrome P450 (CYP) genes from the mouse and human genomes, including nomenclature recommendations for genes, pseudogenes and alternative-splice variants. *Pharmacogenetics.* **14**, 1–18.
13. Guengerich, F. P., and MacDonald, T. L. (1990) Mechanisms of cytochrome P-450 catalysis. *FASEB J.* **4**, 2453–2459.

14. Pelkonen, O., Maenpaa, J., Taavitsainen, P., Rautio, A., and Raunio, H. (1998) Inhibition and induction of human cytochrome P450 (CYP) enzymes. *Xenobiotica.* **28**, 1203–1253.
15. Donato, M. T., Jimenez, N., Castell, J. V., and Gomez-Lechon, M. J. (2004) Fluorescence-based assays for screening nine cytochrome P450 (P450) activities in intact cells expressing individual human P450 enzymes. *Drug Metab Dispos.* **32**, 699–706.
16. Lin, J. H., and Lu, A. Y. (1997) Role of pharmacokinetics and metabolism in drug discovery and development. *Pharmacol Rev.* **49**, 403–449.
17. Hewitt, N. J., Lechon, M. J., Houston, J. B., Hallifax, D., Brown, H. S., Maurel, P., et al. (2007) Primary hepatocytes: current understanding of the regulation of metabolic enzymes and transporter proteins, and pharmaceutical practice for the use of hepatocytes in metabolism, enzyme induction, transporter, clearance, and hepatotoxicity studies. *Drug Metab Rev.* **39**, 159–234.
18. Guillouzo, A., Morel, F., Fardel, O., and Meunier, B. (1993) Use of human hepatocyte cultures for drug metabolism studies. *Toxicology.* **82**, 209–219.
19. Rajan, N., Habermehl, J., Cote, M. F., Doillon, C. J., and Mantovani, D. (2006) Preparation of ready-to-use, storable and reconstituted type I collagen from rat tail tendon for tissue engineering applications. *Nat Protoc.* **1**, 2753–2758.
20. Decaens, C., Durand, M., Grosse, B., and Cassio, D. (2008) Which in vitro models could be best used to study hepatocyte polarity? *Biol Cell.* **100**, 387–398.
21. Rennert, B., and Melzig, M. F. (2002) Free fatty acids inhibit the activity of Clostridium histolyticum collagenase and human neutrophil elastase. *Planta Med.* **68**, 767–769.

Chapter 9

Preclinical Testing of Virotherapeutics for Primary and Secondary Tumors of the Liver

Martina Zimmermann, Timo Weiland, Michael Bitzer, and Ulrich M. Lauer

Abstract

Virotherapy offers a new treatment strategy using oncolytic viruses as self-replicating, tumor-specific agents, which destroy tumors during their natural lytic replication process. To study potential oncolytic viruses, cell culture experiments give basic information about the lytic potential of a virus, measured as cell lysis or decreased viability. For further analysis, animal models are usually employed. As these in vivo experiments are often performed in immunocompromised animals, results have to be interpreted with caution. Therefore, to obtain deeper information of the oncolytic action of specific viruses in a patient's individual context we established a test platform based on human primary tissue slices. In this three-dimensional model, we observed a preferential tumor infection and the penetration of oncolytic measles vaccine virus into deeper cell layers of tumor tissues, which is an essential feature of an effective oncolytic virus.

Key words: Hepatocellular carcinoma, Hepatoma cell lines, Precision-cut liver tissue slices, Sulforhodamine B assay, Lactate dehydrogenase assay, Confocal microscopy

1. Introduction

Hepatocellular carcinoma (HCC) is the fifth most common cancer worldwide with a continuously increasing incidence (1). In about 60–70% of the patients the tumor is detected very late because of lacking symptoms or pain (2). This leads to an inoperable status and palliative therapies for many patients. In addition, particular resistance of these tumors to conventional chemotherapy led to disappointing results (3, 4). Therefore, alternative treatment strategies are urgently required. Virotherapy is a very promising therapy option intended to reach primary tumors or metastasis. The therapeutic effect is caused by oncolytic viruses, which preferentially replicate in tumor cells and destroy them via cell lysis.

Ragai R. Mitry and Robin D. Hughes (eds.), *Human Cell Culture Protocols*, Methods in Molecular Biology, vol. 806,
DOI 10.1007/978-1-61779-367-7_9, © Springer Science+Business Media, LLC 2012

Some virotherapeutics such as measles vaccine virus (MeV), adenovirus (AdV), and vaccinia virus (VACV) already found their way into clinical trials (http://www.clinicaltrial.gov/ct).

To test the oncolytic potency of a virus, tumor cell lines are used as the standard in vitro test system. We focused on tumor cell destruction of hepatoma cell lines PLC/PRF/5, HuH7, Hep3B, and HepG2 by oncolytic MeV and quantified the tumor cell lytic effects via sulforhodamine B (SRB) assay and determination of LDH release into the supernatant as a marker of cellular membrane disintegrity.

To confirm the oncolytic potency of oncolytic MeV in a patient's individual primary tissue context, we established the precision-cut liver slice (PCLS) technology as a patient individual test platform. This method allows the production and cultivation of 200 μm thin tissue slices of primary or secondary tumors of the liver and of healthy liver tissue (5). Therefore, the PCLS technology is able to provide valuable information, being essential for an effective virotherapeutic agent: (I) a tumor preferential infection in a tissue slice containing both, tumor and healthy liver tissue; (II) a penetration throughout the tissue into deeper cell layers.

Here, we present a selection of methods for a quantitative and qualitative analysis of an oncolytic MeV as a potential anticancer drug.

2. Materials

2.1. Cell Culture and Virus Infection

1. Dulbecco's modified Eagle's medium (DMEM), containing stable L-glutamine and 4.5 g/L glucose (Biochrom AG, Berlin, Germany). Store at 4°C.
2. Fetal bovine serum, FBS, heat inactivated at 56°C for 30 minutes. Store at 4°C.
3. Phosphate-buffered saline (PBS, without Ca^{2+} and Mg^{2+}). Store at 4°C.
4. Trypsin EDTA, store at 4°C.
5. Opti-MEM I medium. Store at 4°C.

2.2. LDH Assay

1. Microplate reader (Tecan Genios; Tecan group Ltd., Männedorf, Switzerland).
2. LDH-P mono Kit (Analyticon Biotechnologies AG, Lichtenfels, Germany). Store at 4°C.
3. 0.1% Triton X-100 in PBS. Prepare immediately before use.

2.3 SRB Assay

1. Microplate reader (Tecan Genios; Tecan group Ltd., Männedorf, Switzerland).
2. PBS (without Ca^{2+} and Mg^{2+}). Store at 4°C.
3. Trichloroacetic acid (10% w/v). Store at 4°C.

4. 0.4% SRB in 1% acetic acid. Store at room temperature protected from light.
5. 1% acetic acid.
6. 10 mM Tris base (pH 10.5).

2.4. Precision-Cut Liver Slicing

1. Leica VT1200 S Vibratome (Leica Microsystems, Nussloch, Germany).
2. 8 mm coring tool (Alabama Research and Development, Munford, USA).
3. Razor blades (Personna Medical, Cedar Knolls, USA).
4. Roti Coll 1 glue (Roth, Karlsruhe, Germany).
5. Custodiol transplant medium (Dr. Franz-Köhler GmbH, Alsbach-Hähnlein, Germany).
6. 10× Krebs–Henseleit buffer (KHB) stock containing calcium chloride (25 mM), potassium chloride (50 mM), sodium chloride (1.18 M), magnesium sulfate (11 mM), and potassium dihydrogen phosphate (12 mM). Autoclave and store at 4°C.
7. 1× KHB buffer, always prepare directly before using: dissolve glucose (25 mM), sodium bicarbonate (25 mM) and HEPES (10 mM) in ultrapure water and add to 1× KHB (diluted in ultra pure water). Adjust the pH to 7.4. Do not store for longer. Chill at 4°C.
8. William´s medium E supplemented with 25 mM sterile glucose and 50 μg/mL gentamycin (WEGG). Store at 4°C for a maximum of 4 weeks.

2.5. Immunohisto chemistry and Confocal Analysis

1. 4% formaldehyde solution (Fischar GmbH, Saarbrucken, Germany).
2. PBS (without Ca^{2+} and Mg^{2+}).
3. Tris-buffered saline with Triton (TBST; 150 mM NaCl, 50 mM Tris, 0.02% Triton X-100; pH 7.4).
4. Blocking solution: TBST containing 1% FCS.
5. Primary anti-MeV-N antibody (mouse anti-MeV-N, NP cl. 120, ECACC, Porton Village, UK). The antibody is secreted by hybridoma cells and harvested from the supernatant. The optimal antibody concentration must be determined for every harvested charge. Store at −20°C.
6. Secondary anti-mouse antibody coupled to Alexa Fluor 546 (goat anti-mouse IgG (H+L) A-11003. Molecular Probes/ Invitrogen GmbH, Karlsruhe, Germany). Store at 4°C, protect from light.
7. Nuclear stain: SYTOX green (Molecular Probes/Invitrogen GmbH, Karlsruhe, Germany). Store at −20°C, protect from light. Treat as a potential mutagen and use with appropriate care.

3. Methods

Effective virotherapeutics should fulfill special quality characteristics: (I) the oncolytic virus must be able to infect and effectively eliminate the targeted cancer cells, (II) surrounding non-transformed cells should not be or much less affected by the virus, and (III) the virus must be able to penetrate throughout the tumor tissue to reach as many tumor cells as possible.

3.1. Cell Culture of Hepatoma Cells and Infection with Measles Vaccine Virus Expressing the Green Fluorescent Protein

As the basic level of analysis, tumor cell lines are infected and investigated for the oncolytic efficacy of viruses. Measles vaccine virus expressing the green fluorescent protein (MeV-GFP) is a kind gift of Prof. S. Russell (Mayo Clinic, Rochester, USA).

3.1.1. Cell Culture

In our study we focused on human hepatoma cell lines, namely HepG2 (ATCC), PLC/PRF/5 (ECACC), Hep3B (ECACC) (6, 7), and HuH7 (Riken (8)). All cell lines are routinely tested for mycoplasma contamination and are cultivated in DMEM with 10% FBS at 37°C and 5% CO_2. Confluent cell cultures are split (1:10) using trypsin/EDTA two to three times a week (see Note 1).

3.1.2. Infection of Human Hepatoma Cell Lines

Analysis of viral infections can be done in 6-, 12-, or 24-well plates; the following protocol describes the infection in the 24-well format exemplarily.

1. Cells are seeded at a density of 5×10^4 cells in 500 μL DMEM with 10% FBS per 24 well, ideally in quadruplicates per treatment (see Note 2).
2. Allow the cells to adhere overnight.
3. Equilibrate the required amount of Opti-MEM I at room temperature.
4. Remove the medium, wash the cells once with PBS (37°C) and add 200 μL Opti-MEM I (room temperature) to each well to avoid a drying out of the cells. An infection in a small medium volume increases the chance of a virus cell contact.
5. Rapidly thaw the virus stocks.
6. Prepare the virus solutions with the required virus particles in a volume of 50 μL. For MeV, commonly used MOI (*m*ultiplicity *o*f *i*nfection = virus particles per cell) are in the range of 0.01 and 1. Start with preparing the highest MOI and make serial dilutions.

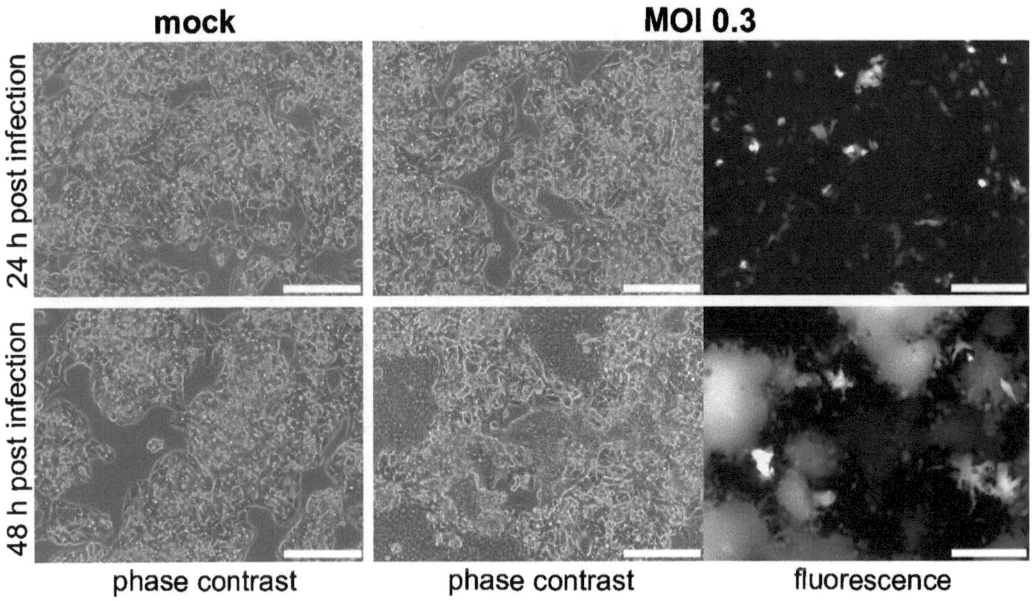

Fig. 1. Phase contrast and fluorescent pictures of HepG2 cells 1 and 2 days post infection with measles vaccine virus (MeV-GFP, MOI 0.3) or mock-treated. Single cell infections are visible 24 h post infection, whereas huge cellular fusions (syncytia) appear at 48 h. Bar represents 200 μm.

7. Add 50 μL of the virus/mock solution to each well, always start with mock-infected control wells (Opti-MEM I without virus) and work from the lowest to the highest MOI.
8. Allow the virus to adhere to the cells at 37°C for 3 h with gentle shaking of the plates every 15 min. This provides optimal virus dissemination and a homogeneous virus infection throughout the cell layer.
9. Remove the inoculum and add 500 μL of fresh DMEM with 5% heat-inactivated FCS.
10. Observe the infection using a fluorescence microscope (enhanced green fluorescent protein (EGFP), excitation maxima at 489 nm, emission maxima at 509 nm, visualize peculiar cell events with phase contrast) and take pictures (see Note 3, Fig. 1).

3.2. Toxicity and Viability Test

To quantify the cellular destruction caused by the virus, SRB and lactate dehydrogenase (LDH) assays are performed. Whereas the SRB assay provides information about cell growth inhibition or just a loss of cell mass, which can be caused by several cell death mechanisms, the LDH assay offers a more detailed interpretation about cell death via cell lysis.

3.2.1 Quantification of Virus Effects via SRB Assay

The SRB assay (9) is used as the standard assay of the US National Cancer Institute (NCI) to screen for all kinds of cytotoxic substances (10, 11). In principle, treated and control cells are fixed and stained

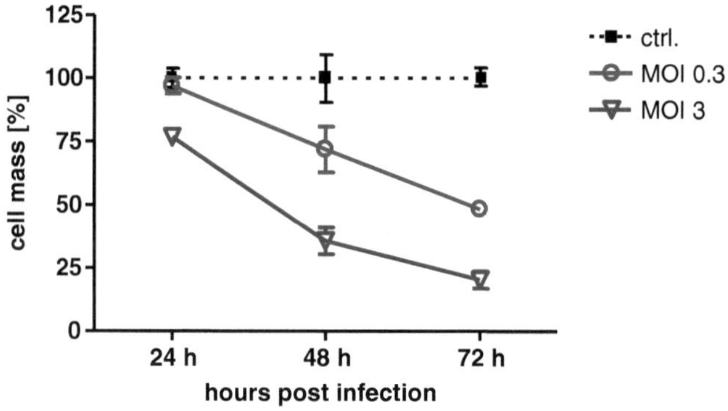

Fig. 2. Exemplary SRB results of MeV-GFP-infected HepG2 cells (mock-infected controls, MOI 0.3 and 3). The cell mass of MOI 3 infected cells decreases earlier and faster than in MOI 0.3 infected cells. Using Triton X-100 as a lysis control, 0.5–4% cell mass remained. Data result from three individual experiments and are shown in mean and SEM.

with the protein-binding SRB dye. By calculating the relative amount of dye bound to a sample, the remaining cell mass of the treated cells can be quantified (see Note 4). The fewer cells remain the more effective the virus is in inducing cell death. SRB cell death results subsequent to infection with MeV-GFP are shown exemplarily in Fig. 2.

1. Assay setup: Cells should be seeded and infected as described before (Subheading 3.1). For this assay the cells are fixed at each time point. As in the LDH assay, an infection with MeV should be observed at least every 24 h for up to 96 h, resulting in multiple wells for every time point, usually 16 identical wells (4 time points in quadruplicates). To compare the protein amount of untreated (mock-infected) cells to virus treated cells, a mock control is essential for every point of time. Therefore, make sure to have separate culture plates for every time point.
2. At each appropriate time point remove the supernatant of the cells and gently wash them twice with ice-cold PBS (see Note 5).
3. Add 500 μL trichloroacetic acid (10% w/v, 4°C) to each well and fix the cells at 4°C for 30 min.
4. Remove the acid (collect and dispose of as hazardous waste) and rinse the plates with tap water for three times very gently not to destroy the remaining cell layer.
5. In a heating-cabinet (40°C) the plates can be dried overnight (at room temperature it takes 1–2 days). Dried plates can be stored indefinitely. It is recommended to store all plates of one assay until the samples of the last time point are fixed.
6. Stain the cell mass with the SRB staining solution (0.4% w/v, the bottom should be covered completely) for 10 min.

7. Discard the staining solution and wash the cells with acetic acid (1% v/v) until no red color can be removed and the washing solution remains clear (two to three times). An equal number of washings must be performed with all samples in one assay. The cell layer will be stained violet now.
8. Dry the plates again until they are completely dry (usually 3–6 h at 40°C). Stained plates should not be stored for more than 48 h.
9. Put the plates on ice and resolve the SRB dye with 500 μL 10 mM tris buffer pH 10.5 for each well.
10. Incubate the plates for 10 min on ice; shaking facilitates the dissolving.
11. Transfer 80 μL of the red colored solution in a transparent flat bottom 96-well plate in duplicates for each sample.
12. Put the plates in a microplate reader (e.g., Tecan Genios) and measure the optical density at 550 nm. The maximum values should not exceed 2 OD units. If higher, dilute the samples to an appropriate volume. All samples of one cell line must be diluted equally for each time point, so that the relation of the treated and control cells remains correct.

Calculate the cell mass of the virus-treated cells in relation to the mock-treated cells to get information on virus induced cell death or proliferation inhibition (see Note 7).

3.3.2. Quantification of LDH Leakage as a Cell Death Marker

LDH is an enzyme which is expressed in nearly all organs and tissues. An increase of LDH in the blood stream is a typical marker for liver diseases such as hepatitis, cirrhosis, liver metastases or cardiac diseases, and cell damage in general. Adapted to in vitro cell culture systems, the quantity of the LDH enzyme released into cell culture supernatant gives information about the extent of cell lysis being caused by oncolytic viruses.

The concentration of LDH can be determined using the method of Wacker et al. (12). Shortly, the LDH enzyme catalyzes the conversion of pyruvate to lactate with a parallel oxidation of NADH to NAD^+. The rate of decrease in NADH (photometric detection at 340 nm) is directly proportional to the amount of LDH in the test solution (see Note 6).

$$\text{pyruvate}+\text{NADH}+\text{H}^+ \xleftrightarrow{\text{LDH}} \text{lacate}+\text{NAD}^+$$

1. Assay setup: Cells are seeded and infected as described before (Subheading 3.1). It is essential to determine the amount of LDH in the supernatant compared to the total amount of LDH (supernatant plus cell-based) in the same well. Therefore, cells must be completely lysed for every time point to be analyzed and cannot be cultured for a longer time period. A typical infection

with MeV should be observed at least every 24 h for up to 72 or 96 h, resulting in the need of employing multiple wells for every time point, usually 16 identical wells (4 time points in quadruplicates). Do not forget to include untreated (mock-infected) wells of each cell line and time point to quantify the basal cell death background over time to distinguish from viral effects. Make sure to have separate culture plates for every time point.

2. It is essential that every well contains exactly 500 μL of medium and that the medium is supplemented with 5% heat-inactivated FBS or less, because active FBS may give a false positive signal during LDH measurement. Prepare one control well with pure medium without cells for background determination.
3. 20 min before measurement, the reaction solution should be prepared: add the buffer (Tris pH 7.5, NaCl, pyruvate) to the dried NADH according to the manufacturer's recommendation and allow to equilibrate at room temperature.
4. A 96-well microplate reader (e.g., Tecan Genios) should be available immediately at the beginning of the reaction. If possible, it should be programmed (e.g., Magellan software) as follows: measure the extinction at 340 nm five times consecutively according to the manufacturer's guidelines. The units per milliliter can be calculated over the extinction value differences over time (ΔE).
5. At any time point, the amount of LDH in the supernatant must be quantified first. Centrifuge the plate at $250 \times g$ for 4 min to pellet cells. Transfer 10 μL of the supernatant of each sample in a well of a clear flat bottom 96-well plate (in duplicates). Additionally, measure the medium background in 10 μL of pure culture medium.
6. With an eight-channel pipet, add 200 μL of the reaction solution per 96-well. Be aware that the reaction starts right then. For this purpose, you should work quickly and accurately. Transfer the 96-well plate into the microplate reader as fast as possible.
7. Start the measurement and observe the curve progression in Magellan. The curves should exhibit a linear decrease; if not, there might be too much or less LDH in the sample. If this is the case try to dilute the sample and measure again. The correlation coefficient should be approximately 1.
8. If the LDH determination of the supernatant was performed successfully, remove the supernatant and add 500 μL of the lysis buffer to the cell layer. Adopt this volume for each cell line because the total amount of LDH varies between cell lines (see Note 6).

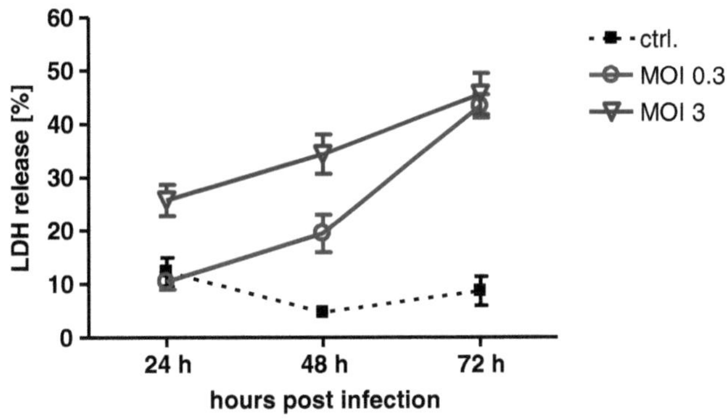

Fig. 3. Exemplary LDH results of MeV-GFP infected HepG2 cells (mock-infected controls, MOI 0.3 and 3). Cells infected with a MOI of 3 show an LDH leakage already after 24 h of incubation, whereas cells infected with a tenfold lower amount of virus (MOI 0.3) exhibit a slower lysis. In combination with SRB results, we also assume a cell death via cell lysis, not only a growth inhibition. Using Triton X-100 as a lysis control, 90–94% cells can be lysed. Data result from three individual experiments and are shown in mean and SEM.

9. Put the plate back in the incubator at 37°C for 10–15 min and let the lysis proceed.
10. If cells are disintegrated (cells appear spherical and swim in the buffer, normally after 10 min in lysis buffer), repeat procedure from **steps 5** to 7.

The relative amount of LDH leakage is calculated as follows:

$$\%\ \text{LDH leakage} = \frac{\text{LDH (supernatant)- background}}{\text{LDH (lysed cell layer)} + (\text{LDH (supernatant)- background})}$$

If the volumes of supernatant and lysis buffer differ, consider this in your calculation. Exemplary results of LDH leakage are depicted in Fig. 3 (see Note 7).

3.4. Precision-Cut Liver Slice Technology

Three-dimensional precision-cut liver tumor slices (PCLS) provide a valuable option to analyze tumor-specific infection and penetration throughout tissue in a patient individual manner. By creating tissue slices, which contain tumor cells as well as normal surrounding liver cells, a selective tumor infection with MeV-GFP can be observed microscopically. More detailed results have already been published using two different slicing machines: the Krumdieck tissue slicer and the newly developed Leica VT1200 S microtome (13, 14).

3.4.1. Generation of PCLS Using the Leica Vibrating Blade Microtome VT1200 S (Leica, Germany)

Human liver resection material must be obtained directly after explantation, and delivery to researchers has to be in consent with the guidelines of the local ethics committee. The slicing process should be performed under sterile conditions and under extended safety for the scientist, which includes the wearing of two pairs of gloves and a mask with a face shield (see Note 8). If a tissue sample is expected, prepare solutions and the Leica VT1200 S tissue slicer as follows (14).

1. Preparation of the tissue slicer: Adjust the stainless steel razor blade (Personna Medical, Stainton, VA, USA) to the 0/0 position using the vibrocheck function according to the manufacturer's guidelines. This step guarantees an optimal horizontal blade oscillation without any quenching of the tissue during the slicing procedure (see Note 9).
2. Prepare the Krebs–Henseleit slicing buffer (KHB): Dilute the 10× KHB stock in sterile water and add a freshly prepared sterile solution containing 25 mM glucose, 25 mM $NaHCO_3$, and 10 mM HEPES. Chill at 4°C.
3. Increase the oxygen concentration in a special incubator to 80% (5% CO_2, 37°C). Primary liver tissue slices are cultivated under a highly oxygenated condition to simulate the intensive blood circulation of this organ as good as possible.
4. Prepare the culture medium: Add glucose and gentamycin to William's E medium (WEGG), fill the required amount of medium in a filter top cell culture flask and aerate it in the 80% O_2 incubator.
5. After resection, store the explanted tissue immediately in ice-cold Custodiol medium and start the slicing procedure within 1 h.
6. Cut the tissue samples into approximately $1.5 \times 1.0 \times 0.5$ cm^3 sized cubes, dry them on a sterile paper for approximately 30 s and fix them onto a magnetic specimen plate (VT1200 S accessory) using Roti Coll 1.
7. Fill the slicing reservoir with ice-cold supplemented KHB and put the magnetic specimen plate on the magnetic plate holder.
8. Adjust the slicing area according to the manufacturer's guidelines and start slicing by generating an even slicing level first.
9. Carry on preparing slices in the continuous mode with the following adjustable settings: knife angle: 15°; sectioning speed: 0.4–1 mm/s; oscillation amplitude: 3 mm; step size: 200 μm; retract: 10 μm; continuous stroke.
10. Transfer the 200 μm sections into a petri dish filled with WEGG medium.
11. To obtain equally sized sections, punch 8 mm diameter cutouts using a stainless steel punch.
12. Transfer the 8 mm precision-cut liver tissue slices (PCLS) into a fresh 24-well plate with 1 mL WEGG medium per well. Cultivate the PCLS at 37°C and 80% O_2, 5% CO_2.
13. Change the medium after 1 h.

3.4.2. Infection of PCLS

The generated PCLS exhibit a thickness of around 200 μm and contain 1×10^6 cells in average. An advantage of the method is the possibility to produce tissue slices containing tumor cells only, normal surrounding liver cells only, or tissue slices containing both, liver and tumor cell areas (Fig. 4). Ideally, the infectability by virus should be observed in all three variants of tissue slices from one patient (see Note 10).

1. As a viability and infectability control, some tissue slices (two to three) are infected with a recombinant marker gene (GFP) encoding adenovirus (AdV GFP, MOI 100) 1 h after slicing. Adenovirus is able to infect nearly all tissues and the expression of the virus-encoded GFP marker appears about 16 h later only in the viable parts of the cultivated tissue slices. In our study, we observed a relation of the infectability with AdV GFP and the feasibility of an infection with MeV. Therefore, employing the AdV GFP test virus we are able to distinguish between tissue slices which are not infectable with MeV and those which are simply dead.

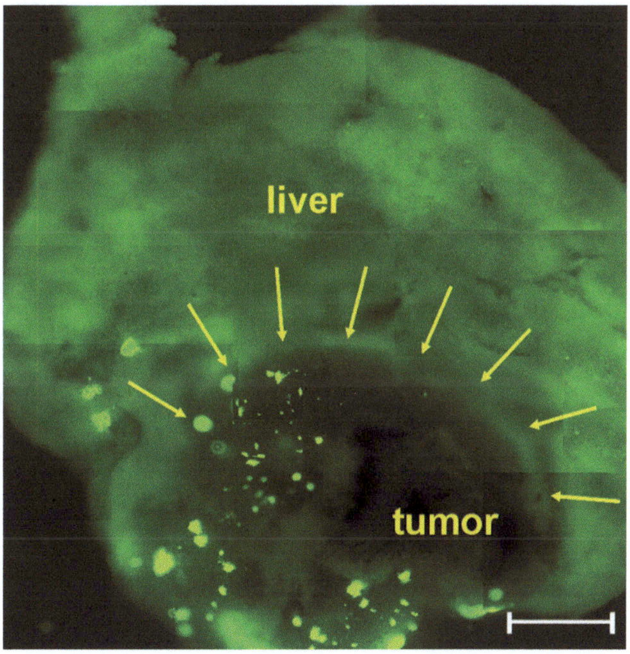

Fig. 4. A mixed, nontumorous/tumorous liver tissue slice infected with MeV-GFP displays a tumor area favouring marker gene expression. A precision-cut liver tissue slice obtained from a patient exhibiting a colorectal liver metastasis was infected with MeV-GFP (MOI of 1) followed by fluorescence microscopy on day 3 post infection (20 single pictures were taken and combined to obtain 1 picture). The tumor area can be discriminated from surrounding normal, nontumorous liver tissue by its dark ground color (tumor margins are depicted by *arrows*); *light dots* represent spots of viral infection visualized by strong expression of the GFP marker gene. Bar represents 1 mm. Picture is reproduced with permission from Spandidos Publications [13].

2. If a profound AdV GFP expression is microscopically visible in the control slices the next day, infect another set of tissue slices with 1×10^6 MeV particles (MOI of 1). Remove the old medium and add the virus solution (diluted in WEGG) in a volume of 500 μL.
3. Put the plates back in the O_2 incubator.
4. Let the infection take place for 3 h, then remove the virus solution and add 1 mL of fresh WEGG medium.
5. Observe the virus dissemination using a fluorescence microscope over a period of 3–4 days (detection of GFP, Fig. 4).

3.4.3. Immunofluorescence Staining for the MeV-N Protein in PCLS

Infections of PCLS with MeV which do not express any visible marker proteins (e.g. GFP) can be stained with antibodies coupled to a fluorochrome. The virus positive regions can then be detected in fluorescent light.

1. Transfer slices in a 48-well plate and fix them with a 4% formaldehyde solution overnight at 4°C.
2. Wash them with PBS three times and block the PCLS with TBST containing 1% FCS for 2 h.
3. Prepare the primary antibody solution: dilute primary mouse anti-measles nucleocapsid protein antibody in TBST. Use appropriate dilutions as titered before.
4. Remove blocking solution and wash the PCLS with TBST.
5. Add the primary antibody in a volume of 200 μL (the tissue slices should be covered completely) and incubate for 1 h.
6. Prepare the secondary antibody solution: dilute goat anti-mouse IgG 1:1,000 in TBST. Add SYTOX green at a final concentration of 1:50,000. Protect from light.
7. Remove the primary antibody solution and wash the PCLS with TBST three times.
8. Add the secondary antibody solution in a volume of 200 μL (the tissue slices should be covered completely) and incubate in the absence of light for 1 h.
9. Remove the secondary antibody solution and wash the PCLS with TBST three times, again.
10. Store the PCLS in PBS at 4°C in darkness until microscopic analysis.

3.4.4. Confocal Microscopy of Measles Vaccine Virus-Infected Tissue Slices

The penetration of virus into deeper cell layers of a PCLS can be explored in stained PLCS (Subheading 3.4.3) using a confocal laser scanning microscope (Fig. 5).

1. Transfer the stained PCLS onto a multiwell slide in a plane orientation.
2. Add PBS so that the PCLS is prevented to dry out, but is immobile.

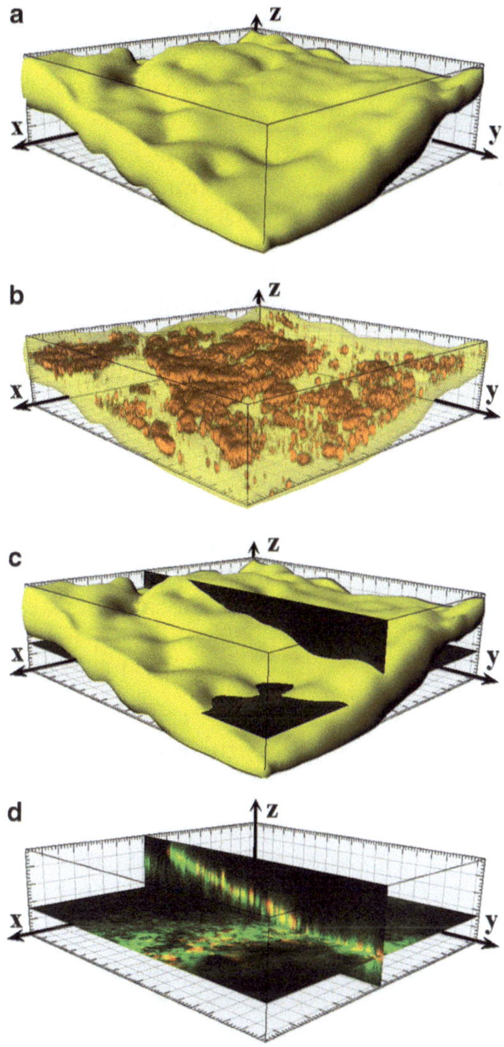

Fig. 5. MeV penetrates into deep layers of tumorous liver tissue slices. A tissue slice from a human donor with colorectal liver metastasis was infected with MeV (Moraten/Edmonston Zagreb strain) (MOI of 1). Formalin fixation and immunofluorescence staining of MeV-N (nucleocapsid) protein were performed 4 days post infection, followed by confocal immunofluorescence microscopy. (**a**) 3D model: the structure of a representative tumorous liver tissue slice (cut-out: *x*/*y* orientation 921 × 921 μm, total height 240 μm, mean tissue slice thickness here: 150 μm) was calculated out of 80 single confocal pictures using the IMARIS 4.0 (Bitplane scientific solutions, Zurich, Switzerland) software. One square represents 50 μm. (**b**) Internal infections: MeV infections (*dark spots*) are visualized in total throughout the tissue. (**c**) Section setting: virtual cross sections along the *y*–*z* plane and plane sections along the *x*–*y* plane were placed into the simulated tissue. (**d**) Infection details: MeV infected cells (indicated by *red*/*yellow* color) are observed even in the depth of the tumorous liver tissue slice; uninfected cells are visible by their unspecific staining with SYTOX green (*green*). Picture is reproduced with permission from Spandidos Publications [13].

3. Adjust a confocal microscope (e.g., Axiovert 200 M) using LSM 510 3.2 SP software (Carl Zeiss MicroImaging GmbH, Goettingen, Germany): perform a z-scan from the upper to the lower surface (~200 μm), which may include around 80 pictures. Alexa Fluor 546 has excitation/emission maxima at 557/572 nm, SYTOX green has excitation/emission maxima at 504/524 nm.

A virtual 3D view of one slice can be calculated with IMARIS 4.0 software (Bitplane scientific solutions, Zurich, Switzerland) (Fig. 5).

4. Notes

1. Some hepatoma cell lines should be treated with a special attention, because of a Hepatitis B surface antigen (HBsAg) production being indicative for the harboring of HBV DNA. Therefore, these hepatoma cell lines are classified into biosafety level 2 (Hep3B, PLC/PRF/5).
2. The tumor cells should be distributed evenly in the cell culture dish with sufficient cell–cell contact to allow cell–cell fusion during viral proliferation.
3. Always observe the infected AND the mock cells in a microscope to notice disseminated virus particles by fluorescence or toxic effects. This should not but may occur during washing procedures. Be aware, that the virus "working reagent" is self-replicating.
4. Each cell line has its own growth behavior; consider this in the SRB assay. Therefore, it may happen that you have to dilute some samples to enable the measurement. Don´t forget to dilute all samples of one cell line, otherwise you cannot evaluate the results.
5. Carefully wash cells with PBS before you fix them for the SRB assay. Some cells tend to detach, if this happens, omit this step. Try this with every cell line before you intent to produce a complete data set.
6. The total LDH amount varies enormously between the cell lines. PLC/PRF/5 cells have a lot, whereas HuH7 and HepG2 have moderate and Hep3B has very low amount of whole cellular LDH enzyme units. You have to be careful to interpret the results and do not forget to include your dilutions in the calculation of the percental leakage.
7. Sometimes, the appearance of syncytia, which are multinuclear giant cells caused by infection with MeV, can lead to "strange" LDH or SRB results. Intact syncytia will not lead to a massive LDH concentration in the supernatant, whereas a fresh syncytial

burst may give very high results. Sometimes, these bursting syncytia still do adhere to the bottom of the well and can give wrong signals in the SRB staining. Consider this in your interpretation.

8. Always be very careful when working with primary patient material, because it may transmit several diseases such as HBV, HCV, HIV, or else.
9. The vibrocheck function of the Leica VT1200 S seems to be essential for an accurate, reproducible slicing technique of liver tissue. We tested several other slicing machines, including older Leica ones, which led to nonsatisfying results. Nearly comparable, but more irregular configured tissue slices, can be prepared with the Krumdieck tissue slicer (Alabama Research and Development, USA) (14).
10. Primary human material may not be in a perfect viable status because of diverse pretreatment strategies. Nevertheless, it reflects the typical recipient material in future clinical studies. Be aware that the PCLS results provide more quality than quantity details of the virus efficacy.

Acknowledgments

We are grateful for grants from the Ministerium für Ernährung und Ländlichen Raum/Ministerium für Wissenschaft, Forschung und Kunst Baden-Württemberg (MLR/MWK Kap. 0802/98174), the Bundesministerium für Bildung, Wissenschaft, Forschung und Technologie (BMBF 01GU503) and from the Deutsche Forschungsgemeinschaft SFB 773, project C3. We thank Prof. S. Russell (Mayo Clinic, Rochester, USA) for providing the MeV-GFP virus, furthermore Prof. Königsrainer (Department of General, Visceral and Transplant Surgery, Tuebingen, Germany), Prof. Falko Fend, and Prof. Bence Sipos (Department of Pathology, Medical University Hospital, Tuebingen, Germany) for providing primary tissue samples. We thank Johanna Lampe and Dr. Susanne Berchtold for proof reading.

References

1. Llovet, J. M., Burroughs, A., and Bruix, J. (2003) Hepatocellular carcinoma, *Lancet 362*, 1907–1917.
2. Bruix, J., and Llovet, J. M. (2002) Prognostic prediction and treatment strategy in hepatocellular carcinoma, *Hepatology 35*, 519–524.
3. Llovet, J. M., Fuster, J., and Bruix, J. (2002) Prognosis of hepatocellular carcinoma, *Hepatogastroenterology 49*, 7–11.
4. Burroughs, A., and Samonakis, D. (2004) Is transarterial chemoembolization an option for all patients with unresectable hepatocellular carci noma?, *Nat Clin Pract Gastroenterol Hepatol 1*, 78–79.
5. Olinga, P., Merema, M., Hof, I. H., de Jong, K. P., Slooff, M. J., Meijer, D. K., and Groothuis, G. M. (1998) Effect of human liver source on the functionality of isolated hepatocytes and liver slices, *Drug Metab Dispos 26*, 5–11.

6. Aden, D. P., Fogel, A., Plotkin, S., Damjanov, I., and Knowles, B. B. (1979) Controlled synthesis of HBsAg in a differentiated human liver carcinoma-derived cell line, *Nature 282*, 615–616.
7. Knowles, B. B., Howe, C. C., and Aden, D. P. (1980) Human hepatocellular carcinoma cell lines secrete the major plasma proteins and hepatitis B surface antigen, *Science 209*, 497–499.
8. Nakabayashi, H., Taketa, K., Miyano, K., Yamane, T., and Sato, J. (1982) Growth of human hepatoma cells lines with differentiated functions in chemically defined medium, *Cancer Res 42*, 3858–3863.
9. Skehan, P., Storeng, R., Scudiero, D., Monks, A., McMahon, J., Vistica, D., Warren, J. T., Bokesch, H., Kenney, S., and Boyd, M. R. (1990) New colorimetric cytotoxicity assay for anticancer-drug screening, *J Natl Cancer Inst 82*, 1107–1112.
10. Rubinstein, L. V., Shoemaker, R. H., Paull, K. D., Simon, R. M., Tosini, S., Skehan, P., Scudiero, D. A., Monks, A., and Boyd, M. R. (1990) Comparison of in vitro anticancer-drug-screening data generated with a tetrazolium assay versus a protein assay against a diverse panel of human tumor cell lines, *J Natl Cancer Inst 82*, 1113–1118.
11. Shoemaker, R. H. (2006) The NCI60 human tumour cell line anticancer drug screen, *Nat Rev Cancer 6*, 813–823.
12. Wacker, W. E. C., Ulmer, D. D., and Vallee, B. L. (1956) Metalloenzymes and myocardial infarction, *New England journal of medicine 225*, 449.
13. Zimmermann, M., Armeanu, S., Smirnow, I., Kupka, S., Wagner, S., Wehrmann, M., Rots, M. G., Groothuis, G. M., Weiss, T. S., Konigsrainer, A., Gregor, M., Bitzer, M., and Lauer, U. M. (2009) Human precision-cut liver tumor slices as a tumor patient-individual predictive test system for oncolytic measles vaccine viruses, *Int J Oncol 34*, 1247–1256.
14. Zimmermann, M., Lampe, J., Lange, S., Smirnow, I., Konigsrainer, A., Hann-von-Weyhern, C., Fend, F., Gregor, M., Bitzer, M., and Lauer, U. M. (2009) Improved reproducibility in preparing precision-cut liver tissue slices, *Cytotechnology 61*, 145–152.

Chapter 10

Isolation and Functional Studies of Human Fetal Gastric Epithelium in Primary Culture

Pierre Chailler, Jean-François Beaulieu, and Daniel Ménard

Abstract

Our understanding of gastric epithelial physiology in man is limited by the absence of normal or appropriate cancer cell lines that could serve as an in vitro model. Research mostly relied on primary culture of gastric epithelial cells of animal species, enriched with surface mucous cells, and devoid of glandular zymogenic chief cells. We successfully applied a new nonenzymatic procedure using *Matrisperse Cell Recovery Solution* to dissociate the entire epithelium from human fetal stomach. Cultures were generated by seeding multicellular aggregates prepared by mechanical fragmentation. We further demonstrate that this simple and convenient technique allows for the maintenance of heterogenous gastric epithelial primary cultures on plastic without a biological matrix as well as the persistence of viable chief cells able to synthesize and secrete gastric digestive enzymes, i.e., pepsinogen and gastric lipase. In wounding experiments, epithelial restitution occurred in serum-reduced conditions and was modulated by exogenous agents. This culture system is thus representative of the foveolus-gland axis and offers new perspectives to establish the influence of individual growth factors and extracellular matrix components as well as their combinatory effects on gastric epithelium homeostasis.

Key words: Human stomach, Human gastric epithelium, Primary culture, Epithelial restitution, Gastric secretion, Chief cells, Pepsinogen, Human gastric lipase, Epidermal growth factor, Transforming growth factor-β, Laminins

1. Introduction

The epithelium of human gastric mucosa which is involved in secretory and digestive functions (see ref. 1 for review) also represents a crucial barrier between a broad spectrum of luminal noxious and immunogenic substances. Distinct epithelial populations located in specific compartments of the gastric pit-gland units mediate these functions. Previous studies from our laboratory (1–5) and others suggest that local growth factors and extracellular matrix (ECM)

Ragai R. Mitry and Robin D. Hughes (eds.), *Human Cell Culture Protocols*, Methods in Molecular Biology, vol. 806, DOI 10.1007/978-1-61779-367-7_10, © Springer Science+Business Media, LLC 2012

proteins may cooperatively regulate epithelial morphogenesis, differentiation, and homeostasis at the level of the stomach. Owing to the absence of a normal or cancer cell line that could serve as an in vitro model to delineate the molecular mechanisms involved, gastric epithelial biology mostly relied on primary culture of epithelial cells freshly isolated from the stomach mucosæ of animal species. The protocols used were based on two different approaches, i.e., microsurgical/scraping techniques and enzymatic treatments. The first requires laboratory expertise and, unless gradient fractionation or elutriation is used to purify glandular cells, these are rapidly overgrown during the first 2 days of culture by dividing mucous cells originating from the pits (foveolæ) of gastric units (6–8). The second approach, which commonly uses a combination of collagenase and hyaluronidase, generates cultures that are initially composed of mucus-secreting cells mainly, with a low proportion of acid-secreting parietal cells (9, 10). Such cultures are also devoid of glandular zymogenic cells, named chief or pepsinogen cells. These represent the major cell type found at the base of gastric glands and they are highly specialized for the secretion of pepsinogens. Their absence in epithelial cultures is particularly deleterious for the advancement of gastric physiology in humans. Indeed, human chief cells not only contribute to the digestion of dietary proteins, but also of triglyceride lipolysis through the secretion of a bile salt-independent and acid-tolerant gastric lipase, which is absent in rodents.

Also consistent with the complex nutritional requirements of epithelial cells, supplements must be added to support the attachment and survival of gastric epithelium in primary culture. Either surgical or enzymatic dissociation procedures yield a crude suspension of single dispersed cells with a few multicellular aggregates that necessarily require the presence of a biological substratum. Hence, the conventional use of fetal bovine serum (FBS) plus fibronectin or collagen-I as supplements appears to optimize the multiplication and differentiation of mucous cells at the expense of other epithelial cell types (11, 12). It is, thus, obvious that primary cultures prepared from total and mixed gastric populations using the above techniques are not representative of the integral gastric epithelium. So far, limited applications have been found for the study of mucous cell proliferation (13–15), mucin biosynthesis (16, 17), and bacterial pathogen binding (18).

In the past, several investigators (19–21) have noted in their extensive review of culture methodologies used for intestinal epithelial cells (IECs) that microexplants or multicellular clumps of epithelial cells have a much greater capacity to adhere and survive on plastic than single dispersed cells. Recently, a new nonenzymatic method based on a procedure to recover cells grown on Matrigel with *Matrisperse Cell Recovery Solution* was used to dissociate the human fetal intestinal epithelium from its surrounding lamina propria (22). We successfully applied this strategy to the human fetal stomach (23, 24) and showed that gastric epithelial cells can easily be dissociated using this commercial agent, as reviewed herein. Primary cultures were generated by seeding viable multicellular aggregates prepared by mechanical fragmentation of the material collected after *Matrisperse* treatment. We further demonstrate that this simple and convenient technique is particularly efficient, allowing for the first time the maintenance of gastric epithelial primary cultures on plastic without a biological matrix, as well as the persistent presence of functional glandular chief cells able to synthesize and secrete gastric digestive enzymes, i.e., pepsinogen and gastric lipase. As exemplified briefly, the new primary culture system will allow one to verify the influence of individual growth factors and ECM components as well as to unravel the specific intracellular signaling pathways that transmit extracellular cues and regulate human gastric epithelial restitution upon wounding.

2. Materials

2.1. Epithelium Isolation

1. Dissection medium: Leibovitz L-15 medium plus gentamicin and nystatin, 40 μg/mL each (all from Gibco BRL/Life Technologies, Burlington, ON, Canada).
2. *Matrisperse Cell Recovery Solution*® (Becton Dickinson Biosciences, Bedford, MA).
3. Buffer: Hank's balanced salt solution (HBSS, Gibco BRL) or phosphate-buffered saline (PBS), 8 g NaCl, 0.2 g/L KCl, 0.73 g/L anhydrous Na_2HPO_4, 0.2 g/L KH_2PO_4, pH 7.4.

2.2. Cell Culture Methodology

1. Culture medium: Mixture of Dulbecco's modification of essential medium (DMEM) and Ham F-12 medium (1:1) supplemented with penicillin (50 U/mL), streptomycin (50 μg/mL) (all from Gibco BRL), and 10% (v:v) FBS (collect Gold FBS from ICN Pharmaceuticals Canada, Montreal, QC, Canada).
2. Plastic multiwall plates, 6 or 24 wells.
3. Alternative media and additives for consideration (see Subheading 3.2.2): DMEM, Ham F-12, RPMI-1640, Opti-MEM (all from Gibco BRL), glutamine, HEPES, insulin, transferrin, and epidermal growth factor (EGF).

2.3. Cell Characterization

2.3.1. Cell Growth

1. Cell counting equipment.
2. [^{3}H]Thymidine, specific activity 80 Ci/mmol (Amersham Canada, Oakville, ON, Canada).

2.3.2. Dye Staining

1. Glass cover slips (round 13 mm, Electron Microscopy Sciences, distributed by Cedarlane, Hornby, ON, Canada).
2. Formaldehyde.
3. 1% Periodic acid.
4. Schiff's reagent (Fisher Scientific, contains basic fuchsin plus sodium metabisulfite).
5. Regaud's fluid: Dissolve 0.48 g of potassium bichromate in 20 mL of distilled water (0.3% solution) and add 5 mL of 37% formaldehyde.
6. Bowie stain solution: Dissolve 1 g of Biebrich scarlet in 250 mL of distilled water and filter using Whatman paper (qualitative 4). Dissolve 2 g of ethyl violet in 500 mL of distilled water. Filter into Biebrich scarlet solution until neutralization (color shifts from red to violet). Filter. Heat-dry the precipitate and dissolve in ethanol to obtain a 1% solution.
7. Acetone.
8. Xylene.
9. Permount cover slip mounting medium (Fisher Scientific).

2.3.3. Immuno cytochemistry

1. Formaldehyde or methanol.
2. Immunocytochemistry quenching solution: 100 mM glycine in PBS, pH 7.4.
3. Immunocytochemistry blocking solution: Either use 2% BSA in PBS or 5% Blotto (Nestlé powdered milk) in PBS.
4. Antihuman antibodies: All are diluted in 0.2% BSA–PBS (see Table 1).
5. Secondary antibodies: FITC-conjugated sheep anti-rabbit and anti-mouse IgG antibodies (Chemicon, Temecula, CA) diluted 1:50 and 1:30, respectively, in 0.2% BSA–PBS.

2.3.4. Western Blotting

1. Homogenization buffer: Tris–HCl, 4% sodium dodecyl sulfate (SDS), 2% β-mercaptoethanol, 20% glycerol, and 0.08% bromophenol blue.
2. Nitrocellulose membranes.
3. Primary antibodies against HGL, Pg5, and keratin-18 (see Subheading 2.3.3 and Table 1).
4. Western-Light Plus Chemiluminescent Detection System (Tropix, Bedford, MA).

2.3.5. Determination of Enzymatic Activities

For these purposes, cells were seeded in 6-well plates and collected by trypsinization after selected intervals along with their corresponding media. Cells from each well were resuspended in 250 μL of either citrate-phosphate buffer (for lipase assay) or glycine–HCl

Table 1
Description of primary antibodies

Category	Antibody	Dilution for use	Source
Growth factor receptors	EGF/TGFα-R mouse monoclonal	1:100	Upstate Biotechnology Inc., Lake Placid, NY
	IGFI-R mouse monoclonal	1:25	Oncogene Research distributed by Cedarlane, Hornby, ON
	HGF-R rabbit polyclonal	1:200	Santa Cruz Biotechnology, Santa Cruz, CA
	KGF-R rabbit polyclonal	1:100	Kindly given by Dr. M. Terada, National Cancer Center, Research Institute, Tokyo, Japan
Zymogen products	Pg5 polyclonal antisera	1:150	From A. DeCaro, F. Carrière, and R. Verger, INSERM, Marseille, France
	HGL polyclonal antisera	1:4,000	From A. DeCaro, F. Carrière, and R. Verger, INSERM, Marseile, France
Intermediate filaments	Epithelial keratin-18 mouse monoclonal	1:2,000	Sigma-Aldrich
	mesenchymal vimentin mouse monoclonal	1:200	Sigma-Aldrich
Junctional proteins	E-cadherin (clone 36)	1:800	Transduction Laboratories, Lexington, KY
	ZO-1 of tight junctions rabbit polyclonal	1:500	Zymed Laboratories, San Francisco, CA

(for pepsin assay), then lysed by freeze–thawing–refreeze cycle in liquid nitrogen.

1. For lipase activity: Glycerol [^{14}C]trioleate, specific activity 50 mCi/mmol (Amersham Canada).
2. Fatty-acid-free BSA (Sigma-Aldrich).
3. Citrate-phosphate buffer pH 6.0: Dissolve 0.1 M citric acid and 0.2 M Na_2HPO_4 in water. Mix 37.4 mL of 0.1 M citric acid solution with 62.6 mL of 0.2 M Na_2HPO_4 to obtain a final pH 6.0 solution.
4. Triton X-100.

5. Methanol:chloroform:heptane (1.41:1.25:1).
6. Carbonate-borate buffer, pH 10.5. This buffer extracts free fatty acids (FFA). Prepare 100 mL each of 0.1 M KOH, 0.1 M Na_2CO_3, and 200 mL of 0.1 M boric acid. Mix KOH and Na_2CO_3 solutions, then add boric acid until pH 10.5 is reached.
7. For pepsin activity: Glycine–HCl buffer, pH 3.0: Dissolve 0.1 M glycine in water and adjust pH with HCl.
8. Acid-denatured and dialyzed membrane hemoglobin: Prepare a 6% hemoglobin (Sigma-Aldrich) solution and dialyze against 0.1 N HCl for 24 h at 4°C. Centrifuge at 10,000 × *g* for 15 min. Collect the supernatant and store 3-mL aliquots at –80°C. Upon thawing, dilute at 2% concentration in glycine–HCl buffer.
9. Trichloroacetic acid.

2.4. Wounding Procedure and Epithelial Restitution

1. Razor blade.
2. Restitution medium, serum-reduced: DMEM/F12 with 4% FBS and 20 mM hydroxyurea.
3. PBS.
4. GIEMSA stain.
5. Restitution quantification.

2.5. Addition of Growth Factors, Pharmacological Agents, and Biological Substratum

1. Human recombinant growth factors: EGF, transforming growth factor-α (TGFα), and insulin-like growth factor-I (IGFI) from Becton Dickinson; transforming growth factor-β1 (TGFβ1) from R&D Systems (Minneapolis MN); and hepatocyte growth factor (HGF) from Calbiochem (La Jolla, CA).
2. Neutralizing rabbit polyclonal pan-specific TGFβ antibody from R&D Systems.
3. Hydroxyurea from Sigma-Aldrich.
4. Pharmacological inhibitor of phosphatidylinositol-3-kinase (Pi3K) activity: LY294002 from Calbiochem.
5. ECM proteins: Collagen-I (purified from rat tail; Becton Dickinson), Matrigel reconstituted basement membrane (Becton Dickinson), laminin-1 (LN-1, Gibco BRL), and laminin-2 or merosin (LN-2, Gibco BRL).

3. Methods

The methods described below outline (1) the procedure of epithelium isolation; (2) the culture methodology; (3) the characterization of cultured cells; (4) the effects of culture supplements, i.e., growth factors and biological substratum; and (5) the restitution assay.

3.1. Dissection and Epithelium Dissociation

3.1.1. Specimens

Tissues from 16 fetuses varying in age from 17 to 20 weeks of gestation (postfertilization ages estimated according to Streeter (25)) were obtained from normal elective pregnancy terminations. No tissue was collected from cases associated with known fetal abnormality or fetal death. Studies were approved by the Institutional Human Subject Review Board. The stomach was brought to the culture room, immersed in dissection medium, and prepared within a few minute at room temperature.

3.1.2. Dissection and Dissociation

The gastric epithelium was dissociated using a new nonenzymatic technique based on a procedure to recover cells grown on Matrigel. As opposed to enzymatic treatment, this technique allows dissociation of epithelial tissue as intact sheets or large multicellular fragments (see Fig. 1).

1. Excise cardia and pyloric antrum regions from the stomach.
2. Cut body and fundus tissues into explants (3×3 mm^2).
3. Rinse twice with dissection medium.

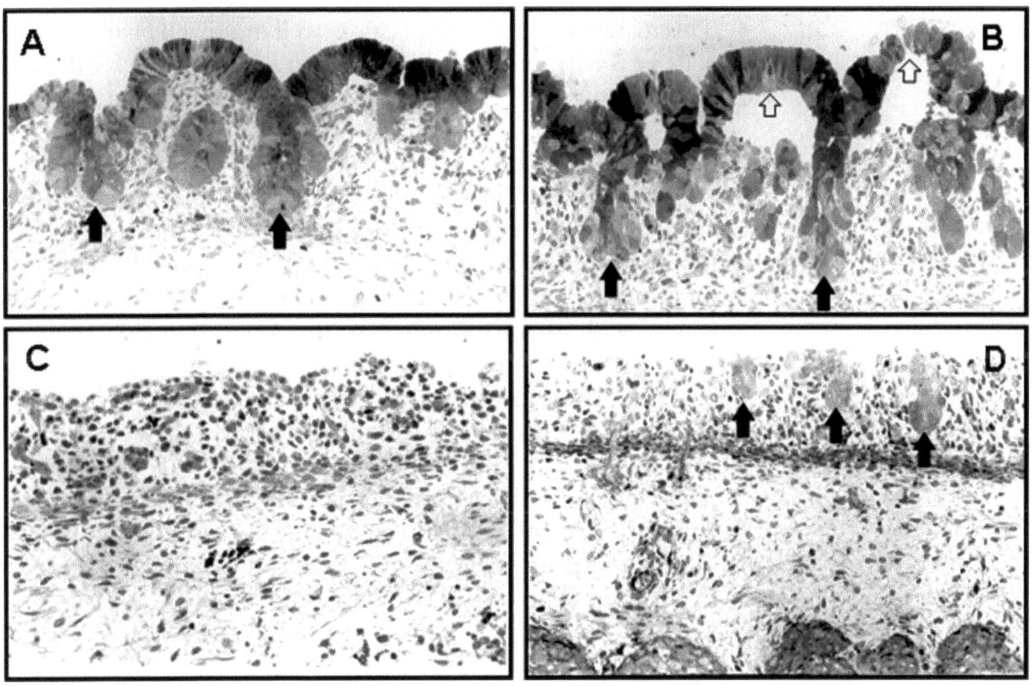

Fig. 1. Dissociation of human fetal gastric epithelium during Matrisperse treatment. (**a**) Optical microscopy of intact explant from 17-week stomach showing the epithelium and the surrounding lamina propria. *filled arrows* indicate the base of forming glands (×20). (**b**) Explant from 17-week stomach treated for 10 h (mid-treatment). *filled arrows* illustrate the initial detachment of surface epithelium in large fragment or intact sheet (×20). (**c**) Explant from 17-week stomach treated for 18 h where epithelial tissue is completely detached (×40). (**d**) Explant from 20-week stomach treated for 20 h. In some samples prepared from older specimens, a few glandular epithelial cells remained not dissociated, as indicated by *filled arrows* (×20). Reproduced from ref. 23.

4. Immerse explants in ice-cold and nondiluted *Matrisperse Cell Recovery Solution* for 16–20 h, depending on the age of each specimen (see Note 1).
5. Dissociate the epithelium by gentle agitation on ice for approximately 1 h (see Note 2).
6. Collect Matrisperse with detached material in a 50-mL tube. Add PBS or HBSS, agitate manually, and collect fluid. Repeat several times (6–8) until no detachment occurs.

3.2. Cell Culture Methodology

The resulting material was then fragmented mechanically into multicellular clumps and seeded in culture vessels.

3.2.1. Seeding

1. Centrifuge the epithelial preparation at low speed, $100 \times g$, 7 min.
2. Resuspend in culture medium.
3. Tissue is fragmented mechanically into multicellular clumps (approximately 10–15 cells) by repeated pipetting cycles: 5–10 cycles with a 5-mL serological pipette and 5–10 cycles with a 1-mL micropipette. Avoid the formation of air bubbles in the suspension. It is important to minimize cell death and single-cell dispersion (see Note 3).
4. This material is seeded in plastic 6-well multiwell plates (5×10^5 cells in 3 mL) or 24-well plates (1.5×10^5 cells in 1 mL) and left undisturbed for at least 24 h to allow attachment.

3.2.2. Culture

Culture medium with nonattached material was discarded after 24 h. Usually, it was necessary to rinse the culture surface with a flow of fresh medium once or twice in order to remove cell debris and attached mucus. Medium was renewed every 48 h thereafter. Cultures were routinely maintained in DMEM/F-12 mixture supplemented with 10% FBS only (no hormones or matrix coating) and characterized after selected intervals using phase-contrast microscopy. After spreading of epithelial colonies, a confluent monolayer of polyhedral and irregularly shaped cells was obtained (see Fig. 2a, b).

The overall quality of primary cultures strictly depended on the initial density of epithelial aggregates. Furthermore, several medium formulations were tested for their efficiency to support cell proliferation and monolayer formation:

1. DMEM.
2. Ham F-12.
3. DMEM/F-12 mixture.
4. RPMI-1640.
5. Opti-MEM.

As expected from cell culture references (26), DMEM/F-12 was optimal for reaching confluence rapidly, i.e., after 3–4 days.

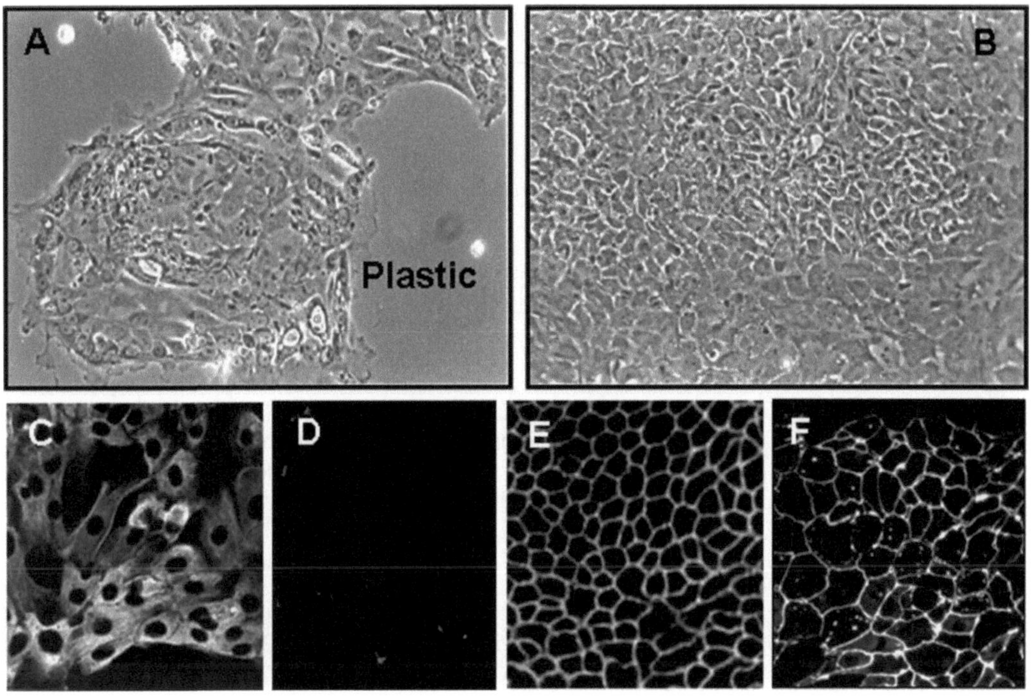

Fig. 2. Epithelial status of human gastric primary cultures. *Upper panel*: Phase-contrast morphology of fetal gastric epithelial cells after 1.5 days (**a**) and after 4 days (**b**) (magnification ×10). *Lower panel*: Indirect immunofluorescence of epithelial keratin-18 (**c**), mesenchymal vimentin (**d**), tight junction ZO-1 (**e**), and E-cadherin (**f**). Adapted from refs. 23, 24.

DMEM alone and Opti-MEM rather supported the formation of more compact epithelial colonies, particularly in the presence of 4 mM glutamine and 20 mM HEPES. Consequently, the adequate medium formulation should be selected according to the necessity of rapid confluence or a more polarized epithelial morphology in specific experimental studies. It is also noteworthy to mention that the concentration of FBS can be reduced to 4–5% when the following growth factors are added: 5 μg/mL insulin, 5 μg/mL transferrin, and 10 ng/mL EGF. Under all previous conditions, a low number of cells spontaneously detached from the culture substratum during the first week of culture without altering the integrity of the monolayer. Thereafter, cell detachment became progressively more intense. Such results may suggest that the cell adhesion kinetics could be ameliorated with the use of ECM protein coatings, as discussed in Subheading 3.4.2.

3.3. Cell Characterization

3.3.1. Cell Growth

Cell number and viability were evaluated after each day by hematocytometer counting and Trypan blue exclusion technique performed on trypsin-dissociated cells (26) (see Fig. 3). The DNA synthesis rate was determined by incubating cells seeded in 24-well plates with 2 μCi/mL of [^{3}H]thymidine during the last 12 h of each interval. Specimens were rinsed, radioactivity incorporated

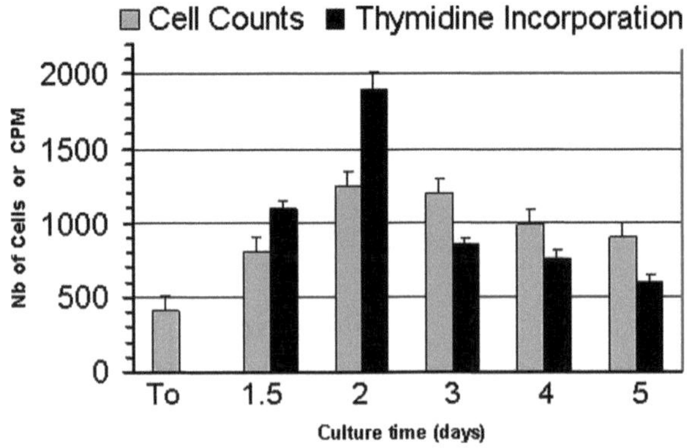

Fig. 3. Growth kinetics. Cell counts were determined at different culture intervals and final results expressed as the number of cells × 10^3 per well. The levels of [^{3}H]thymidine incorporation are in CPM per well. Note that the number of cells and radioprecursor uptake rapidly increase by twofold during the first 48 h. Adapted from ref. 23.

into total DNA was precipitated by trichloroacetic acid treatment and then solubilized with 0.1 M NaOH + 2% Na_2CO_3 for 30 min. One-milliliter aliquots were counted in a β-liquid scintillation system (see Fig. 3).

3.3.2. Dye Staining

In these experiments, suspensions were seeded on sterile glass cover slips deposited in culture vessels. At the end of culture intervals, specimens were rinsed with HBSS, pH 7.4, and then processed. Periodic acid Schiff (PAS) staining revealed that a significant proportion of cultured cells, but not all, were mucous cells (see Fig. 4a). Bowie staining (27) provided the first evidence of the presence of viable chief cells in our primary culture system (see Fig. 4b).

1. For PAS staining, fix specimens in 3.7% formaldehyde for 15 min at room temperature.
2. Incubate in 1% periodic acid during 10 min.
3. Rinse 5 min in running tap water.
4. Incubate 15 min in Schiff reagent.
5. Wash in water, allow to dry completely, and mount with Permount.
6. For Bowie staining, fix specimens 12 min in Régaud's fluid.
7. Rinse 5 min in water.
8. Incubate for 5–10 s in Bowie solution.
9. Rinse in water and PBS.
10. Dehydrate specimens in acetone (two rapid dippings), soak 30 s in acetone:xylene (1:1), differentiate in xylene for 1 min, and mount with Permount.

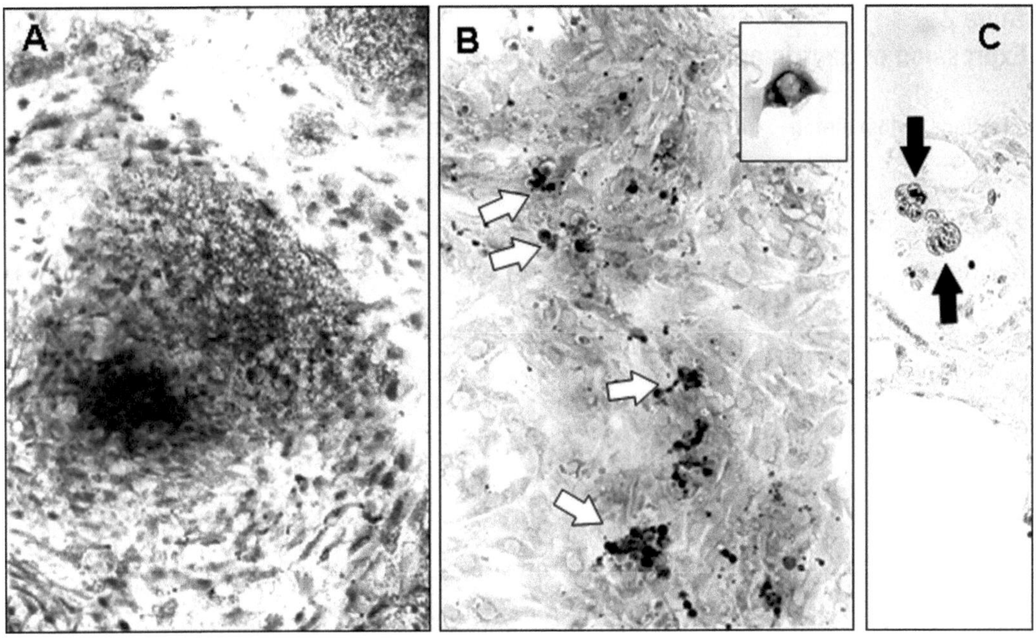

Fig. 4. Histochemical analysis of differentiation markers. (**a**) PAS staining showing intense reactivity in a majority (50–60%) of cells. Dark staining at the center of the specimen reveals the presence of mucus (×16). (**b**) Bowie staining of pepsinogen-containing cells with their cytoplasm appearing dark violet (*white arrows*) (×16). *Inset* illustrates positive cells at higher magnification (×40). (**c**) Bowie staining of parietal cells which appear *pink* (*black arrows*) (×16). Adapted from ref. 23.

3.3.3. Immuno cytochemistry

Several protein markers previously associated with the functional development of human fetal gastric epithelial cells were analyzed using indirect immunofluorescence. Assays were performed on cells cultured on sterile glass cover slips. The epithelial nature of the cells was confirmed by the presence of keratin-18, E-cadherin, ZO-1, and the absence of vimentin staining (see Fig. 2c–f). Table 2 summarizes our data (see refs. 23, 24, 36 for detailed analyses).

1. Fix specimens in fresh 3% formaldehyde for 12 min or in methanol for 10 min.
2. If necessary, free aldehyde residues are quenched by a 15-min incubation in quenching solution.
3. Nonspecific binding is prevented by adding blocking solution for 15 min.
4. Incubate specimens inside a humidified chamber with choice of mouse and rabbit primary antibodies for 60 min.
5. After rinsing, add appropriate dilutions of anti-mouse and anti-rabbit secondary antibodies for 45 min.
6. Rinse and mount on glass slides.
7. Specimens were examined on a Reichert Polyvar microscope equipped for epifluorescence and photographed with Kodak TMAX film (400 ASA).

Table 2
Expression of gastric epithelial markers

Junctions/cytoskeleton	Growth factor receptors	Integrin subunits	Secretory products
E-cadherin	**EGF-R**	**β1**	PAS-pos
ZO-1	**IGFI-R**	**α6**	mucin-6
Keratin-18 (vimentin)	**HGF-R** **KGF-R**	**α3** α2 β4	HGL Pg5

Proteins detected in all cultured cells are written in *bold*. Secretory products were present in the intracellular granules of cell subpopulations, whereas α2 and β4 integrin subunits were absent in a low number of cells. Vimentin immunoreactivity was almost negative (see ref. 23 for detailed results)

3.3.4. Western Blotting

1. Dissociate cultures seeded in 6-well plates.
2. Lyze cells in homogenization buffer.
3. Separate proteins (180–200 μg aliquots) by SDS-PAGE, transfer onto nitrocellulose membranes, and then incubate with primary antibodies against HGL, Pg5, and keratin-18.
4. Membranes are finally processed with the Western-Light Plus Chemiluminescent Detection System.
5. Autoradiograms exposed in a linear range are quantified by densitometry and the signals are normalized to those of the keratin-18 control.

3.3.5. Determination of Enzymatic Activities

Previous studies from our laboratory (cited in ref. 24) have already established that human fetal gastric explants can be maintained fully functional in organ culture for several days. We also unraveled a nonparallelism of synthesis and secretion for both gastric enzymes in the last system: pepsin activity increases over a 5-day period, whereas lipase activity decreases. In order to determine whether isolated gastric epithelial cells (and chief cells, in particular) retain their intrinsic capacity to produce digestive enzymes in primary culture, enzymatic activities were measured in cells and culture fluid as described previously for human fetal stomach (28).

Lipolytic activity attributed to HGL in the presence of triglycerides was determined using glycerol [^{14}C]trioleate as substrate and fatty-acid-free BSA as carrier of released fatty acids (29). Pepsin activity resulting from activation of pepsinogen at acid pH was measured by the method of Anson and Mirsky (30).

1. For lipolytic activity, the assay system contains: 1.2 μmol labeled triglyceride, 10 μmol citrate-phosphate buffer, pH 6.0, 0.1 μmol BSA, 2 μM Triton X-100, and 100 μL of cell homogenate in citrate-phosphate buffer in a final volume of 200 μL.

2. Incubate 60 min at 37°C.
3. Stop reaction and separate free [^{14}C]oleic acid by liquid–liquid partition: add 3.25 mL of methanol:chloroform:heptane (1.41:1.25:1 by volume) and vortex vigorously. Immediately add 1.05 mL of carbonate-borate buffer, pH 10.5, and vortex during 45 s.
4. Centrifuge at 1,800 × *g* for 20 min.
5. Count a 1-mL aliquot of the superior fraction of supernatant in a β-scintillation system.
6. The specific activity of HGL is expressed as nanomoles (nmol) of FFA released per minute per milligram of protein.
7. For pepsin activity, the assay tube contains: 900 μL of glycine–HCl and 100 μL of homogenate in glycine–HCl to which is added 1 mL of 2% hemoglobin as substrate.
8. Reaction is carried out at 37°C for 10 min.
9. Stop with 6.2% trichloroacetic acid, vortex, and keep on ice for 1 h.
10. Centrifuge samples at 20,000 × *g* for 20 min.
11. Quantitate the free amino acid products generated by pepsin activity in the supernatant by spectrometry (280 nm) using a L-tyrosine standard.
12. Pepsin specific activity is expressed in units (μmol/min) per milligram of protein.

Protein content of the homogenates was measured by the method of Lowry et al. (31).

In primary culture, pepsin activity remained fairly constant (see Fig. 5a), whereas intracellular activity of HGL dramatically decreased while its extracellular activity was high on day 3 and reduced on day 5 (see Fig. 5b). This illustrates that the digestive enzymes co-expressed by glandular chief cells are differentially regulated, corroborating previous data obtained in organ culture (3, 28, 32).

3.4. Addition of Growth Factors and Biological Substratum

Homeostasis of digestive epithelia, maintenance of cell polarity, as well as differentiation of specialized cell lineages are governed by cell-to-cell interactions and extracellular signals present in the cell microenvironment. The latter are represented by growth factors and by ECM components such as laminins (LNs), collagen-IV, and heparan sulfate found at the level of the epithelial basement membrane (33–35). Obviously, the current culture system is the first that allows to discriminate their individual effects on the gastric epithelium because it does not contain hormonal supplements, mesenchymal constituents, or a biological substratum.

3.4.1. EGF Supplementation

Growth factors (GFs) known to be trophic for the gastric mucosa such as EGF, TGFα, and IGFs play a major role in gastric physiology

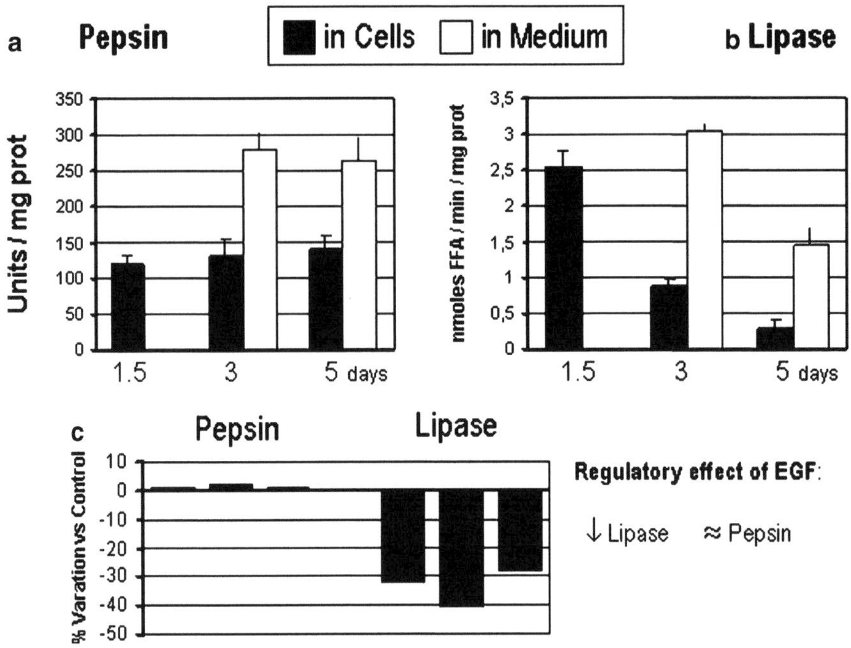

Fig. 5. Measurements of pepsin and lipase activities: variations during culture and upon EGF supplementation. *Upper panel*: synthesis and secretion profiles of Pg5 (**a**) and HGL (**b**). Specific activities were measured in cells and media between culture intervals 1.5, 3, and 5 days. *Lower panel*: (**c**) effect of 100 ng/mL EGF on intracellular lipase and pepsin activities after 1.5 days in three independent experiments. Adapted from refs. 23, 24.

because of their effects on mucosal repair and/or inhibition of gastric acid secretion. Their receptors are ubiquitously expressed along the foveolus-gland axis, they stimulate epithelial cell proliferation in organ culture and they downregulate HGL expression (mRNA, protein, enzyme activity) without affecting Pg5 (refs. 3, 5, 32, and review ref. 24). In total accordance with the last results, EGF triggered the same effects in the current primary culture system (see Fig. 5c): at the 100 ng/mL concentration, it specifically decreased HGL activity after 1.5 days.

3.4.2. Biological Matrix Coatings

Basement membrane LNs (LN-1, LN-2, LN-5, and LN-10) provide cues for cell polarity and promote the expression of tissue-specific genes in differentiating epithelial cells. In addition, their localization along the foveolus-gland axis in the developing human gastric mucosa either suggests a role in differentiation of epithelial cell lineages or gland morphogenesis. TGFβ1 is also one of a few GFs associated with basement membranes and it is now recognized as a regulator of cell–matrix interactions. As described in ref. 36 and illustrated below (see Fig. 6), these mediators had profound effects on cell polarity and HGL expression in gastric primary cultures.

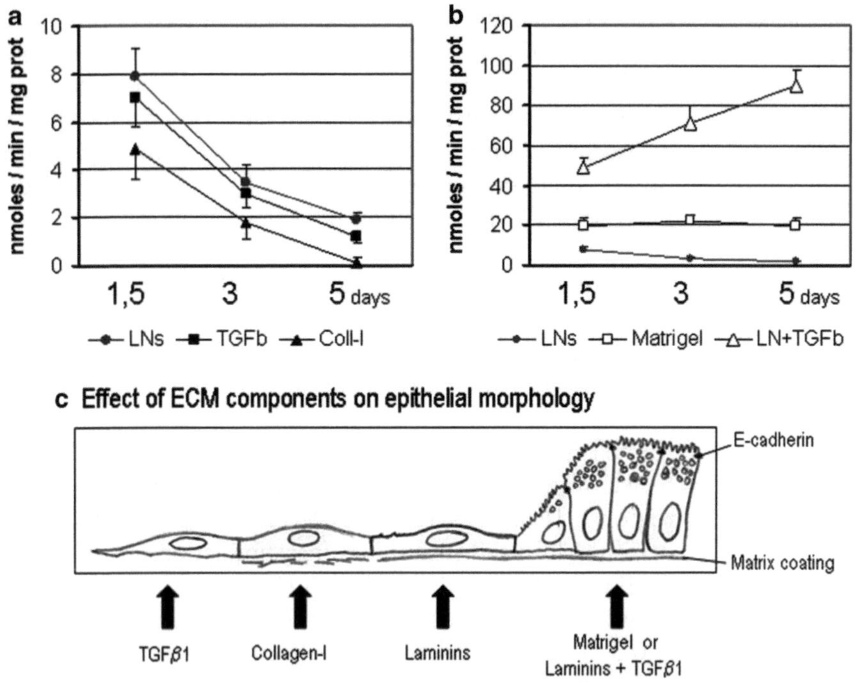

Fig. 6. Regulation of lipase activity by ECM components. In (**a**), specific activities were measured in cells between culture intervals 1.5, 3, and 5 days in the presence of either 5 ng/mL TGFβ1 (plus 0.1% BSA) or thin coatings of LNs (LN-1/LN-2 mixture) and collagen-I (Coll-I). Panel (**b**) shows the comparison of a different scale between the effects of LNs, Matrigel, and a LNs + TGFβ1 combination. Panel (**c**) illustrates the ECM protein bioactivity on gastric epithelial cultures.

1. For this purpose, culture flasks were precoated with ECM proteins: collagen-I (20 μg/mL), LN-1, LN-2, or their mixture (10 μg/mL total), and Matrigel (a complex reconstituted basement membrane, 3.4 mg/mL).
2. Individual ECM components (see Fig. 6a) could not prevent the spontaneous decrease of cellular HGL activity and mRNA that occurs during a 5-day period; stromal-type collagen-I even had a negative influence compared to TGFβ1, and LN-1/ LN-2 mixture (individual LN-1 and LN-2 exerted similar effects; data not shown).
3. By contrast, a thin coating of Matrigel initially increased and then maintained HGL levels (activity) during 5 days (see Fig. 6b). Moreover, a combination of LNs plus TGFβ1, which are purified constituents of Matrigel, was able to upregulate HGL synthesis through a powerful synergism (see Fig. 6b). Matrigel and LNs + TGFβ1 exerted beneficial effects on many aspects of human gastric epithelial morphology such as epithelial cell polarity, accumulation of intracellular granules in apical cytoplasm, and maintenance of E-cadherin at cell–cell contacts (see Fig. 6c and ref. 36 for detailed results).

3.5. Epithelial Repair in Monolayer Culture

3.5.1. Wounding Procedure

Cells were seeded in 6-well plates and allowed to reach confluence. Using a method standardized for rat IEC cultures (37) and then validated for human gastric epithelial cells (38, 39), monolayers were wounded with a sharp razor blade: two wounds per well, each measuring 13-mm wide × approximately 7-mm long. Cultures were rinsed, then further incubated 40 h in serum-reduced medium DMEM/F12 (1:1, v:v) plus 4% FBS (Note 4).

3.5.2. Restitution assay

This experiment was designed to test the effects of GFs which are known inducers of cell proliferation and migration in primary cultures of rabbit, dog, and human gastric epithelial cells (38–42). It was carried in the simultaneous presence of 20 mM hydroxyurea, which inhibits the synthesis of deoxyribonucleoside triphosphates and arrests the cell cycle in G1/S transition, thus allowing to differentiate epithelial restitution (growth-independent spreading and migration responses) from total regeneration. The following dosages are known to exert half-maximal and maximal stimulations: 5–50 ng/mL EGF/TGFα, 3–20 ng/mL HGF, 10–100 ng/mL IGFI. Known for its growth-inhibitory and motility-inducing potential, TGFβ1 was tested at 0.5–5 ng/mL concentrations. Neutralization of TGFβ bioactivity and its possible effect on basal and growth factor-stimulated regeneration was achieved by the simultaneous addition of a pan-specific TGFβ antibody (10 μg/mL). Restitution counts were determined after 40 h in control or supplemented condition.

3.5.3. Quantification of Restitution

At the end of each experiment, specimens were rinsed twice with PBS then contrasted with GIEMSA stain. Restitution response was quantified using phase-contrast microscopy by counting the number of cells across the 13-mm wound edge (razor mark) delineated onto the culture surface. Individual GFs differentially modulated restitution, HGF being the most potent versus experimental control (4% FBS) (Fig. 7a). This assay was useful to demonstrate that GFs acted through a TGFβ1-independent mechanism as opposed to that reported for intestinal epithelial crypt cell (37). Indeed we found no effect of exogenous TGFβ1 and no modulation of HGF stimulation by a neutralizing anti-TGFβ antibody. One can refer to ref. 39 for the validation of TGFβ-RII expression and TGFβ-responsiveness of a luciferase reporter gene. As reported later (43), the primary culture system may serve to decipher the intracellular signaling pathways involved. Inhibition of Pi3K activity with exogenous agent LY294002 significantly attenuated cell migration in the presence or absence of TGFα (Fig. 10.7b), clearly confirming the paramount importance of the Pi3K pathway on human gastric cell migration, and particularly on basal restitution.

Taken together, availability of the whole gastric epithelium in the current primary culture system using a convenient nonenzymatic dissociation technique will certainly contribute to a better understanding of the intrinsic mechanisms by which mucosal growth factors and ECM individually or cooperatively regulate gastric epithelial

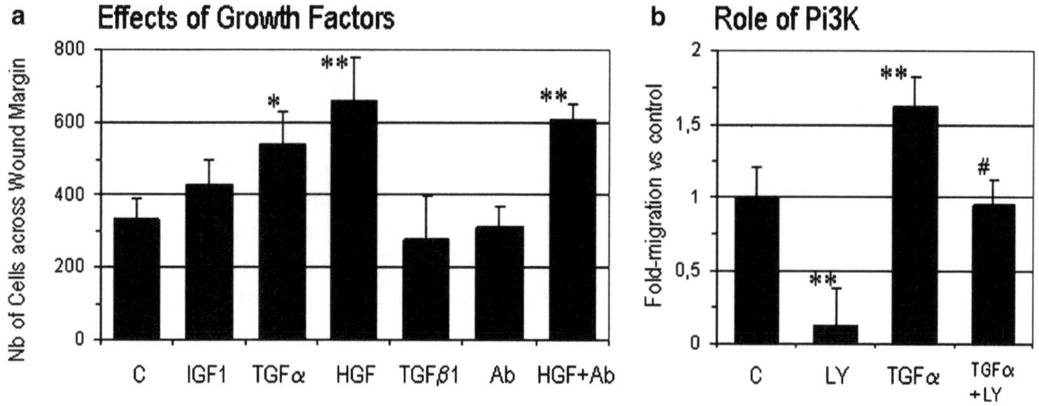

Fig. 7. Regulation of epithelial restitution. (**a**) Restitution of newly confluent primary cultures determined 40 h after wounding with 4% FBS, hydroxyurea plus 100 ng/mL IGFI, 50 ng/mL TGFα, 20 ng/mL HGF, 5 ng/mL TGFβ1, 10 μg/mL anti-TGFβ neutralizing antibody (Ab), or their combination. (**b**) Inhibition of Pi3K activity with LY294002 (LY) in the presence or absence of 50 ng/mL TGFα. *Symbols* represent the significance of statistical variations: *$p < 0.05$ vs. control; **$p < 0.005$ vs. control; #$p < 0.05$ vs. TGFα. Adapted from refs. 39, 43.

functions. More specifically, this technique will allow the identification of signaling cascades involved in cell spreading and cell motility during healing of the gastric epithelium upon wounding. Finally, use of this primary culture system will provide new insight into the study of interactions between the human gastric epithelial barrier and pathogens, luminal noxious, and immunogenic substances.

4. Notes

1. Period of incubation in Matrisperse solution increases with age: 16 h is usually sufficient for a 16-week specimen, 18 h for 17-week, 20 h for 18-week, and 24 h for 20-week stomach. Larger explants will require a longer incubation. It is noteworthy that we noted no significant alteration of culture viability and functionality when incubation was prolonged for a few hours.
2. Gentle agitation on a rocker platform (approximately 30 cycles per min) is recommended and detachment of epithelial sheets is generally visualized after 45–50 min. They appear thin and translucent compared to intact explants. More vigorous manual agitation will accelerate the process at this stage. However, do not agitate the suspension by hand too strenuously in order to prevent the induction of apoptosis (anoikis) in dissociated cells.
3. Hand-to-hand adaptation of the technique aims at minimizing single-cell dispersion. Epithelial cells isolated from gastrointestinal organs attach and survive poorly on plastic compared to multicellular aggregates (19–21). The abundance of damaged cells during initiation of the culture is possibly deleterious and their number

can be reduced by simply performing a 5-min sedimentation at the end of step 3.1.2. Dissection and dissocation the cells/debris that remain in suspension and continue to Subheading 3.2.1.

4. FBS concentration is usually reduced to 0.1% in wounding assays performed with epithelial cell lines, including those originating from human stomach (HGE cell lines; see ref. 38). In primary cultures however, reducing FBS concentration below 4% will cause cell detachment (1 and 2% have been tested) (39). Despite this methodological adaptation, the response of wounded primary cultures remains similar to that of gastric cell lines (39, 43).

Acknowledgments

This work was supported by Canadian Institutes of Health Research (CIHR) grant MOP-36495 to D.M. J.F.B. and D.M. are members of the FRSQ-funded Centre de Recherche Clinique Étienne-LeBel from the CHUS. J.F.B. is the recipient of the Canadian Research Chair in Intestinal Physiopathology.

References

1. Ménard D, Basque JR (2001) Gastric digestive functions In: Delvin EE, Lentze MJ (eds). Gastrointestinal functions, Nestlé Workshop series, vol 46, Vevey/Lippincott Williams & Wilkins, Philadelphia.
2. Tremblay E, Ménard D (1996) Differential expression of extracellular matrix components during the morphogenesis of human gastric mucosa. Anat Rec 245: 668–676.
3. Tremblay E, Monfils S, Ménard D (1997) Epidermal growth factor influences cell proliferation, glycoproteins and lipase activity in human fetal stomach. Gastroenterology 112: 1188–1196.
4. Chénard M, Basque JR, Chailler P et al (2000) Expression of integrin subunits correlates with differentiation of epithelial cell lineages in developing human gastric mucosa. Anat Embryol 202: 223–233.
5. Tremblay E, Chailler P, Ménard D (2001) Coordinated control of fetal gastric epithelial functions by insulin-like growth factors and their binding proteins. Endocrinology 142: 1795–1803.
6. Sanders MJ, Amirian DA, Ayalon A et al (1983) Regulation of pepsinogen release from canine chief cells in primary monolayer culture. Am J Physiol 245: G641-G646.
7. Rattner DW, Ito S, Rutten MJ et al (1985) A rapid method for culturing guinea pig gastric mucous cell monolayers. In Vitro (Cell. Dev. Biol.) 21: 453–462.
8. Chew C (1994) Parietal cell culture: new models and directions. Annu Rev Physiol 56: 445–461.
9. Terano A, Ivey KJ, Stachura J et al (1982) Cell culture of rat gastric fundic mucosa. Gastroenterology 83: 1280–1291.
10. Terano A, Mach T, Stachura J et al (1983) A monolayer culture of human gastric epithelial cells. Dig Dis Sci 28: 595–603.
11. Fukamachi H, Ichinose M, Ishihama S et al (1994) Fetal rat glandular stomach epithelial cells differentiate into surface mucous cells which express cathepsin E in the absence of mesenchymal cells in primary culture. Differentiation 56: 83–89.
12. Fujiwara Y, Arakawa T, Fukuda T et al (1995) Role of extracellular matrix in attachment, migration, and repair of wounded rabbit cultured gastric cells. J Clin Gastroenterol 21(Suppl.): S125–S130.
13. Nakagawa S, Yoshida S, Hirao Y et al (1985) Biological effects of biosynthetic human EGF on the growth of mammalian cells in vitro. Differentiation 29: 284–288.
14. Rutten MJ, Dempsey PJ, Solomon TE et al(1993) TGF-alpha is a potent mitogen for primary cultures of guinea pig gastric mucous epithelial cells. Am J Physiol 265: G361–G369.

15. Takahashi M, Ota S, Shimada T et al (1995) Hepatocyte growth factor is the most potent endogenous stimuklant of rabbit gastric epithelial cell proliferation and migration in primary culture. J Clin Invest 95: 1994–2003.
16. Yoshida S, Kasuga S, Hirao Y et al (1987) Effect of biosynthetic human epidermal growth factor on the synthesis and secretion of mucin glycoprotein from primary culture of rabbit fundal mucosal cells. In Vitro (Cell. Dev. Biol.) 23: 460–464.
17. Boland CA, Kraus ER, Scheiman JM et al (1990) Characterization of mucous cell synthetic functions in a new primary canine gastric mucous cell culture system. Am J Physiol 258: G774–G787.
18. Wagner S, Beil W, Mai UE et al (1994) Interaction between Helicobacter pylori and human gastric epithelial cells in culture: effect of antiulcer drugs. Pharmacology 49: 226–237.
19. Quaroni A, May RJ (1980) Establishment and characterization of intestinal epithelial cell cultures. Methods Cell Biol. 218: 403–427.
20. Kedinger M, Haffen K, Simon-Assman P (1987) Intestinal tissue and cell cultures. Differentiation 36: 71–85.
21. Evans GS, Flint N, Potten CS (1994) Primary cultures for studies of cell regulation and physiology in intestinal epithelium. Annu Rev Physiol 56: 399–417.
22. Perreault N, Beaulieu JF (1998) Primary cultures of fully differentiated and pure human intestinal epithelial cells. Exp Cell Res 245: 34–42.
23. Basque JR, Chailler P, Perreault N et al (1999) A new primary culture system representative of the human gastric epithelium. Exp Cell Res 253: 493–502.
24. Basque JR, Ménard D (2000) Establishment of culture systems of human gastric epithelium for the study of pepsinogen and gastric lipase synthesis and secretion. Microsc Res Tech 48: 293–302.
25. Streeter GL (1920) Weight, sitting head, head size, foot length and menstrual age of the human embryo. Carnegie Inst Contrib Embryol 11: 143–179.
26. Freshney RI (1987) Culture of animal cells: a manual of basic technique. 2nd edn, Wiley-Liss, New York.
27. Bowie DJ (1936) A method for staining the pepsinogen granules in gastric glands. Anat Rec 64: 357–365.
28. Ménard D, Monfils S, Tremblay E (1995) Ontogeny of human gastric lipase and pepsin activities. Gastroenterology 108: 1650–1656.
29. Lévy E, Goldstein R, Freier S et al (1981) Characterization of gastric lipolytic activity. Biochim Biophys Acta 664: 316–326.
30. Anson ML, Mirsky AE (1932) The estimation of pepsin with hemoglobin. J Gen Physiol 10: 342–344.
31. Lowry OH, Rosebrough NF, Farr AL et al (1951) Protein measurement with Folin phenol reagent. J Biol Chem 193: 265–275.
32. Tremblay E, Basque JR, Rivard N et al (1999) Epidermal growth factor and transforming growth factor-α downregulate human gastric lipase gene expression. Gastroenterology 116: 831–841.
33. Beaulieu JF (1997) Extracellular matrix components and integrins in relationship to human intestinal cell differentiation. Prog Histochem Cytochem 31: 1–78.
34. Montgomery RK, Mulberg AE, Grand RJ (1999) Development of the human gastrointestinal tract: twenty years of progress. Gastroenterology 116: 702–731.
35. Podolsky DK, Babyatsky MW (1995) Growth and development of the gastrointestinal tract. In: Yamada T (ed) Textbook of Gastroenterology, 2nd edn, Lippincott, Philadelphia.
36. Basque JR, Chailler P, Ménard D (2002) Laminins and TGF-β maintain cell polarity and functionality of human gastric glandular epithelium. Am J Physiol Cell Physiol 282: C873–C884.
37. Ciacci C, Lind SE, Podolsky DK (1993) Transforming growth factor beta regulation of migration in wounded rat intestinal epithelial monolayers. Gastrotenterology 105: 93–101.
39. Chailler P, Ménard D (2005) Establishment of human gastric epithelial (HGE) cell lines exhibiting barrier function, progenitor and prezymogenic characteristics. J Cell Physiol 202: 263–274.
39. Tétreault MP, Chailler P, Rivard N et al (2005) Differential growth factor induction and modulation of human gastric epithelial regeneration. Exp Cell Res 306: 285–297.
40. Kato K, Chen MC, Nguyen M et al (1999) Effects of growth factors and trefoil peptides on migration and replication in primary oxyntic cultures. Am J Physiol 276: G1105–G1116.
41. Watanabe S, Hirose M, Wang XE et al (1994) Hepatocyte growth factor accelerates the wound repair of cultured gastric mucosal cells. Biochem Biophys Res Commun 199: 1453–1460.
42. Watanabe S, Hirose M, Wang XE et al (2000) J Gastroenterol 35 (Suppl. 12): 65–68.
43. Tétreault MP, Chailler P, Beaulieu JF et al (2007) Epidermal growth factor receptor-dependent PI3K-activation promotes restitution of wounded human gastric epithelial monolayers. J Cell Physiol 214: 545–557.

Chapter 11

Isolation, Characterization, and Culture of Normal Human Intestinal Crypt and Villus Cells

Jean-François Beaulieu and Daniel Ménard

Abstract

The intestinal epithelium is a highly dynamic tissue undergoing constant and rapid renewal. It consists of a functional villus compartment responsible for terminal digestion and nutrient absorption and a progenitor cell compartment located in the crypts that produce new cells. The mechanisms regulating cell proliferation in the crypt, their migration, and differentiation are still incompletely understood. Until recently, normal human intestinal cell models allowing the study of these mechanisms have been lacking. In our laboratory, using fetal human intestines obtained at mid-gestation, we have generated the first normal human intestinal epithelial crypt-like (HIEC) cell line and villus-like primary cultures of differentiated enterocytes (PCDE). In this chapter, we provide a detailed description of the methodologies used to generate and characterize these normal intestinal crypt and villus cell models.

Key words: Human small intestine, Primary culture, Cell line, Sucrase-isomaltase, Keratins, E-cadherin, Gene profiling

1. Introduction

The intestinal epithelium is a highly dynamic tissue being renewed each 3–5 days in the human by a process involving cell generation and migration from the stem cell population located at the bottom of the crypt to the extrusion of the differentiated cells at the tip of the villus (1–3). Its functional unit, the crypt-villus axis, develops relatively early during human ontogeny being established before mid-pregnancy (4, 5). The crypt-villus axis can be defined by typical morphological and functional properties displayed by mature villus enterocytes that distinguish them from crypt cells. For instance, the mature forms of sucrase-isomaltase and apolipoprotein B, lactase-phlorizin hydrolase, and maltase-glucoamylase are markers for functional enterocytes of the villus in the developing and adult

Ragai R. Mitry and Robin D. Hughes (eds.), *Human Cell Culture Protocols*, Methods in Molecular Biology, vol. 806, DOI 10.1007/978-1-61779-367-7_11, © Springer Science+Business Media, LLC 2012

small intestine while the proliferating antigen Ki67 and the specific secretory granule marker MIM-1/39 are only detected in crypt cells (6–8).

It is now obvious that gene expression in intestinal cells must be tightly regulated along the crypt-villus axis in order to control cell proliferation, migration, and differentiation. Our understanding of the precise mechanisms involved is still incomplete but key transcription factors such as Cdx2 and HNF1α have been identified and it is clear that growth factors, hormones, and cytokines as well as cell–cell and cell–matrix interactions (4, 9–13) are participating.

Because no normal intestinal epithelial cell model was available until recently, most of the knowledge about human intestinal cell regulation has been derived from studies conducted on cell cultures generated from experimental animals and human colorectal cancers. The utility as well as limitations of these models are well documented (1, 4, 6, 7, 14). Recent progress in the generation of normal cell models has provided new tools to study human intestinal cell functions without these limitations. In our laboratory, we generated the first normal human intestinal epithelial crypt-like (HIEC) cell line and villus-like primary cultures of differentiated enterocytes (PCDE) (15, 16). HIEC cells exhibit typical crypt cell proliferative and undifferentiated characteristics (15) that make them an excellent model to study the regulation of normal intestinal cell proliferation (17–21) and survival (22–25) as well as the molecular mechanisms leading to the onset of differentiation (26–28). PCDE are fully differentiated enterocytes that can be maintained in primary culture for about 10–12 days (16). They have been successfully used for various purposes such as for studying differentiation (18) and survival (29). In this chapter, a detailed description of the methodologies used to generate and characterize these normal intestinal crypt and villus cell models are provided in order to encourage their use in basic and applied intestinal epithelial biology.

2. Materials

2.1. Tissues

Tissues from fetuses varying in age from 14 to 20 weeks of gestation (postfertilization age estimated according to Streeter (30)) were obtained from normal elective pregnancy terminations. No tissue was collected from cases associated with known fetal abnormality or fetal death. Studies were approved by the Institutional Human Subject Review Board.

2.2. Epithelial Cell Isolation

1. Dissection medium: Leibovitz L-15 medium plus gentamycin and nystatin, 40 μg/mL each (all from Invitrogen, Burlington, ON).
2. Phosphate-buffered saline (PBS): 137 mM NaCl, 10 mM Na_2PO_4, 2.7 mM KCl, 1.75 mM KH_2PO_4, pH 7.4.

3. HEPES buffer: 6.7 mM KCl, 142 mM NaCl, 10 nM HEPES, 1 mM $CaCl_2$, 0.45 mM NaOH, pH 7.4.
4. Thermolysin (protease type X; Sigma-Aldrich, Oakville, ON) solution: 0.005% thermolysin (w/v) in HEPES buffer.
5. *Matrisperse Cell Recovery Solution®* (BD Biosciences, Bedford, MA).

2.3. Cell Culture

1. Complete Dulbecco's modification of Eagle's medium (DMEM) culture medium: DMEM medium supplemented with 20 mM HEPES, 10 mM GlutaMAX (all from Invitrogen), 10 ng/mL of epidermal growth factor (EGF; Invitrogen), and 5% fetal bovine serum (Cellect Gold; MP Biomedicals, Aurora, OH) unless otherwise specified.
2. Complete OPTI-minimum essential medium (OPTI-MEM): OPTI-MEM medium (Invitrogen) supplemented with 20 mM HEPES, 10 mM GlutaMAX, 10 ng/mL EGF, and 4% fetal bovine serum.
3. Plastic petris (Falcon) and LabTek 8-well chamber slides (Nalge Nunc International, Naperville, IL).
4. Rat tail collagen I (Sigma-Aldrich).

2.4. Cell Characterization

2.4.1. Cell Proliferation

1. Cell counter: Z1 Coulter Counter (Beckman, Mississauga, ON).
2. ^{3}H]Thymidine (Amersham, Oakville, ON).
3. 5-Bromo-2-deoxyuridine (BrdU) (Roche, Laval, QC).

2.4.2. Indirect Immunofluorescence

1. Fixation:Ethanol or 2% paraformaldehyde in PBS, pH 7.4 (PFA).
2. Quenching solution: 100 mM PBS-glycine, pH 7.4.
3. Primary anti-human antibodies: all diluted in blocking solution (see Table 1).
4. Secondary antibodies: Sheep FITC-conjugated anti-mouse and anti-rabbit IgG antibodies (Chemicon, Temecula, CA) diluted 1:25 in blocking solution.
5. Blocking solution: 5% (w/v) nonfat dry milk in PBS.
6. Counterstaining: 0.1% Evan's blue solution.
7. Mounting medium: 0.1% *p*-phenylenediamine in 90 mL glycerol and 10 mL PBS, pH 8.5.

2.4.3. SDS-PAGE and Western Blot

1. Laemmli solubilization buffer (2×): 0.125 M Tris–HCl, pH 6.8, 4% sodium dodecyl sulfate (SDS), 20% glycerol, 0.004% bromophenol blue. 5% β-Mercaptoethanol (v/v) is added at the time of lysis.
2. 10% Separating gel: Mix 13.3 mL Acrylamide/Bis (30:0.15, BioRad Laboratories, Mississauga, ON), 20 mL 1.5 M

Table 1
Descriptions of primary antibodies

Category	Antigen	Antibody	Dilution for use		Source
			IF	WB	
Intermediate filaments	Keratin-8	DK80	1/50	1/100	Sigma-Aldrich, Oakville, ON
	Keratin-18	CY90	1/500	1/1,000	Sigma-Aldrich, Oakville, ON
	Keratin-20	K21	1/100	1/200	Kind gift from Dr. A. Quaroni (35)
	Vimentin	3B4	1/10	1/200	Abcam, Cambridge, MA
Junctional proteins	E-cadherin	clone 36	1/500	1/4,000	Sigma-Aldrich, Oakville, ON
	ZO-1	anti-ZO-1	1/2,000		Zymed Lab, San Francisco, CA
Intestinal cell markers	MIM	MIM-1/39	1/1,000	1/5,000	Kind gift from R. Calvert (36)
	Aminopeptidase N	HAPN-1	1/100		Kind gift from Dr. A. Quaroni (37)
	Dipeptidylpeptidase IV	HDPPIV-1	1/100		Kind gift from Dr. A. Quaroni (38)
	Sucrase-isomaltase	HSI-14	1/100	1/1,000	Kind gift from Dr. A. Quaroni (39)

Tris–HCl (pH 8.8), 0.4 mL 10% SDS, 6.1 mL H_2O, 100 μL 10% ammonium persulfate (APS), and 20 μL TEMED.

3. Overlay: Mix 1.25 mL 1.5 M Tris–HCl (pH 8.8), 3.65 mL H_2O, 50 μL SDS 10%, and 50 μL APS.
4. 4% Stacking gel: Mix 2.66 mL Acrylamide/Bis (29:1, BioRad Laboratories), 2.5 mL 1 M Tris–HCl (pH 6.8), 0.2 mL 10% SDS, 14.74 mL H_2O, 100 μL 10% APS, and 10 μL TEMED.
5. Running buffer: Mix 28.8 g glycine, 6.06 g Tris–HCl, and 1 g SDS in 1 L H_2O.
6. Prestained molecular weight markers (Amersham/GE Healthcare, Baie d'Urfe, QC).
7. Transfer buffer: Add 57.60 g glycine, 12.12 g Tris–HCl, and 800 mL methanol to 3,600 mL H_2O and chill to 4°C.
8. Nitrocellulose membrane (Hybond, Amersham/GE Healthcare).
9. 3MM Paper (Whatman).
10. Ponceau S solution: 2 g Ponceau S dissolved in 152 mL of 20% trichloroacetic acid (TCA) and completed to 1 L H_2O.
11. Phosphate-buffered saline (PBS): 4 g NaCl, 0.1 g KCl, 0.72 g Na_2HPO_4, 0.12 g KH_2PO_4, H_2O 0.5 L (final, pH to 7.4).
12. Tween 20 (BioRad).
13. Blocking solution: 5% (w/v) nonfat dry milk in 0.1% PBS–Tween.
14. Washing buffer: PBS–Tween 0.1%.
15. Primary anti-human antibodies: diluted in blocking solution (see Table 1).
16. Secondary antibodies: anti-mouse and anti-rabbit IgG conjugated to horseradish peroxydase (Amersham/GE Healthcare).
17. Enhanced chemiluminescent (ECL) reagents from Amersham/GE Healthcare.

2.4.4. Gene Profiling

1. RNA extraction with TRIzol reagent (Invitrogen).
2. Reverse transcription reaction was performed in the presence of 500 μM dATP, dCTP, and dGTP, 300 μM 5-aminoallyl-dUTP and 200 μM dTTP, 1× first-strand buffer, 10 mM dithiothreitol, and 400 U Superscript II (Invitrogen).
3. cDNA were labeled with either Cy3 or Cy5 (GE Healthcare).
4. cDNA microarrays (SS-H19k8) were obtained from the University Health Network, Microarray Centre, Toronto, ON.
5. Prehybridization solution: 0.1% BSA, 5× SSC, 0.1% SDS.
6. Hybridization solution: 50% formamide, 5× SSC, 0.1% SDS.

3. Methods

The methods described below outline: (1) the procedures for epithelial isolation, (2) the cell culture methodologies for both HIEC and PCDE, and (3) the basic characterization of the cells.

3.1. Dissection

1. Immerse the distal small intestine (ileum) obtained from fetuses (see Note 1) ranging from 17 to 20 weeks in age (see Note 2) in dissection medium within 30 min of pregnancy termination.
2. Cut into 10 mm long explants and if necessary, clean the mesentery from the tissue by splitting them longitudinally and washing in PBS.

3.2. Epithelial Dissociation for the Generation of HIEC and Seeding

1. Wash the explants twice in HEPES buffer.
2. Incubate the explants in thermolysin solution at 37°C for 2–3 h (see Note 2) under constant agitation.
3. Collect detached material by centrifugation in a 15 mL tube and wash twice with HEPES buffer ($100 \times g$, 7 min).
4. Resuspend the pellet in 10 mL complete DMEM containing 4% FBS and seed onto one or two 100-mm culture dishes.
5. Recover medium containing nonadhering cells after 90 min and seed onto new dishes (see Note 3).
6. Change medium every 48 h. At this stage, two types of colonies are readily observable: nonproliferative epithelial colonies (see Fig. 1a) that will degenerate over the next few days and densely packed epithelioid colonies (see Fig. 1b) that rapidly expand to fill most of the plastic surface after 10–15 days. Confluence is reached after 2–3 weeks and coincides with the acquisition of an epithelial-like morphology (see Fig. 1c).

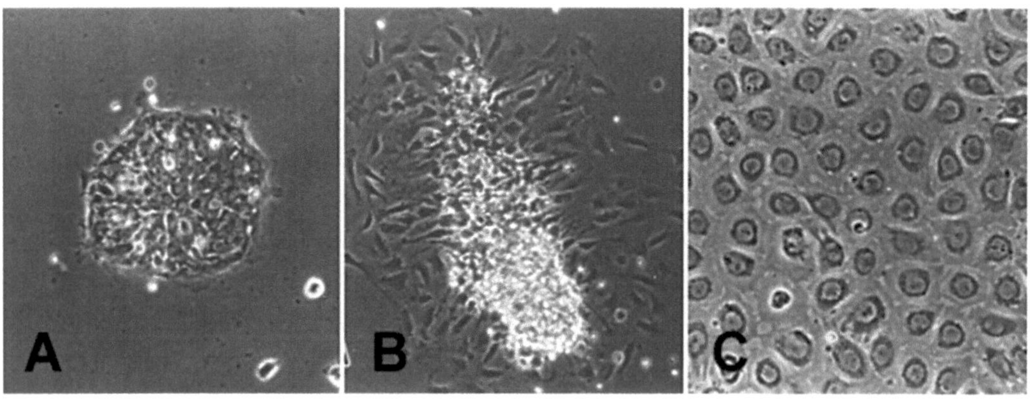

Fig. 1. Morphological features of thermolysin-dissociated human intestinal epithelial cells. Phase-contrast micrographs illustrating the two typical types of colonies present 48 h after the initial plating: (**a**) the nonexpanding epithelial colony that degenerates after a few days and (**b**) the expanding epithelioid colony that rapidly enlarges to eventually reach confluence and exhibit an epithelial-like morphology (**c**). Adapted from ref. 33.

3.3. HIEC Cell Culture

1. Subculture after mild trypsinization (5 min at room temperature) in complete Opti-MEM (see Note 4).
2. Expand cell populations by 1:2 or 1:3 dilution and keep low-passage frozen stocks in liquid nitrogen.
3. Cell cultures can be maintained for up to 28 passages before senescence becomes apparent.

3.4. Epithelial Dissociation for the Generation of PCDE and Seeding

1. Immerse 5–10 explants in 5 mL ice-cold, undiluted *Matrisperse Cell Recovery Solution* for 8–10 h without agitation (see Note 5). The incubation produces little apparent effect on villus structure other than a swollen appearance (see Fig. 2a).
2. Dissociate the epithelium lining (see Fig. 2c) from the underlying mesenchyme (see Fig. 2b) by gentle agitation on ice for approximately 10 min.
3. Collect detached material from Matrisperse by centrifugation in a 15 mL tube and wash twice with PBS (100 × *g*, 7 min). Gently resuspend the pellet in order to reduce fragmentation of the villus cell aggregates. Recover the remaining fragments, free of epithelial cells and wash twice in PBS.

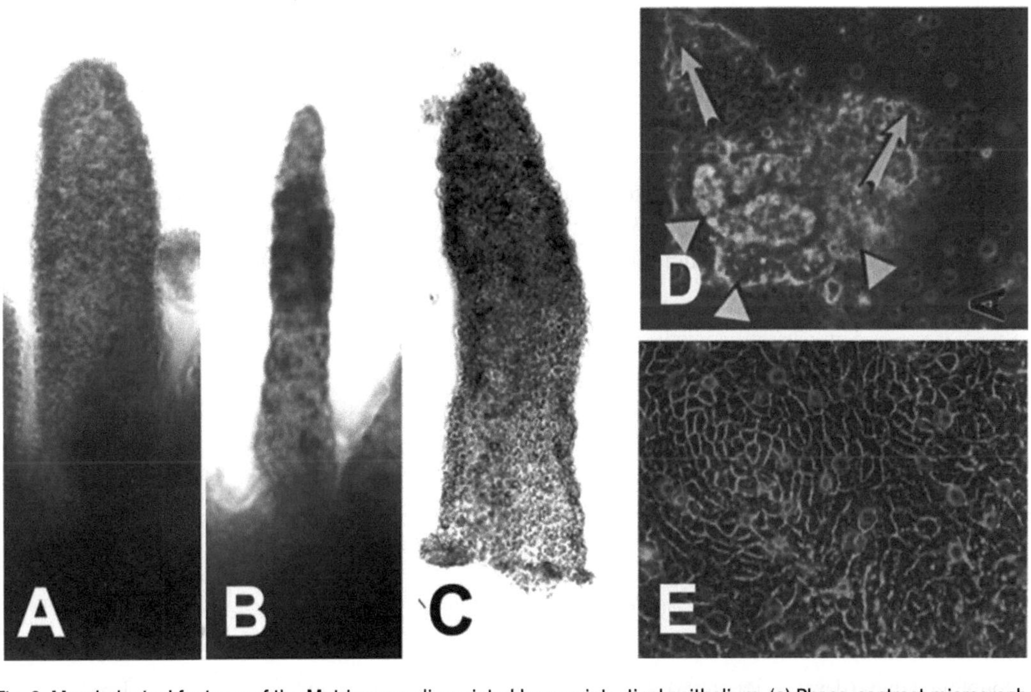

Fig. 2. Morphological features of the Matrisperse-dissociated human intestinal epithelium. (**a**) Phase-contrast micrographs showing the typical swollen appearance of a villus after 9 h of incubation in Matrisperse without agitation. After agitation, the villus stroma (**b**) and the entire epithelial lining (**c**) become separated. Plating villus-like fragments on a collagen substrate results in the attachment and initial spreading of colonies (**d**) that continue to spread until confluency is reached 2–3 days later (**e**). Adapted from ref. 34.

4. Resuspend the pellet in 6 mL complete DMEM and seed onto two to three 60-mm culture dishes and/or 8-well LabTek slides previously coated with a solution of rat tail collagen 1 at a concentration of 50 μg/cm^2. Leave undisturbed for 24 h at 4°C to allow attachment.

3.5. PCDE Culture

1. Discard the culture medium with nonattached material after 24 h. Rinse the culture surface with fresh medium in order to remove cell debris. Renew medium every 48 h thereafter.
2. Characterize the cultures at selected intervals using phase-contrast microscopy. After 24 h, the villus-like fragments have adhered and started to spread (see Fig. 2d). Spreading will continue for an additional 2–3 days until confluence is reached (see Fig. 2e). Confluent PCDE can be maintained for an additional 9–10 days until the first signs of degeneration becomes apparent.

3.6. Cell Characterization

Cell proliferation of HIEC and PCDE is evaluated by various methods as shown in Table 2.

3.6.1. Cell Proliferation

1. For evaluation of doubling time, wash monolayers with PBS, trypsinize and count individual cell suspensions with a cell counter.
2. For evaluation of DNA synthesis, incubate 48 h cultures with 2 μCi/mL (^{3}H]thymidine in complete medium for an additional 12 h. Wash cells twice and precipitate DNA with two subsequent treatments of 10% trichloroacetic acid. Solubilize the DNA in 0.3 M NaOH and neutralize with 1.5N HCl, then evaluate total (^{3}H]thymidine incorporation using a scintillation counter. The relative proliferation index is established by calculating DPI incorporation per 1,000 cells.

Table 2
Cell proliferation data

	HIEC	PCDE	Caco 2/15[a]
DNA synthesis (DPM/1,000 cells) ([^{3}H]thymidine incorporation) (7)	4.57 ± 1.77	0.32 ± 0.08	10.63 ± 2.58
DNA synthesis (% of labeled cells) (BrdU incorporation) (19, 40)	17.5 ± 1.2	n.d.	39.2 ± 5.5
Doubling Times (days) (cell counts) (27, 41)	4	∞[b]	1.5

n.d. not determined
[a]Subconfluent adenocarcinoma Caco-2/15 cells were used as positive control
[b]Doubling time was too long (longer than the culture period itself)

3. Evaluation of DNA synthesis can also be measured using 5-bromo-2-deoxyuridine (BrdU) incorporation. Seed cells onto 35 mm dishes and allow to adhere for 24 h under normal culture conditions. Add BrdU (10 μM final concentration) to the medium for 4 h. Visualize BrdU positive cells according to the manufacturer's instructions and stain nuclei with 10 ng/mL 4′,6-diamidino-2-phenylindole (DAPI). View cells with a microscope equipped for epifluorescence and digital imaging. The proliferation index is established by calculating the ratio of positive BrdU-stained cells over the total number of DAPI-stained cells × 100.

3.6.2. Indirect Immunofluorescence

A number of cell markers associated with human intestinal epithelial cells can be analyzed. Assays are performed on cells cultured on LabTek slides. The epithelial nature of the cells is established by the detection of keratin-8, -18 (see Fig. 3a and Fig. 4a), 19, and 20, E-cadherin (see Fig. 4d), ZO-1 (see Fig. 4c), and the desmosomal component ZK31. Intestinal cell markers such as aminopeptidase N (see Fig. 4f) and dipeptidylpeptidase IV (see Fig. 3b), the crypt

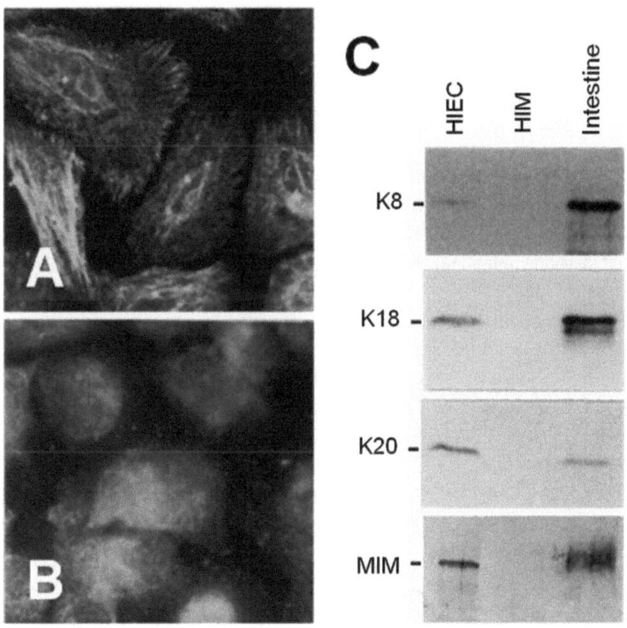

Fig. 3. Immunocharacterization of HIEC cells. Indirect immunofluorescence for the detection of keratin 18 (**a**) and aminopeptidase N (**b**) on HIEC cells. (**c**) Western blot analysis for the detection of keratin-8 (K8), 18 (K18), and 20 (K20) as well as the crypt cell antigen MIM-1/39 (MIM) in HIEC cells. Human intestinal myofibroblasts (HIM) and human intestinal scrapings (Intestine) were used as negative and positive controls, respectively. Adapted from ref. 33.

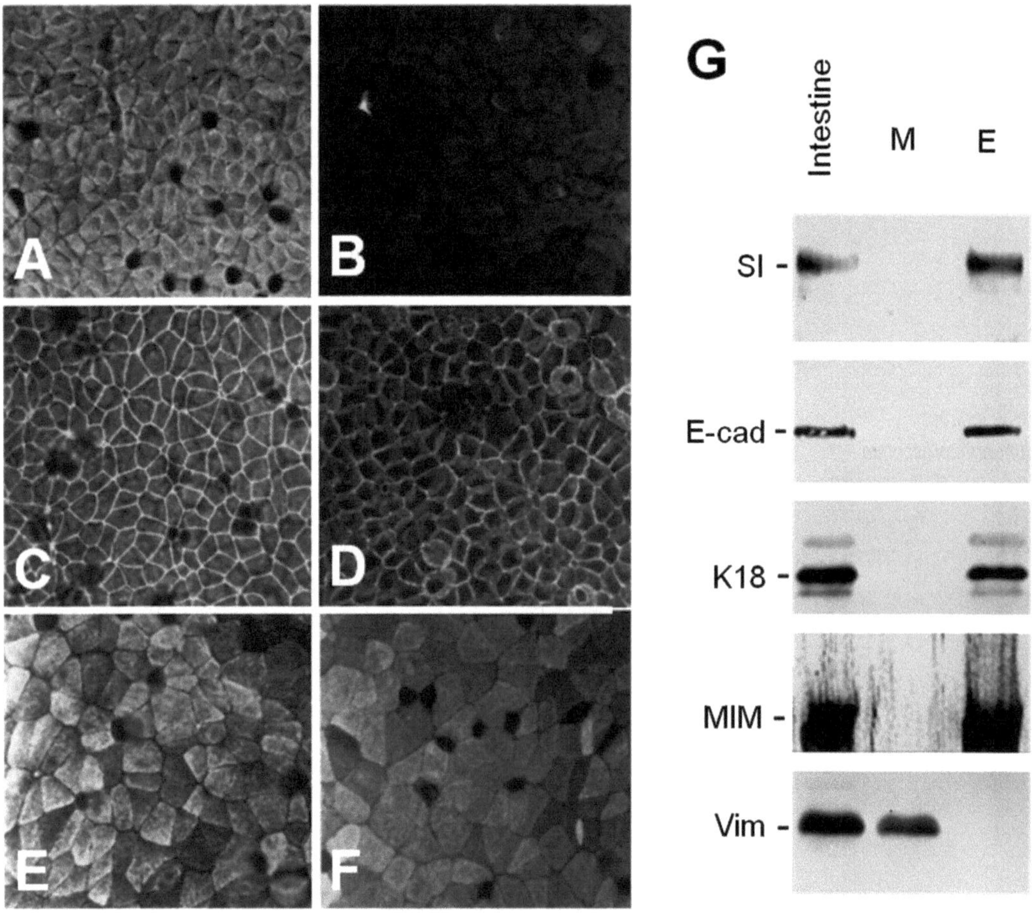

Fig. 4. Immunocharacterization of PCDE. Indirect immunofluorescence for the detection of keratin 18 (**a**), vimentin (**b**), ZO-1 (**c**), E-cadherin (**d**), sucrase-isomaltase (**e**), and aminopeptidase N (**f**). (**g**) Western blot analysis for the detection of sucrase-isomaltase (SI), E-cadherin (E-cad), keratin-18 (K18), the crypt cell antigen MIM-1/39 (MIM), and vimentin (Vim) in the intact intestine (Intestine) as well as in mesenchymal (M) and epithelial (E) fractions used for the generation of PCDE. Adapted from ref. 34.

cell marker MIM-1/39, and villus cell markers sucrase-isomaltase (see Fig. 4e) and lactase-phlorizin hydrolase may also be analyzed.

1. Wash cells grown on 8-well LabTek slides twice with complete culture medium and fix with all of the above antibodies except anti-ZO-1 and MIM-1/39 with ethanol for 10 min at –20°C. Fix cells for anti-ZO-1 and MIM-1/39 detection with 2% paraformaldehyde for 45 min at 4°C. Then wash slides 3×10 min in PBS. Quench with PBS-glycine solution for 45 min at 4°C then wash 3×10 min in PBS.
2. Block reactive sites by incubation with 5% blotto in a humidity chamber for 30 min at room temperature.
3. Remove the blocking solution and add the primary antibody diluted in blocking solution for 1 h at room temperature. Then, wash slides 3×10 min in PBS.

4. Dilute the secondary antibody 1:5 in PBS then centrifuge for 10 min at 10,000 × *g*. Carefully collect the supernatant and further dilute 1:5 in blotto and apply to the cells for 1 h in a humidity chamber at room temperature. Wash slides 3 × 10 min in PBS.
5. Place a drop of mounting medium onto the cells then cover with a cover slip.
6. View cells under a microscope equipped for epifluorescence.

3.6.3. SDS-PAGE and Western Blot

Expression of specific markers associated with human intestinal epithelial cells can be analyzed. The expression of the keratins and E-cadherin confirms the epithelial nature of the cells while MIM-1/39, aminopeptidase N, dipeptidylpeptidase IV, and sucrase-isomaltase can be analyzed to evaluate the expression of intestinal crypt and villus cell markers (see Figs. 3c and 4g).

1. Wash cells twice in PBS 1× before adding 0.5–1.0 mL of 2× Laemmli solubilization buffer containing 5% β-mercaptoethanol. Distribute evenly over the dish using a cell scraper and incubate for 5 min at room temperature. Then transfer to a 2 mL screw-capped microtube and heat for 5 min at 95–100°C.
2. Cool the samples then sonicate (2 × 15 s) and centrifuge at 9,000 × *g* in a microfuge for 10 min. Transfer the supernatants to new tubes and freeze at –80°C for long-term storage. Avoid freeze–thaw cycles.
3. These instructions are based on the use of a GibcoBRL vertical gel electrophoresis apparatus model V-16 (Invitrogen). They are nevertheless adaptable to other formats including mini-gel systems.
4. The glass plates and accessories need to be cleaned using a rinsable detergent and extensively rinsed with distilled water before storage. Assemble the gel-frame with glass plates, clamps, spacers, and combs so that no leaks occur.
5. Prepare a 1.0-mm thick, 10% separating gel by mixing the acrylamide/bis (30:0.15), Tris 1.5 M, SDS 10%, and H_2O as described in Subheading 2. Apply vacuum for 5 min to reduce solubilized oxygen then add APS and TEMED. Mix gently and pour the solution slowly into the gel-frame leaving space for the stacking gel.
6. Overlay the separating gel with water and allow to polymerize for 45 min.
7. Prepare the 4% stacking gel by mixing the acrylamide/bis 30% (29:1), Tris–HCl 1 M, SDS 10%, and H_2O. Apply vacuum for 5 min.
8. Pour off the overlay. Add APS and TEMED to complete the stacking gel solution, gently agitate and pour on top of the separating gel. Insert comb. Polymerization should be complete after 30 min.

9. Assemble the electrophoresis apparatus, fill the top chamber with running buffer then carefully remove the comb and wash the wells with electrophoresis buffer. Fill the bottom chamber with the remaining running buffer and remove any accumulated bubbles from the lower edge of the gel.
10. Prepare samples containing 50–100 μg protein per well. Final volumes should be matched between wells using 1× solubilization buffer.
11. Load the samples and an appropriate protein marker onto the bottom of the well. Install the cover and electrodes. Run the gel for 30–40 min at 70 V to allow the proteins to enter the separating gel. Then, raise the voltage to 100 V for an overnight migration (12 h) or to 170 V for a 3–5 h migration.
12. After gel migration, remove the gel from the glass plates using a piece of Whatman paper and equilibrate the gel by immersion in a small amount of chilled transfer buffer for 15 min.
13. Electro-transfer the proteins that have been separated by SDS-PAGE to a nitrocellulose membranes as follows.
14. Half-fill a tray large enough to hold the opened transfer cassette with chilled transfer buffer, pack the sandwich frame in the following order: sponge-filter paper-nitrocellulose membrane-gel-filter paper and sponge. Remove all bubbles between the various layers and close the sandwich with the clamp.
15. Insert the sandwich in such a way that the gel is oriented toward the negative electrode and the nitrocellulose toward the positive electrode.
16. Run the electro-transfer at 20 V for 30 min then at 100 V for 1.5 h.
17. Once the transfer is complete, the cassette is disassembled and the nitrocellulose membrane is peeled off from the gel, marked for orientation and washed in H_2O for 10 min.
18. Proteins can be stained by soaking the membrane in Ponceau S solution for 1 min then washing twice with H_2O. The molecular weight bands of the protein marker can be labeled with a pencil. If desired, the membrane can be scanned and later used as a loading reference.
19. Wash the membrane twice with PBS for 5 min to remove the Ponceau S and incubate in blocking solution for 60 min.
20. Place the blocked membrane in a plastic bag, add the appropriate antibody diluted in blocking solution and seal the bag. Incubate overnight at room temperature.
21. Wash the membrane 3× for 10 min at room temperature with washing buffer and incubate for 60 min at room temperature with the secondary antibody, diluted 1:5,000 in blocking solution.

22. Wash the membrane 3× for 10 min at room temperature with washing buffer and 3× for 10 min with PBS without Tween 20.
23. ECL reagent is prepared according to the manufacturer's instructions and placed on the membrane between the leaves of a plastic sheet in an X-ray film cassette in a darkroom.
24. Excess ECL reagent is removed and a film is placed over the incubated membrane for a suitable exposure time, typically a few minutes.

3.6.4. Gene Profiling

Gene expression profile analysis can be used to identify the genes associated with human intestinal cell proliferation and/or differentiation in the crypt and villus cell models. The original cluster analysis and the complete data set have been published as complementary data (see ref. 31). An example of differentially expressed clusters of genes between HIEC and cells used to generate PCDE is illustrated in Fig. 5.

1. Extract RNA from samples with TRIzol and store at −80°C. Verify the quality of RNA on agarose gel and by spectrophotometric analysis.
2. For reverse transcription, prime 1 μg of RNA with 6 μg random hexamers (Roche) by heating at 70°C for 10 min, snap cooling on ice for 30 s, and incubating at room temperature for an additional 5–10 min. Perform reverse transcription in the presence of 500 μM dATP, dCTP, and dGTP, 300 μM 5-aminoallyl-dUTP, and 200 μM dTTP, 1× first-strand buffer, 10 mM dithiothreitol, and 400 U Superscript II (Invitrogen) in a volume of 40 μL at 42°C for 3 h to overnight then lyophilize the cDNA.
3. For the labeling reaction, resuspend the cDNA in 5 μL of RNAse-free water. Add 3 μL of 0.3 M sodium bicarbonate, pH 9. Resuspend an aliquot of the NHS ester of Cy3 or Cy5 in 2 μL of DMSO (dye from each tube has been previously dissolved in 72 μL DMSO, divided into 4.5 μL aliquots and lyophilized), add to the reactions, and incubate at room temperature in the dark for 1 h.
4. A reference pool can be produced from a mixture of equimolar aliquots of total RNA from human fetal jejunum, ileum, and colon to obtain a complete representation of the genes spotted on the arrays.
5. The reference pool is labeled with Cy3 dye, while samples to be evaluated are labeled with Cy5 dye (31). Quench the coupling reaction by adding 35 μL of 0.1 M sodium acetate, pH 5.2, and remove unincorporated dye using QIAquick columns. The labeling efficiency is determined by analyzing the whole

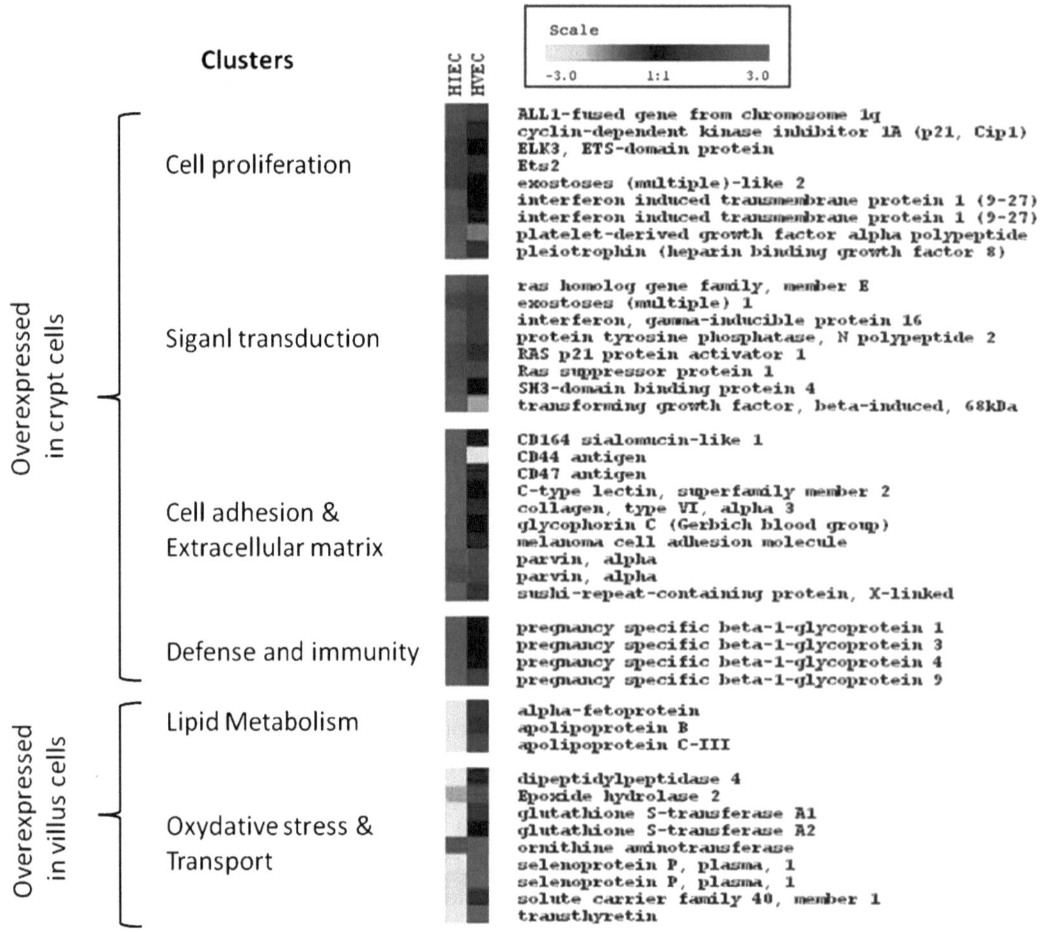

Fig. 5. Gene expression profiles in human intestinal crypt and villus cells. Expression profiles of upregulated genes in HIEC include cell proliferation, signal transduction, cell adhesion, and extracellular matrix as well as defense and immunity clusters while upregulated genes in villus cells include lipid metabolism, oxidative stress, and transport clusters. Adapted from ref. 31.

undiluted sample in a spectrophotometer using a 50 μL microcuvette (Beckman). Lyophilize and store fluorescent cDNA probes.

6. Prehybridize cDNA microarrays representing 19,200 human cDNA clones in 0.1% BSA, 5× SSC, 0.1% SDS for 45 min at 42°C, wash by dipping in deionized water twice and 2-propanol once, and air-dry.
7. Resuspend fluorescent cDNA probes in 30 μL of hybridization solution. Add 20 μg Cot1 DNA and 20 μg poly(A)+DNA to the combined Cy3 and Cy5 samples and denature at 95°C for 3 min, followed by snap cooling on ice for 1 min.
8. Apply room-temperature probes to a prehybridized array, cover with another slide and place in a humidified hybridization chamber (Corning). Hybridize at 42°C for 16 h, followed by

5 min washings in: 1× SSC, 0.2% SDS at 42°C, 0.1× SSC, 0.2% SDS at room temperature, and 0.1× SSC at room temperature, twice.

9. Scan arrays using a ScanArray Express dual-color confocal laser scanner (Perkin Elmer). Collect data in Cy3 and Cy5 channels and store as paired TIFF images.
10. For data analysis, identify spots and subtract local background using the TIGR_Spotfinder 3.1 software (32). A quality control (QC) filter can be used to remove questionable array features. Two criteria for spot rejection are a spot shape that deviates from a circle and a low signal-to-noise ratio. Hybridization intensity data are normalized using iterative mean-log2(ratio)-centering (data range for mean centering ±3 SD) and Lowess procedures (smoothing parameter sets to 33%) using the native Java function of the TIGR MIDAS 4.0 software (Microarray Data Analysis System) (32). Assess statistical significance by one-way ANOVA and apply an adjusted Bonferroni correction to control the false-positive rate ($p<0.05$ is considered significant), and perform hierarchical clustering analysis using the TMEV 3.0 software (TIGR_MultiExperiment Viewer) (32). All software is available at The Institute for Genomic Research (TIGR) website, http://www.tigr.org/.

4. Notes

1. Fetal intestine, although not completely mature as compared to the adult, has many comparable features including a well-defined crypt-villus axis and the expression of numerous cell markers (4, 5). The initial sterility of the fetal intestine significantly contributes to the success in establishing viable intestinal cell cultures.
2. Optimal conditions for the generation of HIEC cells with thermolysin were evaluated according to age (14–20 weeks), segment (jejunum vs. ileum), and incubation time (2–3 h vs. 5–6 h) (33). The best results were obtained with the 17–19 week fetal ileum after a 2–3 h period of incubation.
3. This step was introduced to discard fibroblasts from the epithelial cultures.
4. The use of OPTI-MEM was found to improve the speed of HIEC growth as compared to DMEM.
5. The minimal agitation is to prevent premature separation of the epithelial lining in order to avoid anoikosis. Longer times of incubation were found to reduce the spreading ability of the villus aggregates.

Acknowledgments

The authors thank E. Herring for revision of the manuscript. This work was supported by Canadian Institutes of Health Research grants MOP-57727 and MOP-97836 to J.F.B. J.F.B. is the recipient of the Canadian Research Chair in Intestinal Physiopathology. J.F.B. and D.M. are members of the FRSQ-funded Centre de Recherche Clinique Etienne-LeBel of the CHUS.

References

1. Babyatsky MW, Podolsky DK. (1999) Growth and development of the gastrointestinal tract, In *Textbook of gastroenterology* (Yamada, T., editor, Ed.) 3rd ed., pp 547–584, J.B Lippincott, Philadelphia.
2. Bjerknes M, Cheng H. (2005) Gastrointestinal stem cells. II. Intestinal stem cells, *Am J Physiol Gastrointest Liver Physiol* ***289***, G381–387.
3. Rubin DC. (1999) Small intestine: Anatomy and structural anomalies, In *Textbook of gastroenterology* (Yamada, T., editor, Ed.) 3rd ed., pp 1561–1583, J.B Lippincott, Philadelphia.
4. Menard D. (2004) Functional development of the human gastrointestinal tract: hormone- and growth factor-mediated regulatory mechanisms, *Can J Gastroenterol* ***18***, 39–44.
5. Menard D, Calvert R. (1991) Fetal and postnatal development of the small and large intestine: Patterns and regulation., In *Growth of the gastrointestinal tract: Gastrointestinal Hormones and Growth Factors.* (J, M., and T, S., Eds.), pp 147–162, CRC Press Inc, Boca Raton.
6. Ménard D, Beaulieu JF. (1994) Human intestinal brush border membrane hydrolases, In *Membrane Physiopathology* (Bkaily, G., Ed.), pp 319–341, Kluwer Academic Publisher, Norwell.
7. Pageot LP, Perreault N, Basora N et al. (2000) Human cell models to study small intestinal functions: recapitulation of the crypt-villus axis, *Microsc Res Tech* ***49***, 394–406.
8. Teller IC, Beaulieu J-F. (2001) Interactions between laminin and epithelial cells in intestinal health and disease, *Expert Reviews in Molecular Medicine* http://www-ermm.cbcu.cam.ac.uk/01003623h.htm.
9. Beaulieu JF. (2001) Role of extracellular matrix proteins on human intestinal cell function: Laminin-epithelial cell interactions, In *Gastrointestinal Functions* (E.E., D., and M.J., L., Eds.), pp 59–75, Vevey/Lippincott Williams & Wilkins, Philadelphia.
10. Kedinger M, Duluc I, Fritsch C et al. (1998) Intestinal epithelial-mesenchymal cell interactions, *Ann N Y Acad Sci* ***859***, 1–17.
11. Lussier C, Basora N, Bouatrouss Y et al. (2000) Integrins as mediators of epithelial cell-matrix interactions in the human small intestinal mucosa, *Microsc Res Tech* ***51***, 169–178.
12. Ménard D, Beaulieu J-F, Boudreau F et al. (2006) Gastrointestinal tract, In *Cell Siganling and Growth Factors in Development: From molecules to Organogenesis.* (Krieglstein, K. U. a. K., Ed.), pp 755–790, Wiley-Vch:, Weinheim.
13. Montgomery RK, Mulberg AE, Grand RJ. (1999) Development of the human gastrointestinal tract: twenty years of progress, *Gastroenterology* ***116***, 702–731.
14. Whitehead R, Watson N. (1997) Gastrointestinal cell lines as a model for differentiation., In *The gut as a model in cell and molecular biology* (Halter F, W. D., Wright NA, Ed.), pp 275–290, Kluwer, Dordrecht.
15. Perreault N, Beaulieu JF. (1996) Use of the dissociating enzyme thermolysin to generate viable human normal intestinal epithelial cell cultures, *Exp Cell Res* ***224***, 354–364.
16. Perreault N, Beaulieu JF. (1998) Primary cultures of fully differentiated and pure human intestinal epithelial cells, *Exp Cell Res* ***245***, 34–42.
17. Deschenes C, Alvarez L, Lizotte ME et al. (2004) The nucleocytoplasmic shuttling of E2F4 is involved in the regulation of human intestinal epithelial cell proliferation and differentiation, *J Cell Physiol* ***199***, 262–273.
18. Deschenes C, Vezina A, Beaulieu JF et al. (2001) Role of p27(Kip1) in human intestinal cell differentiation, *Gastroenterology* ***120***, 423–438.
19. Benoit YD, Lussier C, Ducharme PA et al. (2009) Integrin alpha8beta1 regulates adhesion, migration and proliferation of human intestinal crypt cells via a predominant RhoA/ROCK-dependent mechanism, *Biol Cell* ***101***, 695–708.

20. Gagne D, Groulx JF, Benoit YD et al. (2010) Integrin-linked kinase regulates migration and proliferation of human intestinal cells under a fibronectin-dependent mechanism, *J Cell Physiol 222*, 387–400.
21. Schwartz S, Beaulieu JF, Ruemmele FM. (2005) Interleukin-17 is a potent immunomodulator and regulator of normal human intestinal epithelial cell growth, *Biochem Biophys Res Commun 337*, 505–509.
22. Garneau H, Alvarez L, Paquin MC et al. (2007) Nuclear expression of E2F4 induces cell death via multiple pathways in normal human intestinal epithelial crypt cells but not in colon cancer cells, *Am J Physiol Gastrointest Liver Physiol 293*, G758–772.
23. Harnois C, Demers MJ, Bouchard V et al. (2004) Human intestinal epithelial crypt cell survival and death: Complex modulations of Bcl-2 homologs by Fak, PI3-K/Akt-1, MEK/Erk, and p38 signaling pathways, *J Cell Physiol 198*, 209–222.
24. Francoeur C, Escaffit F, Vachon PH et al. (2004) Proinflammatory cytokines TNF-alpha and IFN-gamma alter laminin expression under an apoptosis-independent mechanism in human intestinal epithelial cells, *Am J Physiol Gastrointest Liver Physiol 287*, G592–598.
25. Ruemmele FM, Beaulieu JF, O'Connell J et al. (2002) The susceptibility to Fas-induced apoptosis in normal enterocytes is regulated on the level of cIAP1 and 2, *Biochem Biophys Res Commun 290*, 1308–1314.
26. Benoit YD, Pare F, Francoeur C et al. (2010) Cooperation between HNF-1alpha, Cdx2, and GATA-4 in initiating an enterocytic differentiation program in a normal human intestinal epithelial progenitor cell line, *Am J Physiol Gastrointest Liver Physiol 298*, G504–517.
27. Escaffit F, Pare F, Gauthier R et al. (2006) Cdx2 modulates proliferation in normal human intestinal epithelial crypt cells, *Biochem Biophys Res Commun 342*, 66–72.
28. Escaffit F, Perreault N, Jean D et al. (2005) Repressed E-cadherin expression in the lower crypt of human small intestine: a cell marker of functional relevance, *Exp Cell Res 302*, 206–220.
29. Vachon PH, Harnois C, Grenier A et al. (2002) Differentiation state-selective roles of p38 isoforms in human intestinal epithelial cell anoikis, *Gastroenterology 123*, 1980–1991.
30. Streeter G. (1920) Weight, sitting head, head size, foot length and menstrual age of the human embryo, *Contr Embryol 11*, 143–179.
31. Tremblay E, Auclair J, Delvin E et al. (2006) Gene expression profiles of normal proliferating and differentiating human intestinal epithelial cells: a comparison with the Caco-2 cell model, *J Cell Biochem 99*, 1175–1186.
32. Saeed AI, Sharov V, White J et al. (2003) TM4: a free, open-source system for microarray data management and analysis, *Biotechniques 34*, 374–378.
33. Perreault N, Beaulieu JF. (1996) Use of the dissociating enzyme thermolysin to generate viable human normal intestinal epithelial cell cultures, *Exp.Cell Res. 224*, 354–364.
34. Perreault N, Beaulieu J. (1998) Primary cultures of fully differentiated and pure human intestinal epithelial cells, *Exp Cell Res. 245*, 34–42.
35. Quaroni A, Calnek D, Quaroni E et al. (1991) Keratin expression in rat intestinal crypt and villus cells. Analysis with a panel of monoclonal antibodies, *J Biol Chem 266*, 11923–11931.
36. Calvert R, Millane G, Pothier P et al. (1993) An intestinal secretory protein is found in most glands associated with the gastrointestinal tract: von Ebner's and salivary glands, gallbladder, and pancreas, *J Histochem Cytochem 41*, 1223–1231.
37. Quaroni A, Nichols BL, Quaroni E et al. (1992) Expression and different polarity of aminopeptidase N in normal human colonic mucosa and colonic tumors, *Int J Cancer 51*, 404–411.
38. Daniele B, Quaroni A. (1990) Polarized secretion of diamine oxidase by intestinal epithelial cells and its stimulation by heparin, *Gastroenterology 99*, 1675–1687.
39. Beaulieu JF, Nichols B, Quaroni A. (1989) Posttranslational regulation of sucrase-isomaltase expression in intestinal crypt and villus cells, *J Biol Chem 264*, 20000–20011.
40. Dydensborg AB, Teller IC, Basora N et al. (2009) Differential expression of the integrins alpha6Abeta4 and alpha6Bbeta4 along the crypt-villus axis in the human small intestine, *Histochem Cell Biol 131*, 531–536.
41. Beaulieu JF, Quaroni A. (1991) Clonal analysis of sucrase-isomaltase expression in the human colon adenocarcinoma Caco-2 cells, *Biochem J 280* (Pt 3), 599–608.

Chapter 12

Primary Culture of Human Renal Proximal Tubule Epithelial Cells and Interstitial Fibroblasts

Claire C. Sharpe and Mark E.C. Dockrell

Abstract

Renal physiology and pathology are complex systems that are best studied in whole living organisms. This, however, is often restricted by our desire to limit the number of animal experiments undertaken and to replace them with relevant in vitro models that can be used as surrogates for the system under test. Primary culture cells are derived directly from the relevant tissue and therefore correlate more closely with the system under examination. Although the tissue of origin is not always readily available for culture and cells may quickly change their phenotype after only a few passages, they can be used in many circumstances to validate results obtained from closely related cell lines and to confirm vital protein expression patterns. This chapter outlines methods by which proximal tubular epithelial cells and renal interstitial fibroblasts can be isolated and characterized from human renal nephrectomy tissue.

Key words: Kidney, Primary culture, Proximal epithelial cell, Fibroblast

1. Introduction

The nephron structure immediately adjacent to the glomerulus of the kidney is a tubule formed by a single layer of epithelial cells derived originally from mesenchymal cells. During renal organogenesis, the proximal tubule is part of the nephron formed from the metanephric mesenchyme as a result of intercellular communication between the ureteric bud and the mesenchymal cells. The proximal tubule can be divided structurally into three segments: the convoluted portion consisting of S1, S2, and the straight portion, S3, which descends into the outer medulla. Characteristic structures of the proximal tubule include the microvilli which form the brush border reflecting its absorptive nature and the density of the mitochondria in the cells reflecting the high levels of intracellular energy required for function.

Ragai R. Mitry and Robin D. Hughes (eds.), *Human Cell Culture Protocols*, Methods in Molecular Biology, vol. 806,
DOI 10.1007/978-1-61779-367-7_12, © Springer Science+Business Media, LLC 2012

The proximal tubules are exposed to approximately 150 L/day of glomerular ultrafiltrate; from this, the proximal tubule reabsorbs low-molecular-weight proteins, amino acids, and glucose as well as 90% of filtered phosphate, 80% of filtered bicarbonate, and 60% of filtered chloride. In addition, it also secretes a number of endogenous and exogenous substances as well having endocrine and paracrine functions. There is also increasing evidence that under prolonged stress, proximal tubule epithelial cells (PTECs) can undergo phenotypic change, becoming mesenchymal in nature, allowing them to cross their basement membrane and to contribute to the development of tubulointerstitial fibrosis. Consequently, disorders of the proximal tubule have serious repercussions and the study of its function is of significant importance.

A number of human-transformed PTEC cell lines do exist, notably the SV40-transformed HKC cell line developed by Loraine Racusen in 1997 and the HPV E6/7-transformed HK-2 cell line developed by Ryan et al. (1, 2). Although extremely useful, both of these cells lines have limitations in their phenotypic resemblance to primary PTECs in culture; for example, HKC clone 8 does not express tight junctions while HK-2 cells do not tolerate being grown to confluence. For these reasons, primary PTECs offer a good viable model for the proximal tubule epithelia which can be used entirely independently or to confirm results which have previously been obtained from immortal cell lines.

Interstitial renal fibroblasts, although relatively sparse in the normal kidney, are also important in the pathophysiology of disease. During the process of fibrosis, they increase in number and lay down extracellular matrix and are thus the main perpetrators of scar formation, a process closely correlated with loss of renal function. Fibroblasts obtained from fibrotic kidneys are known to retain a more "active" phenotype than those obtained from normal kidneys even after they have been established in tissue culture (3, 4). To our knowledge, there are no commercially available human renal fibroblast immortal cell lines.

Generally, protocols for the preparation of primary PTECs and fibroblasts are based on preparing suspension of nonglomerular cortical cells and isolating them on the basis of size or density. In addition, renal fibroblasts can be directly cultured from outgrowths from tissue explants. This technique has been described previously in full and is not discussed further in this chapter (5). When using nephrectomy samples, care should be taken to avoid using renal tissue adjacent to tumour and all local ethical permissions should be in place before tissue is obtained.

2. Materials

2.1. Tissue Harvest and Cell Isolation

1. Hank's balanced salt solution (HBSS), pH 7.5: Hank's balanced salts 9.8 g, 1 M HEPES 10 mL, 1 M NaOH (~2 mL to pH 7.5), 5 mL penicillin 10,000 IU/mL, 5 mL streptomycin 10 mg/mL. The solution is made up to 1 L with water and sterilized through a 0.2-μm nitrocellulose filter.
2. Collagenase type IV (Sigma, Poole Dorset, UK), 1 mg/mL in sterile HBSS.
3. Foetal calf serum (FCS) (Sigma, Poole Dorset, UK).
4. Sieves (Endecotts Ltd, UK). 100-μm pore size for method A or 300, 125-, 106-, 70-, and 45-μm pore sizes for method B.
5. Percoll (Sigma, Poole Dorset, UK) made up to a 50% solution with sterile HBSS (containing 0.84 g mannitol and 100 μL 1 M HCl in 30 mL).
6. Round bottom centrifuge tubes (Nalgene).

2.2. Tissue Culture: Proximal Tubule Epithelial Cells

1. Cell culture medium: Dulbecco's modified Eagle's medium (DMEM)/nutrient mixture Ham's F12, (Gibco BRL, Life Technologies, Paisley, UK). Medium is supplemented with 100× ITS [1.0 mg/mL insulin from bovine pancreas, 0.55 mg/mL human transferrin (substantially iron-free), and 0.5 g/mL sodium selenite], 5 mL per 500 mL of medium, and water-soluble hydrocortisone (100 mg hydrocortisone is dissolved in 4.5 mL of culture medium, filter sterilized, and stored frozen in aliquots). 300 μL of the dissolved solution is added to each 500 mL of culture medium. Antibiotics are also added: 5 mL penicillin 10,000 IU/mL, 5 mL streptomycin 10 mg/mL.

2.3. Tissue Culture: Fibroblasts

1. Cell culture medium: DMEM/nutrient mixture Ham's F12. Medium is supplemented with 10% FCS. Antibiotics are also added: 5 mL penicillin 10,000 IU/mL, 5 mL streptomycin 10 mg/mL per 500 ml of medium.

2.4. Immunohistochemistry

1. Phosphate-buffered saline (PBS) tablets.
2. 4% paraformaldehyde solution: Paraformaldehyde 4 g is dissolved in PBS 100 mL containing 100 μM $CaCl_2$ and 100 μM $MgCl_2$ (see Note 1).
3. Anti-vimentin monoclonal antibody (Sigma).
4. Anti-pan-cytokeratin (Sigma).
5. Fluorescein isothiocyanate (FITC)-conjugated secondary antibody (Sigma).
6. Mowiol mounting medium: Glycerol 6 g, Mowiol 4–88 2.4 g (Calbiochem, Nottingham, UK), sterile water 6 mL, 0.2 M Tris,

pH 8.5, 12 mL. The glycerol, Mowiol 4.88, and water are added together in a 50-mL polypropylene centrifuge tube and incubated at room temperature for 2 h. The Tris is then added and the solution is incubated at 50°C for a further 10 min with intermittent shaking. This solution is then centrifuged at 5,000 × *g* for 5 min and the supernatant is aliquoted and frozen at −80°C.

2.5. Functional Assays

1. For γ-glutamyl transferase (γ-GT) staining, make up a freshly prepared and filtered solution containing GMNA (γ-glutamyl-4-methoxy-naphylamide), 2.5 mg/mL, Tris–HCl buffer (0.1 M), pH 7.4, NaCl (0.85%), glycylglycine (0.5 mg/mL), and Fast blue BB (diazotised 4′-amino-2′, 5′-diethoxybenzanilide) (0.5 mg/mL).
2. For leucine aminopeptidase (LAP) staining, the cells make up a freshly prepared and filtered solution containing LMNA (γ-glutamyl-4-methoxy-naphylamide), 4 mg/mL, sodium acetate buffer (0.1 M), pH 6.5, NaCl (0.85%), sodium cyanide (0.02 M), and Fast blue B (tetrazotized-o-dianisidine chloride) (0.5 mg/mL).
3. For cAMP assays: 3-Isobutyl-1-methylxanthine (IBMX), parathyroid hormone [PTH(1–34)], vasopressin, calcitonin (salmon calcitonin), and forskolin.

3. Methods

We describe two separate methods for the isolation of primary tubulointerstitial cells from renal tissue. The first requires the use of an ultracentrifuge to create a density gradient but can yield both PTECs and fibroblasts while the second does not require an ultracentrifuge but yields only PTECs. Once isolated, it is important to adequately characterize cells to ensure that the correct phenotype has been obtained in a reasonably pure culture. Primary PTECs can be characterized by several different methods: (1) morphologically using light microscopy (demonstrating the classic cobblestone appearance and dome formation indicating active transport; Fig. 1) or electron microscopy (demonstrating the presence of a polarized morphology with basolateral nuclei and brush border; Fig. 2); (2) immunohistochemically (staining positive for cytokeratin and negative for vimentin); (3) by enzyme cytochemistry [demonstrating the staining on these cells for brush border enzymes LAP and γ-GT]; and (4) using functional assays, such as differential adenylate cyclase responses to parathyroid hormone and arginine vasopressin. Primary fibroblasts, however, have a more spindle-shaped appearance under light microscopy and stain negative for cytokeratin and positive for vimentin.

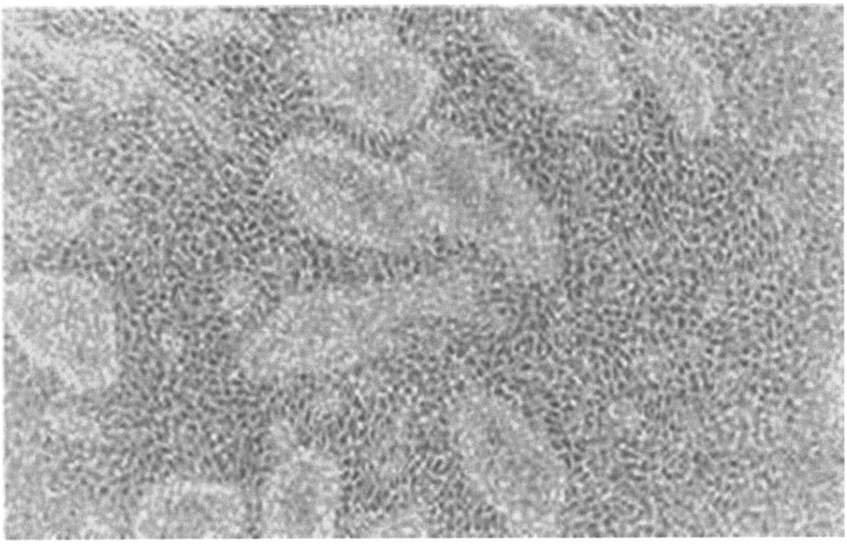

Fig. 1. Cells isolated using method 2, showing the characteristic epithelial cobblestone appearance, and the formation of hemicysts or "domes", indicative of transcellular transport.

3.1. Isolation of PTECs and Fibroblasts from Human Renal Tissue Using a Density Gradient

Fresh renal tissue (approximately 10 g) is obtained from human nephrectomy samples which have been removed for renal cell carcinoma. The tissue should be taken from the pole distant to the tumour and identified as macroscopically tumour-free by a consultant histopathologist (see Note 2). The tissue should be placed directly into sterile, ice-cold HBSS and processed immediately for maximum yield (see Note 3). All procedures should be carried out using aseptic technique in a laminar flow hood where applicable .

1. Place the tissue in a sterile Petri dish and soak with HBSS to remove as much blood as possible. Using autoclaved surgical tools, remove any adherent renal capsule and fat; discard.
2. Cut the outer (paler) cortex from the medulla using scissors and move to a clean Petri dish containing approximately 2 mL of HBSS. Mince the tissue with sharp scissors to form pieces about 1 mm^3 and then wash three times in clean HBSS to remove any remaining blood. Transfer the diced fragments to a pre-warmed 40 mL solution of 1 mg/mL collagenase (type IV) in HBSS in a sterile conical flask. Add a small, sterile magnetic stirrer and place on a stirring block in a water bath at 37°C for 30 min. This results in separation of interstitial and tubular cells from the surrounding tissues and glomeruli (see Note 4).
3. Stop the digestion by adding an equal volume of ice-cold HBSS/5% FCS.

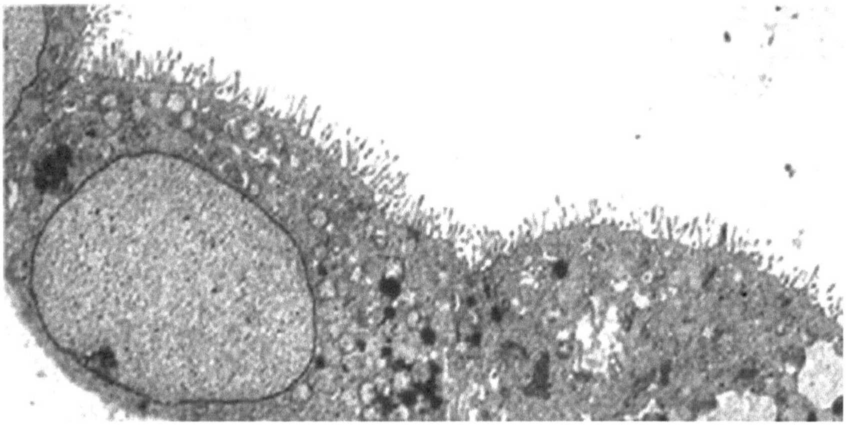

Fig. 2. Electron micrograph of primary human PTEC (courtesy of Dr Alex Pearson).

Remove the glomeruli and any undigested tubular fragments from the digested solution by passing it through a sterilized 100-μm sieve and collect the filtrate.

4. Collect the filtrate into a sterile universal and centrifuge it at 4°C at 200 × *g* for 5 min in a bench-top refrigerated centrifuge to pellet the digested material. Resuspend the pellet in 30 mL of ice-cold HBSS and spin again as above to remove any remaining collagenase solution. Repeat this wash once more.
5. Resuspend the pellet in 30 mL of the cold 50% Percoll solution and load onto a sterile 30-mL round-bottom centrifuge tube and spin at 30,000 × *g* at 4°C for 30 min. We recommend balancing this against a 50% Percoll solution containing marker beads of different colours and weights (also available from Sigma) to demonstrate that the density gradient has been adequately generated by the centrifugation process (Fig. 3). The two tubes must be weighed and equalized using Percoll solution to ensure adequate balancing for spinning at this high speed.
6. After centrifugation, the density gradient should have separated the digested material into four identifiable bands (1–4). The top band (band 1) contains predominantly fibroblasts while band 4, roughly equivalent to the blue marker bead band, contains predominantly proximal tubular cells (Fig. 3). Red blood cells migrate towards the bottom of the tube and are identifiable by their colour. The bands of cells required can be carefully sucked up using a sterile Pasteur pipette.
7. The harvested material should then be diluted in 20 mL ice-cold HBSS and centrifuged at 200 × *g* at 4°C for 5 min to remove any contaminating Percoll. This wash process should be repeated a second time in a pre-weighed universal container.

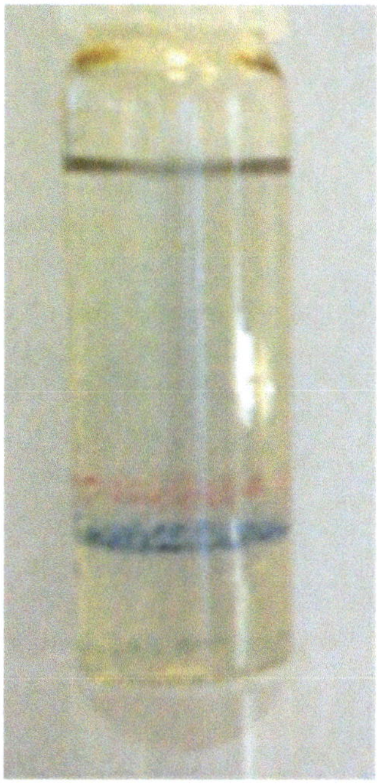

Fig. 3. Density marker beads spun on a Percoll density gradient as described above.

After centrifugation, the supernatant is removed carefully and the tube weighed again to gather the weight of the pellet. The pellet should then be resuspended in 20 mL of culture medium (DMEM Ham's/F12 supplemented with 1 mL ITS/100 mL medium and 300 μL hydrocortisone solution for proximal tubular cells and 10% FCS for fibroblasts) containing antibiotics.

8. The cell suspension is plated onto tissue culture dishes or flasks at an approximate density of 10–20 mg per 35-mm^2 dish. Fibroblasts are plated directly onto tissue culture plastic, whereas the primary proximal tubular cells grow better on dishes pre-coated with collagen type IV. These cultures are then incubated in a standard 5% CO_2 incubator at 37°C undisturbed for 72 h to allow attachment. After this time, the medium is removed and replaced with fresh medium.

3.2. Isolation of PTECs from Human Renal Tissue Using a Sieving Method

1. Renal cortical tissue should be obtained washed and minced as above.
2. The minced pieces of cortex are then mashed through a 300-μm stainless steel sieve using an inverted plunger from a sterile syringe; connective tissue is retained in this sieve. The resulting filtrate is then washed through 125-μm sieve. The red particulate matter retained in this sieve can be identified as glomeruli by microscopic examination.

3. The filtrate is then sequentially washed through sieves of 106-, 70-, and 45-μm pores with ice-cold sterile HBSS. The material retained in the 45-μm sieve should be predominantly tubular fragments which should be transferred to a sterile universal containing ice-cold HBSS and centrifuged at 250 × *g* for 6 min. The supernatant should be discarded and this wash procedure repeated for a second time.
4. The pellet should then be resuspended in a 20-mL solution of collagenase type IV/HBSS (1 mg/mL) and incubated at 37°C for 10 min. This suspension is then centrifuged at 250 × *g* for 6 min and the supernatant discarded. The cell pellet is then washed twice more with sterile HBSS.
5. The cells should then be resuspended in DMEM/Ham's Nutrient Mixture F12 supplemented with 1 mL ITS/100 mL medium and 300 μL hydrocortisone solution and antibiotics then plated onto tissue culture plates coated in type IV collagen as above.

3.3. Subculture of Primary Cells

1. The cells can be trypsinized and passaged (divided 1 in 2) once confluent or near confluence. PTECs only continue to proliferate for 3–4 passages (while retaining phenotype) though the primary fibroblasts continue to grow for 8–10 passages.
2. These cells can be slowly expanded with each passage and frozen in aliquots. Trypsinized cells are pelleted and resuspended in an 80% DMSO/20% FCS solution at a density of 10,000 cells/mL. 1-mL aliquot is pipetted into cryotubes and snap frozen in liquid nitrogen. It should be maintained in a liquid nitrogen store rather than a –80°C freezer if possible.

3.4. Characterizing Cells: Immunohistochemistry

1. For the purposes of immunostaining, the primary culture cells should be grown on glass coverslips in 35-mm tissue culture dishes. Cells are seeded sparsely (1 in 10) in normal growth medium and left for 24 h. The growth medium is then removed and the coverslips should be washed three times in PBS.
2. The cells should be fixed in 4% paraformaldehyde for 10 min and then washed three times in PBS. They are then permeabilized with 0.2% Triton X-100/PBS and blocked with 5% goat serum/PBS (host species for secondary antibody). The primary antibody (anti-vimentin 1:40, anti-pan-cytokeratin 1:400) should be diluted in PBS and 100 μL dropped onto the surface of the coverslip and left at 4°C overnight or room temperature for 1 h (if performing overnight incubation, take care to avoid evaporation of the antibody solution).
3. The coverslips are then washed three times in PBS and 100 μL of FITC-conjugated secondary antibody (1:200) should be added and left for 30 min at room temperature in the dark.

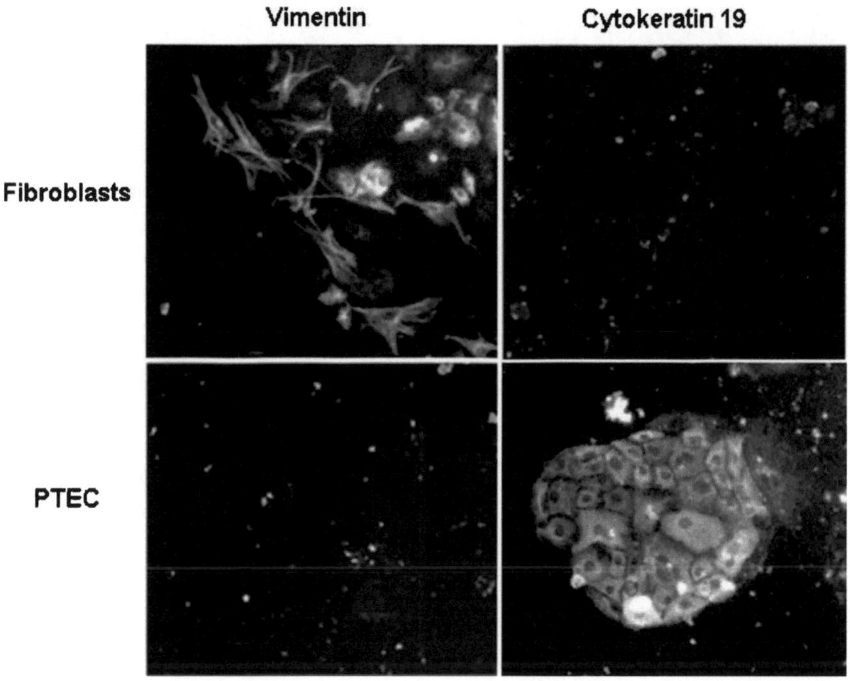

Fig. 4. Primary culture human renal fibroblasts staining positive for vimentin and negative for cytokeratin. Primary culture human renal proximal tubular cells staining positive for cytokeratin 19 and negative for vimentin. The epithelial cells show a characteristic paving stone morphology compared to the more spread-out appearance of the fibroblasts.

The coverslips are again washed three times in PBS and mounted onto glass slides with Mowiol mounting medium and left to dry for 2 h. The cells can then be visualized by conventional fluorescent microscopy (Fig. 4).

3.5. Functional Assays: Brush Boarder Enzymes

The presence of proximal tubular brush border enzymes γ-GT (Fig. 5) and LAP can be confirmed by enzyme cytochemistry as previously described by Rutenburg et al. and Nachlas et al., respectively (6, 7).

1. Cells can be grown on coated chamber slide coverslips or other appropriately treated cell culture surfaces. Best results are achieved by allowing cells to grow to confluence prior to removal of culture medium.
2. For γ-GT staining, the cell monolayers are then incubated for 45 min at 37°C in a freshly prepared and filtered solution containing GMNA (γ-glutamyl-4-methoxy-naphylamide). The cells are then rinsed in NaCl (0.85%) for 2 min and transferred to a 0.1 M copper sulphate solution for 2 min prior to visualization by light microscopy.
3. For LAP staining, cells are incubated at 37°C for 20 min in a freshly prepared and filtered solution containing LMNA (γ-glutamyl-4-methoxy-naphylamide). The cells are then rinsed

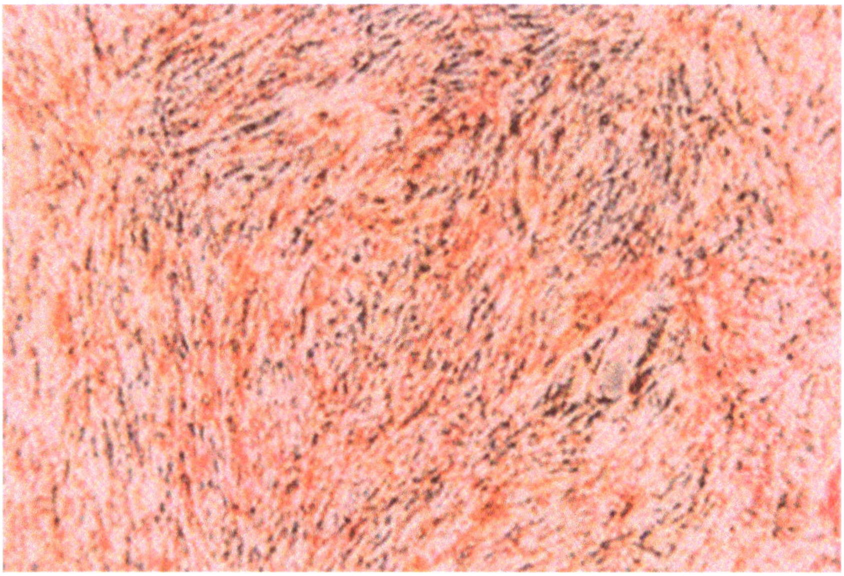

Fig. 5. Light microscopy of γ-GT staining of proximal tubular cells.

in NaCl (0.85%) for 2 min and transferred to a 0.1 M copper sulphate solution for 2 min prior to visualization by light microscopy.

3.6. Functional Assays: Induction of cAMP

Induction of cyclic adenosine monophosphate (cAMP) as a second messenger by distinct G-protein-coupled receptors is characteristic of different segments of the nephron. Typically, the proximal tubule cells show a cAMP response to parathyroid hormone (PTH) but not vasopressin or calcitonin. One may observe an increase of cAMP in response to PTH in cells from the distal segments from the tubule, so use of vasopressin and calcitonin as negative controls is important. An increasing number of assays are available for the measurement of cAMP that avoid the use of radioactivity utilizing different detection platforms, and the method of extraction of cAMP may be determined by the method of measurement; for example, radioimmunoassays have typically used acetate buffers with a pH of 4–5. Consequently, precise details of the assay and the extraction of cellular cAMP are not given here.

1. Cells are grown to confluence in normal culture medium, the medium removed, and the cells rinsed twice in fresh medium.
2. All cultures were incubated in media containing IBMX 1 mM to prevent cAMP breakdown and allow accumulation of $[cAMP]_i$. The cells are treated with vehicle; PTH(1–34), 1 μM; vasopressin 1 μM; calcitonin (salmon calcitonin) 100 nM; or forskolin 25 μM for 30 min at 37°C. Forskolin directly activates adenylyl cyclase and thus acts as a positive control.

3. Typically basal $[cAMP]_i$ levels are in the range 0.25–0.75 pmol/mg of cellular protein depending on cellular conditions and the method of detection. PTH stimulation in the presence of IBMX results in a two- to fourfold rise in 30 min; therefore, any assay used should be able to detect levels across this range of values. Prolonged treatment results in accumulation of higher levels of $[cAMP]_I$, but may also result in cAMP efflux.

4. Notes

1. Paraformaldehyde is toxic and should only be handled in a fume hood.
2. It is best to request that the tissue sample be removed by a histopathologist. This avoids disruption of what may be a clinically important specimen and minimizes sample contamination by tumour cells.
3. Ideally, the nephrectomy sample should be taken directly to the histopathologist as soon as it is available and dissected tissue for culture be placed directly into ice-cold HBSS. The culture process should then be started straightaway for maximum cell yield. If it is too late in the day, the sample can be stored at 4°C overnight, but this will reduce the number of viable cells obtained.
4. The longer the tissue is digested in collagenase, the better the cell separation but the worse the viable cell yield. We have found that 30 min is optimum, but this can be altered according to how active each collagenase solution is found to be.

References

1. Racusen, LC, Monteil, C, Sgrignoli, A, Lucskay, M, Marouillat, S, Rhim, JG & Morin, JP: Cell lines with extended in vitro growth potential from human renal proximal tubule: characterization, response to inducers, and comparison with established cell lines. J Lab Clin Med 129: 318–29, 1997.
2. Ryan, MJ, Johnson, G, Kirk, J, Fuerstenberg, SM, Zager, RA & Torok-Storb, B: HK-2: an immortalized proximal tubule epithelial cell line from normal adult human kidney. Kidney Int 45:48–57, 1994.
3. Muller, GA & Rodemann, HP: Characterization of human renal fibroblasts in health and disease: I. Immunophenotyping of cultured tubular epithelial cells and fibroblasts derived from kidneys with histologically proven interstitial fibrosis. Am J Kidney Dis 17:680–3, 1991.
4. Rodemann, HP & Muller, GA: Characterization of human renal fibroblasts in health and disease: II. In vitro growth, differentiation, and collagen synthesis of fibroblasts from kidneys with interstitial fibrosis. Am J Kidney Dis 17:684–6, 1991.
5. Grimwood, L & Masterson, R: Propagation and culture of renal fibroblasts. Methods Mol Biol 466:25–37, 2009.
6. Nachlas, MM, Monis, B, Rosenbatt, D & Seligman, AM: Improvement in the histochemical localization of leucine aminopeptidase with a new substrate, L-leucyl-4-methoxy-2-naphthylamide. J Biophys Biochem Cytol 7:261–4, 1960.
7. Rutenburg, AM, Kim, H, Fischbein, JW, Hanker, JS, Wasserkrug, HL & Seligman, AM: Histochemical and ultrastructural demonstration of gamma-glutamyl transpeptidase activity. J Histochem Cytochem 17:517–26, 1969.

Chapter 13

Glomerular Epithelial and Mesangial Cell Culture and Characterization

Heather M. Wilson and Keith N. Stewart

Abstract

Culture of isolated kidney glomerular cells has been employed for almost four decades as a tool to dissect pathophysiological effects of individual cell types in renal disease. This chapter aims to highlight in detail the available techniques to isolate, culture, and characterize human glomerular epithelial and mesangial cells. To establish primary culture of these cells, glomeruli are isolated from the cortex of kidney by differential sieving and cellular outgrowths from cultured glomeruli further subcultured in appropriately coated tissue culture plates/flasks. Methods used for characterization of isolated glomerular mesangial and epithelial cells (podocytes) are described as are the phenotypic markers useful for identification. Other sources of isolated glomerular cells such as immortalized cell lines are briefly discussed.

Key words: Mesangial cell, Glomerular epithelial cell, Podocyte, Tissue culture, Kidney glomerulus

1. Introduction

For many years, cultures of homogeneous populations of glomerular cells have been employed in renal research laboratories to aid our understanding of renal physiology and the pathogenesis of glomerular disease at the cellular level. It is now clear that the response of intrinsic glomerular cells to external stimuli plays an important role in glomerular injury (1, 2). Glomerular cells from several mammalian species have been isolated, propagated, and cell lines generated by viral or nonviral oncogenic transformation (3–6).

The renal glomerulus is a complex anatomical structure that contains many different cell types, including visceral and parietal epithelial cells, mesangial cells, and fenestrated endothelial cells.

Ragai R. Mitry and Robin D. Hughes (eds.), *Human Cell Culture Protocols*, Methods in Molecular Biology, vol. 806,
DOI 10.1007/978-1-61779-367-7_13, © Springer Science+Business Media, LLC 2012

These cells have long been recognized as distinct entities, because they occupy defined anatomical locations in vivo and have distinguishable morphological and cytochemical features. Mesangial cells and their matrix form the central stalk of the glomerulus and are part of the functional unit interacting closely with endothelial cells and podocyte epithelial cells. This compartmentalization is, however, lost in culture, as are several of the anatomical characteristics such as endothelial fenestrations and epithelial pedicels (7–10); in addition cultured cells may undergo de-differentiation. Despite these limitations, the study of glomerular cells in culture has proven useful, and valuable information has been obtained about their physiology and pathophysiology. Human glomeruli are usually obtained from the normal pole of kidneys surgically removed from patients with renal carcinoma or from donor kidneys that cannot be used for transplantation for technical reasons but glomeruli have also been derived from autopsy specimens or from fetal kidneys at 8–18 weeks of gestation. Primary cultures are established principally from glomeruli isolated by differential sieving such that they can be rendered virtually free of tubule contamination (10, 11). Thereafter, glomeruli can either be seeded into culture flasks, and after a week, cells can be seen growing out of the glomerular core (5). Alternatively, glomeruli can be dissociated by incubation with an enzyme, such as collagenase before culture (12).

The first cells to emerge from explanted glomeruli are epithelial cells, which have a distinctive "cobblestone" appearance (12) and, for the first 7–10 days of outgrowth, are the most common cell type in the mixed population of glomerular cells. If Bowman's capsule is not stripped from the glomeruli, parietal epithelial cells are the dominant cell type (13); endothelial cells and mesangial cells may present at this stage. Mesangial cells become more evident later in culture and have a stellate appearance. They grow vigorously, in multilayers, whereas epithelial cells grow in a monolayer and are subject to contact inhibition. Mesangial cells, therefore, outgrow the epithelial cells, and after 30 day of growth, the cultures are nearly pure glomerular mesangial cells (14). The difference in growth potential of primary cells in culture can be explained by the "Mosaic Theory" (15), which states that separate populations of cells can be obtained from a mixed population of cells based on their different growth rates or culture requirements.

For most experimental purposes, a homogeneous population of cells is required. It is therefore important to assess the purity of the isolated cells, since even a small population of contaminating cells can affect the experimental results. Homogeneity of cells can be improved by using cloning rings and repeated cloning but the yield of cells is less and they do de-differentiate with passage number. Glomerular cell lines (6, 16–18) have now been established for most of the glomerular cell types that usually results in cellular transformation of cells to a more proliferative and less differentiated

Table 1
Main cell markers that can be used to differentiate between glomerular cell types in vitro

Antibody	Visceral epithelial cells	Parietal epithelial cells	Mesangial cells	Endothelial cells
GLEPP1	+	–	–	–
Vimentin[a]	+	–	+	–
WT-1	+	+	–	–
Cytokeratin 18[a]	–	+	–	–
Cytokeratin 19[a]	–	+	–	–
α-SMA	–	–	+	–
Myosin (smooth)	–	–	+	–
PDGF-Rβ	–	–	+	–
Factor VIII antigen	–	–	–	+
PECAM-1	–	–	–	+

GLEPP1 glomerular epithelial protein 1; (*WT-1*) Wilm's tumor protein-1; α-*SMA* smooth muscle actin; *PDGF-R beta* (CD140b) platelet-derived growth factor receptor beta; *PECAM-1* (CD31) platelet endothelial adhesion molecule-1
+ = positive; – = negative
[a]GVEC and GPEC can de-differentiate in culture and are not always distinguished specifically by vimentin and cytokeratin; cytokeratin can be expressed by both GVEC and GPEC in culture

phenotype compared to normal primary cells but has the advantage in that they have a longer life span.

It is now clear that morphology alone is not sufficiently discriminating to assess cell purity or ensure that a homogenous population of cells has been isolated (19). Antibodies to specific cell-surface and cytoskeletal markers need to be used to confirm both cell identity and purity. Table 1 shows the main markers used to differentiate between the glomerular cell types present in culture (5, 14, 20–23).

The outlined methods are routinely used in our laboratory to obtain and characterize pure populations of adult glomerular epithelial and mesangial cells (see Note 1).

2. Materials

1. Sterile scalpel, sterile petri dishes, plastic sterile 50 mL falcon tubes, tissue culture dishes of flasks.
2. Stainless-steel sieves of mesh size 250, 200, 150 and 106 μm (Endecotts Ltd, London, UK). Sieves should be thoroughly

washed, rinsed in ethanol, dried, and sterilized before use. A diameter size of 7.5 cm or greater is best for processing large amounts of tissue.

3. Wash medium: Rosewell Park Memorial Institute (RPMI) 1640 supplemented with 2 mM L-glutamine, 100 U/mL penicillin, 100 μg/mL streptomycin (all reagents from Invitrogen Ltd, Paisley, Scotland). Store at 4°C for up to 1 month.
4. Culture medium: RPMI 1640 or MEM-D-valine supplemented as in Subheading 1 with the addition of 10% fetal calf serum (Invitrogen Ltd.) and insulin–transferrin–sodium selenite (ITS) medium supplement (Sigma-Aldrich, Dorset, UK). Reconstitute one vial of ITS in 50 mL sterile distilled water, and add 1 mL of this stock to 100 mL of medium to give a final concentration of 5 μg/mL insulin, 5 μg/mL transferrin, and 5 ng/mL sodium selenite. Store the stock solution of ITS at 4°C and protect from light.
5. Fibronectin (Sigma-Aldrich): reconstitute to 1 mg/mL in sterile distilled water, and dilute to 10 μg/mL in RPMI 1640 wash medium. Store at −20°C in aliquots. Coat flasks at a concentration of 5 $\mu g/cm^2$, and leave solution to bind for at least 30 min. Decant excess fibronectin solution. Coated plates/flasks are also commercially available (Vitrogen, Collagen Corp, Palo Alto, USA).
6. Type I collagen (tissue culture grade, Sigma-Aldrich): reconstitute to 1 mg/mL in 0.1 M acetic acid, leave for 1–3 h until it is dissolved. Transfer this to a glass bottle with chloroform at the bottom. Do not shake or stir the collagen after this point. Store in the dark at 4°C. Coat the flasks at a concentration of 10 $\mu g/cm^2$, and leave to bind for 4 h at 37°C or overnight at 4°C. Expose the coated flask to UV radiation if you suspect the solution is not sterile, do not filter sterilize. Collagen I coated tissue culture flasks are commercially available.
7. Phosphate-buffered saline (PBS): 1.5 mM (0.2 g/L) KH_2PO_4, 8.1 mM (1.15 g/L) Na_2HPO_4, 2.7 mM (0.2 g/L) KCl, 140 mM (8.0 g/L) NaCl. Filter sterilize through a 0.22 μm filter before use. Store at room temperature (see Note 2).
8. Trypsin/EDTA 10× (Sigma T-4174): Dilute trypsin/EDTA 1:10 with sterile PBS, and store in aliquots at −20°C.
9. Tris-buffered saline (TBS), pH 7.6. Store at 4°C. Dilute 1:10 in distilled water before use.
10. Lux eight chamber permanox slides from MP biomedicals.
11. Envision detection systems Peroxidase/DAB, Rabbit/Mouse (Dako, Ely, Cambridgeshire, UK). Store at 4°C.
12. Dako REAL antibody diluent (Dako), store at 4°C.

13. Universal Negative control mouse (Dako).
14. Peroxidase-Blocking solution Dako REAL.
15. Scot's tap water substitute: Dissolve 2 g of sodium bicarbonate and 20 g of magnesium sulfate in 1 L of distilled water, and store at room temperature.
16. Hematoxylin.
17. Aqueous mounting medium (Dako).

3. Methods

3.1. Isolation of Whole Glomeruli

Human glomeruli are most often isolated from the nonaffected pole of nephrectomy specimens from patients with renal cell carcinoma (see Notes 3 and 4). Aseptic conditions in a laminar flow hood should be adopted throughout.

1. Place the tissue in a sterile Petri dish, and cover it with RPMI 1640 wash medium; no fetal calf serum should be used during the isolation procedure, since it may initiate clotting due to blood products present in the tissue. Remove surrounding capsule and any fat using a sterile scalpel.
2. Cut the cortex away from the medulla, and chop the cortex into 1–2 mm^2 pieces. Press this through a sieve of mesh size 250 μm, into a sterile Petri dish, using the barrel from a 5 mL syringe. This results in the separation of glomeruli from renal tubules, interstitium, and vasculature. Wash the retained tissue with a generous amount (50–100 mL) of RPMI 1640. Collect the glomerular-enriched filtrate from the Petri dish into sterile 50 mL falcon tubes on ice.
3. Separate the glomeruli from the tubular fragments by passing through a 150 μm sieve with the aid of a 5 ml syringe barrel. This also strips the Bowman's capsule from most of the glomeruli (see Note 5). A further 50 mL of RPMI wash medium are used to rinse the tissue retained on the sieve. Collect the filtrate into sterile containers.
4. Pass the filtrate through a 106 μm sieve, which retains and concentrates the isolated glomeruli. Since there is a substantial volume of filtrate, it is advantageous to hold the sieve above a beaker to allow the filtrate to pass through quickly. Pour the filtrate through the sieve into a sterile funnel inserted in the beaker to prevent any "splashback" that may occur. Rinse the retained glomeruli with approximately 50 mL RPMI 1640 wash medium to eliminate any tubular fragments that may still be present.

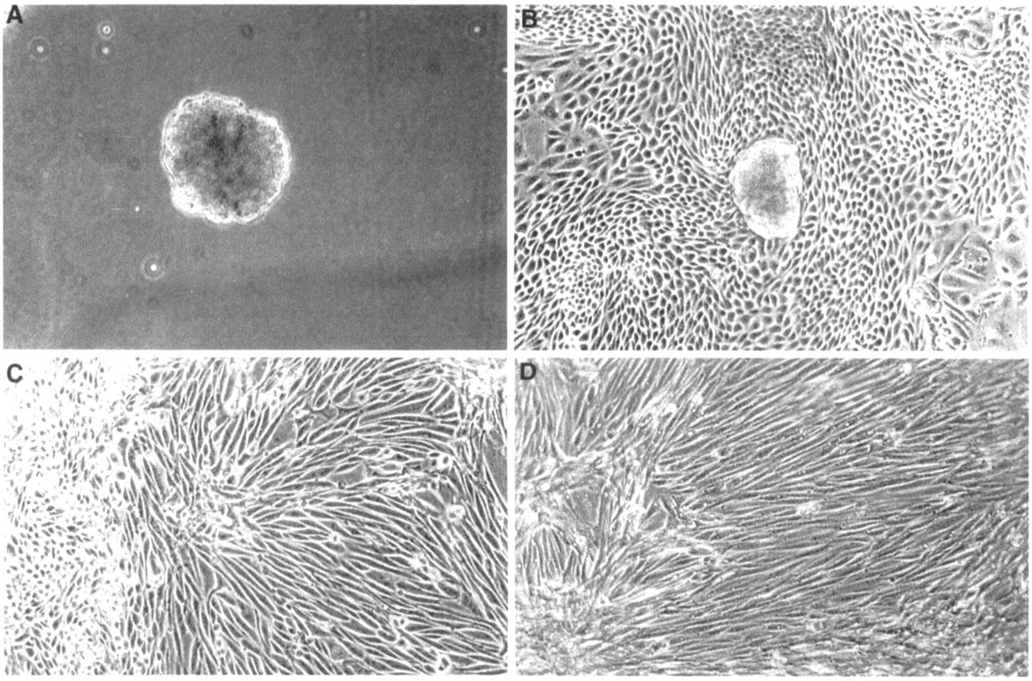

Fig. 1. Glomerular cell culture. (**a**) Decapsulated glomerulus at day 0. (**b**) Cellular growth at day 7; polygonal cells characteristic of epithelial cells. (**c**) Stellate-shaped cells growing in multilayers characteristic of contractile mesangial cells at day 14; a few epithelial cells are still present. (**d**) Mesangial cell outgrowth at day 28; no epithelial cells are evident.

5. Collect the glomeruli retained on the 106 μm sieve by inverting the sieve over a Petri dish and washing with RPMI culture medium. Glomeruli remaining on the sieve should be collected using an automated 1 mL pipette tip and washing. Transfer the glomeruli at a concentration of approximately 15–20 glomeruli/mL to a fibronectin-coated culture flask (see Fig. 1a and Note 6). Fibronectin-coated plates/flasks are also commercially available, see Subheading 2 for details.
6. Culture the glomeruli at 37°C in a 5% CO_2 incubator. Culture flasks/dishes should be left undisturbed for at least 24 h to allow good attachment of glomeruli. Sterility of the media, incubator, and laminar flow hood must be ensured throughout the isolation procedure.

3.2. Isolation of Glomerular Epithelial Cells

There are two types of epithelial cells within the glomerulus, namely, the visceral epithelial cells (GVEC) and the parietal epithelial cells (GPEC). GVEC, better known as podocytes, are located on the outer side of the glomerular basement membrane and are crucial in the urine filtration process where GPEC form the parietal sheet of Bowman's capsule. Podocytes are the most prominent targets in many types of renal injury. It is difficult to differentiate between GVEC and GPEC in culture as phenotypic changes take place. In glomeruli in vivo, GPEC rapidly proliferate and form crescents in proliferative nephritis (24), whereas GVEC rarely undergo cell division. Cytokeratin is present only in the GPEC, whereas vimentin is restricted to GVEC (25). These characteristics are lost in culture and it has been suggested that the GPEC phenotype is adopted (26). When epithelial cells are derived from decapsulated glomeruli they are likely to represent GVEC while GPEC grow at the periphery of glomeruli that have Bowman's capsule present (7, 27). Decapsulated and encapsulated glomeruli are best separated manually under a microscope if purity is to be improved. However, it is unlikely to achieve pure cultures of a specific epithelial cell type without the use of cloning rings. Although not described here, there is now potential to cultivate podocytes from urine as a new and promising method for monitoring the activity of some glomerular-associated diseases such as diabetic nephropathy (28).

The differentiation of human podocytes in culture in vitro is associated with an irreversible growth arrest. To circumvent this problem conditionally immortalized podocyte cell lines have been derived (6, 16), but due to changes induced by this processes and the alteration in cellular physiology due to continuous proliferation, data obtained from these cells should be interpreted with caution.

1. Isolate and culture the glomeruli as described in Subheading 3.1. Once the glomeruli have adhered to the flask, change the medium (RPMI 1640 culture medium) for every 4–5 days.
2. Epithelial cells can be seen growing out of the glomeruli around 7–10 days. This time may vary and generally it takes longer to see cellular outgrowth in glomeruli prepared from kidneys from older patients (see Note 7).
3. Once sufficient epithelial cells are identified (these have a polygonal, cobblestone appearance) (see Fig. 1b), pour off the unbound glomeruli, and rinse the bound glomeruli/cells in PBS. Trypsinize the cells/glomeruli off the flask (see Subheading 3.4) and pass them through a 63-μm sieve. Collect the epithelial cells in the filtrate; glomeruli are retained on the sieve. Rinse the sieve several times with a volume of around 10 mL RPMI culture medium to stop the action of trypsin and ensure maximum epithelial cell recovery.
4. Pellet the cells that have passed through the sieve by centrifugation at 200 × *g* for 5 min, and then resuspend in RPMI culture medium. Plate the cells into tissue culture flasks which have been coated with type I collagen at a concentration of 10 μg/cm^2 (see Subheading 2.6 and Note 8).
5. After approximately 1 week when the epithelial cells have reached confluence (see Note 9), passage using trypsin/EDTA (see Subheading 3.4). Plate the trypsinized cells onto plastic

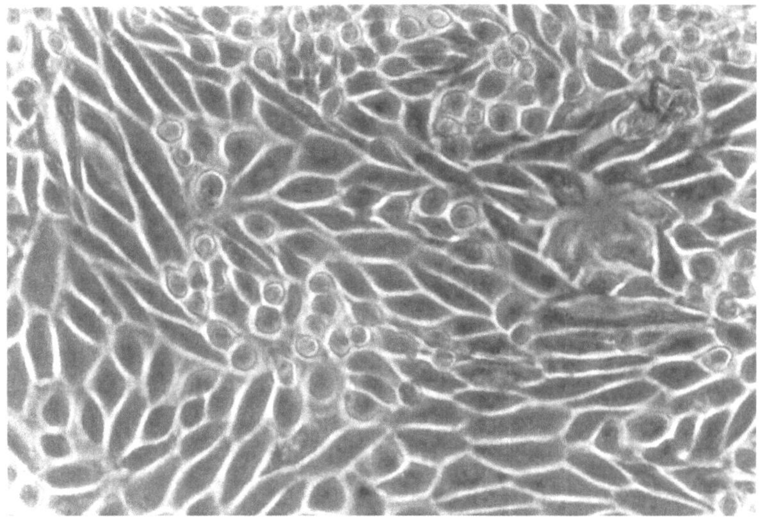

Fig. 2. Epithelial cell preparation grown on type I collagen. At confluence, epithelial cells display a cobblestone-like appearance.

tissue culture grade dishes or flasks without type I collagen at this point. At confluence, the cells will be homogeneous and display a cobblestone appearance (Fig. 2). GVEC tend not to proliferate. Within 2–3 weeks, the subcultured cobblestone cells transform into individual arborized cells and stain for Wilms' Tumor-1 (WT-1) antigen (29).

6. Characterize the cells before use (see Subheading 3.5).

3.3. Glomerular Mesangial Cell Culture

Mesangial cells are the easiest glomerular cells to grow due to their minimal growth requirements and high proliferative capacity. In culture, they show a similar stellate phenotype to that observed in vivo. They synthesize an extracellular matrix that is thought to represent some of the features of the glomerular basement membrane (e.g., collagen). Like glomerular epithelial (and endothelial) cells, a number of laboratories have developed mesangial cell lines stabilized by transfection with viruses. Moreover, primary human mesangial cell cultures are now commercially available (ScienCell Research Laboratories).

1. Glomeruli are isolated and cultured as described in Subheading 3.2. Once the glomeruli have adhered to the flask, the medium should be changed for every 4–5 days.
2. The first cells to grow out of the glomerular core are epithelial cells (Fig. 1b). After a further 2 weeks in culture, a mixed population of glomerular cell types is observed (Fig. 1c). Continue cultures for 2–3 weeks to allow the mesangial cells to overgrow (Fig. 1d). From week 3 onward, grow the cells

in MEM-D-valine culture medium to inhibit any fibroblast growth (see Note 10) (30).

3. Once the cells have grown to confluence, split them 1:1 using trypsin/EDTA (see Subheading 3.4), and continue to culture in uncoated tissue culture grade flasks.
4. Cells should be used between passages 3 and 8 (see Notes 6 and 11) after being fully characterized.

3.4. Trypsinization and Subculture of Adherent Cells

To remove the epithelial or mesangial cells from tissue culture flasks:

1. Pour the spent medium off the cells, and wash out the flask with sterile Ca^{2+} and Mg^{2+}-free PBS.
2. Pour off the PBS, add 5 mL of 1% trypsin/EDTA solution, and then incubate the flask at 37°C for 4–7 min.
3. Tap the flask lightly on the bench to loosen the cells before adding fresh medium. Adding culture medium helps inhibit further trypsinization. Split the cell suspension into two flasks to continue the culture, or centrifuge and count the cells before using for experiments (see Note 12). Clumps of cells (mesangial hillocks) that have formed should be broken down by repeated pipetting.
4. Cell purity increases after the first passage and purity should be confirmed by morphology and immunohistochemical staining although unique and reliable markers for mesangial cells are lacking and identification by this method is based on exclusion criteria.
5. Cells should be subcultured approximately every 5 days, depending on the speed of division. This avoids build up of mesangial hillocks and passage is easier and more efficient.
6. Trypsinized mesangial cells suspension can be resuspended in freezing medium (serum-containing, sterile-filtered media supplemented with 10% (v/v) DMSO as a cryopreservative and frozen in liquid nitrogen for long-term storage) (7).

3.5. Validation of Glomerular Cell Cultures

3.5.1. Cell Morphology

Cell morphology should be assessed throughout the culture period using an inverted microscope with phase-contrast illumination. Cell morphology at different stages of glomerular culture is shown in Fig. 1.

1. Epithelial cells: Epithelial cells are homogeneous in appearance (12), polygonal, and form cobblestone-like monolayers (Fig. 2). They are closely packed and adhere tightly to each other at the edge of the growing monolayer. Epithelial cells are subject to contact inhibition (see Note 8). Original methods for phenotype confirmation of cultured epithelial cells was based on their toxic sensitivity to puromycin aminonucleoside,

however, results were somewhat inconsistent and cells could not be rescued for experimentation after toxicity testing and immunohistochemical characterization is now preferred.

2. Mesangial cells: Mesangial cells are elongated and stellate-shaped, and are not subject to contact inhibition (Fig. 1d). They grow in multilayers to form characteristic hills and valleys when confluent (5, 31).

3.5.2. Envision Horseradish Peroxidase Immunohistochemistry Technique

The morphological assessment of cell cultures should be confirmed by immunohistochemistry using currently available monoclonal antibodies (Mabs) (Table 1).

The Dako Envision horseradish peroxidase (HRP) immunohistochemical technique is a very sensitive method (32) and is particularly useful when using monoclonal antibodies (see Note 13). In this technique, a labeled polymer reagent which has both secondary antibody and HRP enzyme conjugate attached is used to amplify the primary antibody–antigen interaction. An intense immunohistochemical staining with low nonspecific backgrounds can be obtained using the 3,3′diaminobenzidine tetrahyrochloride (DAB) substrate in the Envision staining Kit. Any nonspecific peroxidise activity is blocked by the addition of peroxidise blocking reagent at the start of the staining protocol.

1. The cells to be tested should be passaged, and 0.4 mL of cells at 1×10^6/mL seeded onto 8-well CultureWell™ MultiSlip™ Cell Culture System chamber slides and incubated overnight in a CO_2 incubator.
2. Remove the slides from the incubator, and without removing the upper chamber structure, aspirate tissue culture supernatant. Gently wash the adherent cells twice with 0.8 mL of TBS/well. Discard the final rinse of TBS, and allow the slides to air-dry for at least 20 min at room temperature.
3. Fix the slides by adding 0.3 mL of acetone at 4°C and incubating slides on ice for 10 min. (Note: Do not use more than 0.3 mL of acetone, since this dissolves the upper structure of the 8-well chamber.)
4. Remove the acetone and allow the slides to air-dry at room temperature for 10–15 min.
5. Prepare the antibodies at this stage, diluting them to the optimal concentration (see Table 2) in antibody diluent which has 2.5% normal goat serum added.
6. Rehydrate the slides by adding 0.8 mL of TBS to each well (two washes of 5 min). Ensure that the slide does not dry out at any stage after this point.
7. Remove the TBS from the wells, and add 200 μL of the primary monoclonal antibody diluted to its optimal concentration

Table 2
Source and working dilutions of antibodies used to characterize glomerular cell cultures

Antibody	Clone/source[a]/product code	Dilution
Glomerular epithelial protein-1	5C11/Biogenex/336M	1/100
Vimentin	LN-6/Sigma-Aldrich/V2258	1/200
Wilm's tumor protein-1	6F-H2/Dako/M3561	1/50
Cytokeratin 18	CY-90/Sigma-Aldrich/C8541	1/500
Cytokeratin 19	A53-BA2/Sigma-Aldrich/C6930	1/50
α-Smooth muscle actin	1A4/Sigma-Aldrich/A5691	1/20
Myosin (smooth)	hSM-V/Sigma-Aldrich/M7786	1/500
PDGF-R (CD140b)	PR7212/R&D Systems/MAB1263	1/150
Factor VIII antigen	F8/86/Dako/M0616	1/50
PECAM-1 (CD31)	JC704/Dako/M0823	1/200
Mouse negative control	Dako/N1698	Use neat

[a]Locations of companies: Biogenex (San Ramon CA, USA); Dako (Ely Cambridgeshire, UK); Sigma-Aldrich (Poole, Dorset, UK); AbSerotec (Kiddington, UK) and (Cambridge, UK); R&D Systems (Abington, Oxford, UK)

in antibody diluent (see Note 14). Incubate the slides for 1 h at room temperature.

8. Discard the antibody, and wash the slides three times for 2 min in TBS.
9. To each well, add 200 μL Envision HRP (Rabbit/Mouse) Polymer reagent. Incubate for 30 min at room temperature.
10. Wash the cells in TBS, three times for 2 min each, followed by a rinse in distilled water.
11. Prepare the DAB chromogen solution: from the Envision Kit by adding 20 μL of DAB chromogen to each 1 mL of substrate solution required.
12. Incubate the slides at room temperature for 5–7 min. During this time, a brown color develops if the cells are positive for that marker.
13. Rinse in distilled water.
14. Rinse in tap water, and remove the upper chamber structure from the slides according to the Manufacturer's instructions.
15. Stain the cells lightly in hematoxylin for approximately 10 s, and blue the nuclei in Scot's tap water substitute.
16. Wash the slides well in running tap water.

17. Mount the slides in Faramount aqueous mounting medium with a glass cover slip.
18. Assess slides for positivity (brown) by microscopy, and score on a scale of negative (−) to +++. It is also important to assess the percentage of cells positive for each marker to determine the purity of the epithelial or mesangial cell cultures.

4. Notes

1. The methods described are for isolation of adult human cells, and sieve mesh sizes and culture conditions should be modified if rat cells (or other mammalian cells or human fetal cells) be required (4, 33–36).
2. PBS may be made up from tablet form: PBS tablets (Sigma-Aldrich P4417) and ultrapure sterile water is also a convenient method for the preparation of PBS solution. One tablet is added to 200 mL of sterile water and stored at room temperature.
3. All work undertaken with human tissue must adhere to local Ethical Committee guidelines for confidentiality and consent.
4. Appropriate precautions should be used when handling human tissue, e.g., sterile, disposable gloves should be worn at all times; it is recommended that those engaged in cell isolation procedures should be immunized against hepatitis B.
5. If parietal epithelial cells are required, a 200 μm sieve should be used in place of the 150 μm sieve at step 3 (Subheading 3.1) as Bowman's capsule is retained but tubular contamination is minimized.
6. Two techniques can be used to initiate glomerular cell culture. In addition to the methods which are described in this chapter, glomeruli can be dissociated by incubation with collagenase type I (Sigma-Aldrich C-0130) at a concentration of 1 mg/mL for 20 min at 37°C. After agitation with a Pasteur pipette, glomeruli remnants can be separated from single cells by passing them through a 63 μm sieve. Glomerular fragments and single cells are plated separately for the culture of mesangial and epithelial cells, respectively. Although this improves the plating efficiency of glomeruli, great care must be taken not to "overdigest" the glomeruli, since this may damage and impair viability of epithelial cells.
7. The age of the patient and functional capacity of the tissue will determine how quickly (a) the cells establish themselves in culture (cells from a young, healthy kidney will grow more rapidly) and (b) the number of passages the cells can undergo before reaching senescence.

8. The composition of the extracellular matrix may exert major effects on the phenotypic properties of cells. Attention must be given to the modulatory influences of the matrix on each cell type. Although fibronectin and collagen greatly improve the initial adherence of mesangial and epithelial cells, they can, if necessary, be cultured in the absence of such matrices. Commercially coated type I collagen tissue culture plates are now available and can be used for culturing epithelial cells at the early stages.
9. Epithelial cells are subject to contact inhibition and should be passaged as soon as they reach confluence to reduce cell death. Ideally, cells should be used at, or before, passage 3, since their proliferative activity decreases suddenly around this time. Epithelial cells can sometimes adopt a spindle-like structure after passage; this is usually owing to de-differentiation, and their use for experimentation should be considered carefully. This is distinguished, however, from fully differentiated arborized podocytes (8, 27, 29).
10. One easy and reliable way to check the cultures for fibroblast contamination is by growing cells in medium containing D-valine substituted for L-valine, a condition in which fibroblasts cannot grow. Fibroblasts do not contain the enzyme (D-amino acid oxidase) necessary to convert the D-amino acid to its essential L-form (30).
11. Phenotypic changes may occur in cultured mesangial cells after about ten passages with the loss of angiotensin II receptors. The morphology of the cells may change from stellate-shaped cells to large, flat cells with the development of stress fibers. In addition, cells at high passage number will no longer contract isotonically to vasoactive hormones (37). Changes in matrix due to increased synthesis and decreased degradation (due to reduced matrix degrading enzymes) are also evident after this time.
12. If mesangial or epithelial cell matrix proteins are required, the cells should firstly be dislodged in 1% EDTA/PBS (w/v). The matrix should then be removed in a volume of detergent (e.g., 0.5% SDS) with vigorous scraping using the barrel of a 1-mL syringe.
13. The Envision method in our hands has high sensitivity and low background. However, any standard immunocytochemical method can be used, e.g., Vector Avidin Biotin Complex (ABC) method (Vector, Peterborough, UK) (32).
14. It is recommended that one of the wells should contain negative control antibody such as Universal Negative control Mouse (DAKO code N1698) or a source of nonimmune mouse serum diluted 1 in 20 v/v in antibody diluent. This is to ensure that there is no excessive background staining.

References

1. Sterzel RB, and Lovett DH (1988) Interactions of inflammatory and glomerular cells in the response to glomerular injury, in Contemporary Issues in Nephrology, vol. 18 (Wilson CB, Brenner BM, and Stein JH, eds), Churchill Livingstone, New York, pp. 137.
2. Striker LJ, Doi T, Elliot S and Striker GE (1989) The contribution of glomerular mesangial cells to progressive glomerulosclerosis. *Semin Nephrol* 9, 318–328.
3. Scheinman, J I, Fish AJ, Brown DM, and Michael A J (1976) Human glomerular smooth muscle (mesangial) cells in culture. *Lab Invest* **34**, 150–158.
4. Holdsworth SR, Glasgow EF, Atkins RC, and Thomson NM (1978) Cell characteristics of cultured glomeruli from different animal species. *Nephron* **22**, 454–459.
5. Kreisberg JI and Karnovsky M (1983) Glomerular cells in culture. *Kidney Int* **23**, 439–447.
6. Saleem MA, O'Hare MJ, Reiser J, Coward RJ, Inward CD, Farren T, Xing CY, Ni L, Mathieson PW, Mundel P (2002) A conditionally immortalized human podocyte cell line demonstrating nephrin and podocin expression. *J Am Soc Nephrol* **13**, 630–8.
7. Menè P, Stoppacciaro A (2009) Isolation and propagation of glomerular mesangial cells. *Methods Mol Biol* **466**, 3–17
8. Krtil J, Pláteník J, Kazderová M, Tesar V, Zima T (2007). Culture methods of glomerular podocytes. *Kidney Blood Press Res* **30**, 162–74.
9. Andrews, PM, and Coffey AK (1980) In vitro studies of kidney glomerular epithelial cells. *Scanning Electron Micros* **2**, 179–184.
10. Burlington H and Cronkite EP (1973) Characteristics of cell cultures derived from renal glomeruli. *Proc Soc Ex. Bio. Med* **142**, 143–149.
11. Krakower CA and Greenspon S A (1954) Factors leading to variation in concentration of "nephrotoxic" antigens of glomerular basement membranes. *Arch Pathol* **58**, 401–432.
12. Striker GE and Striker LJ (1985) Biology of disease. Glomerular cells in culture. Lab Inves*t* **53**, 122–131.
13. Norgaard JO (1987) Rat glomerular epithelial cells in culture: parietal or visceral epithelial origin? *Lab Invest* **57**, 277–282.
14. Stewart KN, Roy-Chaudhury P, Lumsden L, Jones MC, Brown PAJ, MacLeod AM, Haites NE, Simpson JG and Power DA (1991) Monoclonal antibodies to cultured human glomerular mesangial cells. Reactivity with normal kidney. *J Pathol* **163**, 265–272.
15. Soukupova M and Holeckova E (1964) The latent period of explanted organs of newborn, adult, and senile rats. Exp Cell Re*s* **33**, 361–367.
16. Delarue F, Virone A, Hagege J, Lacave R, Peraldi MN, Adida C, Rondeau E, Feunteun J, Sraer JD. (1991) Stable cell line of T-SV40 immortalized human glomerular visceral epithelial cells. *Kidney Int* **40**, 906–12
17. Sraer JD, Delarue F, Hagege J, Feunteun J, Pinet F, Nguyen G, Rondeau E (1996) Stable cell lines of T-SV40 immortalized human glomerular mesangial cells. *Kidney Int* **49**, 267–70.
18. Satchell SC, Tasman CH, Singh A, Ni L, Geelen J, von Ruhland CJ, O'Hare MJ, Saleem MA, van den Heuvel LP, Mathieson PW (2006) Conditionally immortalized human glomerular endothelial cells expressing fenestrations in response to VEGF.*Kidney Int* **69**, 1633–40.
19. Fish A J, Michael A F, Vernier, R L, and Brown D. (1975) Human glomerular cells in tissue culture. *Lab Invest* **33**, 330–341.
20. Abbott F, Jones S, Lockwood CM, and Rees A J (1989) Autoantibodies to glomerular antigens in patients with Wegner's granulomatosis. *Nephrol Dial Transplant* **4**, 1–8.
21. Rees AJ (1989) Proliferation of glomerular cells, in *New Clinical Applications in* Nephrology: Glomerulonephritis (Catto GRD, ed.), Kluwer, London, pp. 163–193.
22. Müller GA and Müller C (1983) Characterisation of renal antigens on distinct parts of the human nephron by monoclonal antibodies. *Klin Wochensch* **61**, 893–902.
23. Stewart KN, Hillis G, Roy-Chaudhury P, Brown PAJ, Simpson JG and MacLeod AM (1995) Integrin distribution in normal kidney and cultured human glomerular cells. *Exp Nephrol* **3**, 140–141.
24. Gree DF, Resnick L and Bourgoignie JJ (1992) HIV infects glomerular endothelial and mesangial but not epithelial cells in vitro. *Kidney Int* **41**, 956–960.
25. Holthofer H, Miettinen A, Lehto VP, Lehtonen E and Virtanen I (1984) Expression of vimentin types of intermediate filament proteins in developing and adult human kidneys. *Lab Invest* **50**, 552–559.
26. Weinstein T, Cameron, Katz A and Silverman M (1986) Rat glomerular epithelial cells in

culture express characteristics of parietal not visceral epithelium *J Am Soc Nephrol* **125**, 493

27. Mundel P, Kriz W (1996) Cell culture of podocytes. *Exp Nephrol* **4**, 263–6.
28. Vogelmann SU, Nelson WJ, Myers BD, Lemley KV (2003) Urinary excretion of viable podocytes in health and renal disease. *Am J Physiol Renal Physiol* **1**, F40–8
29. Mundel P, Reiser J, Kriz W. (1997) Induction of differentiation in cultured rat and human podocytes. *J Am Soc Nephrol* **8**, 697–705.
30. Gilbert SF and Migeon BR (1987) D-valine as a selective agent for normal human and rodent epithelial cells in culture. *Cell* **5**, 11–17.
31. Sterzel RB, Loyett DH, Foelimer HG, Perfetto M, Biemesderfer D and Kashgarian M (1986) Mesangial cell hillocks: nodular foci of exaggerated growth of cells and matrix. *Am J Pathol* **125**, 130–140.
32. Bratthauer GL (1999) The peroxidise anti-peroxidase (PAP) method and other all immunologic detection methods. *In: Methods in Molecular Biology* **115**, 191–201.
33. Ringstead S and Robinson GB. (1994) Cell culture from rat renal glomeruli. *Vichows Arch* **425**, 391–398.
34. Harper, PA, Robinson JM, Hoover RL, Wright, TC, and Karnovsky MJ (1984) Improved methods for culturing rat glomerular cells. *Kidney Int* **26**, 875–880.
35. Mendrick DL and Kelly DM (1993) Temporal expression of VLA-2 and modulation of its ligand specificity by rat glomerular epithelial cells in vitro. *Lab Invest* **69**, 690–702.
36. Yaoita E, Kurihara H, Sakai S, Ohshiro K and Yamamoto T (2001) Phenotypic modulation of parietal epithelial cells of Bowman's capsule in culture. *Cell Tissue Res* **304**, 339–349.
37. Kreisberg JL, Venkatachalam MA and Troyer DA. (1985) Contractile properties of cultured glomerular mesangial cells. *Am J Physiol* **249**, F229–F234.

Chapter 14

Culture of Isolated Human Adipocytes and Isolated Adipose Tissue

Kirstin A. Carswell, Mi-Jeong Lee, and Susan K. Fried

Abstract

Adipose tissue (AT) is no longer considered merely as insulation or padding for human organs. It is an endocrine organ in its own right, which includes composite cells with the ability to differentiate into multiple cell lines.

In fact, there is increasing evidence to support the theory that the causation of obesity and its associated metabolic disorders originate at the cellular or tissue level. Adipocyte dysfunction and chronic inflammatory states are able to modulate triglyceride storage and mobilization directly through cytokine and adipokine release.

Significant variability exists between adipocyte isolation and culture techniques which subsequently can impact experimental results. We aim to explain the importance of controlling these variables, to assist tailoring methodological choice towards the investigational outcomes, and modifications of the techniques used accordingly.

The techniques described in this chapter yield cell and adipose tissue which can be utilised in many different ways, including adipose tissue stem cells for differentiation, DNA analysis, RT-PCR, immunohistochemistry, lipolysis, glucose uptake, and LPL activity.

Key words: Adipocytes, Adipose tissue, Cell culture techniques

1. Introduction

1.1. The Role of Adipose Tissue

The functions of adipose tissue (AT) have not been fully elucidated. Adipocytes store energy as triglycerides, the main energy source for the body, and release it in the form of non-esterified fatty acids, when required (1). AT also releases multiple cytokines (2), some with local paracrine effects, e.g. TNF-α and IL-6 which have pro-inflammatory effects (3–5). In addition, adipocytes can respond to and secrete hormones, e.g. leptin and adiponectin as

Ragai R. Mitry and Robin D. Hughes (eds.), *Human Cell Culture Protocols*, Methods in Molecular Biology, vol. 806,
DOI 10.1007/978-1-61779-367-7_14,

well as acylation-stimulating protein and others, which are of paramount importance in the regulation of energy intake and expenditure as well as systemic inflammation and metabolism (6). This suggests that adipose tissue is the largest, and potentially one of the most important, endocrine organ in the body (7).

Adipose-resident stem cells acquired from collagenase digestion of adipose tissue are multipotent with the ability to differentiate into many different cell lines, including adipocytes, bone, cartilage, endothelium, neural (8, 9), and liver cells (10–12). This, combined with their ease of harvesting, suggests their potential as the most important adult stem cell reservoir (13).

1.2. Adipocytes Versus Adipose Tissue

Whether to use adipose tissue or adipocytes in your experiments is a decision which should not be taken lightly. Other cells, especially macrophages can adhere to isolated fat cells (14). Thus, primary adipocytes (fresh or cultured) are useful for acute metabolic studies; however, caution is advised if using isolated cells for other goals.

Organ culture by its nature contains other cells, including preadipocytes, endothelial cells, and fibroblasts and various immune cells (15). Despite this, many believe organ culture to offer substantial benefits over adipocyte culture. It is the preferred method when assessing the long-term regulation of gene expression and adipocyte function within the AT (16–18) as adipocyte-specific gene expressions in the human AT are maintained (19). However, loss of adipocyte-specific gene expression, e.g. GLUT4, and loss of insulin sensitivity have been reported in isolated rodent adipocyte culture (20). Adipocytes isolated from these fragments remain responsive to acute hormonal effects after culture (21, 22). Therefore, studies of the mechanisms involved in the long-term regulation of genes important for studies of insulin resistance, type 2 diabetes mellitus, and obesity, can be performed (17, 23). Encouragingly, this methodology appears to correlate well with in vivo effects (15).

1.3. Regional Variations in Adipocyte Biology

Variations in regional distribution of body fat has been linked with metabolic and cardiovascular disease (24). Increased visceral adiposity in particular has been correlated with hyperlipidaemia, insulin resistance, and T2DM (25). Other studies also suggest heterogeneity among subcutaneous, abdominal, and gluteal depots (26–28).

Significant variations exist not only in AT distribution, but the molecular characteristics of the adipocytes at these depots. Morphological and functional differences have been reported between visceral and peripheral AT (29), e.g. hormone receptor expression, adipokine secretory profile, and expression pattern (30). In addition, depots vary with respect to adipocyte cell size or different cell types present (28, 31, 32). This may be due to different laboratory preparatory techniques or the comparatively small sample sizes used.

Interestingly, visceral adipose tissue have been shown to express more genes encoding pro-inflammatory cytokines than subcutaneous adipose tissue (32–34) which may or may not translate to the protein level. Depot-related differences in adipose-produced molecules have also been reported, including leptin, adiponectin, IL-6, and angiotensinogen (35, 36), with an increase in IL-6 secretion in visceral cells (37) but no difference in TNF-α secretion (33). Variation also has been reported in receptor expression between subcutaneous adipocytes from different sites. Subcutaneous abdominal adipocytes have larger numbers of stimulatory beta adrenergic receptors on the cell surface and higher stimulatory lipolytic activity when compared to gluteal (38) presumed mainly due to higher inhibitory alpha 2 adrenergic receptors in gluteal versus abdominal adipocytes (26, 39).

In response to this, much research is targeted towards identifying the differing roles of adipocytes within their respective depots and potential differences in gender, age, ethnicity, and metabolic disorders (37, 40–42).

2. Materials

2.1. Adipose Tissue Collection and Handling

1. Sterile scalpel and scissors.
2. Sterile 50-cc falcon tubes (see Note 1).
3. Transport medium: 0.9% Saline, Phosphate Buffered Saline (PBS) solution or Medium 199 (M199) (Gibco) with glutamine at room temperature (RT).

   If being used for cell or organ culture, add penicillin (100 U/l), streptomycin (100 μg/ml), and gentamicin (50 μg/ml) to the transport medium.

2.2. Adipocyte Isolation

1. Sterile 50-cc polypropylene tubes.
2. Sterile scissors, forceps, and perforated spoons.
3. 95%O_2:5%CO_2 gas.
4. Parafilm©.
5. 37°C water bath.
6. 240–300 micron nylon mesh (Small parts or Cole-Palmer).
7. Polypropylene funnel (see Note 1).
8. Polypropylene syringes (10 and 20 ml).
9. Elastic bands.
10. Polypropylene tubing (VWR).
11. Polypropylene Erlenmeyer flask.
12. Krebs-Ringer Bicarbonate buffer (KRB) or M199.

13. Fatty acid-free and endotoxin-free bovine serum albumin (BSA) (see Note 2).
14. D-glucose (Sigma).
15. 200 nM Adenosine or N^6-(1-methyl-2-phenylethyl) adenosine (PIA) (Sigma).
16. Collagenase (Type 1) solution: 1 mg/ml Type 1 Collagenase (Worthington) or Liberase Blendzymes (Roche) in KRB (or M199) with 4% BSA (and 200 nM adenosine or PIA) (see Notes 3 and 4).

2.3. Culture Technique

2.3.1. Adipose Tissue Organ Culture

1. AT samples obtained from human subjects (see Note 5).
2. Laminar flow hood.
3. Tissue culture incubator, 5% CO_2 atmosphere.
4. Sterile 50-cc tubes.
5. Sterile scissors, forceps, and perforated spoons.
6. 100-mm-diameter sterile Petri dishes, 60-mm and 35-mm culture dishes.
7. PBS or 0.9% normal saline.
8. Culture medium: M199 (Gibco), 50 μg/ml gentamicin added (see Note 6).
9. Insulin (recombinant human, e.g. Humulin 100 U/ml).
10. Glucocorticoids (dehydrocortisone or dexamethasone (DEX)).
11. 240–300-μm sterile polypropylene or nylon mesh affixed on top of funnel and autoclaved.
12. Sterile pipettes (10 and 25 ml).

2.3.2. Isolated Mature Adipocyte Culture

1. Laminar flow hood.
2. Tissue culture incubator, 5% CO_2 atmosphere.
3. 240–300-μm sterile polypropylene or nylon mesh affixed on top of funnel and autoclaved.
4. Sterile 50-cc tubes (polypropylene or polystyrene).
5. Culture medium: M199 or Dulbecco's modified Eagle medium (DMEM) with Ham's F12 (Gibco) with gentamicin solution (50 μg/ml) ± foetal bovine serum (FBS) or BSA (see Note 6).
6. Insulin (recombinant human, e.g. Humulin 100 U/ml).
7. Glucocorticoids (dehydrocortisone or DEX).
8. Sterile pipettes (10 and 25 ml).
9. 100-mm sterile Petri dishes.
10. Sterile scissors and forceps.

3. Methods

3.1. Adipose Tissue Collection: Adipose Tissues Are Generally Obtained by Surgical Resection or Subcutaneous Adipose Tissue Aspiration

1. Under general or local anaesthesia, the skin is cleaned with aseptic preparation.
2. Skin incisions are made using a scalpel and the required volume of adipose tissue excised using either scalpel or scissors (see Note 7).
3. Careful consideration must be made with respect to specimen extraction in advance (see Note 8).
4. If the adipose tissue is to be needle aspirated, this can be performed directly using a wide-bore (2.5 mm, 1–3-hole cannula) needle into a sterile 50-cc syringe.
5. AT biopsies (obtained through either technique) are placed into capped, sterile 50-cc tubes containing M199 as transport medium (RT).
6. Specimens are transferred to the laboratory as soon as possible, within 30 min.
7. Any surplus tissue can be snap frozen in the operating theatre or clinic and stored at –80°C for future use.

3.2. Adipose Tissue Processing: Sterile Conditions Are Required

1. Perform the following procedures under a laminar flow hood. Mince the tissue into small pieces, approximately 5–10 mg per pieces (1–2 mm^3), using sterile sharp scissors (see Note 9). Adipose tissue obtained from needle aspiration is already fragmented and does not require further mincing.
2. Pour minced tissue through a nylon mesh, affixed to a funnel and placed on top of a 500-ml bottle to capture the waste.
3. Pour room-temperature saline or PBS over the tissue on the funnel to remove broken cell debris and lipid. Several tubes of minced tissue can be combined on the funnel and washed together. Remove any visible blood clots and connective tissues using sterile forceps.
4. Transfer the tissue into a pre-weighed sterile Petri dish using forceps or perforated spoons and weigh the Petri dish with tissue to get tissue weight. Since the tissue is not quite dry, the actual weight of the tissue is overestimated by about 10–20%.
5. The AT can be used for adipocyte isolation or adipose tissue organ culture.

3.3. Adipocyte Isolation

1. The AT from Subheading 3.2 is placed into 50-cc tubes containing 1 mg/ml Collagenase (Type 1) solution. Generally, 2–3 ml of collagenase solution per 1 g adipose tissue is used.

There may be variations in collagenase and BSA obtained from different companies and different batches (see Notes 2 and 3).

2. Place the tubes into the cell culture incubators loosely capped and equilibrate for 15 min.
3. Tightly cap the tubes and incubate in a 37°C water bath for 30–60 min with shaking at 100 rpm until the mixture has a "soupy" consistency (see Note 10). Gently swirl the tubes every 15 min or so and check the degree of digestion.
4. Transfer the mixture into the laminar flow hood and proceed to next steps.
5. Gently pour the mixture through a nylon mesh filter affixed on top of funnel and placed on top of a 50-cc tube (see Note 11). Wash mesh with culture media. Separate the adipocyte fraction from stromal vascular fractions. This can be achieved by removing the lower fraction below adipocytes after centrifuging at low speed (500 × *g* for 1 min) or simply floating adipocytes to the top of tubes. Wash adipocytes with culture media three times (see Note 12).
6. Resuspend the adipocytes in M199 or DMEM:F12 without or with supplementation with BSA or 5% FBS (see Note 6). The volume of fat cell yield is recorded and then diluted in 1:10 ratio.
7. Adipocyte (see Fig. 1) cell size and number should be calculated using established protocols (see Note 13).

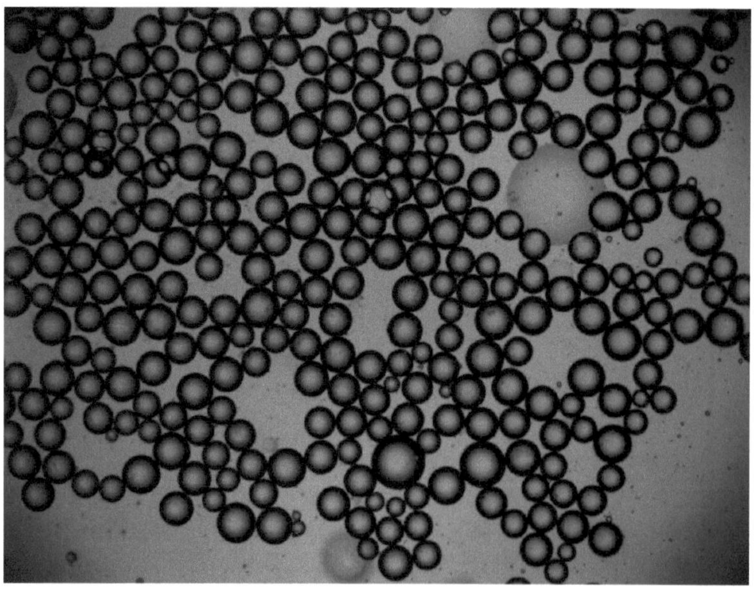

Fig. 1. A representative example of human adipocytes isolated from adipose tissue.

3.4. Culture Technique

This procedure is undertaken in a laminar flow hood.

3.4.1. Adipose Tissue Culture

1. Calculate the potential number of dishes based on the total tissue weight from the procedure 3.2 (~0.5 g/100 mm or smaller if required, keeping the proportion of tissue to media (150–250 mg/60-mm or 50–100 mg/35-mm dishes) at approximately 30 mg/ml[15]) for use in experimentation (see Note 14).
2. Calculate and prepare culture media (see Note 6) with appropriate hormones, chemicals, and other additions added.
3. Distribute the tissue into culture dishes and add pre-warm culture media [15 (100-mm dish), 5–7 ml (60-mm dish), or 2–3 ml (35-mm dish) with appropriate hormones or other additions]. In our experience, the combination of insulin and glucocorticoids maintains adipocyte gene expression similar to freshly obtained tissues.
4. Incubate in cell culture incubator at 37°C under 5%CO_2/95% air atmosphere.
5. Replenish culture media every 2–3 days by aspirating using serological pipettes (see Note 15). This removes adipose tissue metabolites and secretory products (free fatty acids, glycerol and cytokines, etc.) which accumulate during the culture. This also ensures maintenance of hormones that are degraded during long-term culture.
6. Generally, you can maintain AT in organ culture for 7–9 days in vitro (see Notes 16 and 17). It is worth noting that during the early days of cultures (1–2 days), the inflammatory cytokine levels are extremely high while LPL activity is almost null compared to the levels in freshly obtained AT and after days 6–7 of culture their levels are restored.
7. After culture, weigh the amount of tissue in each plate and freeze or use for other experiments.

3.4.2. Adipocyte Culture

This procedure is undertaken in a laminar flow hood.

1. Adipocytes are isolated and resuspended in the culture media using the protocol described in 3.1 with antibiotics added (see Note 6).
2. Resuspend adipocytes ~10% (i.e. 1-ml packed cells into 10-ml culture media) (see Note 6). FBS or BSA can be used for isolated fat cell culture to capture free fatty acids released during the culture (see Note 4).
3. Distribute the adipocyte suspension into culture plates or 50-cc tubes (should be loosely capped during the culture). Mature adipocytes float easily and gentle swirling is required during the dividing process to get equal number of cells per plate or tube.

4. Place the tubes or dishes in a cell culture incubator.
5. Exchange media after first 24 h and then alternate days (see Note 18). Since fat cells float on top of plates or tubes, gently aspirate culture media using 1- or 5-ml pipettes and add new culture media gently avoiding direct shooting into adipocytes or contact with them (see Note 19).

4. Notes

1. Fat cell samples should not be kept in glass vessels as this can result in lysis of mature adipocytes.
2. BSA helps maintain the integrity of adipocytes and prevent breakage.
3. Variability exists between lots in the enzyme activity and performance of the type 1 collagenase used. As such, pre-testing of specific lots should be performed prior to purchasing in bulk for all experiments used in the series.
4. The addition of adenosine or PIA prevents cell breakage and high lipolytic rates.
5. Full research ethics permission must be acquired prior to removal of human tissue specimens. Participants must have given fully informed consent, and appropriate storage facilities must be available.
6. The use of albumin-free conditions or addition of albumin or serum, e.g. 0.5–5% FBS, for culture media is debatable. Supplementation, or not, of serum may impact upon experimental outcome as it contains many growth factors (with inter-lot variability).

   Serum-free media is recommended for AT organ culture since mature adipocytes do not require serum for their maintenance, and BSA or FBS increases cytokine expression while decreasing adipocyte gene expression, such as LPL, during the culture (our unpublished observation). However, in some culture conditions (e.g. addition of free fatty acids), it is necessary to supplement additional BSA or FBS due to deliver to free fatty acids.

   It should also be noted that proliferation of stromal fibroblasts (which can convert to adipocytes) within adipose tissue has been reported in >5% serum-containing media (43).
7. Avoid using diathermy or harmonic scalpel as this can result in damage to the cells being sampled; however, samples aspirated with mini liposuction have been used previously.
8. Should it is necessary to extract the specimen through another adipose tissue depot, e.g. omental AT must pass through subcutaneous AT layer, an Endobag™ or equivalent should be used to avoid contamination at this point.

9. Mincing should be performed using a sharp implement, e.g. scalpel or scissors, to avoid crushing of the cells. This can be performed using a two-handed scissor technique in plastic, conical, 50-cc tubes.
10. The specimen should not be left in the Collagenase solution for longer than 60 min as this reduces adipocytes' viability.
11. This can be performed using end-cut syringes, nylon meshes, and rubber bands. Gently pour the digested AT inside the "end of syringe", place a mesh over the tip, and secure this with an elastic band. While wearing protective clothing, including goggles, the cell solution may then be passed through the mesh into a polypropylene flask using a funnel. Alternatively, the nylon mesh can be affixed on top of funnel and the cell solution poured through this into 50-cc tubes.
12. The adipocytes are washed by pouring media indirectly into the tubes, mixing with polypropylene tubing, and slowly aspirating the supernatant from below the fat cell layer using blunt-end needle connected to a syringe, thereby minimising disruption of the cells. Precursors of adipocytes can be obtained by centrifuging the first aspirate at 500 × *g* for 5 min. Cell pellets can be plated and expanded in α-MEM supplemented with 10% FBS (44, 45). These cells can be differentiated into adipocytes in vitro.
13. Cell sizing can be performed using a Coulter counter or light microscopy. If light microscopy is used, digital images at ×4 magnification can be transferred into Image J software (http://rsb.info.nih.gov.ij/). This software can be used to calculate the average diameter of >200 adipocytes (46). Total triglyceride determination is performed using Dole's lipid extraction and adipocyte cell number is extrapolated as per Di Giralomo (47).
14. For organ culture, it is crucial that the fragments are ~2 mm^3 as necrosis of internal portions of AT has been noted <24 h when pieces exceeded 4–5-mm diameter (48). It is also important to have uniform-sized tissue pieces to minimise your experimental errors. More than 0.5-g tissue/dish (100 mm) decreases maximum LPL activity that can be induced by insulin and DEX and may affect other aspects of adipocyte function (15).
15. When exchanging media, care should be taken to avoid touching the dish. Slowly aspirate the media below the floating adipose tissue fragments using serological pipettes.
16. Time courses may vary depending upon the adipose tissue function being assessed.
17. Basic experimental outcomes using this technique include quantification of the glycerol and adipokine/cytokine content of the media and tissue analysis for LPL activity, mRNA, and protein.
18. This media exchange technique differs from other media exchange techniques as adipocytes have a tendency to float. As

such, the pipette is placed through the media to the bottom of the dish, avoiding touching the base. The media is slowly aspirated and ~0.5–1-ml residual can be left if necessary to facilitate this. Fresh media should be added in a similarly gentle fashion to minimise disruption to the cells.

19. To optimise the insulin response of cells, reduce the glucose concentration in the medium for a few days before experiments.

Acknowledgements

We would like to thank Mr. Ameet G Patel for his generous support throughout this work.

References

1. Arner P (1999) 8. The role of adipose tissue in lipoprotein metabolism. Atherosclerosis 146 Suppl. 1: S11–S12.
2. Trayhurn P, Wood IS (2004) Adipokines: inflammation and the pleiotropic role of white adipose tissue. Br J Nutr 92:347–55.
3. Coppack SW (2001) Pro-inflammatory cytokines and adipose tissue. Proc Nutr Soc 60: 349 -56.
4. Klaus S (2004) Adipose tissue as a regulator of energy balance. Curr Drug Targets 5:241–50.
5. Rajala MW, Scherer PE (2003) Minireview: the adipocyte–at the crossroads of energy homeostasis, inflammation, and atherosclerosis. Endocrinology 144:3765–73.
6. Havel PJ (2004) Section IV: Lipid Modulators of islet function. Update of Adipocyte hormones. Regulation of energy balance and carbohydrate/lipid metabolism. Diabetes 53 (Suppl. 1): S143–S151.
7. Trayhurn P (2007) Adipocyte biology. Obesity reviews 8 (Suppl.1): 41–44.
8. Zuk PA, Zhu M, Mizuno H et al. (2001) Multilineage cells from human adipose tissue: implications for cell-based therapies. Tissue Eng 7: 211–228.
9. Guilak F, Lott KE, Awad HA et al. (2006) Clonal analysis of the differentiation potential of human adipose-derived adult stem cells. J Cell Physiol 206: 229–237.
10. Sgodda M, Aurich H, Kleist S, et al (2007) Hepatocyte differentiation of mesenchymal stem cells from rat peritoneal adipose tissue in vitro and in vivo. Exp Cell Res 313: 2875–2886.
11. Aurich H, Sgodda M, Kaltwasser P, et al (2009) Hepatocyte differentiation of mesenchymal stem cells from human adipose tissue in vitro promotes hepatic integration in vivo. Gut 58: 570–581.
12. Ruiz JC, Ludlow JW, Sherwood S, et al (2010) Differentiated human adipose-derived stem cells exhibit hepatogenic capability in vitro and in vivo. J Cell Phys 225: 429–436.
13. Zuk PA, Zhu M, Ashjian P, et al (2002) Human adipose tissue is a source of multipotent stem cells. Mol Biol Cell 13: 4279–4295.
14. Weisberg SP, McCann D, Desai M, et al (2003) Obesity is associated with macrophage accumulation in adipose tissue. J Clin Invest 112: 1796–1808.
15. Fried SK, Moustaid-Moussa N (2001) Culture of adipose tissue and isolated adipocytes. In: Alihaud G (ed) Methods in Molecular Biology: Adipose Tissue Protocols, Volume 155. Humana Press.
16. Smith U, Jacobsson B (1973) Studies of human adipose tissue in culture. II. Effects of insulin and of medium glucose on lipolysis and cell size. Anat Record 176: 181–183.
17. Smith U (1971) Morphologic studies of human subcutaneous adipose tissue in vitro. Anat Rec 169: 97–104.
18. Robertson JP, Faulkner A, Vernon G (1982) Regulation of glycolysis and fatty acid synthesis from glucose in sheep adipose tissue. Biochem J 206: 577–586.
19. Tan BK, Chen J, Lahnert H, et al (2007) Raised serum, adipocyte, and adipose tissue retinol-binding protein overweight women with polycystic ovary syndrome: effects of gonadal

adrenal steroids. J Clin Endocrinol Metab 92: 2764–2772.

20. Gerrits PM, Olson AL, Pessin JE (1993) Regulation of the GLUT4/muscle-fat glucose transporter mRNA in adipose tissue of insulin-deficient diabetic rats. J Biol Chem 268: 640–644.
21. Appel B, Fried SK (1992) Effects of insulin and dexamethasone of lipoprotein lipase in human adipose tissue. Am J Physiol 262: E695–E699.
22. Wang Y, Fried SH, Petersen RN, et al (1999) Somatotropin regulates adipose tissue metabolism in neonatal swine. J Nutr 129: 139–145.
23. Fried SK, Russel CD, Grauso NL, et al (1993) Lipolprotein lipase regulation by insulin and glucocorticoid in subcutaneous and omental adipose tissues of obese women and men. J Clin Invest 92: 2191–2198.
24. Despres JP, Moorjani S, Lupien PJ, et al (1990) Regional distribution of body fat, plasma lipoproteins, and cardiovascular disease. Arteriosclerosis 10: 497–511.
25. Despres JP, Lemieux I. (2006) Adbominal obesity and metabolic syndrome. Nature 444: 881–887.
26. Wahrenberg H, Lonnqvist F, Arner P (1989) Mechanisms underlying regional differences in lipolysis in human adipose tissue. J Clin Invest 84: 458–467.
27. Arner P (1998) Not all fat is alike.Lancet 351:1301–1302.
28. Wajchenberg BL, Giannella-Neto D, da Silva MER, et al (2002) Depot-specific hormonal characteristics of subcutaneous and visceral adipose tissue and their relation to the metabolic syndrome. Horm Metab Res 34: 616–621.
29. Frayn KN (2000) Visceral fat and insulin resistance – causative or correlative? Br J Nutr 83 (suppl. 1); S71–77.
30. Fruhbeck G (2008) Overview of adipose tissue and its role in obesity and metabolic disorders. In: Yang K (ed) Methods in Molecular Biology: Adipose Tissue Protocols. 2nd edn. Humana Press.
31. Havel PJ (2004) Update on adipocyte hormones: regulation of energy balance and carbohydrate/lipidmetabolism.Diabetes53(Suppl1): S143–S151.
32. Leibel RL, Hirsch J (1987) Site- and sex-related differences in adrenoreceptor status of human adipose tissue. J Clin Endocrinol Metab 64: 1205–1210.
33. Duserre E, Moulin P, Vidal H (2000) Differences in mRNA expression of the proteins secreted by the adipocytes in human subcutaneous and visceral adipose tissue. Biochim Biophys Acta 1500: 88–96.
34. LaFontan M, Berlan M (2003) Do regional differences in adipocyte biology provide new pathophysiological insights? Trends in Pharm Sci 24: 276–283.
35. Albu JB, Curi M, Shur M, et al (1999) Systemic resistance to the antilipolytic effect of insulin in black and white women with visceral obesity Am J Physiol 277: E551–E560.
36. Montague CT, O'Rahilly S (2000) The perils of portliness: causes and consequences of visceral adiposity. Diabetes 49: 883–888.
37. Fried SK, Bunkin DA, Greenberg AS (1998) Omental and subcutaneous adipose tissues of obese subjects release Interleukin-6: depot difference and regulation by glucocorticoid. J Clin Endocrinol Metab 83: 847–850.
38. Engfeldt P, Arner P (1988) Lipolysis in human adipocytes, effects of cell size, age and of regional differences. Horm Metab Res Suppl 19: 26–29.
39. Rosenbaum M, Presta E, Hirsch J et al (1991) Regional differences in adrenoreceptor status of adipose tissue in adults and prepubertal children. J Clin Endocrinol Metab 73: 341–7.
40. Zierath JR, Livingston JN, Thorne A et al (1998) Regional difference in insulin inhibition of non-esterified fatty acid release from human adipocytes: relation to insulin receptor phosphorylation and intracellular signalling through the insulin receptor substrate-1 pathway. Diabetologia 41: 1343–1354.
41. Maeda, K, Okubo K, Shimomura I et al (1997) Analysis of an expression profile of genes in the human adipose tissue. Gene 190: 227–235.
42. Arner P, Hellstrom L, Wahrenberg H et al (1990) Beta-adrenoceptor expression in human fat cells from different regions. J Clin Invest 86: 1595–1600.
43. Cigolini M, Smith U (1979) Human adipose tissue in culture: VII. Studies on the insulin-antagonistic effect of glucocorticoids. Metabolism 28: 502.
44. Zhou X, Li D, Yin J et al (2007) CLA differently regulates adipogenesis in stromal vascular cells from porcine subcutaneous adipose and skeletal muscle. J Lipid Res 48: 1701–1709.
45. Fernandez M, Acuna MJ, Reyes M et al (2001) Proliferation and differentiation of human adipocyte precursor cells: differences between the preperitoneal and subcutaneous compartments. J Cell Biochem 111: 659–664.
46. Sjostrom, Bjorntorp P, Vrana J (1971) Microscopic fat cell size measurements on frozen-cut adipose tissue in comparison with automatic determinations of osmium-fixed fat cells. J Lipid Res 12: 521–530.

47. Di Giralomo M, Mendlinger S, Fertig JW (1971) A simple method to determine fat cell size and number in four mammalian species. Am J Physiol 221: 850–858.
48. Smith U (1972) Studies of human adipose tissue in culture. 1. Incorporation of glucose and release of glycerol. Ant Rec 172: 597–602.

Chapter 15

Primary Culture of Human Adipocyte Precursor Cells: Expansion and Differentiation

Thomas Skurk and Hans Hauner

Abstract

Culture of adipose tissue precursor cells allows gaining insight into the sequential processes involved in adipocyte development. Furthermore, the secretory properties associated with these cellular changes can be studied. Although clonal cell lines are valuable tools for the identification of mechanisms associated with proliferation or differentiation such models do not necessarily represent the complexity of adipose tissue physiology. Primary cell culture systems may be closer to physiology and circumvent some of these restrictions. One advantage is that phenotypic properties of the tissue donor such as gender, age, or body weight are still, at least partially retained in vitro. In primary culture, also differences between various adipose depots can be studied either as a condition per se or in the cellular context of the stromal-vascular (SV) fraction. Furthermore, artificial stressors such as hypoxia and other relevant conditions can be applied to elucidate their functional role. Finally, cultures of human adipose precursor cells may also be used as a screening tool for potential novel drug targets to modulate adipocyte differentiation and biology.

Key words: Tissue culture, Preadipocyte isolation, Collagenase digestion, Stromal vascular fraction, Serum-free culture

1. Introduction

Obesity is a nutritionally related disorder characterized by the expansion of adipose tissue depots. Among others, metabolic disorders like type 2 diabetes mellitus and atherosclerosis are commonly associated with this condition. To study the biology behind adipose tissue expansion the processes of preadipocyte proliferation and differentiation are a subject of research interest. For this purpose, various culture models were established including clonal preadipocyte cell lines from rodent origin such as 3T3-L1 or ob17. These models are now used for more than 30 years, but only recently a human cell strain (SGBS cells) was described (1).

Ragai R. Mitry and Robin D. Hughes (eds.), *Human Cell Culture Protocols*, Methods in Molecular Biology, vol. 806, DOI 10.1007/978-1-61779-367-7_15, © Springer Science+Business Media, LLC 2012

Primary cultures of adipose precursor cells represent another valuable model, as these cells display the individual phenotype of the tissue donor to a certain extent better reflecting the physiological heterogeneity. This technique is based on the isolation of preadipocytes using collagenase as originally described by Rodbell and colleagues (2) which allows a gentle separation of single cells out of a complex matrix composed of proteins, glycoproteins, glycolipids, and mucopolysaccharides. Since that time, the original protocols were repeatedly modified and further developed. As adipocyte metabolism and endocrine function have been shown to differ substantially between humans and rodents (3), only a primary human model facilitates the elucidation of molecular and pathophysiological processes related to human diseases.

Adipose tissue is mainly comprised of adipose precursor cells and mature adipocytes. Additionally, it contains endothelial cells, fibroblasts, and various leukocyte subsets to a variable degree depending on, e.g., BMI and depot. Nevertheless, adipocytes represent up to two thirds of the total cell number and, owing to their enormous cell size make up more than 90% of the tissue volume. Adipocytes are major participants in energy homeostasis of the human body by storing excess energy in the form of triglycerides and by releasing fatty acids to meet the energy demand of the body depending on substrate balance and hormonal regulation (4, 5). In addition, adipocytes exert multiple auto-, para-, and endocrine functions; they secrete numerous signaling factors either constitutively or in dependence of triggering factors such as ER-stress, hypoxia, or inflammatory mediators. Metabolites secreted by adipocytes may be involved in the regulation of energy homeostasis and a variety of neuroendocrine, metabolic, and immune functions (6). Besides the homeostatic hormone leptin, chemokines like the macrophage migration inhibitory factor (MIF), the γ-interferon inducible protein 10 (IP-10), as well as angiotensinogen, plasminogen activator inhibitor-1 (PAI-1), and IL-6 are some of the products secreted by adipocytes (7, 8). During adipose tissue enlargement, most of these “adipokines” are increasingly produced contributing to a state of chronic, low-grade inflammation. This condition has been suggested to build a common soil for pathologies such as insulin resistance/type 2 diabetes (TNF-α), atherosclerosis (IL-6, low adiponectin), hypertension (angiotensinogen), and enhanced thrombogenesis (PAI-1).

As adipose tissue is easily accessible, either during surgery or by needle biopsy it is possible to study specific questions concerning human adipocyte function. Also liposuction material can be used without loss of viability depending on the aspiration technique. In principle, three methods for the study of human adipocyte function are possible in primary culture (1) culturing adipose tissue explants (9); (2) culturing freshly isolated mature adipocytes (after collagenase digestion) (9); and (3) culturing and in vitro differentiation of the stromal cell fraction of adipose tissue. Whereas (1) and (2) exhibit methodological disadvantages – e.g., hypoxia in the explants can alter adipocytes gene expression (10) and freshly isolated adipocytes easily disrupt, in vitro differentiated preadipocytes offer some advantages (1). These cells can be cultured under defined conditions (in serum-free medium) for a longer duration (up to 1 month), (2) effects of external factors, i.e., from the circulation, can be excluded, and (3) the process of adipocyte development can be studied. Disadvantages include the long duration of adipocyte development (14–20 days) and variation in adipose differentiation from donor to donor and from experiment to experiment.

This chapter describes the technique to isolate primary human adipocyte precursor cells and to differentiate them into adipocytes. In addition, it is demonstrated how cell number can be expanded if only little starting material is available (e.g., from biopsies).

2. Materials

2.1. Buffers and Media

1. *Basal medium*: Dulbecco's modified Eagle's medium (DMEM)/Ham's F12 containing phenol red (see Note 1). DMEM/Ham's F12 nutrient mixtures can be purchased ready to use or may be prepared from powder by dissolving DMEM/F12-mixtures (see Note 2). Media are supplemented with final concentrations of 33 μM for biotin and 17 μM for D-pantothenate, respectively. The pH must be adjusted to 7.4 and in case of powder preparations the medium must be sterilized by filtration (see Note 3).
2. *Phosphate-buffered saline (PBS)*: 10 mM KH_2PO_4, 10 mM Na_2HPO_4, 2.7 mM KCl, and 0.137 mM NaCl, pH 7.4. The buffer must be sterilized by filtration and may be kept at 4°C for up to several months.
3. *Collagenase solution*: Collagenase powder (e.g., Worthington CLS type 1) is dissolved in PBS containing 2% bovine serum albumin (BSA, fraction V) at pH 7.4 to obtain a final concentration of 250 U/ml (see Note 4).
4. *Erythrocyte lysis buffer*: 155 mM NH_4Cl, 5.7 mM K_2HPO_4, 0.1 mM EDTA at pH 7.3. When sterilized by filtration it can be kept at 4°C for several months.
5. *Inoculation medium*: *Basal medium* supplemented with 10% fetal bovine serum (FBS, not heat inactivated) and 1% PS (10,000 U penicillin G and 10 mg/ml streptomycin) (see Note 5).
6. *Maintenance culture medium*: *Basal medium* with 66 nM human insulin, 1 nM triiodo-L-thyronine, 10 μM/ml human transferrin (stock solution is 1 mg/ml H_2O), and 1% PS. The stock solutions are sterilized by filtration (except that in EtOH) and kept at −20°C for several months in aliquots.

They are never refrozen after thawing. Once thawed, they can be kept for 1 week at 4°C.

7. *Preadipocyte differentiation medium: Maintenance culture medium* supplemented with 0.5 mM isobutylmethylxanthine (IBMX), 100 nM hydrocortisone, and 2 mM rosiglitazone (from Axxora, Lörrach, Germany; #ALX-350-125-M025). The 20-mM IBMX stock solution is prepared in H_2O and slightly alkalized with Na_2CO_3. This stock can be kept after sterile filtration at 4°C for several months. For hydrocortisone a stock solution of 0.1 mM in 50% EtOH is prepared. The 2-mM stock solutions of troglitazone or rosiglitazone are dissolved in dimethyl sulfoxide (DMSO). The stock solutions of both rosiglitazone (troglitazone) and hydrocortisone are kept at –20°C for up to 3 months. *Preadipocyte differentiation medium* is always freshly prepared.
8. *Proliferation medium: Basal medium* supplemented with 8.7 μM insulin, 100 ng/ml EGF, 1 ng/ml FGF, and 2.5% FBS (referred as PM4) (see Note 6).
9. *Oil-red-O staining solution:* 0.5 g of this dye is dissolved in 100 ml 99% isopropanol giving a 0.5% solution. 6 ml of this stock solution is mixed with 4 ml H_2O, resulting in a 0.3% solution ready to use.

All media and buffers are kept in autoclaved glass bottles. Sterile filtration is done through 0.2 μm pore size filters. Stock solution aliquots of factors added to the medium are kept in regular sterile plastic tubes.

2.2. Additional Materials Required

1. Laminar air-flow hood.
2. Incubator with 5% CO_2 humidified atmosphere.
3. Autoclavable sharp long-handled surgical scissors (15–18 cm), forceps, small scissors (10 cm).
4. Sterile filters (polypropylene) with pore size of 70 and 150 μm, respectively.
5. Centrifuge for 50-ml tubes (200 × *g*).
6. Inverse light microscope.
7. Neubauer chamber and Trypan blue (0.4% solution).
8. Temperature-controlled shaking water bath.

3. Methods

Adipose tissue specimens from the abdominal region [subcutaneous (sc) and intra-abdominal (omental or visceral)] are usually obtained from elective or laparoscopic abdominal surgery (hernia, gall stones, gynecology, gastric banding for morbid obesity, and so on).

Fat reduction surgery (e.g., body lift) is particularly suitable to obtain larger amounts of fat (up to 1 kg and more), whereas tissue from laparoscopic interventions and subcutaneous needle biopsies result only in a limited amount of specimens (1–10 g). Also fat from liposuction may be suitable (11), if low pressure and no vaso-active hormones are used during the procedure. Using liposuction and biopsy material, there may be a high contamination with blood cells. In general, depending on the site of operation, age, or gender of the subject the number of stromal-vascular cells can vary considerably. Yields between 100,000 and 700,000 cells/g can be obtained (unpublished observations). Tissue collection and the procedures must be approved by an ethics committee and informed consent must be obtained from the patient/subject prior to tissue sampling.

3.1. Tissue Collection

1. Fat tissue may be crudely prepared in the surgical room to remove skin and other nonadipose tissue material, such as gland tissue in case of mammary fat. Sometimes it may be necessary to collect samples as fresh as possible (e.g., for RNA extraction, histology). Then, it may be necessary to immediately shock freeze the sample in liquid nitrogen.
2. Consecutively, adipose tissue must be immediately transported to the laboratory in sterile containers containing *basal medium* (see Note 5). Preparation should be started within 1.5 h after removing the tissue from the patient. Although viability of the tissue and cell yield during preparation is best when directly starting the isolation procedure, it is sometimes also possible to store the adipose tissue samples (crudely prepared into small pieces) in basal medium overnight at 4°C. We have observed that the adipose tissue differentiation capacity is similar in preadipocytes isolated immediately or stored overnight at 4°C (unpublished observations).

3.2. Cell Isolation

1. For cell isolation the adipose tissue is placed in a Petri dish with PBS to prevent drying. Remaining visible connective tissue and blood vessels are removed as good as possible as these tissues may be a source for unwanted cellular contamination (e.g., circulating leukocytes, endothelial cells, etc.). Isolated fat lobules are kept in PBS in separate Petri dishes until the next step (Fig. 1, 3.2.1).
2. The adipose tissue pieces are transferred into 50-ml tubes and cut with surgical scissors until a papescent texture is reached. The samples are incubated in collagenase solution (3 ml solution/ml or gram adipose tissue). It is important not to fill the tube completely (45 ml max), as cells require oxygen and collagenase has to surround the tissue pieces. Tubes are then tightly closed and horizontally incubated under water in a 37°C shaking water bath for 90 min (60–70 bpm).

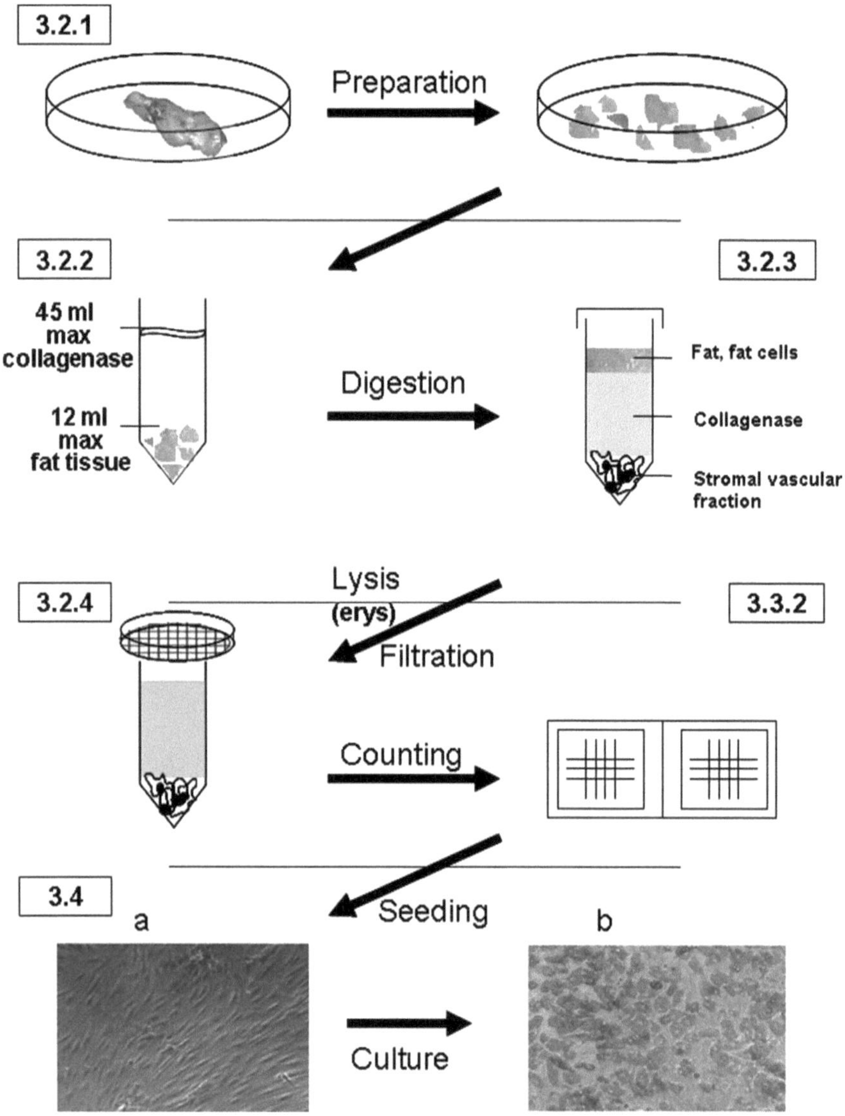

Fig. 1. Schematic diagram showing the process of isolation, culture, and differentiation of human adipocyte precursor cells.

3. Following, the tubes are centrifuged at $200\times g$ for 10 min at room temperature. Thereafter, the supernatant containing free fat from damaged cells, mature adipocytes, and the collagenase solution is removed. As the pellet which contains the preadipocytes (Fig. 1, 3.2.3) should not dry, 1 ml solution should remain on the cells.
4. Pellets from several tubes can be pooled and incubated in erythrocyte lysis buffer (1:9; v:v) (see Note 7) for not more than 10 min. The suspension is filtered through a 150-μm filter and centrifuged again as above (see Note 8). The supernatant is discarded and the pellet is resuspended in an

appropriate volume of *basal medium* (5–20 ml, depending on the expected yield of preadipocytes).

5. The suspension is once again filtered through a cell strainer (70 μm) to remove aggregates of cells and collagen.

3.3. Determination of Cell Number

1. For determination of the cell number 50 μl of the suspension is diluted in Trypan blue (1:2). In case a high cell number is expected, the aliquot can be diluted in *basal medium* first.
2. 10 μl of the solution is transferred to a Neubauer chamber to assess Trypan blue exclusion in the cells from four quadrants to measure the percentage of vital cells, which appear light or colorless. Blue shining cells are considered to be dead. This procedure is repeated and the mean value is used to calculate total stromal cell number. From the mean value counted, the total cell number is calculated.
3. The remaining cells are centrifuged again and resuspended in an appropriate amount of *inoculation medium* (e.g., 150,000 cells/ml for seeding 1 ml in a 4.5-cm^2 well) for subsequent seeding into their respective culturing vessels (see Note 9). From now on cells should be kept in a humidified atmosphere (37°C, 5% CO_2).

3.4. Cell Inoculation and Attachment

1. Cells can be either seeded for immediate experiments or plated for proliferation to achieve a higher cell number. The advisable inoculation density without proliferation is 30,000–50,000 cells/cm^2 to achieve optimal differentiation. A lower inoculation density may lead to a reduced differentiation capacity. Furthermore, it has to be tested which plate format is appropriate. For example, 24-well plates may lead to higher cell loss during the culturing period due to shearing forces according to the medium changes. We mostly use 6-, 12-well plates, or culture flasks (25 and 75 cm^2) to reduce these shearing forces.
2. For optimal cell attachment, it is recommended that the cells are kept for 16–24 h in the *inoculation medium*. Longer exposure of the cells to this medium is associated with an increased mitogenic activity due to the high amount of FBS and may reduce the differentiation capacity considerably (12).

3.5. Cell Propagation

The limiting factor for many applications (e.g., tissue engineering/reconstruction) is the low yield of stromal cells from adipose tissue when starting with small amounts of tissue. This prompted attempts to stimulate stromal cell proliferation without losing differentiation capacity. Preadipocytes can be proliferated in different ways. Use of the fibroblast growth factor (FGF) alone results in only a moderately reduced preadipocyte differentiation capacity (13). Recently, a new medium formulation was described which effectively promotes cell proliferation and differentiation in cultured adipose precursor cells (PM4) (14). This PM4 medium uses a combination of FGF, endothelial growth factor (EGF), insulin, and FBS and does not adversely influence differentiation capacity. For both options, cells are inoculated at a density of approximately 10,000/ cm^2 in the usual *inoculation medium* for 16–24 h. Then, cells are washed twice with PBS. In the case of culture with FGF alone cells are fed with *maintenance medium* with reduced hydrocortisone (10 nM), supplemented with recombinant bFGF (hu-bFGF; 1 nM, stock 1 μM in PBS). In case of PM4, *proliferation medium* supplemented with the growth factors is added to the cells after washing. With both methods media are changed every other day until confluence is reached (5–6 days) and differentiation is subsequently induced as usual. Proliferated preadipocytes are treated like cells directly seeded after isolation (see below).

3.6. Preadipocyte Differentiation

1. After cell adhesion or proliferation to confluence, cells are washed twice with prewarmed PBS to remove nonattached cells, detritus, and traces of serum.
2. Then, cells are incubated in a serum-free, hormone-supplemented medium (*preadipocyte differentiation medium*) for 72 h to induce adipose conversion (see Note 10).
3. Afterward, medium is changed to *maintenance medium* and differentiation is followed until the maximum is reached (usually reached within 16 days) (see Note 11). This culture medium is renewed every 3–4 days (see Note 12). Visible lipid accumulation starts within 6–8 days under these conditions. The average differentiation rate is 50–60%, but up to 90% can be reached depending on the donor characteristics (gender, age, BMI, etc.). Differentiated preadipocytes are filled with lipid droplets (see Note 13) and have developed a characteristic morphology with multilocular lipid droplets (Fig. 1, 3.4b). Differentiated preadipocytes exert the functional characteristics of "mature" adipocytes (e.g., expression of PPARγ and genes of lipid synthesis and mobilization, secretion of leptin and adiponectin, etc.). The percentage of differentiated cells should be determined by counting differentiated and undifferentiated cells in five randomly selected areas (mm^2) under the light microscope using phase contrast. Alternatively, differentiation capacity can be assessed by enzymatic measurements of the lipogenic oxidoreductase glycerol-3-phosphate dehydrogenase (GPDH; 1.1.5.3), which serves as a marker enzyme for terminal differentiation (see Note 14).

3.7. Oil-Red-O-Staining

Differentiation may also be followed by the lipophilic stain Oil-Red-O. Cultured cells are fixed with a 10% formaldehyde solution for 2 h. Thereafter, they are washed with PBS and subsequently incubated with a 0.3% *Oil-Red-O solution* (400 μl/4.5 cm^2 well). After 1 h, the staining solution is removed and cells are incubated with 60% isopropanol (500 μl/4.5 cm^2 well) for 5 min. Finally, cells are washed twice with PBS. Stained cultures can also be used for the extraction of the dye with 100% isopropanol to measure lipid content in a photometer at 518 nm or for observation or photographs using an inverted light microscope.

4. Notes

Various problems may appear during the isolation and culture of cells from the stromal cell fraction. As an example, the differentiation capacity can vary considerably from experiment to experiment. Several technical aspects influencing differentiation were already mentioned in this chapter. However, this variation may also be due to inherent characteristics of the donors. For example, we have shown that mammary preadipocytes differentiate less well with increasing BMI. A strong influence is also exerted by the age of the donor (15). Therefore, some details may help to identify and solve these complications:

1. Phenol red is a dye serving as a pH indicator. Waste products or bacterial contamination results in an acidification and turns the color into yellow. It should be noted that in some assays (e.g., luminometric assay to measure glycerol release from the cells), phenol red interferes with the assay.
2. Cave: under certain circumstances $NaHCO_3$ must be added to the powder preparations as recommended by the manufacturer. Usually, a final concentration of 15 mM is recommended.
3. This medium can be kept for 1 month at 4°C. Media should be stored in aliquots to avoid warming–cooling cycles in order to prevent the degradation of L-glutamine and other sensitive components of the medium. To reduce cytotoxicity due to ammonia formation from L-glutamine a stable preparation (GlutaMAX™, Invitrogen, Carlsbad, CA, USA) may be appropriate for critical applications.
4. Each lot of collagenase has to be tested for efficiency of the digestion, cell yield, and for cytotoxicity. Additionally, also each lot of BSA must be tested for cell cytotoxicity to optimize cell yield. The collagenase solution should be freshly prepared before each preadipocyte isolation and must be sterilized by filtration through a bottle top filter (0.2 μM pore size). Nevertheless, surplus collagenase solution may also be stored at −20°C and used after thawing at 37°C although in this case enzyme activity of the collagenase may be reduced. A too long

exposure of tissue to the collagenase solution may result in a poor attachment rate and reduced spreading of the cells. Different batches of collagenase may differ substantially in activity. When a new batch of collagenase is needed, this batch should be tested in terms of cell yield and cytotoxicity and compared to the old batch. Collagenase of the same batch must be used throughout a whole series of experiments. The collagenase with the best activity should be chosen and larger quantities of the batch should be purchased and adequately stored. To achieve comparable results between collagenase batches, concentration or incubation time in the shaking water bath can be adapted.

5. This medium is intended to improve cell attachment. FBS is kept in sterile aliquots at −20°C and never refrozen after thawing. The inoculation medium can be stored at 4°C for up to 2 weeks. Every lot of FBS has to be tested in terms of cell adherence and cytotoxicity. During tissue collection in the surgery room it is not always possible to work under controlled sterile conditions. Nevertheless, it is mandatory to reduce potential contamination sources, e.g., by the use of sterile surgical instruments. To minimize the risk of contamination a PS (penicillin/streptomycin) solution is widely used in the basal media for transportation or cell culture. Also gentamicin (50 μg/ml) can also be used. This aminoglycoside antibiotic protects against Gram-negative and Gram-positive bacteria as well as against mycoplasma. But, it is noteworthy to mention that gentamicin may exhibit metabolic effects. Additionally, if tissue has to be transported outside a hospital, look for local regulations.
6. Although higher concentrations of FBS were shown to reduce adipose differentiation capacity this mixture of different growth factors was shown to retain this capacity or even enhance it presumably by promoting selective proliferation of adipose precursor cells (14).
7. Erythrocytes may interfere with preadipocyte attachment and should therefore be eliminated using the erythrocyte lysis buffer described above. This buffer is a *mildly* hypotonic solution and makes the erythrocytes burst, because the plasma membrane of these cells is freely permeable to water in contrast to preadipocytes or other cell types. This buffer does not interfere with preadipocytes attachment and differentiation, but should be washed off thoroughly.
8. Adipose tissue obtained from the visceral region is densely vascularized and contamination with endothelial cells in the stromal cell fraction is very likely. The proportion of endothelial cells should be reduced to less than 5% of the total cell number because differentiation can be adversely influenced. Therefore, after the first filtration through a 150-μm sieve

a second filtration using a sieve with a pore size of 30 μm is performed. Endothelial cells re-aggregate rapidly and are therefore retained by this filter (16). However, some contamination by endothelial cells cannot be avoided. If precursor cells from different depots are compared, all samples must undergo the same filtration procedure. (The preceding filtration step through 70 μm can be omitted in this case.)

9. Direct inoculation in *preadipocyte differentiation medium* will cause a low cell adhesion rate (30–40%) due to the lack of FBS. The differentiation capacity of the attached preadipocytes on the other hand may be higher in this case. It is also possible to precoat the culture dishes with, e.g., fibronectin (pre-incubation of the dishes with 0.02 mg/ml fibronectin overnight at 37°C). Thereafter, cells can be inoculated in these wells under serum-free conditions (before seeding, the wells are rinsed with basal medium). Cell adhesion is high, but cannot fully replace the effect of serum from the inoculation medium. However, precoating with fibronectin does not negatively influence the differentiation capacity of the preadipocytes.

10. To maximize preadipocyte differentiation, the nonselective phosphodiesterase inhibitor IBMX and/or the thiazolidinedione rosiglitazone are used in the differentiation medium for the first 3 days (an exposure longer than 3 days does only slightly improve differentiation). One can also use other thiazoldinediones such as troglitazone, pioglitazone, ciglitazone, or other class members. They all stimulate preadipocytes differentiation by activating the transcription factor PPARγ (peroxisome proliferator-activated receptor-γ) that acts as master regulator of adipogenesis by controlling the expression of adipocyte-specific genes.

11. To facilitate lipogenesis from glucose, the *basal medium* is supplemented with biotin and pantothenate and it is continuously present throughout the culture period. Furthermore, 100 nM hydrocortisone is also present because it has, like insulin and triiodo-L-thyronine, positive effects on the metabolic function and lipid accumulation in preadipocytes (17). However, hydrocortisone foremost has an adipogenic activity by controlling the expression of adipocyte-specific genes. Slightly higher rates of preadipocyte differentiation may be reached by adding dexamethasone (1 μM) instead of hydrocortisone to the medium.

12. It is recommended to change the medium very gently and not more often than every 2–4 days. After lipid accumulation, the differentiated preadipocytes may detach more and more easily from the culture dish. Therefore, experiments should not be planned for more than 4 weeks.

13. A multilocular accumulation of lipid droplet is characteristic for in vitro differentiated human adipocytes. Full differentiation

is usually seen after 12–16 days. However, cells can be cultured for another 10–14 days after complete differentiation, but the multiple lipid droplets will grow in size and may finally form one central lipid droplet to give a monovacuolar appearance.

14. Some compounds are dissolved in EtOH or DMSO. A combination of such solvents may lead to (sub)toxic concentrations and affect preadipocyte function considerably. It is important to note that a final concentration of 0.1% DMSO and 0.1% EtOH should not be exceeded. When a compound is dissolved in DMSO or EtOH, control cultures should contain the same concentration of DMSO or EtOH, respectively.

References

1. Wabitsch M., Brenner R.E., Melzner I. et al. (2001) Characterization of a human preadipocyte cell strain with high capacity for adipose differentiation. Int. J. Obes. Relat. Metab. Disord. 25, 8–15.
2. Rodbell M. (1964) Metabolism of isolated fat cells. I. Effects of hormones on glucose metabolism and lipolysis. J. Biol. Chem. 239, 375–380.
3. Ailhaud G. and Hauner H. (1998) Development of white adipose tissue. In: Bray GA, Bouchard C, James WPT, eds. Handbook of Obesity. 1st ed. New York, USA. Marcell Dekker Inc. 359–378.
4. Carey G.B. (1998) Mechanisms regulating adipocyte lipolysis. Adv. Exp. Med. Biol. 441, 157–170.
5. Ramsay T.G. (1996) Fat cells. Endocrinol. Metab. Clin. North Am. 25, 847–870.
6. Trayhurn P., Wood I.S. (2005) Signalling role of adipose tissue: adipokines and inflammation in obesity. Biochem. Soc. Trans. 33, 1078–1081.
7. Skurk T., Herder C., Kräft I., Müller-Scholze S., Hauner H., Kolb H. (2005) Production and release of macrophage migration inhibitory factor from human adipocytes. Endocrinology 146, 1006–1011.
8. Crandall D.L., Growling T.M., Busler D.E., Antrilli T.M. (2000) Release of PAI-1 by human preadipocytes and adipocytes independent of insulin and IGF-1. Biochem. Biophys. Res. Commun. 279, 984–988.
9. Fried S.K., Moustaid-Moussa N. (2001) Culture of adipose tissue and isolated adipocytes. Methods Mol. Biol. 155, 197–212.
10. Trayhurn P., Wang B., Wood I.S. (2008) Hypoxia in adipose tissue: a basis for the dysregulation of tissue function in obesity? Br. J. Nutr. 100, 227–235.
11. Sommer B., Sattler G. (2000) Current concepts of fat graft survival: histology of aspirated adipose tissue and review of the literature. Dermatol. Surg. 26, 1159–1166.
12. Entenmann G., Hauner H. (1996) Relationship between replication and differentiation in cultured human adipocyte precursor cells. Am. J. Physiol. 270, C1011-6.
13. Hauner H., Röhrig K., Petruschke T. (1995) Effects of epidermal growth factor (EGF), platelet-derived growth factor (PDGF) and fibroblast growth factor (FGF) on human adipocyte development and function. Eur. J. Clin. Invest. 25, 90–96.
14. Skurk T., Ecklebe S., Hauner H.(2007) A novel technique to propagate primary human preadipocytes without loss of differentiation capacity. Obesity (Silver Spring) 15, 2925–2931.
15. van Harmelen V., Skurk T., Röhrig K., Lee Y.-M., Halbleib M., Aprath-Husmann I., Hauner H. (2003) Effect of BMI and age on adipose tissue cellularity and differentiation capacity in women. Int. J. Obes. Relat. Metab. Disord. 27, 889–895.
16. Björntorp P., Karlsson M., Pertoft H., Pettersson P., Sjöström L., Smith U. (1978) Isolation and characterization of cells from rat adipose tissue developing into adipocytes. J. Lipid Res. 19, 316–324.
17. Hauner H., Entenmann G., Wabitsch M., Gaillard D., Ailhaud G., Negrel R., Pfeiffer E.F. (1989) Promoting effect of glucocorticoids on the differentiation of human adipocyte precursor cells in a chemically defined medium. J. Clin. Invest. 85, 1663–1670.

Chapter 16

Primary Culture of Ovarian Cells for Research on Cell Interactions in the Hormonal Control of Steroidogenesis

Jerzy F. Galas

Abstract

Ovary is a polymorphic complex structure in which the cells are arranged in two essential endocrine mini glands: the follicle (F) and the corpus luteum (CL). Their secretory function creates an optimal milieu for growth, maturation, and selection of follicles and oocytes competent for ovulation.

Monoculture of isolated ovarian cells has identified the secretory potential of the different cell types functioning in this complex gland in vivo. Primary culture of isolated ovarian cells is a good tool for the investigation of cell interactions and its impact on steroidogenesis, dynamics of steroidogenic enzymes, hormone receptors, changes in the cytoskeleton in granulosa cell populations, regulatory mechanisms, and the intracellular pathways of gonadotropin signaling in steroidogenic ovarian cells. Since the granulosa cell number recovered during isolation from the follicle is substantial, this makes primary culture fairly easy and enables many kinds of studies in vitro. Ovarian cells are highly differentiated and express characteristic functional specificity dependent on the dynamics of the sexual cycle. This is important so as not to produce artifacts in vitro.

Key words: Isolated ovarian cells, Primary culture, Granulosa, Theca, Corpus luteum, Coculture

1. Introduction

Ovarian cells from humans and animals express characteristic specificity, not observed in other cell types. This has to be taken into consideration when planning a culture. Their function changes in time dependent on the stage and length of the sexual cycle. Thus, the tissue culturer has to be familiar with the timing of these events. GC within the follicle are arranged in three different zones: (1) mural, a compact layer of endocrine cells attaching the basal membrane and separating granulosa compartment from that of theca cells (ThC), (2) antral compartment composed of loosely arranged cells surrounding antrum of the follicle, and (3) cumulus

Ragai R. Mitry and Robin D. Hughes (eds.), *Human Cell Culture Protocols*, Methods in Molecular Biology, vol. 806,
DOI 10.1007/978-1-61779-367-7_16, © Springer Science+Business Media, LLC 2012

compartment, most internally situated, containing the oocyte occupying the center of this zone. The cells of these areas differ both morphologically and functionally. GC secrete mainly progesterone. Indeed the cells contain enzyme aromatase (Cyp 19) which, however, is inactive until exposed to aromatizable androgens (androstendione or testosterone) which are substrates for conversion of progesterone to estradiol.

Theca interna constitutes an internal layer of the follicular wall. It is composed of connective tissue stroma, blood capillaries and groups of large steroidogenic cells synthesizing androgens. In a majority of mammals and humans, ThC lack aromatase except in the pig and Rhesus monkey. However, the activity of this enzyme in ThC of these two species is much weaker than that in GC and does not significantly influence levels of estradiol secretion. Theca cells synthesize androgens, which penetrate the basal membrane and stimulate the activity of aromatase. High preovulatory level of follicular estrogens is a physiological trigger of a cascade of processes resulting in ovulation. Growth and maturation of follicles and oocytes as well as simultaneous timing of follicular secretory dynamics during the sexual cycle are controlled centrally by hypothalamic gonadotropin releasing hormone (GnRH) and hypophyseal follicle stimulating hormone (FSH), luteinizing hormone (LH) and prolactin, and locally by paracrine factors and prostaglandins (PG).

Ovarian cells are also useful material for studying follicular cell differentiation in vitro. Androgens apart from involvement in estradiol synthesis by the follicle, take active part in follicle and oocyte differentiation. Androgen receptors (AR) mediate this process. GC of primary follicles and the oocyte express a strong immunopositive reaction for AR (Fig. 1a). In the course of follicle growth and maturation the number of AR declines. Thus, GC of mature antral follicle and the oocyte show very weak or zero immunopositive response. In some species follicles of animals kept in long day conditions mature faster than in those kept in short day conditions. In

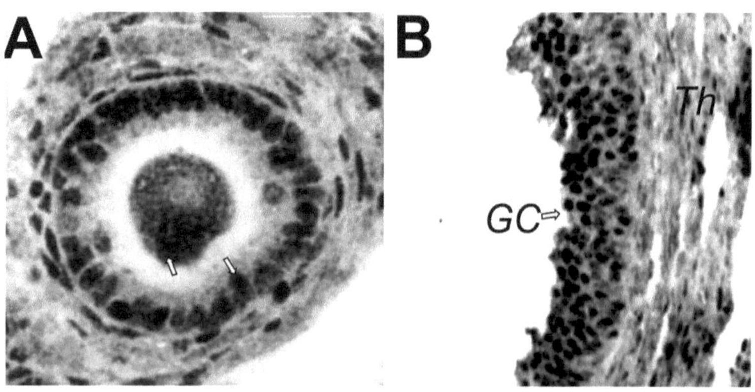

Fig. 1. (**a**) Primary ovarian follicle of short day bank vole showing androgen receptors in GC and the oocyte. (**b**) A wall of short day, large follicle. GC express very strong immunopositive AR reaction (*arrow*) (1).

Fig. 1b the wall of a large antral short day follicle is shown. GC express very strong immunoreaction. This suggests that differentiation of short day follicles is markedly delayed and their differentiation is also photoperiod regulated (1, 2).

Coculture is a cultivation of two or more ovarian cell types first separated and then inoculated into a common dish. GC and ThC in primary coculture "remember" their arrangement in the follicle in vivo. Thus, they express the mutual position they used to occupy in the follicle. In a monolayer GC localize as central colonies with ThC surrounding them concentrically. This is a kind of a flat pseudofollicle, since in this system spatial and vectorial contacts between cells are lacking (Fig. 2g) (4).

CL is a structure which develops from the follicle after ovulation as a result of morphological and biochemical remodeling of the follicular wall. It is composed of two cell types small of theca and large of granulosa origin. According to (5), large CL cells derive first from the GC, but in the course of CL maturation part of them undergoes differentiation into small cells (Fig. 2h). Progesterone is the main steroid secreted during luteal phase of the sexual cycle. Hormonal function of CL shows a specific life span marked by several distinct phases: early (1–3 days), maturing (4–6 days), mature (7–9 days), and regressing. During the life span of porcine (6), bovine (5), ovine (7), and human CL (8), the number of hormonally active luteal cells gradually declines along with the decrease in P4 secretion. In regressing CL, the majority of cells are nonsteroidogenic fibroblastic cells (Fig. 3). During early and mature phases, luteal cells are equipped with receptors for gonadotropic hormones while in the late mature phase these cells gain receptors for prostaglandin $F_{2\alpha}$ which when bound induces luteolysis of CL bringing its endocrine termination (6, 7). This information is essential for tissue culturers of ovarian cells who have to keep in mind the specificity and functional dynamics of CL when collecting cells for culture. Sexual cycles: estrous cycles in the majority of animals and menstrual cycles in human and primates dominate function of three ovarian cells types. Therefore, the tissue culturer of these cells has to know that: (1) ovarian cells function in vitro in the same manner as they do in vivo at the time of cell isolation from the ovary; (2) in primary culture the differentiated state of GC is maintained up to 8 days but only up to 4 days of ThC and up to 3 days of early CL. After that time cells dedifferentiate; (3) cells have to be collected from follicles and CL of the same phase of the cycle; (4) cells have to be harvested from ovaries of the same age; (5) from follicles of the same size (diameter), sorted in separate groups of small, medium, or large, because all three follicle types coexist in the ovary, but their physiological and hormonal statuses are different; (6) cells should be collected in the same phase of the sexual cycle or in the same phase of the life span of CL; (7) from animals kept in the same photoperiod (2, 3) (Fig. 4a, b).

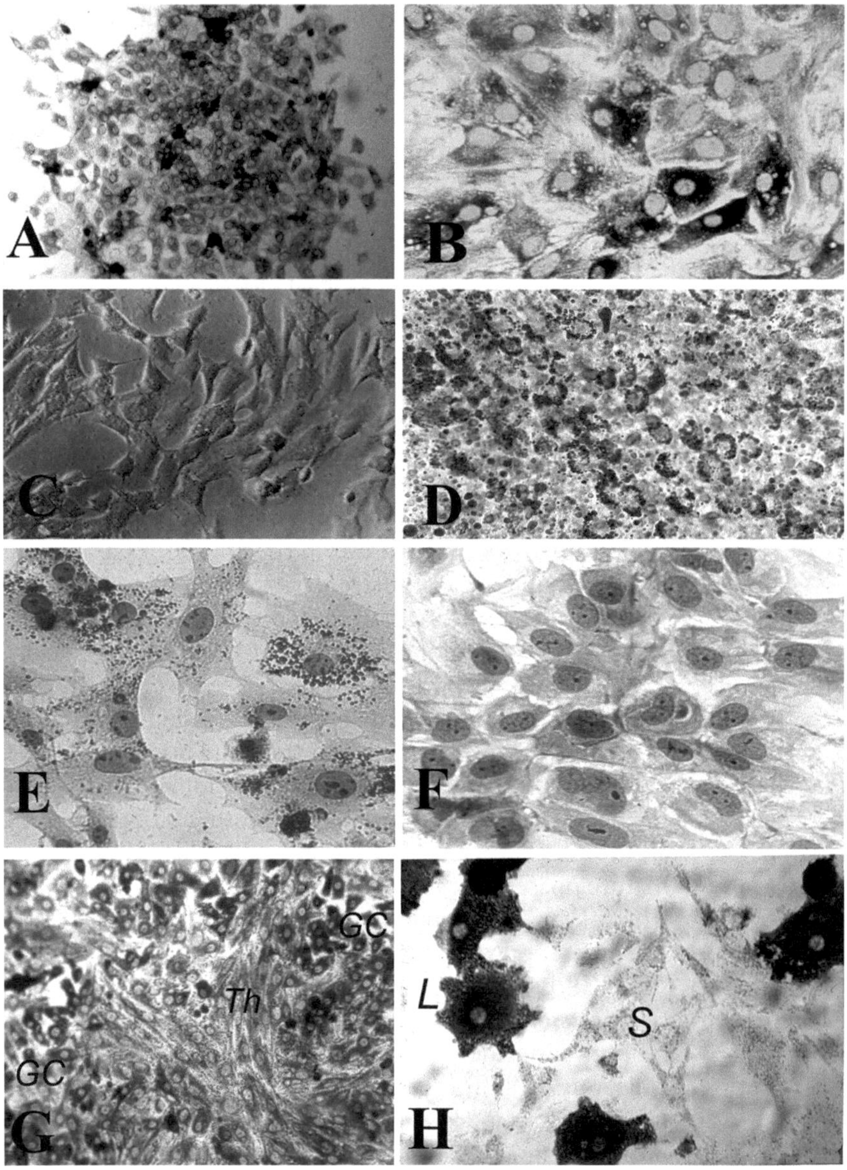

Fig. 2. (**a**) A colony of GC after 2 days in culture (×120); (**b**) a *monolayer* of porcine GC in 4-day culture. Cells stained by histochemical assay for activity of 3β-hydroxysteroid dehydrogenase. The cells show different intensity of the reaction; (**c**) a *monolayer* of live bank vole GC. Nomarski optics; (**d**) suspension of porcine ThC stained with oil red O showing lipid droplets as *black* granules in cell plasma; (**e**) a *monolayer* of porcine ThC after 2-day culture. See fine lipid droplets in cell plasma; (**f**) a *monolayer* of subcultured ThC, passage 52 (×120); (**g**) a *monolayer* of ThC and GC coculture in 4-day culture. In a *monolayer* GC localize as central colonies with ThC surrounding them concentrically. This is a kind of flat pseudofollicle since in this system spatial and vectorial contacts between cells are lacking (×180); (**h**) a *monolayer* of early CL cells. Note small (S) and large (L) cells (×140) [(**a**, **f**, **h**) unpublished; (**b**, **d**, **e**) in *Cell and Tissue Culture* (in Polish: Hodowla komórek i tkanek S. Stoklosowa ed. PWN, Warszawa, 2006) in color insert; (**c**) Galas et al. (3); (**g**) Stoklosowa et al. (4)]

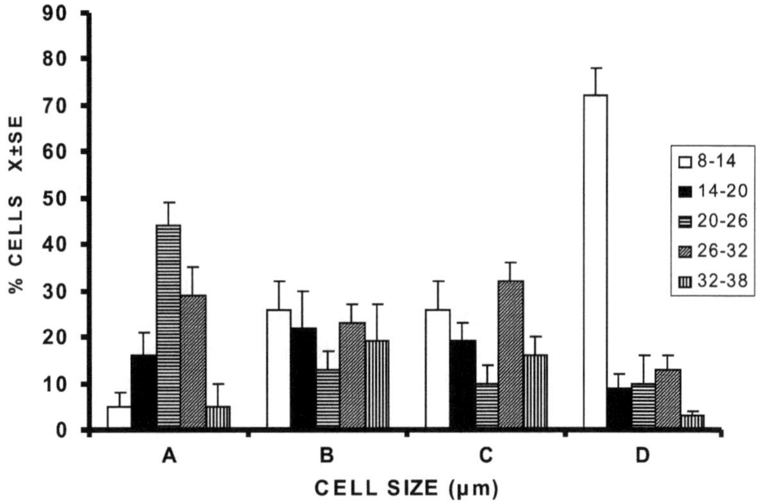

Fig. 3. Histogram of cell composition (μm) in consecutive four phases of life span of porcine CL: (**a**) Early, (**b**) maturing, (**c**) Mature, (**d**) regressing phases. *Empty bars* – not steroidogenic cells; *black* and *striped bars* luteal steroidogenic cells (size from 20 to 38 μm) [the figure in *Cell and Tissue Culture* (in Polish): Hodowla komórek i tkanek, S. Stoklosowa (ed.) PWN, Warszawa, 2006, p. 441].

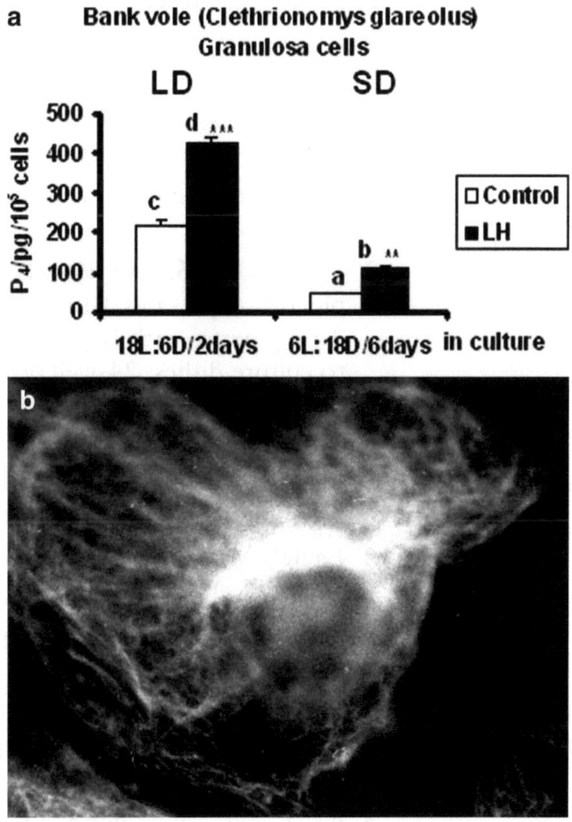

Fig. 4. (**a**) Difference in progesterone secretion by LH stimulated GC in long day (LD) and short day (SD) photoperiod (3). (**b**) Cytoskeletal protein arrangement in short day photoperiod granulosa cell (J. Galas, unpublished).

Cells cannot be collected from animals treated with hormones or pharmaceutical drugs in vivo for breeding, commercial or therapeutic purposes. Ovarian function of treated animals is significantly changed, and this is why they cannot be used as control research material unless the objective of research is drug or hormone action on ovarian cell function.

Ovarian follicles of large animals: cow, mare, and specially pig are a rich source of large number of GC making it possible to plan large experiments in culture. Pig ovary is especially useful for cell harvest because its interstitial compartment is scarce (see Note 1). This chapter describes protocols for human and animal cell culture, as the techniques are readily interchangeable.

2. Materials

2.1. Culture of Porcine GC

1. Porcine ovaries in proestrus stage (see Note 2).
2. Medium 199 without phenol red; supplemented with 10% calf serum.
3. Phosphate-buffered saline (PBS).
4. Calf serum; antibiotics penicillin (120 IU/ml), streptomycin (10 μg/ml), and mycostatin (10 IU/ml).
5. Petri dishes, 60–100 cm diameter; platinum or stainless steel loops (5–7 mm diameter) fixed in a glass or metal handle.
6. Small and medium anatomical and surgical forceps, small scissors.
7. Syringes (1, 5, and 10 ml); beakers (25, 50, and 100 ml); pipettes (2, 5, and 10 ml), preferably glass; Pasteur pipettes; 2-ml pipettes with large orifice for dispensing cell suspension to culture dishes, 24-well plastic plates; round cover slips.
8. Hemocytometer.

2.2. Culture of Rat GC

1. Large, proestrus, ovarian follicles of the rat (range of 0.6–1.5 mm diameter) (see Notes 1 and 3).
2. Eagle's MEM without phenol red.
3. All items as Subheading 2.1, except for antibiotics.
4. Cuvette with disinfecting bath, 70% alcohol and iodine.

2.3. Culture of Human GC

1. Follicle aspirates of women submitted to IVF technique.
2. Media: M199 with Earle's salts (6), Dulbecco Eagle's MEM, Ham's-F12 mixture, without phenol red.
3. Synthetic medium SSR2 (Medi-Cult, 2004) with 0.1%/ml calf serum.

4. Beakers, pipettes, syringes. Other equipment as used in Subheading 2.1 (see Notes 4 and 15).

2.4. Theca Interna Cell Culture

1. Proestrous porcine follicles excised from at least ten large ovaries.
2. Medium M199 and antibiotics as used in Subheading 2 for cultivating GC see Subheading 2.1, trypsin, 0.25% solution in Ca^{++}, Mg^{++} free PBS; hyaluronidase 10 IU/ml PBS; fetuin (1 mg/ml), calf serum.
3. Syringes 2, 5, and 10 ml to measure medium and serum.
4. Surgical knife No 11; small and medium surgical and anatomical forceps.
5. Trypsinization flask 25 ml with Teflon-coated magnetic bar.
6. Two Erlenmeyer flasks, 25 ml; 2-ml pipette with large orifice, glass funnel.
7. Three-layer sterile gauze pads fitting diameter of the funnel.
8. 24-Well plastic plates with round cover slips or Leighton tubes (glass or plastic) with cover slips specially cut to fit growth surface of the tube; Petri dishes 60, 100-mm diameter.

2.5. Coculture of Granulosa and Theca Cells

1. Separate suspensions of GC and ThC in culture medium with 10% calf serum.
2. Culture plates with 2-ml wells with cover slips of adequate diameter.
3. 5-ml Pipettes with large orifice. Other items as in Subheadings 2.1 and 2.4 (see Note 5).

2.6. Primary Culture of Luteal Cells

2.6.1. Mechanical Dispersion of Luteal Cells

1. Excised early or other phases corpora lutea (see Note 6).
2. Medium M199, calf serum, DNAse (250 IU/ml/PBS).
3. Stainless steel sieve; glass rod; injection needles 2, 1, 0.8, and 0.4 mm diameter.

2.6.2. Dispersion by Trypsin

1. Excised corpora lutea.
2. 0.25% Trypsin solution in Ca^{++}, Mg^{++} free PBS.
3. Medium M199, calf serum or 0.5 ml fetuin, 1 mg/ml. Other materials as in Subheading 2.4.

2.7. Culture of Human Corpus Luteum Cells

1. Excised human CL of the cycle (see Note 15).
2. Simple media mainly M199 or Eagle's MEM with Earle's salts buffered with 25 mmol HEPES.
3. Calf serum or at least 0.1–2% BSA.
4. Collagenase II or IV in 1 mg/ml PBS (see Note 7). Other items as in Subheading 2.6.

2.8. Dispersion of Cells with Chelating Agents

1. Mg^{++} and Ca^{++} free salt solution is needed to disperse cells with chelating agents such EDTA, or for trypsinization. The phosphate buffer is used to maintain pH over short periods of time and does not depend upon shifts of CO_2.
2. 10× stock solution; NaCl 80 g, KCl 3 g, Na_2HPO_4 0.73 g, KH_2PO_4 0.20 g, glucose 20.0 g.
3. Working solution; dilute one in ten with sterile distilled water, dispense in suitable amounts.

3. Methods

GC, ThC, and CL cells change functionally from hour to hour from day to day along with the phases of sexual cycle. Thus, selection of tissue material for culture cannot be done arbitrarily.

Phases of the cycle are the same in animals and humans, differences concern menstrual phase (human) and length of the cycle. Coculture of GC and ThC is of special value being a model of cell interaction in the follicle. Mixing GC with ThC in a proportion of 4:1 is important because it reflects the proportion of these two cell types in vivo. Thus, GC–ThC coculture is a simple in vitro model of the follicle. This structure is flat since in monolayer three-dimensional (3D) contacts of cells are lacking. An in vitro model of skin organotypic coculture that could easily be applied to organotypic coculture of GC and ThC was reported (9). Such organotypic model of the follicle would enable spatial and vectorial arrangement of mutual cell contacts similar to that in vivo situation. Several basic primary culture methods comprised in this chapter will help to obtain reliable results enabling to clarify some processes occurring in the complex ovarian structure in vivo.

3.1. Culture of Porcine GC

1. Porcine ovaries are collected at slaughter and placed into a 1-l jar containing cold PBS with a double dose of antibiotics, penicillin (240 IU/ml), PBS, streptomycin (20 μg/ml), mycostatin (20 IU/ml) and are brought as fast as possible to the laboratory.
2. Ovaries are rinsed in three 1-l jars with cold PBS supplemented sequentially with double, one and a half and single dose of penicillin (180 IU/ml), 15 μg/ml streptomycin and (15 IU/ml) mycostatin and a single dose of penicillin (120 IU/ml), 10 μg/ml streptomycin, and 10 IU/ml mycostatin.
3. Ovaries are placed, one by one, on a Petri dish with M199 alone and diameter of follicles is measured.
4. Using small scissors large follicles (1–1.5 cm diameter) are excised as well as follicles of other size dependently on experiment;

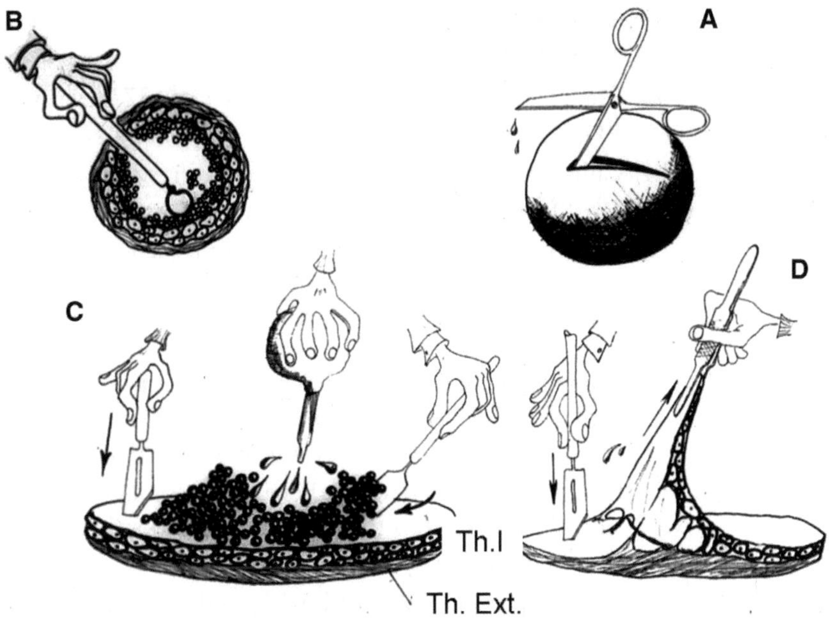

Fig. 5. Isolation of GC from the mature porcine follicle: (**a**) cutting follicle wall; (**b**) collecting GC from the follicle using a stainless steel loop; (**c**) collecting GC that were not removed in (**b**); (**d**) detaching ThC layer from Th externa; drawing by Marek Stoklosa (10).

excised follicles are placed in a 10-ml beaker with fresh culture medium alone and single dose of antibiotics.

5. With small scissors, an incision is made in the follicle wall sufficient for introducing a metal loop inside the follicle.
6. By moving the loop around the internal side of the follicular wall, granulosa cells are gently scraped and transferred into a 20-ml beaker containing serum M199 alone; this action is repeated several times (Fig. 5a, b) (see Note 8).
7. Using a Pasteur pipette cell clumps and aggregates are dispersed by vigorous pipetting to obtain single cell suspension.
8. Cells are rinsed two times with medium alone and followed with 10-min centrifugation ($90 \times g$).
9. The pellet is suspended in 10-ml medium alone and dispersed by energetic pipetting.
10. 0.1-ml sample of the suspension is taken to count cells and establish their viability.
11. Suspension is diluted with culture medium supplemented with (10%) calf serum to a concentration of 5×10^5 or 1×10^6 cells per 1-ml medium. Such inocula result in a successful growth of cells (see Note 9).
12. Round cover slips are placed into 24 wells of a plastic plate and using 2-ml glass pipette with a large orifice 1-ml cell suspension

is inoculated; simultaneously pipetting is continued to prevent cell sedimentation.

13. The plate is transferred to an incubator (37°C with 5% CO_2 in air).
14. Cultures are controlled under the inverted microscope after 6, 12, and 24 h and then every other day.
15. Medium is changed after 24 h and then every second day; medium is collected from each well into separate small flasks and stored at –20°C for further hormone analyses.

3.2. Culture of Rat GC

1. Animals are killed by cervical dislocation and sterilized in a bath of 70% alcohol with 0.5 ml iodine/100 ml alcohol; abdomen is cut along the *linea alba*, ovaries exposed and dissected.
2. Ovaries are placed on a Petri dish containing Eagle MEM alone, cleaned from the attached fat and stripes of connective tissue; and rinsed two times in medium.
3. Bottom part of a Petri dish with medium alone is placed under the stereomicroscope, ovaries are transferred one by one into a Petri dish.
4. The diameter of follicles is measured using a micrometer mounted in the ocular of the microscope.
5. Follicles are segregated and selected for ones of the same size to be used in the experiment.
6. With small sharp scissors, the oblique end of an intradermal injection needle (gauge 26) is cut and mounted on a 1-ml syringe; 0.1 ml of medium alone is taken up in the syringe.
7. The follicular wall is punctured with the needle, the follicle is squeezed gently with anatomical forceps, GC flow out in the form of a semitransparent milky cloud of cells; this operation is repeated several times until cells stop flowing out.
8. Cells are collected by aspiration into a syringe or Pasteur pipette and transferred into a small 10-ml beaker containing medium alone; cells are suspended by energetic pipetting and the procedure continued as in Subheading 2.1 (see Notes 3 and 10).

3.3. Culture of Human GC

1. After oocytes are removed, aspirates are retrieved from follicular fluid, centrifuged (5 min) in cold PBS with 0.1% BSA, rinsed several times (all on ice), each followed by centrifugation (500 × *g*).
2. Supernatants of aspirates are discarded to remove the majority of blood cells.
3. Pellet is suspended in serum free Medium 199 with Earle salts, 25 mM HEPES, 0.1% BSA, penicillin 500 IU/ml of medium, 500 μg/ml streptomycin and 5% glutamine (8, 11).

4. 0.1-ml Aliquot of suspension is taken for measuring cell viability and for cell counting in a hemocytometer.
5. Culture medium is prepared: M199 (phenol red free) supplemented either with 5% calf serum or synthetic serum replacement SSR2 (Medi-Cult) (12).
6. Cells are diluted with culture medium to a concentration of 1×10^4 or 1×10^5 cells per 0.5 ml (8, 13).
7. 0.5-ml Suspension is distributed into 1-ml wells of 24-well plates and transferred to the incubator (in humidified air with 5% CO_2, 37°C).
8. Cultures are examined every day under the inverted microscope (see Notes 11 and 15).

3.4. Primary Culture of Porcine ThCs

1. Large follicles are dissected from proestrous ovaries and processed in the same way as follicles used for GC collection (see Subheading 2.1, item 5).
2. Using stainless steel or platinum loops, GC are removed from follicles.
3. To loosen the mural layer of GC, 0.5 ml or more, – depending on follicle size, hyaluronidase (10 IU/ml PBS for 1–3 s) is injected into the follicle, which then is immediately rinsed several times with fresh PBS.
4. The follicle is cut in two halves and spread in a Petri dish with granulosa layer facing upwards; with a vigorous stream of PBS or medium alone the theca layer is rinsed several times to remove mural GC attached to the theca and checked under a stereoscope whether all GC are detached. If not, then tissue layers are vortexed 2–3 times, 10 min each.
5. Th layer is spread on the bottom of a Petri dish, covered with PBS; one end of Th layer is caught with small surgical forceps and fixed by pressing the tissue with the sharp end of a knife to the bottom of the dish.
6. The end of the opposite side of theca interna layer is caught with small surgical forceps and with an energetic pull detached from the layer of connective tissue theca externa (Fig. 5c, d) (14).
7. Isolated thecas are transferred into a clean Petri dish and minced in small volume of PBS using small sharp scissors.
8. Using a pipette, minced tissue is transferred into trypsinization flask containing Ca^{++}, Mg^{++} free PBS with (0.25%) trypsin and the flask is placed on a magnetic stirrer (37°C) for 30 min.
9. Suspension of released cells is filtered through four layers of gauze pads into a 150-ml Erlenmeyer flask.
10. Suspension obtained after 30 min trypsinization should be discarded, since it contains fragments of connective tissue not affected by the enzyme and few cells are released.

11. Tissue is transferred back from the funnel into the trypsinization flask, fresh trypsin solution added, vortexed for 15 min; released cells are filtered through gauze pads into a sterile 100-ml Erlenmeyer flask; several drops of serum or 0.5 ml fetuin 1 mg/ml are added and the flask is placed on ice.
12. Trypsinization is repeated further after 15 and 10 min.
13. After each trypsinization cell suspensions are pooled into the Erlenmeyer flask on ice, several drops of serum are added to inhibit trypsin activity, trypsinization is repeated for the next 10 min twice; all the time trypsinized tissue is examined under a stereoscope to check how many cells are still entangled in the digested tissue.
14. Pooled cell suspensions are centrifuged, 10 min at $90 \times g$.
15. The pellet is suspended in 10- or 5-ml medium alone (depends on the efficiency of trypsinization) (see Note 12).
16. Cell viability is checked, cells are counted, final cell suspension is diluted to a concentration of 2.5×10^5 per 1-ml medium with 10% serum and 1-ml suspension dispensed into each well of a plastic plate.
17. Aliquots of cell suspension ready to be cultured is taken for staining with oil red O, to detect lipid droplets within cells. Put one drop of cell suspension on a slide and check under the microscope. Usually suspension is very rich in ThC heavily stained with oil red O (Fig. 2d).
18. ThC are attached to the substrate after 12 h and in two days colonies of epithelial-like cells are formed (Fig. 2e) (see Note 13).

3.5. Coculture of Granulosa and Theca Cells

1. 1 ml of GC suspension (1×10^6 GC/ml of culture medium) and 1 ml of ThCs (2.5×10^5/ml) in culture medium are dispensed into each 2-ml well.
2. Cocultures are transferred to a humidified incubator with controlled CO_2 (5% in air, 37°C).
3. Medium is changed after 24 h and then every second day up to day 6 in culture; cocultures usually survive longer than cultures of ThC alone.
4. Collected media are frozen at −20°C for further analyses (see Note 5).

3.6. Primary Culture of Porcine CL Cells

3.6.1. Dispersion of Luteal Cells Without Enzyme Treatment

1. Using sharp scissors and surgical forceps, the connective tissue capsula of CL in a phase of the life span of your choice is removed and discarded (see Note 14).
2. CL tissue is minced into tiny fragments.
3. Using the glass rod, the fragments are rubbed through a sieve, rinsed vigorously on the sieve with PBS letting cells flow into a beaker placed under the sieve.

4. Cell suspension is taken up into a 5-ml syringe and then forced out through injection needles of sequential 2, 1, 0.8 and 0.4 mm in diameter to disperse cell aggregates and obtain single cell suspension; centrifuge suspension at 90 × *g* for 10 min.
5. Pellet is treated with 3 ml of DNAse solution (30 IU/ml PBS for 6 min) to remove damaged cells and cell nuclei.
6. Pellet is rinsed three times, centrifuged, cells are counted and diluted to an inoculum of 3–5 × 10^5 cells/ml of culture medium.

3.6.2. Dispersion of Luteal Cells by Trypsin

Maturing and mature CL are dispersed using trypsin solution. Collagenase can be used for dispersing CL of regressing phase, which has rich and compact connective tissue component.

1. CL is decapsulated, luteal tissue minced into tiny fragments, rinsed three times with PBS supplemented with antibiotics (penicillin, 100 IU/ml, streptomycin 0.1 mg/ml, and amphotericin B, 0.25 μg/ml).
2. Using a pipette with a large orifice, tissue is transferred into a 150-ml trypsinization flask.
3. Trypsin solution is added in a proportion of 1:3 (v/v) tissue: trypsin.
4. Tissue is trypsinized 1× 30 min and then 3× 15 min on a magnetic stirrer at 37°C.
5. After each trypsinization, tissue is filtered through four layers of gauze pads in the funnel; released cells are collected into 100-ml Erlenmeyer flask; several drops of serum added and the flask is placed on ice (see Subheading 2.4).
6. Cell suspension is rinsed with medium alone and centrifuged two times.
7. After the last trypsinization, 3 ml of DNAse (300 IU/ml medium alone) is added to pooled suspensions, shaken for 7 min, rinsed three times with PBS and centrifuged each time.
8. Cell viability is determined, cells are counted and diluted to a concentration of 3–5 × 10^5 cells per 1 ml of culture medium.
9. 1 ml of suspension is distributed into each well; every day, cultures should be evaluated under an inverted microscope.
10. Medium is changed every other day and frozen at −20°C until steroid analyses.
11. CL cells attach to the substrate after 24 h and in 2 days a *monolayer* of large and small cells is formed (Fig. 2h) (see Note 14).

3.7. Culture of Human Corpus Luteum Cells

1. Dispersion of human CL is similar to porcine and bovine corpora lutea (13), see Subheading 2.6 by cutting tissue into fragments and digestion in collagenase II or IV solution (1 mg/ml PBS, in calcium, magnesium free PBS) on a magnetic stirrer at 37°C, 1× 30 min, 2× 15 min each.

2. Released cell suspensions are pooled in one Erlenmeyer flask, kept on ice to diminish enzyme activity and the whole pool is then centrifuged and rinsed several times until all collagenase is removed.
3. After each enzyme treatment, pellet is rinsed 3–4 times and the last pellet is suspended in fresh PBS.
4. Tissue is filtered through four layers of gauze pads into a 100-ml Erlenmeyer flask placed under the funnel and then put on ice.
5. Cell viability is determined, cells are counted and suspension diluted to a concentration of 3–5 × 10^5 cells per 1 ml of culture medium.
6. 1 ml of cell suspension is distributed into each well; cultures are examined every day under an inverted microscope.
7. Medium is changed every other day and is frozen at −20°C until steroid analyses.
8. CL cells are attached to the substrate after 24 h and in 2 days a *monolayer* of large and small cells is formed (see Note 15).

3.8. Several Simple Methods to Enable Steroidogenic Cell Identification

Several simple, routine methods are included to enable steroidogenic cell identification as well as distribution of androgen receptors in follicular cells and tissues. Androgens play an important role in follicular differentiation and oocyte maturation in the earlier stages of folliculogenesis.

3.8.1. Detection of Lipid Droplets and Their Cellular Distribution in Follicular and Luteal Cells

1. Prepare a saturated solution of oil red O: 0.5–1.0 g of oil red O dissolved in 100 ml of 98% isopropyl alcohol (stock solution). For working solution, mix 6 ml of stock solution with 4 ml distilled water to obtain 60% isopropyl alcohol; filter working solution though one layer of nylon gauze prior to each staining procedure.
2. Cover slip with monolayer is submerged in 60% isopropyl alcohol for 15–20 s.
3. Culture is placed in working solution of oil red O for 15–20 s.
4. Rinsed several times.
5. Counterstained with Mayer hematoxylin.
6. Rinsed in tap water and then in distilled water.
7. Cover slip with monolayer is mounted in glycerine gel.

3.8.2. Histochemical Assay for Detection of Activity of 3-Beta-Hydroxysteroid Dehydrogenase

1. Prepare: Phosphate buffer, pH 7.0 and medium for detection of enzyme activity.
2. Medium: Dissolve in 1 ml DMF (dimethylformamide), 5 mg of DHA (dehydroepiandrosterone), 13 mg of NAD, 4 mg of

NBT (nitroblue tetrazolium salt), and 10 mg of PP (amide of nicotinic acid).

3. To fix cells, monolayer is immersed in a small flask with 10% buffered formalin for 30 s.
4. Culture is transferred to phosphate buffer for 5 min.
5. Culture is placed in the medium for 2–3 h in the incubator.
6. Cultures are transferred to 10% buffered formalin plus 50% ethyl alcohol for 10 min; mounted in glycerol gel. DHA is the donor of hydrogen ions, NAD transfers ions on tetrazolium salts reducing them to dark blue deposits (formazan) in cell cytoplasm. The number of formazan fine granules is a measure of intensity of the histochemical reaction and proves the steroidogenic nature of cells. Semiquantitative analysis can be done cytophotometrically.

3.8.3. Immunocytochemical Assay for Detection of Androgen Receptor (AR): Part A

1. Collect medium from culture wells and freeze it.
2. Rinse cultured GC 3× with TBS at −20°C.
3. Using ice-cold methyl alcohol fix cells in −20°C for 10 min and then in 100% ice-cold acetone for 6 min at −20°C.
4. Rinse cultures 3× in TBS.
5. Incubate cells in 3% H_2O_2 for 20 min to block activity of endogenous peroxidise.
6. Rinse cells 3× in TBST.
7. Block unspecific reactions by cell incubation in 20% NGS (normal goat serum) in Tris-buffered saline (TBS) pH 7.4 with Tween-20 (0.5 ml) per 500 ml of TBS to prevent the nonspecific binding of the secondary antibody for 40 min in a humid chamber.
8. Incubate cells with polyclonal rabbit anti-human AR antibody at a dilution 1:100 in TBST.
9. Transfer cultures to a humid chamber for 24 h at 4°C.

3.8.4. Immunocytochemical Assay for Detection of Androgen Receptor (AR): Part B

1. Warm cells up to room temperature and rinse 3× in TBST.
2. Incubate cells with the biotinylated secondary goat anti-rabbit IgG at a 1:300 dilution in TBST for 90 min in room temperature.
3. Rinse cells in TBST.
4. Incubate cells with the standard avidin-biotinylated horseradish peroxidase complex at 1:100 dilution for further 40 min in dark.
5. Rinse in TBST and in TBS.
6. To visualize antibody bound to cells, 3,3′-diaminobenzidine tetrachloride (TBS buffer pH 7.4 containing 0.001% H_2O_2;

0.05% diaminobenzidine and 0.07% imidazole) as a substrate was added for 4 min; prepare DAB fresh prior to staining procedure.

7. In control cultures, the primary antibody is omitted or substituted by irrelevant IgG.
8. Reaction is stopped by rinsing cells in distilled water.
9. Cover slips with cells are mounted in DPX.
10. The same procedure can be used for immunodetection of 3β-HSD (hydroxysteroid dehydrogenase) incubating cultures with a polyclonal antibody anti-recombinant mouse 3β-HSD (a gift from Professor Anita H. Payne from Stanford University Medical Center) at a 1:1,000 dilution overnight at 4°C instead of antibody used in step 2.

3.8.5. Viable Cell Count Using Trypan Blue Exclusion Test

1. Use cell suspension with cell concentration 10^6 cells/ml medium.
2. Mix one drop of cell suspension with one drop of Trypan blue (0.4% in PBS).
3. Transfer one drop of the mixture into the chamber of hemocytometer for 1–2 min sharp. Do not prolong this time because viable cells can be damaged and can stain.
4. Count total number of cells and separately number of cells stained.
5. Count percent of unstained viable cells.

4. Notes

1. Normal untreated pig has ovaries, which generate many follicles of various sizes, ranging from 3–5 mm diameter (small follicles), 6–8 mm diameter (medium follicles), and up to 1–1.5 cm (large follicles). The most reliable material for culture would be ovaries of animals with controlled estrous cycle. However, such animals are rarely accessible. Sometimes, it is possible to get ovaries from animals with hormonally controlled estrous cycles at experimental farms and some research laboratories of veterinary schools.

   When working with slaughter-house animals one has to diagnose the phase of the cycle according to the criteria published (10). To establish phases of the estrous cycle in small laboratory animals (rat), which have short cycles (3–5 days), technique of taking daily vaginal smears is necessary.
2. Antibiotics are particularly important when working with the slaughter-house material.

3. GC attach to the substrate during 12–24 h. First, they form colonies of healthy epithelial-like cells (Fig. 2a) and after 2 days on cover slips form a regular *monolayer* (Fig. 2b, c). After 6–8 days, culture becomes confluent. At this point, the culture is suitable for subculture. If not passaged, the culture should not be continued because confluent culture tends to grow in several layers (multilayer). The lowest layer becomes hypooxygenated, cells die and the multilayer ends up in detaching from substrate. Excision, isolation, and collection of rat GC is carried out under the stereomicroscope in strict sterile conditions without antibiotics.
4. GC of large animals are good material for learning techniques of separation and culture of ovarian cells. Instead, human GC are obtained from follicles from women submitted to IVF. In this situation, the number of cells recovered is lower and part of them is apoptotic. Moreover, GC after gonadotropin stimulation are luteinized so, their hormonal status is different from that of untreated follicles. Rarely, GC of women with normal cycles can be collected when women are submitted to surgical treatment of hemorrhage or tubal ligation (Fig. 6a, b). In addition, puncturing follicles in vivo results in significant intrafollicular bleeding and contamination of aspirates with erythrocytes and other types of blood cells. Cultures can start only after patients give written consent and the protocol of the study is approved by the local ethics committee (8, 15).
5. Coculture of GC and ThC cells showed that for the successful course of the sexual cycle the interaction of these two cell types is essential. Thanks to this system and to organotypic cultures, spatial models providing 3D contacts between two or more cell types offer a research tool whose function mimics the function of an organ in vivo; Coculture of granulosa and theca cells is the simplest model of the ovarian follicle in vitro, which helped to clarify mechanisms of the appearance of preovulatory LH surge and other processes in the ovary in vivo (16). When starting coculture, it is preferable to prepare in one experiment separate suspensions of both follicular cell types from the same set of follicles. Help of a graduate student or a technician familiar to cell culture would be appreciated.

   Important: Prior to setting up coculture of GC and ThC, it is necessary to maintain GC and ThC in a cell ratio of 4:1 since in a batch of ten proestrous follicles the number of recovered GC is $1–2 \times 10^6$ GC cells but only $2–5 \times 10^5$ ThC cells per 1-ml medium. Only coculture of cells mixed in this proportion produces steroid hormones in amounts comparable to the secretion of a follicle in vivo. When the cells are mixed, e.g., in a proportion of 1:1, then the results obtained will be an artifact.

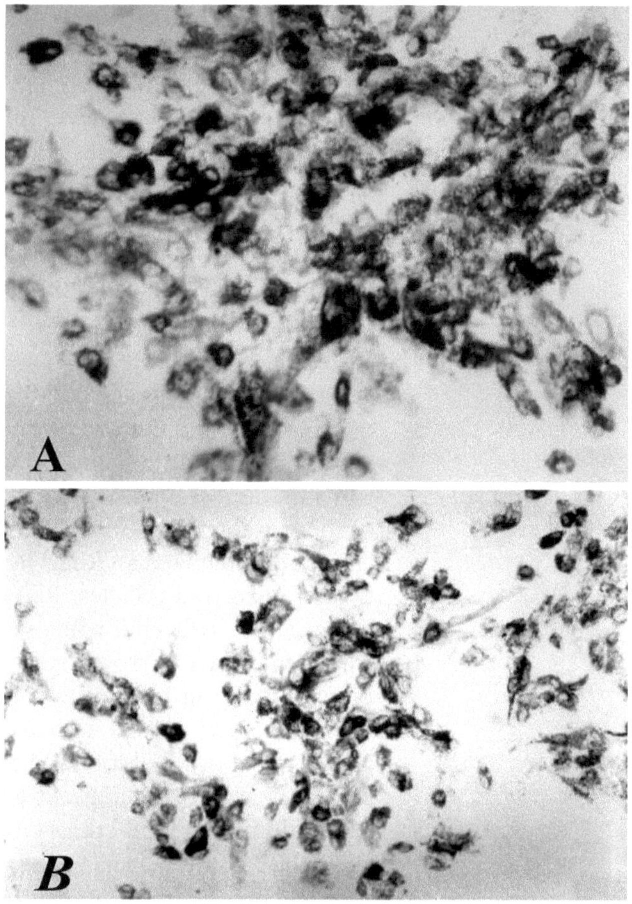

Fig. 6. (**a**) A colony of human GC after 3-day culture (×150). Cells were collected from normally cyclic women. Granulosa cells are viable. Culture was submitted to a cytochemical assay for activity of Δ^5, 3β-hydroxysteroid dehydrogenase. The cells show intensive steroidogenicity. Courtesy of Professor Jana Skrzypczak, Medical University in Poznań, Poland (unpublished). (**b**) 3-day culture of human granulosa cells collected from serous cystic ovary (×120), cytochemically tested for activity of steroid dehydrogenase. Microphotograph accessible thanks to courtesy of Professor Jana Skrzypczak, Medical University in Poznań, Poland (unpublished).

6. In pregnant females CL life span is much longer. This is connected with increased demand for progesterone necessary for maintaining pregnancy in good health. From the middle of the mature phase, the number of large luteal cells gradually declines while a certain amount of small cells remains throughout the regressing phase. At that time increased level of estrogen appears. This may be connected with the involvement of small luteal cells in stimulation of a wave of follicle proliferation for the next ovarian cycle. Strict life span of CL is a limiting factor in luteal cell proliferation and hormonal function. In spite of many efforts it was impossible to extend CL life span as well as to subculture luteal cells of the cycle. CL is a temporary

highly differentiated organ which does not have cellular equipment to remodel into another one, as it has with GC and ThCs (Fig. 2f). Regression phase is termination of CL existence in a particular sexual cycle.

7. Using so-called luteinized GC, either after cell isolation from the follicle, when GC spontaneously "luteinize" or luteinized GC as a result of gonadotropin treatment do not represent a model relevant to physiological cell composition and function of the true CL, which is composed of two luteal cells types of different origin (11, 17). Normal physiological CL is obtained from ovaries of regularly cycling women who were not given hormonal therapy. They were infertile from nonendocrine disorders and were submitted to surgical treatment, e.g., fibroids or tubal ligation.
8. Take care not to disturb the theca layer to avoid contamination of GC suspension with fibroblasts present in theca interna tissue; GC can also be isolated by aspiration using 1- or 2-ml syringe with a small volume of medium, but such material is of worse quality since many loose antral GC are apoptotic.
9. For serum-free culture cover growth substrate with fibronectin 25–50 μg/ml collagen II or preferably IV (10 μg/ml); use medium 199 with 0.1% BSA or with artificial serum (SSR2). Serum can be also omitted when you culture GC on a feeder layer produced by cells of 3T3 line. Better results are obtained when using a feeder layer of mouse embryo cells.
10. Rat GC collected from large (1.5 mm) follicles (Fig. 7) (18) grow and function successfully in monolayer culture in media supplemented with 10% calf serum. Moreover, steroidogenic cells have a reservoir of lipids in their cytoplasm in the form of numerous droplets which can be visualized by staining *monolayers* with oil red O.

    Serum concentration can be reduced down to 2%. Unfortunately, ovarian cells perform poorly in the absence of serum. It is recommended to grow GC starting with higher cell density inoculates. Cells secreting favorable factors elaborate their own culture substrate and biological milieu in which they grow successfully. There are efforts to eliminate serum. Artificial serum, SSR2 (Medi-Cult) is commercially available. It can be a substitute and is often used in culture of human ovarian cells. However, before starting a culture, set up several cultures in medium supplemented with artificial serum and compare results with cultures in medium with natural serum. Growth surface covered with collagen, fibronectin or extracellular matrix stimulates efficiently cell growth in serum free condition.
11. Ovarian cells differ from blood cells by sedimentation rate so, after transferring cell suspension into a Petri dish, preincubation

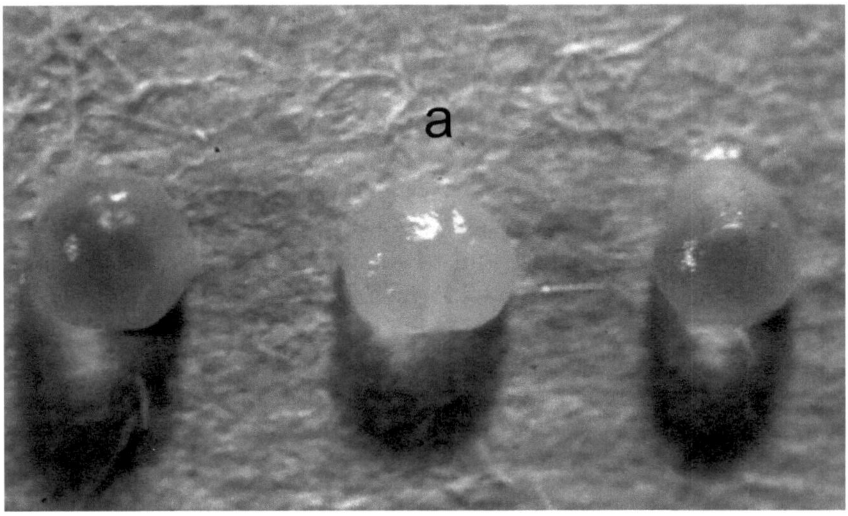

Fig. 7. Isolated preovulatory rat follicles. Foggy follicle (**a**) in the middle is atretic; healthy follicles are transparent (×80) [*Cell and Tissue Culture* (in Polish: Hodowla komórek i tkanek S. Stoklosowa (ed.) PWN, Warszawa, 2004) in color insert Fig.A.2.A].

for 30 min on ice, helps to remove a large part of the blood cells. Some authors use Percoll density gradient for cell separation. From our own experience, we know that blood cells, except for macrophages, do not interfere with cell growth and do not affect the results. Ovarian cells attach earlier to the substrate than blood cells, so the latter are gradually removed at each medium change (11, 15, 19).

12. Isolation of ThC is more laborious and difficult because cells can only be released by enzymatic dispersion resulting in the loss of certain number of ThC entangled in thecal stroma, as well as in a damage of some membrane receptors by collagenase. From own experience is known that gentle trypsinisation (1× 30 min, 2× 15 min, and 1× 10 min) does not affect membrane receptors. Two hour preincubation showed that cells recover and respond to gonadotropin stimulation successfully. Trypsin use has an advantage over collagenase because it is easy to block its activity immediately in cell suspension by the addition of several drops of serum. Collagenase cannot be blocked, it has to be removed by several cell rinsings followed by centrifugation which requires more time and causes cell stress. It is always preferable to check tissue material in several drops of PBS or medium alone under the stereoscope prior to setting up culture. However, e.g., bovine theca, which has a rich and compact stroma, must be dispersed by collagenase. Note that whole theca layer is composed of two layers (1) theca interna (Th.I) of endocrine nature and (2) connective tissue theca externa (Th.Ext). When separating theca interna from externa

care has to be taken not to contaminate theca interna cell suspension with large number of theca externa fibroblasts. So the Th.Ext should be discarded. Theca interna layer is transparent, so under higher magnification of a stereoscope theca cells embedded in the stroma are visible.

13. Suspension obtained from trypsinization, although rich in cells, is not homogeneous. The majority, 70–80%, are ThC, 6–8% are GC, and around 20% are nonsteroidogenic cells (erythrocytes, leukocytes, endothelial cells, macrophages, and fibroblasts). Interestingly fibroblasts contained in this suspension do not overgrow cultures and do not affect steroid secretion. To obtain pure theca cells flow cytometry or cell separation in Percol gradient are recommended (20).

14. Porcine corpora lutea are large 1–1.8 cm in diameter, intension molecules dark pink in maturing and mature phases, turning yellowish with the progress of the life span. At the end of the regressing phase copora lutea called *corpora albicantia* turn white and are composed mainly of connective tissue. Two cell types of CL differing in progesterone and estradiol secretion can be separated by Percoll gradient or by an immunomagnetic method, MACS or by flow cytometry (20).

15. *Final note*: In general isolated ovarian cells grow very well in primary culture, multiply and secrete steroid hormones, adhesion molecules and growth factors. They favor simple media mainly M199 or Eagle's MEM. The only demand is at least 5% serum. Without serum cells starve from the lack of the substrate for synthesis of steroid hormones. Proliferation and steroid secretion declines. Good knowledge of their change with time in terms of hormonal function along with the phases of sexual cycle is very important. Comparison of results of human, bovine, and porcine cycling CL showed that the course of the life span, hormonal dynamics and that of luteal cell composition are very similar. The difference can be only in the length of sexual cycles (5). Monocultures of T, G, CL and coculture significantly differ in basal progesterone secretion as well as in sensitivity to gonadotropins [LH or prolactin (PRL)] stimulation. The cells function as individual entities expressing their own secretory potential. Monoculture of GC can hardly be considered granulosa lutein cells. Without androgen substrate cells will continue secreting progesterone independently of cycle fluctuations. Compare the difference between progesterone secretion by GC alone and by corpus luteum (Fig. 8). The author of the chapter does not have experience in culturing human luteal cells, but is convinced that these cells could be cultured successfully according to culture method for porcine luteal cells with minor modifications.

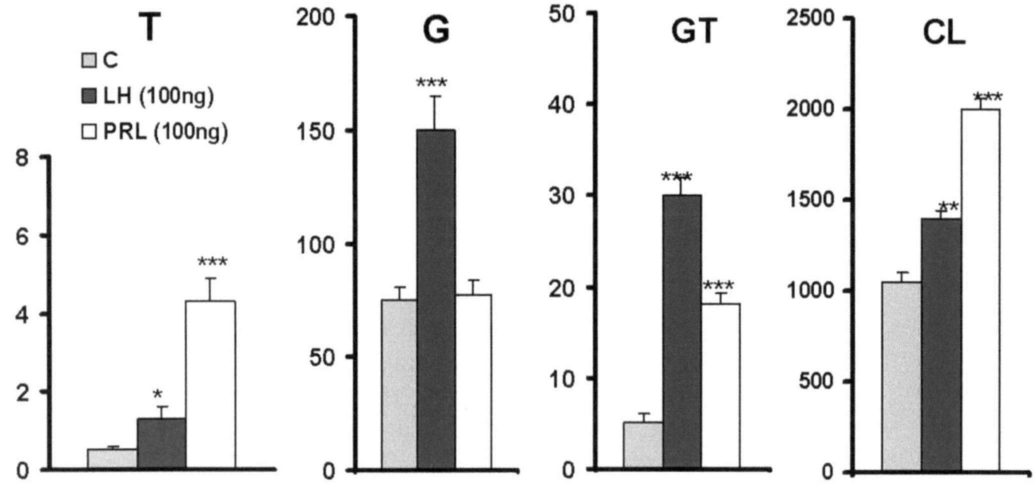

Fig. 8. Progesterone production by monocultures of theca (T), granulosa (G), corpus luteum (CL), and coculture (GT). Cells were treated with LH or prolactin (PRL), C control. The figures shows differences in the response of four ovarian cell types to gonadotropins (unpublished).

Acknowledgments

Advice of Professor S. Stoklosowa from the Department of Endocrinology, Jagiellonian University was greatly appreciated.

References

1. Galas, J., Slomczynska, M., Pierściński, A, (2003) Effect of photoperiod on distribution patterns of androgen receptors and steroid hormone concentrations in ovaries of bank voles. *Acta histochem.* 105, 175–181.
2. Galas, J., Knapczyk, K., Slomczynska, M., Kwaśnik, A. (2007) Distribution of androgen receptor and steroid hormone concentrations in ovaries of immature bank voles: Effect of photoperiod. *Acta histochem.* 109, 437–445.
3. Galas, J., Clarke J.R., Soltys, Z., Stoklosowa,S. (2005) Effect of photoperiod on cultured granulosa cell of the bank vole *Cletrionomys glareolus. Comp. Biochem. Physiol.* 140(2), 187–194.
4. Stoklosowa,S., Gregoraszczuk, E.L., Channing, C.P. (1982) Estrogen and progesterone secretion by isolated cultured porcine thecal and granulosa cells. *Biol. Reprod.* 26, 943–952.
5. Lei, Z.M., Chegini, N., Rao, C.H. (1991) Quantitative cell composition of human and bovine corpora lutea from various reproductive states. *Reproduction,* 44, 1148–1156.
6. Gregoraszczuk, E. L. (1992) Interrelationship in steroid hormone secretion and morphological changes of porcine corpora lutea at various periods of luteal phase. *Endocrine Regulations.* 26, 189–194.
7. Fitz, T.A., Mayan,M.H., Sawyer H.R., Niswender, G.D. (1982) Characteris of two steroidogenic cell types in the ovine corpus luteum. *Biol. Reprod.,* 27, 703–711.
8. Fish,B., Margara, R.A., Winston, R.M.L., Hillier, S.G. (1989) Cellular basis of luteal steroidogenesis in the human ovary. *J. Endocrinol.,* 122, 303–311.
9. Mass-Szabowski N., Stark H.J., Fusenig N.E. (2000) Keratinocite growth regulation in defined organotypic cultures through Il-1 induced keratinocyte growth factor expression in resting fibroblasts. *J. Invest. Dermatol.* 114, 1075–1084.
10. Channing, C.P., Ledwitz–Righby, F. Methods for assessing hormone mediated differentiation of ovarian cells in culture and short term incubation. In: Colowick, S.P.,

Kaplan, N.O, Methods in Enzymology, XXXIX, Part D., Hartman, J.B., O'Malley, B., (eds), Academic Press, New York, London, (1975).

11. Karamouti, M., Kaltitsaris, A, Vamvakopaulos, N., Kotlios, G., Messinis J.E. (2008) Growth hormone, insulin-like growth factor I, and leptin interaction in human cultured lutein granuloza cell steroidogensis. *Fertility Sterility* 90, 1444–50.
12. Fingenschau, Y.,Sandsfjord, J.A., Yousef, M.I., Fuskevag, O.M., Sveinbjornson, B., Bertheussen, K. (1997) A simplified serum-free method for preparation and cultivation of human granulose luteal cells. *Human Reproduction* 12, 523–531.
13. Taketani, T., Yamagata, Y., Takasaki, A., Matsuoka, A., Tamura, H., Sugino, N. (2008) Effects of growth hormone and insulin-like growth factor-1 on progesterone production in human luteinized granulosa cells. *Fertility and Sterility* 90, 744–48.
14. Stoklosowa, S. (1982) Tissue culture of gonadal cells. *Acta biol. Acad. Sci.hung.* 33(4), 367–379.
15. Alila, Rogo, K. O., Gombe S. (1987) Effects of prolactin on steroidogenesis by human luteal cells in culture. *Fertility and Sterility* 47, 947–955.
16. Stoklosowa, S. (2001) Three dimentional tissue and organ models *in vitro*: their application in basic and practical research. *Folia Histochem. Cytobiol.* 39, 91–96.
17. Gregoraszczuk, E.L. (1995) Effect of FSH on progesterone secretion by porcine large and small luteal cells isolated from early, developing corpora lutea. *Exp. Clin. Endocrinol.* 103, 272–274.
18. Cell and tissue culture (in Polish): Hodowla komórek i tkanek S. Stoklosowa (ed) PWN, Warszawa, 2006
19. Kmicikiewicz, I., Wojtusiak, A., Bilinska, B. (1999) The effect of testicular macrophages, macrophages-conditioned medium and interleukin-1 on bank vole Leydig cell steroidogenesis. *Exptl. Clin. Endocrinol. Diabetes.* 107, 262–271.
20. Freshney, R.I. Culture of Animal Cells. Wiley-Liss N. Y (2004).

Chapter 17

Human Vascular Smooth Muscle Cell Culture

Diane Proudfoot and Catherine Shanahan

Abstract

Human vascular smooth muscle cells (VSMCs) in culture are an important tool in understanding how VSMCs function and contribute to vessel wall contraction as well as disease. In this chapter, we describe methodologies that enable the investigator to culture large numbers of proliferative VSMCs. These VSMCs are heterogeneous and vary in size, shape, and proliferative capacity depending on the disease state and location of the vessel of origin. Therefore, we also describe techniques to validate their identity as *bone fide* VSMCs. Briefly, the methods include information on how to dissect the blood vessel to remove the medial layer containing VSMCs, as well as methods on how to propagate these cells, by either allowing VSMCs to migrate from the explanted medial tissue or by enzymatically dispersing the cells from the tissue. Both methods are suitable for culturing VSMCs derived from most vessel types with modifications of the enzyme dispersal method suitable for the isolation of microvessel VSMCs. An important feature of VSMCs in culture is that they lose many of their in vivo contractile properties and so model disease-associated VSMCs in the vessel wall rather than a non-proliferative contractile cell. To overcome this limitation, we also describe alternate methods that enable the study of cultured VSMCs in their contractile state by allowing the VSMCs to remain within an intact vessel ring. Overall, these procedures enable the investigator to undertake a diverse array of experimental assays on cultured VSMCs.

Key words: Human vascular smooth muscle cells, Arterial, Isolation, Identification

1. Introduction

1.1. Location of VSMCs in the Blood Vessel Wall

Vascular smooth muscle cells (VSMCs) are a major component of arteries, veins, and the microvasculature. Their main function is to maintain vascular tone through co-ordinated vasoconstriction and vasodilation, thereby regulating blood pressure and flow. The blood vessel wall is made up of three distinct layers with slight variations in the composition of these layers between individual arteries and veins. The layer exposed to the circulating blood is the tunica intima, which is covered on the lumenal side by a single flattened sheet of

Ragai R. Mitry and Robin D. Hughes (eds.), *Human Cell Culture Protocols*, Methods in Molecular Biology, vol. 806, DOI 10.1007/978-1-61779-367-7_17, © Springer Science+Business Media, LLC 2012

endothelial cells lying on a basement membrane composed of collagen type IV. In some arteries, such as the aorta and coronary arteries, there are also VSMCs in the intima. The intima provides a smooth, non-thrombogenic surface and acts as a permeability barrier to cells and macromolecules. Beneath the intima is the internal elastic lamina, which is a sheet of fenestrated elastic fibres. The layer below the internal elastic lamina is the tunica media which contains longitudinal, spirally arranged VSMCs interspersed with elastin fibres and an extracellular matrix rich in collagen I, fibronectin, and proteoglycans. The VSMCs are surrounded by a basement membrane containing collagen IV, laminin, and heparan sulphate. The outermost layer is the tunica adventitia, which is separated from the media by the external elastic lamina, and contains fibroblasts surrounded by collagen, proteoglycan matrix, vasa vasorum, and innervation (1, 2).

1.2. VSMC Phenotype and Importance in Disease

In the normal vessel media, the VSMCs exist in a contractile phenotype. However, VSMCs are not terminally differentiated and in response to vascular injury, such as in atherosclerosis, VSMCs de-differentiate, migrate, and proliferate, leading to their accumulation in the intima. VSMCs can also synthesise extracellular matrix and this plasticity is essential for vascular repair at sites of injury. These properties also contribute to a number of other vascular pathologies, including hypertension, re-stenosis after coronary angioplasty, bypass grafting, and small artery disease associated with rejection after heart transplantation. Thus, VSMCs in the media are phenotypically distinct from their intimal counterparts. In general, intimal VSMCs have lower levels of contractile proteins, contain fewer myofilaments, more synthetic organelles (e.g. sarcoplasmic reticulum and mitochondria), and, most importantly, they express a number of gene markers not normally expressed in the media (3–5). This property of VSMCs to undergo phenotypic change was considered as a key pathogenic factor by Ross and Glomset in 1976 when they proposed the "response to injury" hypothesis for the development of atherosclerosis (6, 7). In atherosclerosis, VSMCs are found in the often enlarged intima in differing quantities, depending on the type and stage of the lesion. VSMCs were previously considered to be detrimental as their presence in the intima led to narrowing of the lumen. However, more recently, it has been suggested that the stability of the lesions, that is their propensity to rupture and initiate a broad range of clinical symptoms, is dependent on the VSMCs within the lesion. Stable plaques contain abundant VSMCs and matrix, which form a thick fibrous cap above a lipid core. In contrast, plaques which are unstable and more likely to rupture contain few VSMCs, many inflammatory cells (most of which are macrophages), and a thin cap (8, 9). Therefore, although intimal VSMCs are associated with vascular disease, they may be beneficial in some circumstances in that they add stability to plaques by forming a protective cap rich in extracellular matrix. Furthermore, VSMCs may also contribute to lipid accumulation, calcification, and cell death within atherosclerotic plaques (10). Understanding the mechanisms that regulate VSMC phenotypic change in association with specific vascular diseases is essential for modulation of the disease process.

1.3. VSMC Heterogeneity

Human VSMCs are isolated from the normal medial layer of blood vessels by removing the endothelial layer, peeling and chopping the underlying sheets of VSMCs, and either dispersing the cells enzymatically or allowing VSMCs to migrate from the explanted pieces of tissue (see Subheading 3 below). With either method, the VSMCs undergo phenotypic change and de-differentiate and proliferate to become "synthetic" cells. These synthetic cells have broad similarities to cells found in the intima in vivo and have therefore been extensively used to model disease-associated VSMCs. Morphology of VSMCs can be analysed by electron microscopy; cells with a "contractile" phenotype have a heterochromatic nucleus and abundant actin and myosin filaments, compared with "synthetic" VSMCs with a euchromatic nucleus and prominent sarcoplasmic reticulum and Golgi complex (11). However, evidence from studies of human vascular disease in vivo has demonstrated that the definition of VSMCs as either "contractile" or "synthetic" is simplistic. VSMC heterogeneity has been observed in human blood vessels in vivo in different vascular beds in the sub-intimal layer, different layers within the media, as well as in atherosclerotic plaques (12, 13). Tissue culture models of these differences are not as well-characterised as those in other species (14, 15). Indeed, human VSMCs in culture can display heterogeneous characteristics, including different cell shape, size, protein expression, and growth pattern (16). For example, human VSMCs isolated from young arteries contain populations of cells with higher proliferative rates than VSMCs from adult arteries (17). Human VSMCs cultured from the vessel intima were reported to exhibit a higher natural rate of apoptosis in culture compared with cells derived from the normal medial layer of the same blood vessel (18). We have also found that human arterial VSMCs cultured from normal media can exist in different forms judged by morphology. These include spindle-shaped VSMCs which, approaching confluence, retract and form a classic "hills and valleys" morphology that can further develop into spherical multicellular nodules (shown in Fig. 1); large, rounded, senescent cells; small, relatively fast-growing cells; and cells that are contact inhibited with a cobblestone-like appearance at confluency. Some of these different morphological types can coexist in cultures obtained from the same tissue explant. However, many of these morphologies are not stable with time in culture, and specific gene markers of different human VSMC sub-populations in vitro have yet to be identified.

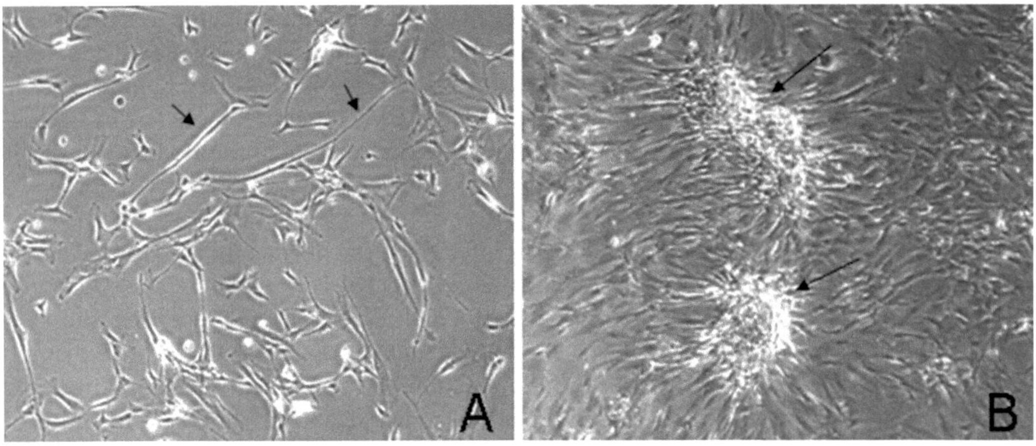

Fig. 1. Human VSMCs in culture. (**a**) Sub-confluent VSMCs showing typical *spindle-shaped* morphology often with long projections connecting adjacent cells (*arrows*). (**b**) Confluent VSMCs showing typical hills and valleys morphology. Larger "hills" form into nodules (*arrows*).

It should be noted that human VSMC cultures that form multicellular nodules have been used as a model for studying the mechanisms and regulation of vascular calcification as these nodular cells spontaneously calcify after about 30 days in culture (19).

It is most appropriate to use human VSMCs whenever possible as animal models do not adequately mimic human disease, particularly atherosclerosis. Studies have also demonstrated major differences in gene expression between human and rodent VSMCs (20, 21). The use of human VSMCs depends on the availability of human tissue and success in growing cultures. In the following sections, we concentrate on the methods for isolation, culture, and identification of VSMCs from a number of different vascular sources in man.

2. Materials

2.1. Media

1. Media commonly used for the successful culture of human VSMCs include medium 199 and DMEM buffered with 3.7 g/L $NaHCO_3$ and 5% CO_2, stored at 4°C. Antibiotic supplements recommended are 100 IU/ml penicillin, 100 mg/ml streptomycin, and 250 ng/ml amphoteracin B (the latter is important to avoid fungal contamination in the early stages of culture). The medium is also supplemented with fresh 4 mM of L-glutamine. Antibiotics and glutamine can be stored as concentrated stock solutions at −20°C. The medium containing supplements is stable at 4°C and should be used within a month.

2. For tissue digestion, a serum-free medium containing 3 mg/ml collagenase with or without 1 mg/ml elastase (both from Sigma) is used.
3. Digested tissues are separated by sieving through a 70 μm filter (Falcon).
4. The sera used for supplementing the medium are commonly derived from foetal bovine sources, although it may be possible to use human autologous serum if the VSMCs are obtained from patients having certain surgical procedures, such as bypass grafting. Serum is used at concentrations of either 10% (for passaged cells) or 20% (for explant cultures). The serum should be heat inactivated at 56° to destroy complement. The concentrations of some components of serum can differ vastly between batches; therefore, serum batch testing is essential. Supplementing the serum with growth factors is not usually necessary. Some companies now supply human VSMCs together with media, serum, and supplements that have been pre-tested and guarantee optimal cell growth for the researcher. Some serum-free systems have been tested, where it is necessary to use defined supplements to maintain cell viability (16). However, cultures must first be established in the presence of serum, which is then gradually replaced with serum-free medium.
5. Human VSMCs are frozen in 10% DMSO and 20% FBS.

2.2. Culture Materials

1. Human VSMCs adhere and grow on most tissue culture plastic and glass vessels. We recommend the use of 6-well plates (Falcon) for the adherence of medial explants and 25, 75, or 150 cm^2 flasks (Corning) for the passaging of cultures. Human VSMCs also grow well on tissue culture-treated glass slides and tissue culture inserts for co-culture experiments.
2. Culture dishes can be pretreated with extracellular matrix proteins to test the effects of attachment factors on cell adhesion and proliferation.
3. Cells can be removed from the cell suspensions using magnetic beads with specific cell surface ligands (Dynal).
4. Cells are detached in freshly thawed 0.5 mg/ml trypsin, 0.2 mg/ml EDTA solution at 37°C (ideally, in the tissue culture incubator).
5. Boyden chambers have been used to measure VSMC migration in response to various factors (22), and specialised devices that can measure mechanical strain in human VSMC cultures have also been utilised (23).
6. Flexible dishes can also be used to measure the effects of mechanical forces on human VSMCs (FlexCell International).

3. Methods

3.1. Obtaining Blood Vessels

VSMCs obtained from different blood vessels may exhibit different phenotypic properties in vitro. The aorta is made up of VSMCs that originate from neural crest in the aortic arch and from mesenchymal precursors in the thoracic segment, whereas coronary VSMCs may be derived from a different embryonic source (24). Aortic samples are obtained from organ donors and other blood vessels are obtained after surgical procedures, such as carotid endarterectomy and saphenous vein bypass grafts. However, the choice of blood vessel used normally depends on local availability. We have also cultured VSMCs from the microvasculature using human placenta as a source. These cells are often referred to as pericytes. Placental tissue is readily available and can also be used as a source of umbilical artery and vein VSMCs. We would recommend using material freshly obtained at surgery and placing the material in serum-free medium at 4°C until use. Ideally, the material should not be stored for more than 24 h.

3.1.1. Explant Method

Explant culture of medial tissue is the simplest and most reliable method of obtaining VSMC cultures.

1. The vessel is washed in serum-free medium, cut open longitudinally using forceps and scissors, and the lumenal side of the vessel is freed of endothelium by gently scraping the surface with a scalpel blade and washing with serum-free medium. If the vessel is diseased and contains fatty streaks or atherosclerotic plaques, these areas should be avoided so that VSMCs are cultured from normal areas of media.
2. The media, which contains layers of VSMCs, is now exposed and these layers can be easily removed by peeling using forceps. The external elastic lamina separates the last layer of VSMCs from the adventitia, providing an easily recognisable physical barrier between VSMCs and fibroblasts.
3. The medial layers containing VSMCs are placed in a separate Petri dish and cut carefully using a scalpel blade into approximately 1 × 1-mm pieces (see Note 1).
4. The explanted pieces are then carefully placed into six-well plates and it is essential that the explants adhere to the tissue culture plastic to ensure VSMC migration from them (see Note 2). To achieve this, either a sterile coverslip is placed on top of the explants and covered with medium or the explants are given only a small amount of medium for the first 24 h of culture, which discourages movement/floating of the explants. Cells should start to grow out radially from the explants within 2–3 weeks of culture, with medium change every 3–4 days (see Note 3).

3.1.2. Enzyme Dispersion

Enzyme dispersion of the medial layer does not yield as many cells as the explant method, but the advantage is that the whole population of VSMCs is represented, rather than selecting for those capable of migrating from the tissue. In addition, the cells isolated by enzyme dispersion may represent a phenotype more closely related to that in the vessel wall because the time to isolation is much quicker (24 h as opposed to 2–3 weeks). This method is also useful for isolating mRNA from the medial layer of VSMCs.

1. The vessel is prepared to the stage of tissue chopping, as described above for explants.
2. The tissue pieces are then placed in a sterile conical flask and incubated in serum-free digestion medium.
3. This incubation is carried out in a shaking water bath at 37°C overnight, with occasional dispersion using a wide-mouthed pipette.
4. If the tissue pieces are not completely dispersed overnight, the cells which have dispersed are removed by centrifugation (1,000 rpm for 5 min) or by sieving through a 70-μm filter and the remainder left to disperse further. Ideally, the dispersed cells should be collected over a time course to maximise the yield of viable cells. When dealing with a small piece of tissue or a small blood vessel, such as a coronary artery, the incubation should be carried out for much shorter times since long exposure to collagenase and elastase may harm the cells.
5. The dispersed cells should be centrifuged and resuspended in fresh medium before plating out at approximately 8,000 cells/cm^2.
6. Cells isolated from small arteries are few in number and should be plated in 24- or 96-well plates to ensure an adequate cell density. Because human VSMCs require close association with other cells to maintain survival and growth, we have not been successful in single-cell cloning by limiting dilution of VSMCs.

3.1.3. Isolation of Microvascular VSMC

Obtaining VSMCs from very small blood vessels initially requires the isolation of microvessels from the surrounding tissue. We have isolated microvascular SMCs (or pericytes) from human placenta and these can be obtained fresh from maternity units.

1. A central section of placenta, which is rich in villi and distant from large blood vessels and the outer membrane, is dissected and washed thoroughly in serum-free medium.
2. The tissue is then manually chopped into small pieces and placed in serum-free medium containing 3 mg/ml collagenase (without elastase).
3. After incubation at 37°C in a shaking water bath for 3 h, the digested material contains single cells and microvessels.

4. To separate the microvessels from the cells in suspension, the material is sieved through a 70-μm filter and washed in serum-free medium.
5. The microvessels, which remain on the sieve, are then removed and placed into 25-cm^3 flasks in medium containing 20% serum.
6. After 5–6 days, pericytes as well as endothelial cells migrate from the adherent microvessels. Without special supplements to encourage endothelial cell growth, the endothelial cells die after two rounds of trypsinisation, leaving pure pericyte cultures. Alternatively, endothelial cells can be selectively removed from the cultures by plating established cells onto dishes coated with l-leucine methyl ester. Due to differences in esterase expression, this substance is toxic to endothelial cells but not to pericytes (25).

3.1.4. Isolation of VSMCs from Diseased Blood Vessels

Isolation of VSMCs from atherosclerotic plaques is complicated by the presence of several other cell types, such as macrophages and lymphocytes. Depending on the type of diseased vessel, VSMCs may be abundant (e.g. in re-stenotic vessels) or scarce (in lipid-rich atherosclerotic plaques with necrotic cores). Explant culture of VSMC from diseased tissue is possible, although a number of different cell types may grow from the explanted material. However, VSMCs are the most likely cell type to survive beyond primary culture. If cell dispersion is used, undesired cell types can be removed from the cell suspensions using magnetic beads. In our hands, specimens from very small arteries, such as coronary atherectomies of re-stenotic vessels (typically, 1-mm thick and up to 10 mm in length), are too small to prepare tissue explants. Instead, successful isolates of VSMCs from these arteries can be obtained by digestion in 3 mg/ml collagenase and 1 mg/ml elastase for 1 h in a shaking water bath and plating cells in 96-well plates as only small numbers of cells are generated, typically 30,000 cells per 100-mg tissue. It should be noted that although it is possible to culture VSMCs from diseased vessels, including the aorta and carotid endacterectomy specimens, the rate of success in obtaining VSMCs using diseased tissue is very low. In addition, if VSMCs are successfully explanted, they often have a very limited lifespan in culture. Another major problem with VSMC isolation from diseased tissue is contamination by other mesenchymal cells, such as pericytes and fibroblasts. It is, therefore, important to check for VSMC-specific markers in the cell cultures obtained (see Subheading 3.2).

3.1.5. Culture of VSMCs Within Intact Blood Vessels

Given the problems outlined above regarding the phenotypic changes VSMCs undergo in vitro, from a contractile to a synthetic cell, it may be useful in some circumstances to culture VSMCs within the intact vessel wall. Under these circumstances, VSMCs retain their contractile phenotype and do not proliferate. This culture procedure is suitable for small calibre arteries from 2 to 5 mm in diameter.

1. Vessels are gently stripped of adventitia and cut into 1-mm rings with a scapel.
2. The vessel rings are placed in serum-free or serum-containing media, free floating in the well of a 24-well plate.
3. Medium is changed twice a week and under these conditions vessel rings can be kept for up to 2 weeks in serum-free conditions and longer in the presence of serum (although phenotypic change of the VSMCs needs to be analysed after longer periods).
4. Rings are harvested using sterile forceps or with a pipette and then processed for immunhistochemistry (26, 27).

3.2. Identification of VSMCs

Because of the heterogeneity and adaptive modulation displayed by VSMCs in culture, a number of criteria must be used to ascertain that cultured cells are indeed VSMCs. As described above, this is particularly important for cells derived from atherosclerotic vessels as some cell types, such as myofibroblasts and primitive mesenchymal cells, share expression of a number of smooth muscle contractile proteins with VSMCs. There are in fact no definitive markers to differentiate between these cell types and VSMCs; therefore, a "battery" of markers must be used, together with knowledge of the origin of the cells.

3.2.1. Cell Markers

The smooth muscle cell protein α-smooth muscle (α SM) actin has been universally used to identify VSMCs (28). However, some studies have reported that endothelial cells and adventitial fibroblasts can express α SM actin under certain culture conditions (29, 30). Therefore, additional smooth muscle markers must be used to identify cells in vitro. Most VSMC markers are contractile proteins and are commonly detected using antibodies and immunohistochemistry, which requires very few cells grown in multiwell chamber slides or on coverslips and fixed in paraformaldehyde. Alternatively, Western analysis is performed on cell lysates. Common VSMC markers include smooth muscle myosin heavy chain (SMMHC), calponin, SM22 α, desmin, h-caldesmon, metavinculin, and smoothelin.

To confirm that the cultured cells are VSMCs, positivity for α SM actin and at least one other smooth muscle contractile protein is required (e.g. calponin or SM22 α). It may also be appropriate to confirm that cultures are von Willebrand negative if there is concern about endothelial cell contamination. Markers for microvascular VSMCs (pericytes), such as 3G5, have been described, but there is debate about the specificity of these markers (31, 32). The high-molecular-weight melanoma-associated antigen is another marker for pericytes in vivo (33).

3.2.2. VSMC Contraction

Contractility is a unique feature of VSMCs that is not shared by other cell types present in blood vessels. However, measuring contractility in cultured cells is not straightforward as they lose many of their contractile properties with time in culture. Most studies of VSMC contraction involve culture of blood vessel ring segments that retain contractile properties which can be measured by treatment with agonists, such as noradrenaline or KCl, and subsequent measurement of contractile force and/or intracellular free calcium ion concentration (34). However, measurement of contraction of freshly dispersed VSMCs in culture is feasible and has been observed in rabbit and human VSMCs (35, 36). In one report, human umbilical vein VSMCs were grown in scaffolds containing rings of cultured VSMCs in which various vasoconstrictor and vasodilator agents were successfully tested (37). Additionally, calcium transients can be measured in human-cultured VSMCs (38).

3.2.3. Multicellular Nodule Formation

The capacity for human VSMCs to form multicellular nodules at post-confluence (see Fig. 1) has been well-described for arterial VSMCs and pericytes (16, 19, 39, 40). We have found that human arterial VSMCs (aortic, tibial, and femoral) and human placental pericytes spontaneously form nodules in culture in which calcium deposits occur after approximately 30 days (19). This property of VSMCs exists in most of the isolates we have prepared. The only VSMC phenotype where we did not observe nodule formation was in the cobblestone cells which appear to be contact inhibited. Although nodule formation is not unique to VSMCs (mesenchymal cells with osteoblastic properties may also form nodules), this property distinguishes human VSMCs from fibroblasts.

3.3. Passaging Cultures

When primary explants or dispersed cell cultures become dense in culture, they should be separated and re-plated by trypsinisation.

1. The cells are trypsinised by washing gently three times in calcium-free balanced salt solution and incubated in detachment medium.
2. After approximately 3 min, the cells begin to detach from the tissue culture vessel and when the dish is tapped lightly, the cells are released into the medium. This can be checked under the light microscope (see Note 4).
3. Fresh medium containing serum is then added to the wells to inactivate the trypsin. If coverslips have been used to encourage adherence of explants, it is likely that some cells will remain on the coverslip. To extract these, the coverslips can be inverted using sterile forceps and the washing and trypsinisation steps are repeated.
4. It is highly recommended that the cells are plated out relatively densely, covering about 40% of the culture dish. This density encourages the growth of human VSMCs (see Note 5).

5. For maintenance of cultures, it is recommended that the medium is changed every 2 days. The typical passage survival time for normal, medial VSMCs is up to ten passages, although pericyte cultures do not often survive beyond passage five. For VSMCs isolated from diseased vessels, the passage survival time is lower, typically three passages. The cell cycle time for human VSMCs varies between isolates (typical doubling time is ~44 h) and on approaching senescence, the cells become large and rounded and may also produce large amounts of debris.

Human VSMCs are stored frozen. DMSO inhibits VSMC growth; therefore, cells should be thawed quickly and either centrifuged at 1,000 rpm for 5 min to remove the DMSO or the medium changed as soon as the cells have adhered to the culture flask.

4. Notes

1. It is important that the tissue pieces are not sheared when cutting as this can reduce efficiency of outgrowth of VSMCs from the explants. If large samples are used, a mechanical tissue chopper can be used to generate small tissue explants.
2. A lack of adherence of medial explants results in a lack of outgrowth of VSMCs. Aside from placing a coverslip on top of the explants to encourage adherence, another successful method is to limit the volume of media covering the explants so that they have less chance of floating. A small volume of medium that just covers the explants should be applied and incubated overnight. The cultures should then be gently topped up with fresh medium with minimal disturbance to the explants. The explants should be left alone as much as possible as minimal disruption of early explant cultures is essential to encourage migration of VSMCs from the explants.
3. It is noteworthy that some groups have reported that different layers of the aortic media contain different types of VSMCs (13). It may, therefore, be appropriate to culture different layers of medial VSMCs separately.
4. Some human VSMC isolates detach from the tissue culture plastic/cell matrix more rapidly than others. It is, therefore, advisable to check the cells under the microscope after 1 min for signs of detachment.
5. To maintain healthy growing human VSMC cultures, they should not be plated out too sparsely. Additionally, they are unlikely to become completely confluent as they tend to form close associations with other cells and form hills/valleys and

nodules. If human VSMC cultures are kept too long in culture before passaging, they can become difficult to disperse upon trypsinisation and can detach from the dish as a sheet rather than a suspension of individual cells. Therefore, to keep cultures growing optimally, it is advised to aim to have a density between 40 and 80%.

References

1. Chamley-Campbell, J., Campbell, G. R., and Ross, R. (1979) The smooth muscle cell in culture. *Physiol Rev* 59, 1–61.
2. Ross, R. (1986) The pathogenesis of atherosclerosis – an update. *N Engl J Med* 314, 488–500.
3. Schwartz, S. M., deBlois, D., and O'Brien, E. R. (1995) The intima. Soil for atherosclerosis and restenosis. Circ Res 77, 445–65.
4. Jonasson, L., Holm, J., Skalli, O., Gabbiani, G., and Hansson, G. K. (1985) Expression of class II transplantation antigen on vascular smooth muscle cells in human atherosclerosis. J Clin Invest 76, 125–31.
5. Hansson, G. K., Jonasson, L., Holm, J., and Claesson-Welsh, L. (1986) Class II MHC antigen expression in the atherosclerotic plaque: smooth muscle cells express HLA-DR, HLA-DQ and the invariant gamma chain. Clin Exp Immunol 64, 261–8.
6. Ross, R., and Glomset, J. A. (1976) The pathogenesis of atherosclerosis (first of two parts). N Engl J Med 295, 369–77.
7. Ross, R., and Glomset, J. A. (1976) The pathogenesis of atherosclerosis (second of two parts). N Engl J Med 295, 420–5.
8. Libby, P. (2000) Coronary artery injury and the biology of atherosclerosis: inflammation, thrombosis, and stabilization. *Am J Cardiol 86*, 3J–8J; discussion 8J–9J.
9. Davies, M. J. (1995) Acute coronary thrombosis – the role of plaque disruption and its initiation and prevention. *Eur Heart J 16 Suppl L*, 3–7.
10. Iyemere, V. P., Proudfoot, D., Weissberg, P. L., and Shanahan, C. M. (2006) Vascular smooth muscle cell phenotypic plasticity and the regulation of vascular calcification. J Intern Med 260, 192–210.
11. Thyberg, J., Hedin, U., Sjolund, M., Palmberg, L., and Bottger, B. A. (1990) Regulation of differentiated properties and proliferation of arterial smooth muscle cells. Arteriosclerosis 10, 966–90.
12. Babaev, V. R., Bobryshev, Y. V., Stenina, O. V., Tararak, E. M., and Gabbiani, G. (1990) Heterogeneity of smooth muscle cells in atheromatous plaque of human aorta. Am J Pathol 136, 1031–42.
13. Glukhova, M. A., Kabakov, A. E., Frid, M. G., Ornatsky, O. I., Belkin, A. M., Mukhin, D. N., Orekhov, A. N., Koteliansky, V. E., and Smirnov, V. N. (1988) Modulation of human aorta smooth muscle cell phenotype: a study of muscle-specific variants of vinculin, caldesmon, and actin expression. Proc Natl Acad Sci USA 85, 9542–6.
14. Frid, M. G., Dempsey, E. C., Durmowicz, A. G., and Stenmark, K. R. (1997) Smooth muscle cell heterogeneity in pulmonary and systemic vessels. Importance in vascular disease. Arterioscler Thromb Vasc Biol 17, 1203–9.
15. Bochaton-Piallat, M. L., Ropraz, P., Gabbiani, F., and Gabbiani, G. (1996) Phenotypic heterogeneity of rat arterial smooth muscle cell clones. Implications for the development of experimental intimal thickening. Arterioscler Thromb Vasc Biol 16, 815–20.
16. Dartsch, P. C., Weiss, H. D., and Betz, E. (1990) Human vascular smooth muscle cells in culture: growth characteristics and protein pattern by use of serum-free media supplements. Eur J Cell Biol 51, 285–94.
17. Fujita, H., Shimokado, K., Yutani, C., Takaichi, S., Masuda, J., and Ogata, J. (1993) Human neonatal and adult vascular smooth muscle cells in culture. Exp Mol Pathol 58, 25–39.
18. Bennett, M. R., Littlewood, T. D., Schwartz, S. M., and Weissberg, P. L. (1997) Increased sensitivity of human vascular smooth muscle cells from atherosclerotic plaques to p53-mediated apoptosis. Circ Res 81, 591–9.
19. Proudfoot, D., Skepper, J. N., Shanahan, C. M., and Weissberg, P. L. (1998) Calcification of human vascular cells in vitro is correlated with high levels of matrix Gla protein and low levels of osteopontin expression. Arterioscler Thromb Vasc Biol 18, 379–88.
20. Shanahan, C. M., Weissberg, P. L., and Metcalfe, J. C. (1993) Isolation of gene markers of differentiated and proliferating vascular smooth muscle cells. *Circ Res* 73, 193–204.
21. Newman, C. M., Bruun, B. C., Porter, K. E., Mistry, P. K., Shanahan, C. M., and Weissberg, P. L. (1995) Osteopontin is not a marker for proliferating human vascular smooth muscle cells. Arterioscler Thromb Vasc Biol 15, 2010–8.

22. Rauch, B. H., Millette, E., Kenagy, R. D., Daum, G., Fischer, J. W., and Clowes, A. W. (2005) Syndecan-4 is required for thrombin-induced migration and proliferation in human vascular smooth muscle cells. J Biol Chem 280, 17507–11.
23. Cheng, G. C., Briggs, W. H., Gerson, D. S., Libby, P., Grodzinsky, A. J., Gray, M. L., and Lee, R. T. (1997) Mechanical strain tightly controls fibroblast growth factor-2 release from cultured human vascular smooth muscle cells. Circ Res 80, 28–36.
24. Gittenberger-de Groot, A. C., DeRuiter, M. C., Bergwerff, M., and Poelmann, R. E. (1999) Smooth muscle cell origin and its relation to heterogeneity in development and disease. Arterioscler Thromb Vasc Biol 19, 1589–94.
25. Lee, C. S., Patton, W. F., Chung-Welch, N., Chiang, E. T., Spofford, K. H., and Shepro, D. (1998) Selective propagation of retinal pericytes in mixed microvascular cell cultures using L-leucine-methyl ester. Biotechniques 25(3), 482–8, 490–2, 494.
26. Shroff, R. C., Shah, V., Hiorns, M. P., Schoppet, M., Hofbauer, L. C., Hawa, G., Schurgers, L. J., Singhal, A., Merryweather, I., Brogan, P., Shanahan, C., Deanfield, J., and Rees, L. (2008) The circulating calcification inhibitors, fetuin-A and osteoprotegerin, but not matrix Gla protein, are associated with vascular stiffness and calcification in children on dialysis. Nephrol Dial Transplant 23, 3263–71.
27. Shroff, R. C., McNair, R., Skepper, J. N., Figg, N., Schurgers, L. J., Deanfield, J., Rees, L., and Shanahan, C. M. (2010)Chronic mineral dysregulation promotes vascular smooth muscle cell adaptation and extracellular matrix calcification. J Am Soc Nephrol 21(1): 103–12.
28. Skalli, O., Ropraz, P., Trzeciak, A., Benzonana, G., Gillessen, D., and Gabbiani, G. (1986) A monoclonal antibody against alpha-smooth muscle actin: a new probe for smooth muscle differentiation. *J Cell Biol* 103, 2787–96.
29. Arciniegas, E., Sutton, A. B., Allen, T. D., and Schor, A. M. (1992) Transforming growth factor beta 1 promotes the differentiation of endothelial cells into smooth muscle-like cells in vitro. J Cell Sci 103 (Pt 2), 521–9.
30. Shi, Y., O'Brien, J. E., Jr., Fard, A., and Zalewski, A. (1996) Transforming growth factor-beta 1 expression and myofibroblast formation during arterial repair. Arterioscler Thromb Vasc Biol 16, 1298–305.
31. Nayak, R. C., Berman, A. B., George, K. L., Eisenbarth, G. S., and King, G. L. (1988) A monoclonal antibody (3 G5)-defined ganglioside antigen is expressed on the cell surface of microvascular pericytes. J Exp Med 167, 1003–15.
32. Shepro, D., and Morel, N. M. (1993) Pericyte physiology. Faseb J 7, 1031–8.
33. Schlingemann, R. O., Rietveld, F. J., de Waal, R. M., Ferrone, S., and Ruiter, D. J. (1990) Expression of the high molecular weight melanoma-associated antigen by pericytes during angiogenesis in tumors and in healing wounds. Am J Pathol 136, 1393–405.
34. Maguire, J. J., and Davenport, A. P. (1999) Endothelin receptor expression and pharmacology in human saphenous vein graft. Br J Pharmacol 126, 443–50.
35. Shirinsky, V. P., Birukov, K. G., Sobolevsky, A. V., Vedernikov, Y. P., Posin, E., and Popov, E. G. (1992) Contractile rabbit aortic smooth muscle cells in culture. Preparation and characterization. Am J Hypertens 5, 124S–130S.
36. Li, S., Sims, S., Jiao, Y., Chow, L. H., and Pickering, J. G. (1999) Evidence from a novel human cell clone that adult vascular smooth muscle cells can convert reversibly between noncontractile and contractile phenotypes. Circ Res 85, 338–48.
37. L'Heureux, N., Stoclet, J. C., Auger, F. A., Lagaud, G. J., Germain, L., and Andriantsitohaina, R. (2001) A human tissue-engineered vascular media: a new model for pharmacological studies of contractile responses. Faseb J 15, 515–24.
38. Ewence, A. E., Bootman, M., Roderick, H. L., Skepper, J. N., McCarthy, G., Epple, M., Neumann, M., Shanahan, C. M., and Proudfoot, D. (2008) Calcium phosphate crystals induce cell death in human vascular smooth muscle cells: a potential mechanism in atherosclerotic plaque destabilization. Circ Res 103, e28–34.
39. Gimbrone, M. A., Jr., and Cotran, R. S. (1975) Human vascular smooth muscle in culture. Growth and ultrastructure. Lab Invest 33, 16–27.
40. Schor, A. M., Allen, T. D., Canfield, A. E., Sloan, P., and Schor, S. L. (1990) Pericytes derived from the retinal microvasculature undergo calcification in vitro. J Cell Sci 97 (Pt 3), 449–61.

Chapter 18

Culture of Human Endothelial Cells from Umbilical Veins

Richard C.M. Siow

Abstract

The present protocol offers an economical option for the isolation and culture of human endothelial cells for vascular cell biology research due to the non-invasive collection procedure being devoid of ethical concerns and ease of the isolation technique, expansion and maintenance under standard cell culture conditions. The human umbilical vein endothelial cell (HUVEC) model is useful for any research on general properties of human endothelium, but as these cells are of foetal and venous origin, other sources could be more appropriate models for studies on specific pathological areas, for example, atherosclerosis or cancer angiogenesis. Nevertheless, HUVEC still represent the most simple and available human vascular cell type widely used in biomedical research.

Key words: Umbilical cord, Endothelial cell, Cryopreservation, von Willebrand factor

1. Introduction

Advances in vascular endothelial cell biology over the past three decades have made a major impact on our understanding of the mechanisms underlying many pathophysiological conditions, from cardiovascular disease to cancer and inflammation. In addition, the endothelium has become a major target of human clinical therapeutics as endothelial dysfunction contributes significantly to hypertension, coronary artery disease, atherosclerosis, and tumour progression (1). Endothelial cells can alter their phenotype and contribute to atherosclerotic lesion formation and progression, not only in response to inflammatory mediators but also to mechanotransduction elicited by mechanical and fluid shear forces (2, 3). Recent studies employing gene expression profiling have shown significant endothelial cell diversity at different sites in the vascular tree with differences in the genotype of large vessel and microvascular endothelium with specific arterial and venous gene expression

Ragai R. Mitry and Robin D. Hughes (eds.), *Human Cell Culture Protocols*, Methods in Molecular Biology, vol. 806,
DOI 10.1007/978-1-61779-367-7_18, © Springer Science+Business Media, LLC 2012

profiles being identified (4, 5). It has become evident that the tissue microenvironment surrounding blood vessels can control endothelial cell phenotype and recent proteomic studies have identified endothelial cell proteins restricted to specialised membrane microdomains which are expressed in specific tissues (6). Much of the growth in the field of endothelial cell biology has arisen from studies employing in vitro culture of human umbilical vein endothelial cells (HUVECs) as they can be derived from a readily available source of human tissue (umbilical cords), from which a relatively pure population of cells can be easily isolated and grown (7). The therapeutic impact of advances in understanding the molecular biology of the endothelium over the past three decades has mainly focused on agents that regulate angiogenesis to target tumour growth and metastasis (8). In addition, regulation of the synthesis and bioactivity of endothelium-derived nitric oxide has also received much interest for the development of cardioprotective strategies against atherosclerosis (9, 10). In this chapter, a protocol is provided for the isolation, culture, and characterisation of HUVEC. The technique has been adapted from the original description by Jaffe et al. (11), and has now been widely adopted in vascular cell biology laboratories. This protocol will yield confluent primary HUVEC cultures that can be obtained within a week and reliably subcultured for up to 4–5 passages for expanding cell number, cryopreservation, characterisation with endothelial specific makers such as von Willebrand factor (vWF), lectin, or uptake of acetylated low-density lipoprotein (12), and for experimental procedures in culture media containing low serum and without additional growth factors.

2. Materials

2.1. Culture Medium and Solutions for Isolating and Maintaining HUVEC Cultures

1. The most commonly used culture medium in HUVEC culture is Medium 199 (M199) but other commercially available growth media are also suitable (see Note 1). The following additions to basal M199 (without glutamine or bicarbonate) are necessary prior to use: (final concentration) 5 mM L-glutamine, 40 mM bicarbonate, 100 U/ml penicillin, 100 μg/ml streptomycin, and 20% (v/v) foetal calf serum (see Note 2). Sterile stocks of these additional components are prepared and stored as frozen aliquots as described below. The complete medium can be stored at 4°C for up to 1 month and is prewarmed to 37°C prior to use in routine cell culture. Medium without the FCS component is used during the initial isolation procedures and the FCS content can be reduced as necessary following seeding of cells for experimental procedures.

2. For maintenance of proliferating HUVEC cultures, further addition of endothelial cell growth factor (ECGF, Sigma), diluted in serum-free M199 containing heparin, is necessary. Stocks solutions (concentration: 1 mg/ml ECGF, 4.5 mg/ml heparin) are filtered through a 0.5-μm prefilter and a 0.22-μm sterile filter connected in series and can be stored as 5 ml aliquots for up to 3 months at −20°C. Stocks are diluted in serum containing M199 to provide final concentrations of ECGF (20 μg/ml) and heparin (90 μg/ml). This component can be omitted from the medium following seeding of cells for experimental procedures.
3. Hank's balanced salt solution (HBSS) is used for umbilical cord specimen collection. The following additions, from sterile stock solutions, are necessary (final concentration in HBSS): 100 μg/ml gentamycin, 0.025 M HEPES, 20 mM bicarbonate, and 0.001% phenol red. The final HBSS cord collection medium can be stored as aliquots at 4°C for up to 3 weeks.
4. L-Glutamine (200 mM) stock solution: dissolve 5.84 g L-glutamine in 200 ml tissue culture-grade deionized water and sterilise by passing through a 0.22-μm filter. Aliquots (5 ml) can be stored at −20°C for 3 months.
5. Bicarbonate (4.4%, 0.52 M)–Phenol red (0.03%) solution: dissolve 44 g $NaHCO_3$ and 30 mg phenol red in 1,000 ml tissue culture-grade deionized water, and sterilise by autoclaving for 10 min at 115°C. Aliquots of 15 ml are stored at 4°C for up to 3 months.
6. Penicillin and streptomycin stock solution (80× concentrate): dissolve 480 mg penicillin (G sodium salt) and 1.5 g streptomycin sulphate in 200 ml tissue culture-grade deionized water and sterilise by passing through a 0.5-μm pre-filter and a 0.22-μm filter. Aliquots of 5 ml are stored at −20°C and one aliquot used in 400 ml of medium.
7. Gentamycin solution (80× concentrate): dissolve 750 mg gentamycin sulphate in 100 ml tissue culture-grade deionized water and sterilise by passing through a 0.22-μm filter. Aliquots of 5 ml are stored at −20°C and one aliquot used in 400 ml of HBSS.
8. HEPES solution (1 M): dissolve 47.6 g of HEPES in 200 ml tissue culture-grade deionized water and sterilise by passing through a 0.22-μm filter. Aliquots of 5 ml can be stored at −20°C and two aliquots used in 400 ml of HBSS.
9. Sterile Dulbecco's phosphate-buffered saline (PBS).
10. Trypsin solution (2.5%): trypsin from porcine pancreas (Sigma) is dissolved (2.5 g/100 ml) in PBS and sterilised by passing through a 0.22-μm filter. Aliquots of 10 ml are stored at −20°C.

11. EDTA solution (1%): EDTA disodium salt is dissolved (500 mg/50 ml) in tissue culture-grade deionized water and sterilised through a 0.22-μm filter. Aliquots of 5 ml are stored at 4°C.
12. Trypsin (0.1%)–EDTA (0.02%, 0.5 mM) solution is prepared by adding 10 ml Trypsin (2.5%) and 5 ml EDTA (1%) to 250 ml sterile PBS. This solution is pre-warmed to 37°C before use to detach cells from culture flasks and stored at 4°C for up to 2 months.
13. Gelatin solution (1%): dissolve gelatin (Sigma G9382) in tissue culture-grade deionised water and sterilise by autoclaving for 20 min. Aliquots can be stored at 4°C for up to 3 months.
14. Collagenase, Type II (Sigma): dissolve collagenase in serum-free medium (0.5 mg/ml) on ice. Particulate material is removed by filtering the solution through a 0.5-μm pre-filter and then sterilise by passing through a 0.22-μm filter. This enzyme solution can be stored for long term as 5–10 ml aliquots at −20°C until use in the proportion of 10 ml per 20 cm of umbilical cord.

2.2. Reagents for Characterising HUVEC Using Immunofluorescence Microscopy

1. Monoclonal antibody against human von Willebrand factor (Sigma).
2. Normal rabbit serum (Sigma).
3. Fluorescein isothiocyanate or Alexa Fluor-488 conjugated rabbit anti-mouse antibody (Dako).
4. Methanol (100%).

2.3. Equipment

All procedures should be carried out in a Class II laminar flow safety cabinet using aseptic technique. Dissection equipment should be thoroughly washed and kept sterilised by immersion in 70% ethanol or by autoclaving at 121°C for 20 min.

1. 25 and 75 cm^2 tissue culture flasks. Culture vessels should be coated with warmed gelatin solution (~6 ml for a 75-cm^2 flask) for ~10 min prior to use. The gelatin is removed by aspirating using a sterile glass Pasteur pipette before the addition of cell suspensions. Flasks and plates can be stored for a week at 4°C after coating.
2. Sterile glass Pasteur, 5 and 10 ml pipettes.
3. Sterile 50 and 15 ml centrifuge tubes.
4. Sterile scalpel blades, scissors, watchmakers forceps, strong thread, and cannulae.
5. Lab-Tek chamber slides (Nunc).

3. Methods

3.1. Collection of Tissue Samples

In this laboratory, human endothelial cells are routinely isolated from umbilical veins. Umbilical cords should be obtained with informed consent and local ethics approval. As human tissues should be treated as potentially hazardous, it is recommended that those engaged in isolating HUVEC should be immunised against hepatitis B as a precaution. To ensure sterility and operator safety, all work should be performed in a Class II biological safety cabinet using sterile techniques. As soon as possible after delivery, the whole umbilical cord is placed in the HBSS collection medium and stored at 4°C. Cords that are collected in this way can be used for HUVEC isolation up to 48 h after delivery (see Note 3).

3.2. Isolation and Primary Culture of Endothelial Cells

1. All cell culture media and solutions are prepared and warmed to 37°C in a water bath.
2. Sufficient collagenase solution is made up (or defrosted) for the lengths of cords to be processed.
3. The cord is removed from the collection solution, and paper towel (see Note 4) is used to blot the cord for removal of blood and excess medium.
4. Make a clean transverse cut across one end of the cord to expose the two umbilical arteries and the umbilical vein. The latter can be identified by its thinner wall and larger, stretchable lumen.
5. Insert a cannula into the vein, and remove tissue and arteries to expose about 1 cm of vein. It is important to remove a short section of the adjacent arteries; otherwise, it may be difficult to tie the ligature tightly enough, and the cannula may slip from the vein during cell isolation. The cannula is secured by tightly tying it to the vein, above the barb, with strong thread.
6. The process is repeated at the other end of the cord. Very long cords can be divided to give two shorter lengths of about 20 cm.
7. Fill a 20-ml syringe with warmed PBS and attach it to the cannula at one end, and then use this solution to flush the vein free from blood, clots, and so forth, collecting the eluate in a beaker.
8. Repeat the procedure from the other end (gentle massage may facilitate the removal of clots). Before expelling all the wash solution, occlude the cord at one end, and check for leaks from perforations made when blood samples were taken. Any holes found can be closed using suitable "crocodile" clips.
9. Attach an empty syringe to one cannula, and then use another syringe to fill the vein with sufficient pre-warmed collagenase solution to ensure that the lumen is distended.

10. Loosely cover the cord in cling-film or aluminium foil (see Note 4) and, with the syringes still firmly attached, place the cord in an incubator at 37°C for 10 min.
11. Remove the cord from the incubator, and massage it gently to assist the detachment of cells from the vessel wall.
12. Apply suction with one syringe while exerting gentle pressure along the cord in the direction of flow to draw out as much of the cell suspension eluate as possible in to the syringe, which is then transferred in to a sterile 50 ml centrifuge tube.
13. Leaving one syringe still attached to the cord, take up 20 ml of pre-warmed sterile PBS into a syringe, and use this to flush free any further loosened cells, working the solution backward and forward between the two syringes a few times. Draw off as much of this second cell suspension as possible, and add to the same 50 ml tube containing the first eluate.
14. Pellet the cells by centrifugation at 1,000 rpm (~150 × *g*) in a bench top centrifuge for 5 min at room temperature.
15. Carefully aspirate or decant the supernatant and resuspend the cells in 5 ml pre-warmed serum containing culture medium.
16. Transfer the cell suspension to a gelatin-coated 25 cm^2 tissue culture flask, and place in an incubator at 37°C in a humidified atmosphere of 5% CO_2/95% air. At this point, the flask will contain large numbers of erythrocytes in addition to endothelial cells (see Note 5).
17. The next day all the medium is aspirated from the culture flask using a sterile glass Pasteur pipette, the cells washed with pre-warmed serum-free medium to remove any residual erythrocytes and non-adherent cells, and 5 ml of fresh serum containing culture medium is added.
18. Every second day, cells should be fed by removing and discarding the medium in the flask and replacing it with fresh warmed serum containing culture medium. The cells should reach confluence in 5–7 days and are then ready for use or subculture to obtain larger numbers of cells.

3.3. Subculture of Endothelial Cells

1. Once a confluent monolayer has been attained in a 25-cm^2 flask, the HUVEC can be subcultured (passaged) into further 25 or 75 cm^2 flasks, plates, or dishes (see Note 6). The culture medium is removed and cells are washed twice with pre-warmed sterile PBS to remove traces of serum.
2. Pre-warmed trypsin/EDTA solution (0.5 ml for 25 cm^2 flask or 1 ml for a 75-cm^2 flask) is added to cover the cells and the flask incubated at 37°C for up to 4 min. The cell layer is periodically examined under the microscope to ensure cells have fully detached. This can also be facilitated by vigorous tapping of the side of the flask to dislodge cells and disassociate aggregates (see Note 7).

3. Serum containing culture medium (~5–7 ml) is added to stop the action of the trypsin which can reduce cell viability through prolonged exposure. The cell suspension is then drawn up and down a sterile Pasteur or 5 ml pipette 4–6 times to further break up any cell aggregates.
4. The cell suspension is then transferred into new culture flasks at a dilution ratio of ~1:3 and sufficient serum containing medium added to the new flasks (5 ml in a 25-cm^2 and 12 ml in a 75-cm^2 flask). The flasks are returned to the 37°C, 5% CO_2 humidified incubator and the serum containing culture medium changed every 2 days.
5. The subcultured cells, which are now at passage 1, should again become confluent within 4–7 days. If required, cells can be further subcultured by dividing the contents of a 75-cm^2 flask between three new flasks or in other culture plates/dishes. Further subculture beyond passages 4–5 is not advisable, since the cell growth rate may decline and phenotypic changes become evident.
6. For experimental protocols, once HUVEC monolayers are confluent, the growth medium can be replaced with either low (1% v/v FCS) serum or serum-free medium for up to 24 h before cells exhibit an altered apoptotic phenotype due to the absence of growth factors.

3.4. Characterisation of Endothelial Cells

Confluent monolayers of vascular endothelial cells exhibit the characteristic "cobblestone" morphology in culture (11, 12). Isolated primary cells can be positively identified by their positive immunostaining with antibodies against von Willebrand factor (vWF) and uptake of acetylated low-density lipoproteins, which have been shown to be a specific markers for endothelial cells (7), and negative immunostaining for smooth muscle α-actin, a smooth muscle cell-specific marker (13). The following procedure, described here in brief, can be used to identify HUVEC by their positive vWF staining using fluorescein isothiocyanate (FITC) or Alexafluor-488 labelled secondary antibodies. Some cultures may be contaminated with smooth muscle cells during the initial isolation. The elongated morphology of SMC is characteristic using low power microscopic examination and these cultures should be discarded.

1. Endothelial cells are subcultured into Lab-Tek chamber slide wells and characterised after 48 h.
2. The culture medium is removed from the wells and cells gently washed three times with serum-free culture medium before being fixed with ice-cold methanol (100%) for 45 s, and then further washed three times with ice-cold PBS.
3. Cells are then incubated with a mouse monoclonal anti-vWF antibody at 1:100 dilution in PBS for ~60 min at room temperature. As a negative control, some cells are incubated with PBS only at this stage.

4. The primary antibody or PBS is then removed and cells washed three times with PBS and incubated for 20 min with normal rabbit serum at 1:50 dilution.
5. After a single wash with PBS, cells are then further incubated for 60 min at room temperature with FITC or Alexa Fluor-488 conjugated rabbit anti-mouse IgG diluted 1:100 in PBS.
6. Finally, cells are washed three times with PBS and viewed under a microscope equipped for epifluorescence with appropriate fluorescence filters. Images are captured using a CCD camera.

3.5. Cryopreservation

HUVEC can be cryopreserved at passage 2 with a recovery rate exceeding ~50% of viable cells which do not appear to be adversely affected by freezing; however, they may exhibit an initial reduced proliferation rate on thawing. The following protocol is suggested, however, other techniques of cryopreservation are also available.

1. Cells at passage 2 should be detached from a 75-cm^2 culture flask as described in Subheading 3.3.
2. Following centrifugation of the cell suspension for 5 min at 1,000 rpm (~150 × *g*), the supernatant is aspirated and the cell pellet resuspended in serum containing culture medium (~1.5 ml) with an additional 10% (v/v) dimethyl sulfoxide (DMSO) and transferred to a suitable cryovial (Nunc).
3. To facilitate a slow freezing rate (~1°C/min), the cryovial is then placed at 4°C for 30 min, transferred to −20°C for 30 min and at −80°C for 2 h before being immersed into liquid nitrogen for long-term (~6 months to 1 year) storage. Alternatively a "Mr. Frosty" cryovial freezing chamber (Nalgene) containing isopropanol is placed at −80°C to facilitate the slow cooling procedure.
4. To thaw cells, the cryovial should be rapidly warmed to room temperature in a water bath and the cell suspension (~1.5 ml) transferred to a new gelatincoated 25 cm^2 culture flask. Pre-warmed serum containing culture medium (~10 ml) is added and cells incubated overnight at 37°C in a 5% CO_2 humidified incubator to facilitate cell attachment, with a change in medium to fresh serum containing medium the next day.

4. Notes

1. Other "endothelial cell optimised" culture media are commercially available, such as Endothelial Cell Growth Media (Promocell or Lonza) which are based on MCDB131 medium.

These media may help to promote more rapid growth of cells, but it should be noted that they may not contain antibiotics, have lower concentrations of serum and include components which could alter HUVEC gene expression, such as insulin and vascular endothelial growth factor.

2. Alternatively, M199 containing 10% (v/v) foetal calf serum (Sigma) and 10% (v/v) new born calf serum (Sigma) can be used to reduce long-term costs. Serum batches should be tested prior to bulk purchase to ensure they support consistent cell proliferation.
3. For maximal HUVEC yield and viability, cells should be isolated from the umbilical cord as soon as possible after child birth. This also reduces the risk of infections arising on initial isolation of cells since the tissues are often handled and excised under non-sterile conditions.
4. Paper towels, cling film, and aluminium foil should be sterilised using 70% ethanol spray to reduce the risk of bacterial and fungal infections.
5. Should infections frequently occur following isolation, additional antibiotics, and fungicides can be supplemented to the culture medium for the initial 24 h following isolation and then the medium replaced. Gentamycin (20 μg/ml) and amphotericin B (2 μg/ml) are commonly used and aliquots can be stored at -20°C as 2 mg/ml and 200 μg/ml stocks respectively.
6. HUVEC cultures left confluent more than 2–3 days will start to detach from the flask and cease to be viable. Primary cultures that start to proliferate but fail to become confluent in 7 days can be sometimes stimulated to grow if detached and seeded in a new culture flask.
7. The trypsin/EDTA solution should be pre-warmed to 37°C only immediately prior to use and not left in a heated water bath for extended periods to prevent loss of activity. Cells should not need incubation with trypsin/EDTA at 37°C for longer than 5 min to detach from the flask and this may indicate that a fresh solution should be prepared.

Acknowledgements

The endothelial cell and vascular biology research in this laboratory is supported by the British Heart Foundation and Heart Research UK. We are grateful to the midwives at the Maternity Unit, St. Thomas' Hospital, Guy's, and St. Thomas' NHS Foundation Trust, for their routine collection of human umbilical cords.

References

1. Tammela, T., Enholm, B., Alitalo, K., and Paavonen, K. (2005) The biology of vascular endothelial growth factors, *Cardiovasc. Res.* 65, 550–563.
2. Ando, J. and Yamamoto, K. (2009) Vascular mechanobiology: endothelial cell responses to fluid shear stress, *Circ. J.* 73, 1983–1992.
3. Davies, P. F. (2009) Hemodynamic shear stress and the endothelium in cardiovascular pathophysiology, *Nat. Clin. Pract. Cardiovasc. Med.* 6, 16–26.
4. Conway, E. M. and Carmeliet, P. (2004) The diversity of endothelial cells: a challenge for therapeutic angiogenesis, *Genome Biol.* 5, 207.
5. Chi, J. T., Chang, H. Y., Haraldsen, G., Jahnsen, F. L., Troyanskaya, O. G., Chang, D. S., Wang, Z., Rockson, S. G., van de, R. M., Botstein, D., and Brown, P. O. (2003) Endothelial cell diversity revealed by global expression profiling, *Proc. Natl. Acad. Sci. USA* 100, 10623–10628.
6. Oh, P., Li, Y., Yu, J., Durr, E., Krasinska, K. M., Carver, L. A., Testa, J. E., and Schnitzer, J. E. (2004) Subtractive proteomic mapping of the endothelial surface in lung and solid tumours for tissue-specific therapy, *Nature* 429, 629–635.
7. Nachman, R. L. and Jaffe, E. A. (2004) Endothelial cell culture: beginnings of modern vascular biology, *J. Clin. Invest* 114, 1037–1040.
8. Loges, S., Roncal, C., and Carmeliet, P. (2009) Development of targeted angiogenic medicine, *J. Thromb. Haemost.* 7, 21–33.
9. Forstermann, U. and Munzel, T. (2006) Endothelial nitric oxide synthase in vascular disease: from marvel to menace, *Circulation* 113, 1708–1714.
10. Napoli, C., de, N. F., Williams-Ignarro, S., Pignalosa, O., Sica, V., and Ignarro, L. J. (2006) Nitric oxide and atherosclerosis: an update, *Nitric. Oxide.* 15, 265–279.
11. Jaffe, E. A., Nachman, R. L., Becker, C. G., and Minick, C. R. (1973) Culture of human endothelial cells derived from umbilical veins. Identification by morphologic and immunologic criteria, *J. Clin. Invest* 52, 2745–2756.
12. Baudin, B., Bruneel, A., Bosselut, N., and Vaubourdolle, M. (2007) A protocol for isolation and culture of human umbilical vein endothelial cells, *Nat. Protoc.* 2, 481–485.
13. Skalli, O., Ropraz, P., Trzeciak, A., Benzonana, G., Gillessen, D., and Gabbiani, G. (1986) A monoclonal antibody against alpha-smooth muscle actin: a new probe for smooth muscle differentiation, *J. Cell Biol.* 103, 2787–2796.

Chapter 19

Human Peripheral Blood Mononuclear Cell Culture for Flow Cytometric Analysis of Phosphorylated Mitogen-Activated Protein Kinases

Athanasios Mavropoulos, Daniel Smyk, Eirini I. Rigopoulou, and Dimitrios P. Bogdanos

Abstract

Lymphocyte activation and fine tuning of downstream signaling circuits for the regulation of cytokine expression are critical for a successful immune response. Hence, technical protocols permitting simultaneous testing of these attributes in peripheral blood lymphocytes are of paramount importance. Phospho-specific flow cytometry is a novel methodology that detects phosphorylation of signaling effectors in multiple, rare cellular populations within peripheral blood. In addition, it allows the quantification of phosphorylation levels for signaling proteins within each single cell, and therefore is superior compared to traditional biochemical approaches, such as Western blotting. One such important signaling pathway within immune cells is the p38 MAPK pathway involved in the regulation of cytokine expression, cell proliferation and apoptosis. In this chapter, we provide technical instructions for culturing human peripheral blood lymphocytes for simultaneous monitoring of p38 MAPK phosphorylation and associated cytokine expression, especially in rare cell populations, such as NK and NKT.

Key words: Peripheral blood, Protein kinases, Flow cytometry, Cell signaling

1. Introduction

Peripheral blood leukocyte analysis has thus far been limited to simple characterization of immune phenotypes and intracellular cytokine expression (1). For more elegant functional studies involving phoshorylation of signaling cascades, prolonged culture of primary cells is necessary in order to achieve maximal cell densities. However, extensive culturing periods, often in the presence of interleukin (IL)-2 and other growth factors, can affect or bias signaling analyses that depend on interaction of such cell receptors

Ragai R. Mitry and Robin D. Hughes (eds.), *Human Cell Culture Protocols*, Methods in Molecular Biology, vol. 806, DOI 10.1007/978-1-61779-367-7_19,

and phosphorylation of intracellular proteins (2, 3). Alternatively, when immortalized cell lines such as Jurkat and NK-92 cells are the culture systems of choice, the fundamental differences in precise signaling regulation between cell lines and primary cells have to be taken into account (4).

The p38 mitogen-activated protein kinase pathway (5) is initiated by cellular stresses or inflammatory cytokines, and regulates cytokine gene expression in part through stabilization of mRNA transcripts (6). This pathway is also important for leukocyte survival, proliferation, and differentiation (7). The kinase is fully activated by dual phosphorylation of threonine-180 and tyrosine-182 residues. There is a plethora of commercial antibodies that recognize the active form of p38 with great selectivity. The mainstream techniques for p38 detection include Western blotting, immunoprecipitation, and confocal immunofluorescence, which especially detects the active p38 kinase translocated to the nucleus (8). These applications, however, involve large populations of homogenous cells and the data obtained is not fully quantitative. It is also not possible to precisely correlate signaling events to a particular subcellular phenotype. Therefore, complex and mixed cell populations, such as peripheral blood mononuclear cells (PBMCs), present a significant challenge to the researcher wishing to study signaling pathways within human NK cells that constitute 15–20% of total PBMCs (9, 10).

The necessity to circumvent these difficulties, together with the evolved technology and availability of reagents, paved the way for the development of alternative protocols based on more sensitive and cell-specific methods, such as flow cytometry (11–14). The emergence of phospho-specific flow cytometry technology delivers extremely rapid, sensitive, and fully quantifiable observations. Moreover, it allows multiparametric analysis of samples containing mixed subpopulations, such as PBMCs (15). This is feasible upon successful combination of fluorochrome-conjugated antibodies against surface markers, such as anti-CD3 and anti-CD56, and intracellular epitopes such as phospho-p38 (9). It follows that upon availability of powerful flow cytometers equipped with multiple lasers and different fluorochrome-conjugated reagents, it is possible to monitor multiple intracellular and/or surface epitopes simultaneously. Therefore, direct analysis of rare populations becomes feasible, as does multiparametric detection of several epitopes within those cells. A typical example would be analysis of both p38 and ERK phosphorylation within rare human NKT cells.

In this chapter, we describe optimized methodology for the successful application of phospho-specific flow cytometry (Fig. 1) in order to detect phosphorylated p38 MAPK within innate immune cells, such as NK and NKT (9). In addition, we provide a new technical extension of the currently available protocols permitting simultaneous flow cytometric measurement of p38 MAPK

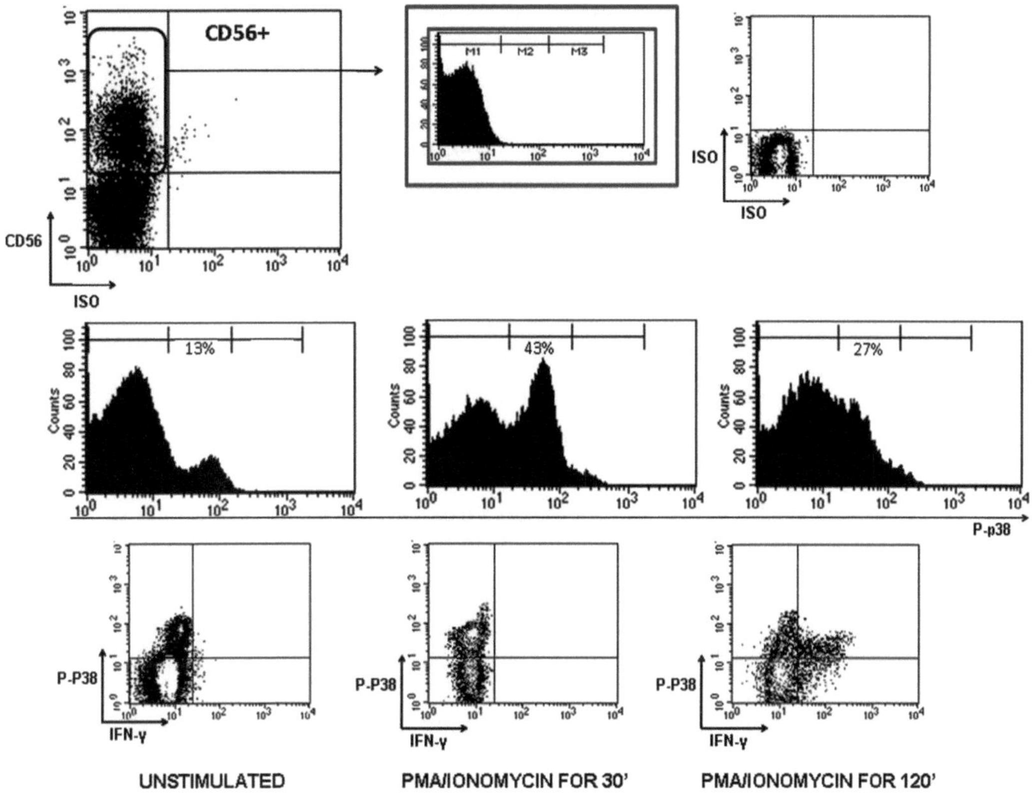

Fig. 1. Isolated PBMCs treated with 50 ng/mL PMA and 1 μM Ionomycin. Cells were harvested at consecutive time intervals and stained using an antibody against the surface epitope CD56. The CD56+ cell fraction was subgated and the percentage of cells positive for phosphorylated p38 MAPK are shown in the histograms. Multiparametric analysis showing p38 MAPK phosphorylation and IFN-γ expression was also conducted.

phosphorylation and intracellular cytokine production. The detection of double-positive (p38 activated and cytokine producing) cells provides direct proof of p38 dependence, and can complement experiments using p38-specific inhibitors to examine whether there is an impact on cytokine expression (16).

2. Materials

2.1. General Equipment

1. Biological safety cabinet certified for Level II handling of biological material.
2. Incubator set at 37°C with 5% CO_2 in air and >95% humidity (e.g., Thermo).
3. Low-speed centrifuge (e.g., Beckman TJ-6) equipped with biohazard containers.
4. Inverted phase-contrast microscope for cell evaluation and preliminary analysis.

5. Neubauer hemocytometer.
6. Vortex.
7. Pipet aid.
8. Micropipettors: 20, 200, and 1,000-μL sterile tips (e.g., Gilson).
9. Culture supplies: 1-, 5-, and 10-mL sterile pipettes; 15- and 50-mL sterile polypropylene tubes.
10. 6- and 24-well tissue culture plates.
11. 12 × 75-mm polystyrene round-bottom tube.
12. Sterilizing equipment.
13. Water purification equipment (if needed).
14. 10-mL syringes (sterile). Syringe-driven filter (0.22-μm pore size, 16 mm in diameter, sterile).
15. Cryovials and passive freezing container (cell cooler or Nalgene "Mr. Frosty™").
16. Cytometer FACSCalibur™ and FACSFlow™, FACSRinse™, FACSClean™ solutions and CaliBrite fluorescent particles all from BD Biosciences.

2.2. Tissue Culture Media, Supplements, and Other Reagents

1. Phosphate-buffered saline (PBS): 140 mM NaCl, 27 mM KCl, 7.2 mM Na_2HPO4, 14.7 mM KH_2PO_4, pH 7.4.
2. Tris-base saline (TBS): 50 mM Tris–HCl and 150 mM NaCl, pH 7.4.
3. Fetal bovine serum (FBS): Sera should be aliquoted and stored at −20°C (see Note 1).
4. Penicillin–streptomycin solution. 100× stock solution. Should be aliquoted and stored at −20°C (see Note 2).
5. L-Glutamine (e.g., Gibco, Invitrogen Ltd) 200 mM stock solution. Should be aliquoted and stored at −20°C.
6. Trypan blue dye (0.4% w/v trypan blue in PBS filtered to remove particulate matter] or eosin stain (0.14% w/v in PBS; filtered) for determination of cell viability.
7. Tissue culture-grade dimethyl sulfoxide (DMSO; e.g., Sigma) stored at room temperature. Freezing medium, freshly prepared and chilled on ice, consisting of 90% FBS and 10% DMSO.
8. RPMI 1640 complete growth medium: RPMI supplemented with 5% FBS, 2 mM glutamine, 100 IU penicillin, and 100 μg/mL streptomycin. Prepare a 500-mL bottle under sterile conditions and store at 4°C for up to 1 month (see Note 3).
9. Methanol absolute.
10. *Para*-formaldehyde solution (e.g., Sigma): 10% *Para*-formal dehyde in PBS or TBS (pH 7.2) (see Note 4).
11. 2% bovine serum albumin (BSA) in TBS.

12. Ficoll-Paque™ Plus, density 1.077 g/mL (e.g., Amersham BioScience). Store at room temperature.
13. Ethylenediaminetetracetic acid (EDTA).
14. Recombinant human cytokine stock solutions: IL-12 (20 ng/mL), IL-18 (25 ng/mL) (see Note 5).
15. Phorbol 12-myristate 13-acetate (PMA) (50 ng/mL) and 1 μM ionomycin.
16. Human peripheral blood (PB) samples buffy coat fraction. Cells are collected using heparin as the anticoagulant following procedures and handling precautions approved by the institution.
17. Saponin (see Note 6).
18. Appropriate normal serum: Goat, rat, mouse, and human.
19. Sodium azide.
20. GolgiPlug™ Brefeldin A containing solution (BD Biosciences).
21. Primary fluorophore-conjugated antibodies. Alexa 647 dye (Molecular Probes, Invitrogen) and phycoerythrin (PE)-conjugated phospho-p38 MAPK (T180/Y182)-specific antibodies were utilized together with fluorescein isothiocyanate (FITC)-conjugated against surface CD3, FITC-conjugated against hIFN-γ, all obtained from BD Biosciences. PE-Cy5-conjugated antibody against CD56 was obtained from Beckman Coulter Inc.
22. SB203580 p38 MAPK inhibitor (Calbiochem). Dissolve in sterile DMSO, aliquot, and store at −20°C.
23. Wash buffer: 2% BSA/0.1% sodium azide/1 mM EDTA in TBS and store at 4°C. TBS and/or PBS can be used to prepare this solution. Generally, TBS is used instead of PBS, where phospho-specific antibodies are used.

3. Methods

3.1. Human Peripheral Blood Mononuclear Cell Isolation

1. Measure and record peripheral blood starting volume.
2. Dilute (optional) with an equal volume of RPMI/1% FBS and mix well by pipetting up and down 4–5 times.
3. Add 15 mL of Ficoll per 50-mL conical tube.
4. Slowly layer 30 mL of the cell suspension per 50-mL tube on the surface of the density gradient medium by resting the tip of the pipette against the side of the tube. It is important not to expel too quickly to avoid mixing the cell suspension into the Ficoll that leads to poor cell recovery.

5. Centrifuge at room temperature for 21 min at 400 × *g* with the brake "off."
6. Using a sterile pipette, carefully remove the cells from the interface between the plasma/medium layer and the Ficoll. Transfer cells to another 50- or 15-mL tube and dilute cell suspension with RPMI/2% FBS.
7. Mix well by pipetting up and down 4–5 times and then centrifuge for 7 min at 300 × *g* with the brake "on."
8. Carefully decant off supernatant, leaving ~0.5 mL of medium on the cell pellet.
9. Resuspend the cells, fill tube with RPMI/1% FBS, and mix well by pipetting up and down 4–5 times. Centrifuge tube(s) for 7 min at 300 × *g* with the brake "on."
10. Carefully decant off supernatant. Add 1–3 mL of RPMI/1% FBS to the cell pellet and make a single-cell suspension by pipetting up and down 4–5 times.
11. Measure and record sample volume. Perform a nucleated cell count using trypan blue exclusion and a Neubauer chamber. A dilution of 1:20 is usually suitable. Calculate and record the cell concentration. Cell viability should exceed 95%.
12. Isolated PBMCs can be aliquoted into separate cryogenic vials, kept at −80°C for 1 day, and then transferred and stored in the vapor phase of liquid nitrogen (see Note 7).

3.2. Stimulation, Fixation, and Permeabilization Methodology

1. For signaling analysis, remove a cryotube (10^7 PBMC) from liquid nitrogen and thaw in a 37°C water bath (see Note 7). Alternatively, freshly isolated PBMCs can be directly processed.
2. Wash cells in serum-free RPMI-1640 to remove the freezing medium, count in 0.2% (w/v) trypan blue to confirm that dead cells are less than 5%, and pellet by centrifugation (5 min at 300 × *g*, brake "on").
3. Resuspend at 10^6 cells/mL in RPMI complete culture medium.
4. Seed PBMCs in 6- or 24-well plates and allow them to rest at 37°C in a CO^2 incubator for at least 1 h before stimulation.
5. All stimulations should be initiated at baseline and cells should be kept in a 37°C CO^2 incubator to allow signal transduction and phosphorylation.
6. To determine basal levels of phosphorylation, unstimulated cells should be maintained and fixed in parallel with stimulated cells.
7. For intracellular cytokine detection, such as IFN-γ, cells should be treated with Brefeldin A (GolgiPlug™, BD Biosciences) during stimulation to block cytokine secretion.

8. For p38 MAPK inhibition experiments, the p38 MAPK inhibitor is added 15′ prior to stimulation at a final concentration of 1–5 μM within the dose-range response previously reported to have complete inhibitory effects on MAPK and IFN-γ expression by PBMC.
9. Stimulate cells by adding PMA plus Ionomycin or IL-12 plus IL-18 at the indicated final concentrations for 30, 60, 120, and 240 min.
10. For fixation and optimal preservation of phosphorylated p38 MAPK in cells, add at the indicated times of the time course 200 μL of 10% *para*-formaldehyde pre-warmed at 37°C directly in the cell culture medium (1 mL) to give a final concentration of 2% and place the cells at 37°C for 10–15 min.
11. For efficient cell permeabilization, two different strategies are recommended. Wash cells, centrifuge, and then vortex mix the pellet thoroughly in slowly added 1 mL of 75–80% (v/v) methanol in TBS. Incubate for at least 30 min on ice and then store at -20°C (up to 3–4 weeks) until further use (see Note 8).
12. Alternatively, cells can be permeabilized using 200 μL of 0.5% BSA/0.1% sodium azide/0.1% saponin in TBS for 30 min. Saponin-treated cells should be always washed in TBS/0.5% BSA/0.1% saponin (see Note 9).

3.3. Intracellular Phospho-Specific Flow Cytometry

1. *Para*-formaldehyde-fixed, methanol-permeabilized cells should be washed twice to ensure complete removal of methanol and rehydrated for 1 h by addition of 1 mL of TBS-based wash buffer, followed by gentle resuspension, and then centrifugation (see Note 10).
2. After pelleting, cells are resuspended in 100 μL of blocking reagent containing 20% (v/v) human plasma in TBS followed by 30′ incubation at room temperature to allow blocking of Fc receptors. Alternatively, cells can be incubated with FcR blocking reagent (e.g., Miltenyi Biotech).
3. Wash cells, pellet, and resuspend in 50 μL 2% BSA/TBS (w/v). Add appropriate amounts of fluorochrome-conjugated antibodies to all samples and leave at room temperature to stain for 1 h (see Note 11).
4. To confirm the specificity of the phospho-p38 antibody, flow cytometry experiments should be performed in the presence of p38 phospho T180/Y182 or non-phospho-peptide competitors (Abcam, Cambridge, UK). The specific antibody for phospho-p38 MAPK has been also validated by direct comparison with Western blotting (see Note 12).
5. Use an aliquot of cells as autofluorescence control and a second stained with the appropriate isotype control IgG antibodies. Incubations should be for 1 h at room temperature. Insignificant

background staining was observed using appropriate control FITC-conjugated, PE-, PC5-, and APC-conjugated antibodies.

6. Flow cytometric analysis can be performed on an FACS Calibur (Becton Dickinson Mountain View, CA) using logarithmic amplification. The cytometer is first calibrated using CaliBrite fluorescent beads. The instrument should be always compensated for optical spillover using three setup samples: one unlabeled, one with the isotype control, and one containing cells labeled with each of the fluorophores used.
7. Create a gate around the total lymphocyte population based on forward and side scatter intensities. Collect 100,000 cells within the lymphocyte gate for each sample and use BD CellQuest software (BD Bioscience) for data acquisition and off-line analysis. For each of the gated populations, the percentage of gated cells and the geometric mean fluorescence intensity should be analyzed.

4. Notes

1. Purchased FBS for cell culture applications should be certified mycoplasma-free and endotoxin low/negative. FBS can be heat inactivated by incubation at 56°C for 30 min to inactivate complement. Heat-inactivated FBS can be cooled overnight at 4°C and then aliquoted under sterile conditions for long-term storage at −20°C.
2. Antibiotics are not necessary in the culture of mammalian cells. However, they protect against bacterial or fungal contamination and are thus recommended for use by inexperienced personnel. It is however desirable, that the antibiotics are not included in the culture media to minimize the emergence of antibiotic-resistant bacterial strains. It is also advisable to culture primary cells without antibiotics to eliminate the likelihood of biological influences of these agents on cell behavior.
3. Culture media include phenol red as a pH indicator. Repeated usage results in a shift in the pH and a change in the color from red to light purple. Most cells display optimal growth within a defined physiological pH range. If the pH does change, the media should be discarded and freshly prepared. The same should be done in the case that the culture medium looks cloudy since this is an indicator of bacterial contamination.
4. Dissolve the desirable amount of *para*-formaldehyde powder, e.g., 10 g in 100 mL of TBS in a fume hood. *Para*-formaldehyde is toxic and volatile, therefore avoid breathing vapors from dissolved *para*-formaldehyde or powder. Heat to 70°C while

stirring (monitor temperature with thermometer and avoid boiling because it can volatilize and pose a serious hazard for respiratory system). Add 50–100 μL of 2 N NaOH to clear the solution. Remove the flask from the heat, let it cool, and adjust to pH 7.2 with HCl. Filter and store at room temperature.

5. Cytokines are usually available from suppliers, such as R&D Systems, in dry lyophilized powder forms. Reconstitute according to manufacturer's instructions. For human IL-18, reconstitute at 25 μg/mL aliquot and store at −80°C for up to 6 months.

6. Prepare a 10% saponin stock solution by mixing 10 g of saponin with 100 mL of TBS. Place at 37°C with mild stirring until saponin has completely dissolved. Sterile filter and store at 4°C. Saponin concentrations for optimal permeabilization range from 0.1 to 0.5%. The process is reversible and therefore antibody cocktails need to be prepared in saponin-based buffers.

7. Cryovials containing primary cells should be placed in passive freezing containers (i.e., Nalgene "Mr. Frosty") and stored at −80°C overnight before transferred to liquid nitrogen for long-term storage. Alternatively, cells can be kept in a Styrofoam rack for overnight storage. Caution should be taken when removing vials from the liquid nitrogen tank since there is a possibility that liquid nitrogen is trapped into the vial and the pressure generated while the vial warms might cause it to explode. Always wear a face shield and gloves when removing frozen vials of cells. Take care to immerse only the bottom half of the vial into the water bath to prevent leakage of water into the vial. Once the cells have nearly thawed, remove the vial from the water bath. Rinse with 70% ethanol to decontaminate it prior to culture procedures.

8. Methanol permeabilization is a prerequisite for optimal detection of phospho p38 MAPK, and also allows efficient intracellular cytokine determination (9) in contrast to conventional saponin-based permeabilization protocols, which accommodate proper intracellular cytokine analysis but do not permit p38 MAPK or other phospho-specific analysis (14, 17). In saponin-treated cells, however, labeling of surface proteins is optimal. Methanol compromises the labeling of surface epitopes and this varies among available commercial antibodies.

9. Saponin-based methodology additionally requires the use of phosphatase and protease inhibitors and ice-cold buffers to eliminate enzymatic activity that can deteriorate phosphorylation events.

10. The epitopes on several surface proteins are detectable only after extensive rehydration (17, 18). Otherwise, they can be easily compromised and removed especially after high-concentration methanol fixation. Typically, staining for surface antigens needs

to be evaluated using saponin- and methanol-containing buffers. Yet, the two-step staining approach for saponin (i.e., first stain for surface markers, and then permeabilize and stain for intracellular markers) should be generally avoided when methanol is utilized. In this chapter, we describe for the first time that the one-step staining approach after rehydration and receptor blocking is efficient also for optimal surface epitopes and intracellular cytokine detection.

11. The antibody concentrations necessary for saponin-based methods are higher than those for methanol permeabilization. Typically, phospho-specific antibodies stain at 0.5–1 μg/10^6 cells for saponin-based protocols, and 0.1–0.5 μg/10^6 cells for methanol protocol (17).

12. Successful evaluation of phospho specificity can be achieved by various approaches. Either outcompete phospho-antibody with phospho-peptide or treat samples with phosphatase cocktails after activation. Alternatively, generate kinase-deficient or knock-out cells. It is highly recommended to verify phosphorylation conditions by traditional immunoblot analysis (9, 19, 20).

References

1. Loza, M. J., Faust, J. S., and Perussia, B. (2003) Multiple color immunofluorescence for cytokine detection at the single-cell level. *Mol Biotechnol* 23, 245–58.
2. Cheng, L. E., Ohlen, C., Nelson, B. H., and Greenberg, P. D. (2002) Enhanced signaling through the IL-2 receptor in CD8+ T cells regulated by antigen recognition results in preferential proliferation and expansion of responding CD8+ T cells rather than promotion of cell death. *Proc Natl Acad Sci USA* 99, 3001–6.
3. Bessoles, S., Fouret, F., Dudal, S., Besra, G. S., Sanchez, F., and Lafont, V. (2008) IL-2 triggers specific signaling pathways in human NKT cells leading to the production of pro- and anti-inflammatory cytokines. *J Leukoc Biol* 84, 224–33.
4. Wang, K. S., Ritz, J., and Frank, D. A. (1999) IL-2 induces STAT4 activation in primary NK cells and NK cell lines, but not in T cells. *J Immunol* 162, 299–304.
5. Saklatvala, J. (2004) The p38 MAP kinase pathway as a therapeutic target in inflammatory disease. *Curr Opin Pharmacol* 4, 372–7.
6. Clark, A. R., Dean, J. L., and Saklatvala, J. (2003) Post-transcriptional regulation of gene expression by mitogen-activated protein kinase p38. *FEBS Lett* 546, 37–44.
7. Dodeller, F., and Schulze-Koops, H. (2006) The p38 mitogen-activated protein kinase signaling cascade in CD4 T cells. *Arthritis Res Ther* 8, 205.
8. Perfettini, J. L., Castedo, M., Nardacci, R., Ciccosanti, F., Boya, P., Roumier, T., Larochette, N., Piacentini, M., and Kroemer, G. (2005) Essential role of p53 phosphorylation by p38 MAPK in apoptosis induction by the HIV-1 envelope. *J Exp Med* 201, 279–89.
9. Mavropoulos, A., Sully, G., Cope, A. P., and Clark, A. R. (2005) Stabilization of IFN-gamma mRNA by MAPK p38 in IL-12- and IL-18-stimulated human NK cells. *Blood* 105, 282–8.
10. Vivier, E., Nunes, J. A., and Vely, F. (2004) Natural killer cell signaling pathways. *Science* 306, 1517–9.
11. Far, D. F., and Rossi, B. (2000) Measurement of protein tyrosine phosphorylation in T-cell subsets by flow cytometry. *Methods Mol Biol* 134, 301–6.
12. Montag, D. T., and Lotze, M. T. (2006) Rapid flow cytometric measurement of cytokine-induced phosphorylation pathways [CIPP] in human peripheral blood leukocytes. *Clin Immunol* 121, 215–26.
13. Perez, O. D., and Nolan, G. P. (2002) Simultaneous measurement of multiple active kinase states using polychromatic flow cytometry. *Nat Biotechnol* 20, 155–62.
14. Krutzik, P. O., Irish, J. M., Nolan, G. P., and Perez, O. D. (2004) Analysis of protein phosphorylation and cellular signaling events

by flow cytometry: techniques and clinical applications. *Clin Immunol* 110, 206–21.

15. Grammer, A. C., Fischer, R., Lee, O., Zhang, X., and Lipsky, P. E. (2004) Flow cytometric assessment of the signaling status of human B lymphocytes from normal and autoimmune individuals. *Arthritis Res Ther* 6, 28–38.
16. Koprak, S., Staruch, M. J., and Dumont, F. J. (1999) A specific inhibitor of the p38 mitogen activated protein kinase affects differentially the production of various cytokines by activated human T cells: dependence on CD28 signaling and preferential inhibition of IL-10 production. *Cell Immunol* 192, 87–95.
17. Perez, O. D., Krutzik, P. O., and Nolan, G. P. (2004) Flow cytometric analysis of kinase signaling cascades. *Methods Mol Biol* 263, 67–94.
18. Irish, J. M., Hovland, R., Krutzik, P. O., Perez, O. D., Bruserud, O., Gjertsen, B. T., and Nolan, G. P. (2004) Single cell profiling of potentiated phospho-protein networks in cancer cells. *Cell* 118, 217–28.
19. Haas, A., Weckbecker, G., and Welzenbach, K. (2008) Intracellular Phospho-Flow cytometry reveals novel insights into TCR proximal signaling events. A comparison with Western blot. *Cytometry A* 73, 799–807.
20. Aerts, N. E., Ebo, D. G., Bridts, C. H., Stevens, W. J., and De Clerck, L. S. (2009) Flow cytometric analysis of phospho-p38 mitogen-activated kinase (MAPK): p38 MAPK does not mediate the effect of adalimumab on peripheral T cell cytokine production in rheumatoid arthritis. *Cytokine* 47, 178–84.

Chapter 20

Human CD4+CD25^{high}CD127$^{low/neg}$ Regulatory T Cells

Haibin Su, Maria Serena Longhi, Pengyun Wang, Diego Vergani, and Yun Ma

Abstract

CD4+CD25^{high}CD127$^{low/neg}$ regulatory T cells (Tregs) play a critical role in the maintenance of peripheral tolerance and in controlling the development of autoimmune diseases. A combination of surface and intracellular markers, namely, CD25, CD39/CD73, CD62L, CD45RO, CD127, glucocorticoid-induced tumor necrosis factor receptor (GITR), CTLA-4, and the forkhead/winged helix transcription factor (FOXP3), has been used to characterize Tregs. Tregs suppress T effector responses mainly in a direct cell–cell contact manner. However, other mechanisms independent from this manner cannot be excluded entirely. It has been shown that Tregs can undergo limited expansion in vitro after the stimulation of TCR in the presence of exogenous cytokines, e.g., IL-2. Expanded Tregs retain their suppression function. Human Tregs have demonstrated their great potential to be used as a therapeutic intervention in preventing graft rejection and treating autoimmune diseases. In this chapter, we have given a review on how to characterize, isolate, expand Tregs and assess their suppressive functions.

Key words: Regulatory T cells, Autoimmunity, Immunnoregulation, Tolerance

1. Introduction

Among several subsets of T cells, which have (1) the ability of immunoregulation and immunosuppression, such as natural killer T cells (NKT), T regulatory cell 1, γδ T cells, and CD4+CD25+CD127$^{low/neg}$ T cells (Tregs), Tregs have stood out and been the focus since their identification in 1995 by Sakaguchi and colleagues, who reported that Tregs constitutively express CD25 (IL-2 receptor α chain) and can suppress autoreactive T cells in vivo (2). CD4+CD25+ Tregs represent a minor population with CD4+ T cells, being 5–10% in healthy humans and mice (3–5). Removal of these cells in early life results in the development of

Ragai R. Mitry and Robin D. Hughes (eds.), *Human Cell Culture Protocols*, Methods in Molecular Biology, vol. 806, DOI 10.1007/978-1-61779-367-7_20, © Springer Science+Business Media, LLC 2012

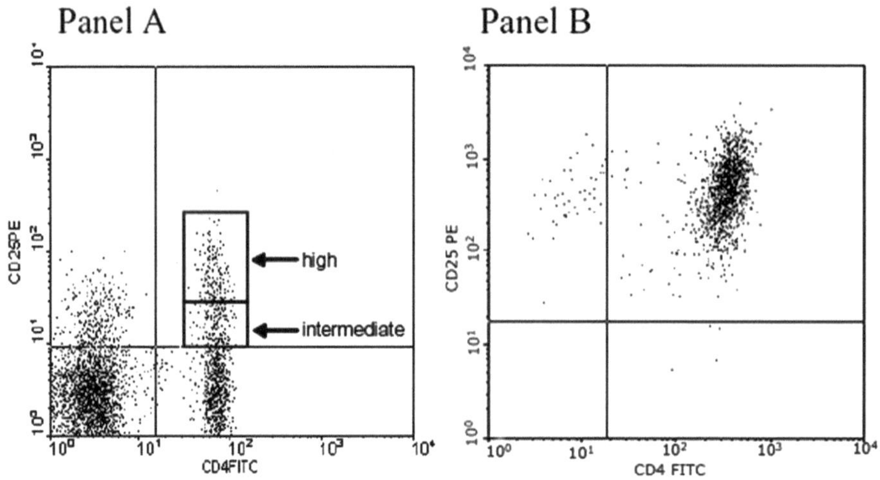

Fig. 1. CD25^{int} and CD25^{high} T cells within CD4+CD25+ T cells. *Dot plot* of anti-CD25 PE fluorescence (*Y*-axis; log scale) vs. anti-CD4 FITC fluorescence (*X*-axis; log scale). (**a**) CD4+CD25^{int} and CD4+CD25^{high} within lymphocytes are indicated by *arrows*. (**b**) Purified CD4+CD25^{high} cells by FACS sorting. Data are obtained from a representative healthy subject.

various T-cell-mediated murine autoimmune diseases, such as autoimmune thyroiditis, gastritis and insulin-dependent diabetes (6), while reconstitution of the CD4+CD25+ Tregs prevents these pathological conditions (2). In patients with autoimmune diseases, CD4+CD25+ Tregs are either markedly reduced in number or impaired in function, or show both impairments (7–10). Depletion of CD4+CD25+ T cells in humans leads to marked clonal expansion of autoreactive T cells in vitro (11). In addition to suppress proliferation and proinflammatory cytokine production of CD4+CD25− T effectors in vitro, human Tregs can modulate the responses of CD8+ T cells and NKI cells (12, 13). In patients with autoimmune hepatitis, we observed both numeric and functional impairment of Tregs (14).

In mouse, FACS profiles show that CD4+CD25+ cells constitute a distinct population that can be easily distinguished from the CD4+CD25− cells. In human, CD4+CD25+ T cells exhibit a continuous expression of CD25, which can be sub-divided into CD25^{int} and CD25^{high} T-cell populations. However, there is no unanimously consensus as to where the boundary between CD25^{int} and CD25^{high} should be put (Fig. 1).

1.1. Biologic Characteristic of Human CD4+CD25+CD127$^{low/neg}$ Regulatory T Cells

1.1.1. Phenotype

Identification of phenotype of Tregs is important in terms of characteristic and therapeutic manipulation of Tregs. Because the CD25^{int} and CD25^{high} cells exhibit suppressive function at different levels (15), it is not entirely clear which subpopulation of CD25+ T cells should be considered as “true” Tregs before identifying other surface and intracellular makers for human Tregs listed below.

CD39/CD73: CD39 is expressed on all Foxp3+ T cells in mice and a subset of naturally occurring Tregs in humans (16, 17). Tregs from CD39$^{-/-}$ mice had reduced suppressive capacity in vitro and failed to prevent allograft rejection in vivo (18), suggesting that hydrolysis of ATP and/or generation of adenosine via the expression of CD39 and CD73 represents an important mechanism of immunoregulation. CD4+CD25^{high}CD39+, which are predominantly Foxp3+, suppressed IL-17 production, whereas CD4+CD25^{high} CD39− T cells produced IL-17 in patients with multiple sclerosis (19).

CD45RO: Tregs isolated from peripheral blood can be classified as CD45RO+ and CD45RA negative (20, 21). Although both cell subsets have inhibitory function, CD45RO+ cells exhibit five-fold more suppression and only 20% of the proliferative ability compared to the of CD45RO− subset (22). However, cord blood-derived CD4+CD25+CD45RA+ T cells are suppressive as CD45RA negative cells (23). Therefore, separation of "true" Tregs based on expression of surface marker CD45 isoform is not adequate.

CD62L: Both mouse and human CD4+CD25+ Tregs can be divided into subsets according to the cell-surface expression of CD62L. Though both subsets express FOXP3 and are anergic, the CD62L+ population is more potent on a per cell basis, and proliferates and maintains suppressive function better than the CD62L− population and undivided CD4+CD25+ Treg (24, 25).

CD127$^{low/neg}$: CD127 is the α chain of IL-7 receptor. CD127$^{low/neg}$ has become a feasible marker for isolation of "true" Tregs. It was shown by both Seddiki et al. and Liu et al. that there is a significant correlation between FOXP3 and CD127$^{low/neg}$ phenotype, expression of CD127 being inversely correlate with Tregs suppressive function (12, 13). Assessing CD127 expression can be used to assist the isolation of CD25^{high} cells (Fig. 2).

Glucocorticoid-induced tumor necrosis factor receptor (GITR): GITR is expressed mainly on the surface of CD4+CD25+ Tregs, CD4+CD25+CD8− thymocytes, and activated T cells. Blockade of GITR by administrating specific antibody results in development of organ specific autoimmune diseases in mice, indicating induction of autoimmunity by functionally interfering with Tregs (26). Since CD4+ T cells also express GITR after activation (27), the use of GITR as a surface marker to isolate CD4+CD25+ Tregs is not appropriate.

CTLA-4: It is induced by T-cell activation (28). The studies investigating CTLA-4 function in Tregs in vitro are contraversial. For example, CD4+CD25+ Tregs-mediated suppressive function to T effectors was abolished by the blockade of CTLA-4 on CD4+CD25+ Tregs in vitro (29). However, CD4+CD25+ Tregs from CTLA-4-deficient mice still retain their inhibitory capacity in vitro, although they suffer from a lethal lymphoproliferation and

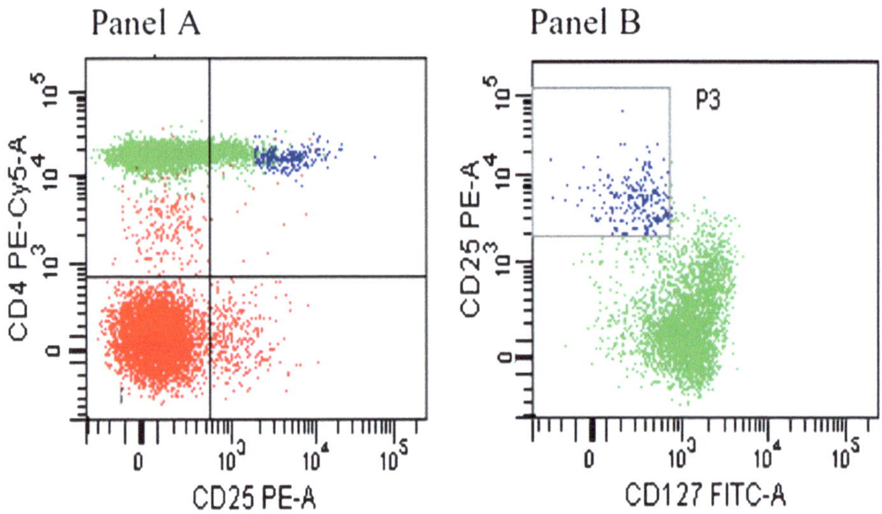

Fig. 2. Flow cytometric analysis of CD4+CD25^{high}CD127$^{low/neg}$ Tregs in a representative healthy subject. CD4+CD25^{high}CD127$^{low/neg}$ Tregs are analyzed by flow cytometry using FACS Diva version 6.1.2 software. PBMCs are stained with PE-CY5 conjugated anti-CD4, PE-conjugated anti-CD25, and FITC-conjugated anti-CD127 monoclonal antibodies. CD4+CD25^{high} Tregs are shown in *upper right quadrant* (**a**) in *blue color*, CD4+CD25^{int} and CD25 negative T cells are shown in *green color*. (**b**) Total CD4 T cells are gated. Those CD4+CD25^{high} T cells in *blue* appear to be CD127$^{low/neg}$.

autoimmune syndrome (30), indicating that CTLA-4 expression is not absolute necessary for the suppressive function of CD4+CD25+ Tregs.

Forkhead/winged helix transcription factor (FOXP3) (31, 32): Expression of FOXP3 by Tregs positively correlates with their regulatory properties (33). Its mutation results in the autoimmune disease of the scurfy mouse (34) and the human neonatal immune dysregulation, polyendocrinopathy, enterophahy X-linked (IPEX) syndromes (35, 36). However, FOXP3 expression is not expressed by CD4+CD25+ T cell only, but also expressed in a proportion of CD4+CD25− and CD8+ T cells (30, 37). Since FOXP3 is expressed in the nucleus, therefore require cell fixation and permeabilization for detection, it cannot be used to purify live Tregs for functional study (38).

1.1.2. Main Mechanism of Suppressive Function of CD4+CD25^{high} Tregs

The mechanism of suppression of Tregs has not been fully understood. The most consensus view is that CD4+CD25^{high} Tregs execute suppression by direct cell–cell contact manner in vitro. In a transwell experiment (30), we found that direct contact is necessary for Tregs to suppress proliferation and IFN-γ production by CD4+CD25− and CD8+ T effectors in healthy subjects and also in patients with AIH. Moreover, in both populations, direct contact of Tregs with their targets leads to increased secretion of regulatory cytokines IL-4, IL-10, and TGF-β, suggesting a mechanism of linked immunosuppression.

Several studies have demonstrated that anti-IL-4, anti-IL-10, and anti-TGF-β were unable to restore proliferation of CD4+ and CD8+ T cells cultured in direct contact with Tregs, suggesting that Treg suppression function may be irrelevant to the action of these regulatory cytokines. The role of TGF-β is controversial, with some studies showed no effect on Tregs after blocking it however, and others studies reporting that neutralizing TGF-β reduces suppression (39, 40).

2. Materials

1. Dynabeads regulatory CD4+CD25+ T-cell kit (Dynal – Invitrogen, UK).
2. Isolation buffer: Ca^{2+}- and Mg^{2+}-free phosphate-buffered saline (PBS) supplemented with 0.1% BSA and 2 mM EDTA.
3. Fetal calf serum (FCS) (Sigma, UK).
4. Human AB serum (Sigma, UK).
5. RPMI1640 (Invitrogen, UK).
6. Anti-CD4 FITC-conjugated, anti-CD25 PE-conjugated, and other fluorescence-conjugated MAbs (BD Biosciences, UK).
7. X-VIVO 15 medium (Cambrex, UK).
8. Penicillin/streptomycin (Invitrogen, UK).
9. L-Glutamine (Invitrogen, UK).
10. 4-(2-Hydroxyethyl)-1-piperazine ethanesulfonic acid (Sigma, UK).
11. Amphotericin B (Invitrogen, UK).
12. T-cell expander (CD3/CD28 Dynabeads, Dynal – Invitrogen, UK).
13. Phorbol 12-myristate 13-acetate (PMA) (Sigma, UK).
14. Human recombinant IL-2 (PeproTech, UK).
15. Trypan blue (Sigma, UK).
16. Paraformaldehyde (BD Biosciences, UK).
17. Ionomycin (Sigma, UK).
18. Brefeldin A (Sigma, UK).
19. Cytoperm/Cytofix solution (BD Biosciences, UK).
20. [^{3}H]-Thymidine (PerkinElmer, UK).
21. Anti-human-IL-4 (clone 34019.111), anti-human-IL-10 (clone 23738), and anti-human-TGF-β (clone 9016) neutralizing antibodies (BD Biosciences, UK).

3. Methods

3.1. Isolation of CD4+CD25^{high} Regulatory Cell from Peripheral Blood

This chapter is focuses on isolation of human CD4+CD25^{high} Tregs from peripheral blood mononuclear cells (PBMCs), which can be separated from peripheral blood by Ficoll-Paque gradient centrifugation (method for PBMCs separation is not included). Two major methods have been developed to isolate Tregs, mainly immuno-magnetic beads and FACS sorting. We have tested magnetic beads from Dynal Invitrogen and Miltenyi Biotec, Germany and obtained comparable results. Purity higher than 95% can be achieved by magnetic beads from both biocompanies. This chapter describes the application of Dynal beads (see Note 1).

3.2. Isolation of CD4+CD25^{high} Tregs by Dynal Magnetic Beads

Two sequential steps are included in this method. First, total CD4+ T cells are selected by negative selection, using a cocktail of antibodies to CD14, CD56, CD19, CD8, CD235a, and CD45RA and depletion beads coated with an Fc-specific human immunoglobulin G4 antibody; second, CD25+ cells are isolated by positive selection using immunomagnetic beads coated with anti-human CD25 antibodies. A summary of each step is given below. It is based on manufacture's instruction.

1. A cocktail of antibodies is added to PBMCs resuspended in isolation buffer containing FCS. Cells are then incubated in the dark at 4°C for 20 min. After washing, depletion beads are added. Cells are incubated for 15 min at room temperature with rolling and tilting.
2. The reaction tube is then placed in a magnetic stand for 3 min to remove all the non-CD4+ cells. Transfer supernatant that containing CD4+ T cells to a new tube. Repeat this step to increase the purity of CD4+ cells.
3. Re-suspend CD4+ T cells in isolation buffer and adjust the concentration to 1.5×10^7 cells/ml, and add 200 μl Dynabeads CD25 per 1.5×10^7 CD4+ cells.
4. After incubation at 2–8°C for 25 min, the tube is placed in the magnetic stand to get supernatant containing CD4+CD25− T cells (effectors). Re-suspend bead-bound cells and place the tube in magnetic stand again to remove residual non-CD4+CD25+ as completely as possible (see Note 2).
5. Finally, bead-bound cells (CD4+CD25+ T cells) are re-suspended in RPMI1640 with 1% FCS and add 80 μl Detach beads. After incubating at room temperature for 45 min, the tube is placed in the magnet. The supernatant that contains CD4+CD25+ Tregs can be harvested now.
6. CD4+CD25^{high} Tregs and CD4+CD25− effector T cells can be used in downstream application because all the cells are

beads free. However, Tregs purity should be checked by FACS before proceed to next experiment.

3.3. Isolation of CD4+CD25^{high} Tregs by FACS Sorting

Tregs can be isolated directly from either total PBMC or CD4+ T cells isolated by commercially available CD4+ T-cell isolation kits. The advantage of this method is that is simple and fast (see Note 3).

1. Total PBMCs or CD4+ cells are re-suspended in RPMI1640 supplemented with 1% FBS at a concentration of 1×10^7 cells/well. Then, anti-CD4 FITC and anti-CD25 PE-conjugated MAb (20 μl/10^7 cells) are added if PBMCs are used anti-CD25 PE-conjugated MAb is added if purified CD4+ T cells are used (20 μl/10^7 cells) only is added for CD4+ T cells. When a smaller number of cells are used for sorting, the amount of antibodies can be scaled down (see Note 4).
2. Incubate the mixture of cells and antibodies at 4°C for 30 min in the dark.
3. Wash the cells twice with RPMI containing 1% inactivated FCS, and re-suspend cells at a final concentration of 2×10^7 cells/well.
4. To avoid cell cluster blocking FACs sorter, the cells should be passed through a cell strainer (30 μm pore size) before transferring into a sterile FACS tube. Keep the cells on ice until the FACS sorter is ready.
5. Prepare collection tubes for positive and negative fractions. Set the gate for positive selection to include only the CD4+CD25^{high} population and collect the whole CD4+CD25− population into the tube for negative fraction.
6. When separation is complete, centrifuge the two fractions, re-suspend the cells in X-VIVO 15 medium supplemented with 5% human AB serum and 100 U/ml penicillin/ 0.1 mg/ml streptomycin, and count.
7. Check the purity of the collected fractions before proceed to next experiment.

3.4. Culturing and Expansion of Human CD4+CD25^{high} Tregs

CD4+CD25^{high} Tregs are known to be anergic (41). Because of their poor ability to expand and their small number in peripheral blood, it is a challenge to explore the therapeutic potential of Tregs.

Up to date, it has been shown that bulk population of Tregs can undergo expansion in vitro after stimulation of TCR in the presence of exogenous cytokines and expanded Tregs retain their suppression function when co-cultured with T responder cells (20, 42). Our group has successfully expanded CD4+CD25^{high} Tregs in health and also in autoimmune hepatitis, using anti-CD3/anti-CD28 T cell expander and a high dose of IL-2, the cytokine critical for Tregs survival and growth. However, CD4+CD25^{high} Tregs obtained from AIH, though retain their suppression function, are less effectively expanded than those from healthy controls in terms of their fold expansion and ability of suppression (43).

In vitro, CD4+CD25^{high} Tregs must be activated through TCR to be operationally suppressive (4, 44–46). Once activated, Tregs suppress both CD4+ and CD8+ T effectors by either nonantigen-specific or antigen-specific fashion (47).

Overall, more evidence supports the notion that human CD4+CD25^{high} Tregs have great potential to be used as a therapeutic intervention in treating autoimmune diseases. In the last decade, simple and reproducible methods to isolate "true" Tregs have attempted.

1. 2×10^4 cells/well are cultured in 96-well plate with RPMI1640, supplied with 2 mM L-glutamine, 25 mM [4-(2-hydroxyethyl)-1-piperazine ethanesulfonic acid], 100 U/L penicillin, 0.1 mg/ml streptomycin, 2.5 μg/ml amphotericin B, and 10% FCS at 37°C and 5% CO_2, anti-CD3/anti-CD28 T cell expander (four beads per cell) and 300 U/ml recombinant IL-2 should be added.
2. Replace the culture medium with fresh one containing 100 U/ml recombinant IL-2 every 3 days. Cell viability and expansion should be assessed weekly by Trypan blue exclusion.
3. Transfer the cells into 24 well, then 12-well plate when the cell number is higher than 3×10^6 per ml (see Note 5).

3.5. Assessment of Suppressive Function of CD4+CD25^{high} Tregs

3.5.1. Phenotype Analysis

Human Tregs should be defined using a combination of staining for the surface and intracellular markers mentioned in Subheading 2 (see Note 6). For detection of surface markers and analysis by FACS, the following procedures should be carried out:

1. Harvest $1–5 \times 10^5$ cells from culture and wash twice with PBS/1% FCS at room temperature.
2. Re-suspend cells in cold PBS/1% FCS and add 2 μl fluorescence-conjugated antibodies per 1×10^5 cells.
3. Incubate the cells in the dark for 20 min at 4°C. Then wash twice with PBS/1% FBS.
4. Add 300 μl PBS/1% FCS and acquire the cells by FACS. If the samples are not analyzed immediately, they should be kept in PBS containing 1% paraformaldehyde at 4°C (see Note 7).

3.5.2. Proliferation Assay

Proliferation assay has been used effectively to assess the suppressive function of human CD4+CD25^{high} Tregs. It can be carried out by co-culturing Tregs with CD4+, CD8+ T effectors. In order to maintain Treg suppressive function in vitro, (anti-CD3/CD28) expander and IL-2 should add into the coculture. This polyclonal

Stimulus can also able to induce target T-cell proliferation. Furthermore, neutralization assays may be performed in parallel by adding neutralizing antibodies, such as anti-IL-4, anti-IL-10, anti-IL-17, anti-IFN-γ, and anti-TGFβ, into the coculture in order to define whether the neutralization of these cytokines affects Tregs function or not.

1. CD4+CD25− or CD8+ T effectors (1–5 × 10^5 cells/well) are plated alone or with Tregs at various ratios, e.g., Tregs: T effectors at 1:1, 1:2, 1:4, 1:8, 1:16, and 1:32. For stimulation, T-cell expander (four beads per 100 cells) and low-dose IL-2 (30 U/ml) will be added into 96-well plate. The cells are cultured in a final volume of 200 μl complete medium.
2. Fresh IL-2 can be added every 3 days.
3. After a 5-day culture, 100 μl of supernatant are harvested from each well for cytokine measurement, and then 0.25 μCi [^{3}H] thymidine is added to each well.
4. The cells are cultured for further 16–18 h and then harvested. The amount of incorporated [^{3}H] thymidine is determined by a β counter (Canberra Packard Ltd, UK). Percentage inhibition can be calculated as: 1 − (cpm in the presence of Tregs/cpm in the absence of Tregs) × 100.

3.5.3. Intracellular Cytokine Staining

Cytokines secreted by Tregs in the culture supernatant can be detected by ELISA, but the pitfall of this assay is that cytokines produced by target cells cannot be excluded. For this reason, we prefer to choose intracellular cytokine staining, which can be used for the simultaneous analysis of surface markers of Tregs and intracellular cytokines at single-cell level.

1. On the last day of coculture, cells are exposed, PMA (10 ng/ml) and Ionomycin (500 ng/ml) in complete medium. At the same time, Brefeldin A (10 μg/ml) is added to block cytokine releasing. In vitro stimulation of cells is usually required for the detection of cytokines by flow cytometry, otherwise the cytokine levels could be too low to be detected.
2. Harvest cells after incubation for 5 h at 37°C and 5% CO_2 (see Notes 8 and 9).
3. After washing twice with cold PBS/1% FCS, cells are stained with antibodies directed to surface markers (CD4, CD25, etc.).
4. Re-suspend cells in Cytoperm/Cytofix solution and incubate them for 30 min in the dark to permealize and to fix them.
5. Stain the cells with fluorescence-conjugated cytokine antibodies and incubate for in the dark 30 min at 4°C.
6. Wash the cells once with PBS/1% FCS and analyze the cytokine profiles by flow cytometry.

4. Notes

1. It has been reported that the percentage of CD4+CD25^{high} Tregs isolated by magnetic beads varies from 6 to 30% among PMBCs (45). The ratio of bead volume to cell number used in isolation may influence Treg yield.
2. It is impossible to remove all CD4+CD25− T cells in this way. The purity of CD4+CD25^{high} Tregs varies between blood samples and it should be carefully checked before starting downstream in vitro experiments. Although the purity of CD4+CD25^{high} Tregs can be improved by using Miltenyi CD4+CD25+CD127$^{dim/-}$ regulatory T-cell isolation kit II, the final yield of Tregs may be small unless a big number of PBMCs are available as start point. In addition, the users should keep in mind that time required by Miltenyi method is longer than using Dynabeads.
3. The expected yield of CD4+CD25^{high} cells is usually less than 2% of the total PBMC and 5% of the CD4+ T cells, and a purity of at least 90%. This method requires expensive instrumentation (FACS sorter) and the running cost is also high. In addition, the chance of Treg contamination is also much higher than that of using magnetic beads. Therefore, downstream functional assays might not be carried out.
4. The quantity of antibodies used for staining is based on the standard recommendation of BD Biosciences and it might not be optimal for isolation of Tregs. The final quantity of antibody used should be justified through titration, especially when antibodies obtained from biocompanies other than BD.
5. It is necessary to examine surface markers (CD4, CD25) and FOXP3 on the cultured cells frequently because some of CD4+CD25^{high} T cells may lose CD25 marker. In addition, staining for CD127 should be performed to make sure that the expanded Tregs are CD127$^{low/neg}$.
6. For the highly specific marker, FOXP3, intracellular staining should be used. The technique is similar to that described in Subheading 3.5.3.
7. In order to get the most reliable result, the maximum time for preserving cells in PBS containing 1% paraformaldehyde at 4°C should not exceed 48 h. This option is subject to the fluorochromes used, e.g., APC-Cy5 can disassociate in the presence of formaldehyde. APC-H7 may be selected to replace APC-Cy5.
8. The reagents and duration of stimulation should be optimized. For example, the best time for the detection of most cytokines

is 6 h, whereas induction of IL-10 may need at least 24 h stimulation.

9. For detection of intracellular cytokines, it is necessary to block the release of cytokines with reagents such as Monensin or Brefeldin A during the last 5 h of stimulation.

References

1. Alyanakian, M. A., S. You, D. Damotte, C. Gouarin, A. Esling, C. Garcia, S. Havouis, L. Chatenoud, and J. F. Bach. 2003. Diversity of regulatory CD4+ T cells controlling distinct organ-specific autoimmune diseases. *Proc Natl Acad Sci USA 100:15806.*
2. Sakaguchi, S., N. Sakaguchi, M. Asano, M. Itoh, and M. Toda. 1995. Immunologic self-tolerance maintained by activated T cells expressing IL-2 receptor alpha-chains (CD25). Breakdown of a single mechanism of self-tolerance causes various autoimmune diseases. *J Immunol 155:1151.*
3. Baecher-Allan, C., J. A. Brown, G. J. Freeman, and D. A. Hafler. 2001. CD4+CD25high regulatory cells in human peripheral blood. *J Immunol 167:1245.*
4. Shevach, E. M., R. S. McHugh, C. A. Piccirillo, and A. M. Thornton. 2001. Control of T-cell activation by CD4+ CD25+ suppressor T cells. *Immunol Rev 182:58.*
5. Ng, W. F., P. J. Duggan, F. Ponchel, G. Matarese, G. Lombardi, A. D. Edwards, J. D. Isaacs, and R. I. Lechler. 2001. Human CD4(+) CD25(+) cells: a naturally occurring population of regulatory T cells. *Blood 98:2736.*
6. Asano, M., M. Toda, N. Sakaguchi, and S. Sakaguchi. 1996. Autoimmune disease as a consequence of developmental abnormality of a T cell subpopulation. *J Exp Med 184:387.*
7. Kukreja, A., G. Cost, J. Marker, C. Zhang, Z. Sun, K. Lin-Su, S. Ten, M. Sanz, M. Exley, B. Wilson, S. Porcelli, and N. Maclaren. 2002. Multiple immuno-regulatory defects in type-1 diabetes. *J Clin Invest 109:131.*
8. Crispin, J. C., A. Martinez, and J. Alcocer-Varela. 2003. Quantification of regulatory T cells in patients with systemic lupus erythematosus. *J Autoimmun 21:273.*
9. Habibagahi, M., Z. Habibagahi, M. Jaberipour, and A. Aghdashi. 2011. Quantification of regulatory T cells in peripheral blood of patients with systemic lupus erythematosus. *Rheumatol Int 31:1219.*
10. Ehrenstein, M. R., J. G. Evans, A. Singh, S. Moore, G. Warnes, D. A. Isenberg, and C. Mauri. 2004. Compromised function of regulatory T cells in rheumatoid arthritis and reversal by anti-TNFalpha therapy. *J Exp Med 200:277.*
11. Danke, N. A., D. M. Koelle, C. Yee, S. Beheray, and W. W. Kwok. 2004. Autoreactive T cells in healthy individuals. *J Immunol 172:5967.*
12. Liu, W., A. L. Putnam, Z. Xu-Yu, G. L. Szot, M. R. Lee, S. Zhu, P. A. Gottlieb, P. Kapranov, T. R. Gingeras, B. Fazekas de st Groth, C. Clayberger, D. M. Soper, S. F. Ziegler, and J. A. Bluestone. 2006. CD127 expression inversely correlates with FoxP3 and suppressive function of human CD4+ T reg cells. *J Exp Med 203:1701.*
13. Seddiki, N., B. Santner-Nanan, J. Martinson, J. Zaunders, S. Sasson, A. Landay, M. Solomon, W. Selby, S. I. Alexander, R. Nanan, A. Kelleher, and B. Fazekas de St Groth. 2006. Expression of interleukin (IL)-2 and IL-7 receptors discriminates between human regulatory and activated T cells. *J Exp Med 203:1693.*
14. Longhi, M. S., Y. Ma, D. P. Bogdanos, P. Cheeseman, G. Mieli-Vergani, and D. Vergani. 2004. Impairment of CD4(+)CD25(+) regulatory T-cells in autoimmune liver disease. *J Hepatol 41:31.*
15. Baecher-Allan, C., and D. A. Hafler. 2006. Human regulatory T cells and their role in autoimmune disease. *Immunol Rev 212:203.*
16. Enjyoji, K., J. Sevigny, Y. Lin, P. S. Frenette, P. D. Christie, J. S. Esch, 2nd, M. Imai, J. M. Edelberg, H. Rayburn, M. Lech, D. L. Beeler, E. Csizmadia, D. D. Wagner, S. C. Robson, and R. D. Rosenberg. 1999. Targeted disruption of cd39/ATP diphosphohydrolase results in disordered hemostasis and thromboregulation. *Nat Med 5:1010.*
17. Borsellino, G., M. Kleinewietfeld, D. Di Mitri, A. Sternjak, A. Diamantini, R. Giometto, S. Hopner, D. Centonze, G. Bernardi, M. L. Dell'Acqua, P. M. Rossini, L. Battistini, O. Rotzschke, and K. Falk. 2007. Expression of ectonucleotidase CD39 by Foxp3+ Treg cells: hydrolysis of extracellular ATP and immune suppression. *Blood 110:1225.*
18. Deaglio, S., K. M. Dwyer, W. Gao, D. Friedman, A. Usheva, A. Erat, J. F. Chen, K. Enjyoji, J. Linden, M. Oukka, V. K. Kuchroo, T. B. Strom, and S. C. Robson. 2007. Adenosine generation

catalyzed by CD39 and CD73 expressed on regulatory T cells mediates immune suppression. *J Exp Med 204:1257.*
19. Fletcher, J. M., R. Lonergan, L. Costelloe, K. Kinsella, B. Moran, C. O'Farrelly, N. Tubridy, and K. H. Mills. 2009. CD39+Foxp3+ regulatory T Cells suppress pathogenic Th17 cells and are impaired in multiple sclerosis. *J Immunol 183:7602.*
20. Taams, L. S., J. Smith, M. H. Rustin, M. Salmon, L. W. Poulter, and A. N. Akbar. 2001. Human anergic/suppressive CD4(+)CD25(+) T cells: a highly differentiated and apoptosis-prone population. *Eur J Immunol 31:1122.*
21. Dieckmann, D., H. Plottner, S. Berchtold, T. Berger, and G. Schuler. 2001. Ex vivo isolation and characterization of CD4(+)CD25(+) T cells with regulatory properties from human blood. *J Exp Med 193:1303.*
22. Jonuleit, H., E. Schmitt, M. Stassen, A. Tuettenberg, J. Knop, and A. H. Enk. 2001. Identification and functional characterization of human CD4(+)CD25(+) T cells with regulatory properties isolated from peripheral blood. *J Exp Med 193:1285.*
23. Wing, K., S. Lindgren, G. Kollberg, A. Lundgren, R. A. Harris, A. Rudin, S. Lundin, and E. Suri-Payer. 2003. CD4 T cell activation by myelin oligodendrocyte glycoprotein is suppressed by adult but not cord blood CD25+ T cells. *Eur J Immunol 33:579.*
24. Fu, S., A. C. Yopp, X. Mao, D. Chen, N. Zhang, D. Chen, M. Mao, Y. Ding, and J. S. Bromberg. 2004. CD4+ CD25+ CD62+ T-regulatory cell subset has optimal suppressive and proliferative potential. *Am J Transplant 4:65.*
25. Bach, J. F., and L. Chatenoud. 2001. Tolerance to islet autoantigens in type 1 diabetes. *Annu Rev Immunol 19:131.*
26. Shimizu, J., S. Yamazaki, T. Takahashi, Y. Ishida, and S. Sakaguchi. 2002. Stimulation of CD25(+)CD4(+) regulatory T cells through GITR breaks immunological self-tolerance. *Nat Immunol 3:135.*
27. McHugh, R. S., M. J. Whitters, C. A. Piccirillo, D. A. Young, E. M. Shevach, M. Collins, and M. C. Byrne. 2002. CD4(+)CD25(+) immunoregulatory T cells: gene expression analysis reveals a functional role for the glucocorticoid-induced TNF receptor. *Immunity 16:311.*
28. Asseman, C., and M. von Herrath. 2002. About CD4pos CD25pos regulatory cells. *Autoimmun Rev 1:190.*
29. Caramalho, I., T. Lopes-Carvalho, D. Ostler, S. Zelenay, M. Haury, and J. Demengeot. 2003. Regulatory T cells selectively express toll-like receptors and are activated by lipopolysaccharide. *J Exp Med 197:403.*
30. Longhi, M. S., M. J. Hussain, R. R. Mitry, S. K. Arora, G. Mieli-Vergani, D. Vergani, and Y. Ma. 2006. Functional study of CD4+CD25+ regulatory T cells in health and autoimmune hepatitis. *J Immunol 176:4484.*
31. Fontenot, J. D., M. A. Gavin, and A. Y. Rudensky. 2003. Foxp3 programs the development and function of CD4+CD25+ regulatory T cells. *Nat Immunol 4:330.*
32. Khattri, R., T. Cox, S. A. Yasayko, and F. Ramsdell. 2003. An essential role for Scurfin in CD4+CD25+ T regulatory cells. *Nat Immunol 4:337.*
33. Wang, J., A. Ioan-Facsinay, E. I. van der Voort, T. W. Huizinga, and R. E. Toes. 2007. Transient expression of FOXP3 in human activated nonregulatory CD4+ T cells. *Eur J Immunol 37:129.*
34. Brunkow, M. E., E. W. Jeffery, K. A. Hjerrild, B. Paeper, L. B. Clark, S. A. Yasayko, J. E. Wilkinson, D. Galas, S. F. Ziegler, and F. Ramsdell. 2001. Disruption of a new forkhead/winged-helix protein, scurfin, results in the fatal lymphoproliferative disorder of the scurfy mouse. *Nat Genet 27:68.*
35. Bennett, C. L., J. Christie, F. Ramsdell, M. E. Brunkow, P. J. Ferguson, L. Whitesell, T. E. Kelly, F. T. Saulsbury, P. F. Chance, and H. D. Ochs. 2001. The immune dysregulation, polyendocrinopathy, enteropathy, X-linked syndrome (IPEX) is caused by mutations of FOXP3. *Nat Genet 27:20.*
36. Scholz, C., K. T. Patton, D. E. Anderson, G. J. Freeman, and D. A. Hafler. 1998. Expansion of autoreactive T cells in multiple sclerosis is independent of exogenous B7 costimulation. *J Immunol 160:1532.*
37. Correale, J., and A. Villa.(2010) Role of CD8+ CD25+ Foxp3+ regulatory T cells in multiple sclerosis. *Ann Neurol 67(5):625.*
38. Roncador, G., P. J. Brown, L. Maestre, S. Hue, J. L. Martinez-Torrecuadrada, K. L. Ling, S. Pratap, C. Toms, B. C. Fox, V. Cerundolo, F. Powrie, and A. H. Banham. 2005. Analysis of FOXP3 protein expression in human CD4+CD25+ regulatory T cells at the single-cell level. *Eur J Immunol 35:1681.*
39. Levings, M. K., R. Sangregorio, C. Sartirana, A. L. Moschin, M. Battaglia, P. C. Orban, and M. G. Roncarolo. 2002. Human CD25+CD4+ T suppressor cell clones produce transforming growth factor beta, but not interleukin 10, and are distinct from type 1 T regulatory cells. *J Exp Med 196:1335.*
40. Annunziato, F., L. Cosmi, F. Liotta, E. Lazzeri, R. Manetti, V. Vanini, P. Romagnani, E. Maggi, and S. Romagnani. 2002. Phenotype, localization, and mechanism of suppression of CD4(+)

CD25(+) human thymocytes. *J Exp Med 196:379.*

41. Akbar, A. N., L. S. Taams, M. Salmon, and M. Vukmanovic-Stejic. 2003. The peripheral generation of CD4+ CD25+ regulatory T cells. *Immunology 109:319.*
42. Levings, M. K., R. Sangregorio, and M. G. Roncarolo. 2001. Human cd25(+)cd4(+) t regulatory cells suppress naive and memory T cell proliferation and can be expanded in vitro without loss of function. *J Exp Med 193:1295.*
43. Longhi, M. S., F. Meda, P. Wang, M. Samyn, G. Mieli-Vergani, D. Vergani, and Y. Ma. 2008. Expansion and de novo generation of potentially therapeutic regulatory T cells in patients with autoimmune hepatitis. *Hepatology 47:581.*
44. Thornton, A. M., and E. M. Shevach. 1998. CD4+CD25+ immunoregulatory T cells suppress polyclonal T cell activation in vitro by inhibiting interleukin 2 production. *J Exp Med 188:287.*
45. Takahashi, T., Y. Kuniyasu, M. Toda, N. Sakaguchi, M. Itoh, M. Iwata, J. Shimizu, and S. Sakaguchi. 1998. Immunologic self-tolerance maintained by CD25+CD4+ naturally anergic and suppressive T cells: induction of autoimmune disease by breaking their anergic/suppressive state. *Int Immunol 10:1969.*
46. Piccirillo, C. A., and E. M. Shevach. 2001. Cutting edge: control of CD8+ T cell activation by CD4+CD25+ immunoregulatory cells. *J Immunol 167:1137.*
47. Mallone, R., S. A. Kochik, H. Reijonen, B. Carson, S. F. Ziegler, W. W. Kwok, and G. T. Nepom. 2005. Functional avidity directs T-cell fate in autoreactive CD4+ T cells. *Blood 106:2798.*

Chapter 21

Human Chondrocyte Cultures as Models of Cartilage-Specific Gene Regulation

Miguel Otero, Marta Favero, Cecilia Dragomir, Karim El Hachem, Ko Hashimoto, Darren A. Plumb, and Mary B. Goldring

Abstract

The human adult articular chondrocyte is a unique cell type that has reached a fully differentiated state as an end point of development. Within the cartilage matrix, chondrocytes are normally quiescent and maintain the matrix constituents in a low-turnover state of equilibrium. Isolated chondrocytes in culture have provided useful models to study cellular responses to alterations in the environment such as those occurring in different forms of arthritis. However, expansion of primary chondrocytes in monolayer culture results in the loss of phenotype, particularly if high cell density is not maintained. This chapter describes strategies for maintaining or restoring differentiated phenotype by culture in suspension, gels, or scaffolds. Techniques for assessing phenotype involving primarily the analysis of synthesis of cartilage-specific matrix proteins as well as the corresponding mRNAs are also described. Approaches for studying gene regulation, including transfection of promoter-driven reporter genes with expression vectors for transcriptional and signaling regulators, chromatin immunoprecipitation, and DNA methylation are also described.

Key words: Chondrocyte, Type II collagen, Aggrecan, Monolayer culture, Suspension culture, Alginate, Agarose, PolyHEMA, Three-dimensional scaffolds, Collagen scaffolds, Transfections, Luciferase reporter plasmids, Adenovirus-mediated expression, Chromatin immunoprecipitation assay

1. Introduction

The mature articular chondrocyte embedded in the cartilage matrix is a resting cell with no detectable mitotic activity and a very low synthetic activity (1). The markers of mature articular chondrocytes are type II collagen (COL2A1), other cartilage-specific collagens IX (COL9) and XI (COL11), and the large aggregating proteoglycan aggrecan (ACAN) (Table 1). Chondrocytes also synthesize a number of small proteoglycans such as biglycan and decorin and

Ragai R. Mitry and Robin D. Hughes (eds.), *Human Cell Culture Protocols*, Methods in Molecular Biology, vol. 806, DOI 10.1007/978-1-61779-367-7_21, © Springer Science+Business Media, LLC 2012

Table 1
Proteins synthesized by mature chondrocytes

Collagens
Type II
Type IX
Type XI
Type VI
Type XII
Type XIV
Type XVI
Type XVII
Proteoglycans
Aggrecan (with link protein and hyaluronic acid)
Versican
Perlecan
Biglycan
Decorin
Asporin
Fibromodulin
Lumican
PRELP (proline/arginine-rich and leucine-rich repeat protein)
Chondroadherin
Lubricin, superficial zone protein (SZP), or proteoglycan (PRG)-4
Other noncollagenous proteins (structural)
Cartilage oligomeric matrix protein (COMP; thrombospondin-5)
Thrombospondin-1 and -3
Cartilage matrix protein (matrilin-1); matrilin-3
Fibronectin
Tenascin-C
Cartilage intermediate layer protein (CILP)
Other noncollagenous proteins (regulatory)
S-100
Chondromodulin-I (SCGP) and -II
Glycoprotein (gp)-39, YKL-40
Matrix Gla protein (MGP)
CD-RAP (cartilage-derived retinoic acid-sensitive protein)
Bone morphogenetic proteins (BMP) 2, 7, 13 (GDF-6), 14 (GDF-5)
Transforming growth factor β
Membrane-associated proteins
Integrins (α1β1, α2β1, α3β1, α5β1, α6β1, α10β1, αvβ3, αvβ5)
Anchorin CII (annexin V)
CD44
Syndecan-3
Discoidin domain receptor-2 (DDR2)

The collagens, proteoglycans, and other noncollagenous proteins in the cartilage matrix are synthesized by chondrocytes at different stages during development and growth of cartilage and the mature articular chondrocyte may have a limited capacity to maintain and repair some of the matrix components, particularly collagens. Regulatory proteins are secreted by the chondrocytes and may be stored in the matrix. Membrane-associated proteins permit specific interactions with extracellular matrix proteins

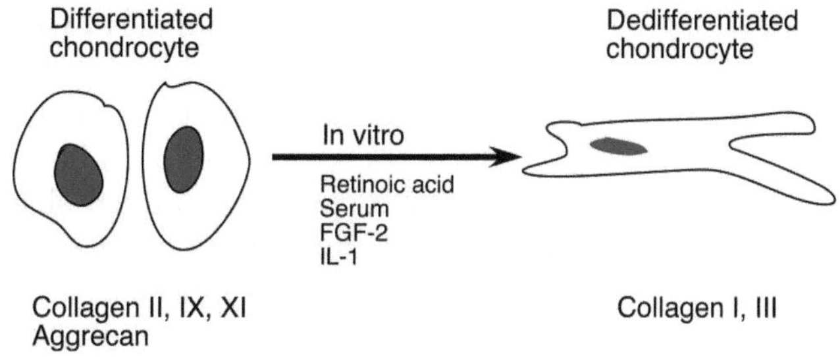

Fig. 1. Schematic representation of the "switch" from the differentiated to the dedifferentiated chondrocyte phenotype that occurs during culture and in response to certain cytokines and growth factors in vitro. The extracellular matrix genes that are differentially expressed are indicated.

other specific and nonspecific matrix proteins both in vivo and in vitro. As the single cellular constituent of adult articular cartilage, chondrocytes are responsible for maintaining the cartilage matrix in a low turnover state of equilibrium. In normal adult articular cartilage, the turnover of collagen occurs with a half-life of greater than 100 years (2, 3). The half-life of aggrecan subfractions is in the range of 3–24 years, whereas the glycosaminoglycan constituents on the aggrecan core protein are more readily replaced (4). Furthermore, normal chondrocyte metabolism in situ occurs in low oxygen tension and is remote from a vascular supply. Thus, it is not surprising that changes in expression of these cartilage matrix constituents occur when the chondrocytes are isolated and placed in monolayer culture, where they increase synthetic activity by several orders of magnitude.

Primary cultures of articular chondrocytes isolated from various animal and human sources have served as useful models for studying the mechanisms controlling responses to growth factors and cytokines (see, for review, refs. 5–7). Early attempts to culture chondrocytes were frustrated by the tendency of these cells to "dedifferentiate" in monolayer culture and their inability to proliferate in suspension culture where cartilage-specific phenotype could be maintained (8–10). High-density monolayer cultures maintain the cartilage-specific phenotype until they are subcultured, although gene expression of type II collagen is generally more labile than that of aggrecan (11–13). Monolayer chondrocytes maintain a rounded, polygonal morphology (Fig. 1), but there may be a progressive change to a fibroblast-like morphology with passage of time, especially after subculture with acquisition of some but not all characteristics of the fibroblast phenotype, such as type I and type III collagen gene expression. This dedifferentiation can be accelerated by plating the cells at low densities or by treatment with cytokines such as interleukin-1 (IL-1) (14, 15) or retinoic acid and appears to be associated with the increased expression of genes involved in cell proliferation, such as cyclin D1 (16). Furthermore, the substrate on which the chondrocytes are plated can influence the differentiation capacity of articular chondrocytes (17).

Since the stability of the phenotype of isolated chondrocytes is critically dependent on cell shape and cell density (18, 19), high-density micromass cultures are useful if sufficient numbers of chondrocytes can be isolated (20, 21), particularly for studying proteoglycan biosynthesis (22). It is also possible to expand the cultures through a limited number of subcultures and "redifferentiate" the cells in fluid or gel suspension culture systems, where the chondrocytes regain morphology and the cessation of proliferation is associated with increased expression of cartilage-specific matrix protein. Culture systems that support chondrocyte phenotype include suspension culture in spinner flasks (23), in dishes coated with a nonadherent substrates (24–26), in pellets or micromasses (27, 28), and in three-dimensional matrices such as collagen gels (29), agarose (12, 30, 31), alginate (32, 33), or collagen sponges (34, 35). Serum-free defined media of varying compositions, but usually including insulin, have also been used, frequently in combination with the other culture systems mentioned above (36).

The use of chondrocytes of human origin has been problematical, since the source of the cartilage cannot be controlled, sufficient numbers of cells are not readily obtained from random operative procedures, and the phenotypic stability and proliferative capacity in adult human chondrocytes are lost more quickly upon expansion in serial monolayer cultures than in cells of juvenile human (11) or embryonic or postnatal animal origin (37, 38). Alternatively, explant cultures of human, but usually bovine, articular cartilage where the chondrocytes remain encased within their own extracellular matrix have been used as in vitro models to study cartilage biochemistry and metabolism (39). However, many experimental manipulations are done more easily using isolated chondrocytes. Later studies focused on adult human articular chondrocytes as target cells for immortalization, using immortalizing antigens such as SV40-TAg (40–42), human papilloma virus type 16 (HPV-16) early function genes E6 and E7 (43), and telomerase (44). Strategies that maintain high cell density and decrease cell proliferation must also be applied to immortalized chondrocyte cell lines, since stable integration of immortalizing genes disrupts normal cell cycle control but does not stabilize expression of the type II collagen gene (45, 46).

This chapter focuses on strategies for isolation, culture, and characterization of isolated human articular chondrocytes and also describes approaches for using different chondrocyte culture systems for evaluating chondrocyte phenotype and studying the regulation of chondrocyte functions.

2. Materials

2.1. Isolation and Culture of Human Chondrocytes

1. Growth medium for chondrocytes: Dulbecco's modified Eagle's medium (DMEM)/Ham's F-12 1:1 mixture, with L-glutamine (Mediatech, Inc. Manassas, VA). Add 10% fetal calf serum (FCS) immediately before use. Selection of lots of FCS that maintain chondrocyte phenotype (see Note 1) is recommended.
2. Phosphate-buffered saline (PBS) Ca^{2+}- and Mg^{2+}-free (Mediatech, Inc. Manassas, VA).
3. Trypsin–EDTA solution: 0.05% trypsin and 0.02% EDTA in Hank's balanced salt solution without $Ca^{2+}$and Mg^{2+} (Gibco Invitrogen, Carlsbad, CA). FCS and trypsin–EDTA are stored at −20°C, but should not be refrozen after thawing for use.
4. Serum substitutes for experimental incubations: Nutridoma-SP (Roche Applied Science, Indianapolis, IN) is provided as sterile concentrate (100×, pH 7.4; storage at 15–25°C protected from light). Dilute 1:100 (v/v) with sterile DMEM/Ham's F-12, without FCS, and use immediately. ITS + Universal Culture Supplement Premix (BD Biosciences, San Jose, CA); provided as a concentrated stock sterile aqueous solution (storage at 2–8°C). Dilute the stock solution 1:100 (v/v) in serum-free DMEM/Ham's F-12 to give final concentrations of 12.5 μg/mL human insulin, 12.5 μg/mL human transferrin, and 12.5 μg/mL selenious acid for experimental purposes.
5. Enzymes for cartilage digestion:
   (a) Pronase (Roche Applied Science, Indianapolis, IN): Prepare freshly in DMEM/Ham's F12 with 10% FCS at a final concentration of 1 mg/mL and sterilize by filtration using a 0.22-μm filter.
   (b) Collagenase P (Roche Applied Science, Indianapolis, IN): Prepare freshly in DMEM/Ham's F12 with 10% FCS at a final concentration of 1 mg/mL and sterilize by filtration through a sterile 0.22-μm filter.

2.2. Suspension and Three-Dimensional Culture Systems for Chondrocytes

1. Agarose-coated dishes (25): Weigh out 1 g of high melting point agarose in an autoclavable bottle and add 100 mL of dH_2O to make a 1% solution. Autoclave with cap tightened loosely, allow to cool to approximately 55°C, and pipette quickly into culture dishes (1 mL/3.5-cm well of 6-well plate, 3 mL/6-cm dish, or 9 mL/10-cm dish). Allow the gel to set at 4°C for 30 min and wash the surface 2 or 3 times with PBS. Plates may be used immediately or wrapped tightly with plastic or foil to prevent evaporation and stored at 4°C.

2. Poly-HEMA (poly-2-hydroxyethyl-methacrylate)-coated dishes (26): Prepare a 10% (w/v) solution by dissolving 5 g of poly-HEMA (Sigma-Aldrich, St. Louis, MO) in 50 mL of ethanol in a sterile capped bottle or centrifuge tube. Leave overnight at 37°C with gentle shaking to dissolve polymer completely. Centrifuge the viscous solution for 30 min at 2,000 × *g* to remove undissolved particles. Dilute the stock solution to 1% (1 mL of 10% polyHEMA in 9 mL ethanol). Layer the 1% poly-HEMA solution on dishes at 0.3 mL/well of 6-well plate or 0.9 mL/6-cm dish and leave with lids in place to dry overnight in a tissue culture hood. Expose open dishes to bactericidal ultraviolet light for 30 min to sterilize.
3. Agarose: Autoclave 2% (w/v) low gelling temperature agarose in dH_2O, cool to 37°C, and dilute with an equal volume of 2× DMEM containing 20% FCS either without cells or with a chondrocyte suspension.
4. Alginate (Keltone LVCR, NF; ISP Alginates Inc., San Diego, CA, USA). Low-viscosity (LV) alginate is used generally. Request LVCR for more highly purified preparation:
   (a) Prepare 1.2% (w/v) solution of alginate in 0.15 M NaCl.
   (b) Dissolve alginate in a 0.15 M NaCl solution, heating the solution in a microwave until it just begins to boil. Swirl and heat again two or three times until the alginate is dissolved completely. (*Caution*: Do not autoclave.) Allow the solution to cool to about 37°C and filter-sterilize. Filtering when warm permits the viscous solution to pass through the filter.
   (c) Prepare 102 mM $CaCl_2$ and 0.15 M NaCl solutions in tissue culture bottles and autoclave.
   (d) Prepare 55 mM Na citrate, 0.15 M NaCl, pH 6.0, filter-sterilize, and store at 4°C. Make fresh weekly.
5. Three-dimensional (3D) scaffolds: Several types of scaffolds are available commercially, including the following:
   (a) Gelfoam® (Pharmacia & Upjohn, Kalamazoo, MI), sterile absorbable collagen sponge, purchased as sponge-size 12–7 mm (box of 12).
   (b) BD™ Three-Dimensional Collagen Composite Scaffold (BD Biosciences): Contains a mixture of bovine type I and type III collagens and is provided as 3D scaffolds with 48-well plates.
   (c) BD™ Three-Dimensional OPLA® Scaffold (BD Biosciences): Contains a synthetic polymer synthesized from D,D,-L,L-polylactic acid and is provided as 3D scaffolds with 48-well plates.
6. Recovery of cells from scaffolds: Cell lysis solution: 0.2% v/v Triton X-100, 10 mM Tris–HCl, pH 7.0, 1 mM EDTA.

7. Collagenase solution: 0.03% (w/v) Collagenase (bacterial, clostridiopeptidase A; Worthington Biochemical Corp., Freehold, NJ) in Hank's balanced salt solution (HBSS).

2.3. Analysis of Matrix Protein Deposition and Synthesis

2.3.1. Alcian Blue Staining

1. 2.5% Glutaraldehyde (diluted from 50% solution; Sigma-Aldrich, St. Louis, MO) in 0.4 M $MgCl_2$ and 25 mM sodium acetate, pH 5.6.
2. Alcian blue 8GX (Sigma-Aldrich, St. Louis). Dissolve in the 2.5% glutaraldehyde solution to give final concentration of 0.05%. Filter the solution through Whatman paper (or coffee filter).
3. Washes: 3% acetic acid solutions without and with 25 and 50% ethanol.

2.3.2. Collagen Typing

1. L-(5-^{3}H)proline (1 mCi/mL; specific activity >20 Ci/mmol) at 25 μCi/mL in serum-free culture medium supplemented with 50 μg/mL ascorbate and 50 μg/mL β-aminopropionitrile fumarate (β-APN). Filter-sterilize 10× solution of ascorbic acid and β-APN (5 mg of each dissolved in 10 mL of serum-free culture medium), dilute in medium at 1/10 (v/v) to give the volume required for the incubation, and add 25 μL of (^{3}H) proline per mL using a sterile pipette tip.
2. Pepsin–acetic acid solution: Dissolve 2 mg of pepsin in 1 mL dH_2O, then add 58 μL of glacial acetic acid per each mL of solution and cool on ice.
3. Reagents for SDS-PAGE (47) and autoradiography:
   (a) Gel sample buffer (GSB): 0.1 M Tris–HCl, pH 7.6, 3% (w/v) SDS, and 16% (v/v) glycerol.
   (b) Loading dye: 1% (w/v) Bromophenol blue (BPB; sodium salt; Sigma).
   (c) Tris–glycine/SDS–5% gradient polyacrylamide or 7–15% gradient gels and Laemmli buffer system (47).

2.3.3. Proteoglycan Synthesis

1. Biosynthetic labeling of proteoglycans: (^{35}S)sodium sulfate (2 mCi/mL; specific activity >1,000 Ci/mmol). Add to culture medium at 50 μCi/mL.
2. 7 M urea.
3. DE52 columns, 3.5 cm × 12 cm containing 4 mL of DEAE-cellulose (Whatman, Hillsboro, OR).
4. Guanidine extraction buffer: 4.0 M guanidine HCl, buffered with 50 mM sodium acetate, and containing 10 mM disodium EDTA. Add immediately before use 100 mM 6-aminocaproic acid, 2.5 mM benzamidine HCl, 5 mM *N*-ethylmaleimide, and 0.25 mM phenylsulfonyl fluoride (PMSF) from 100× stock solutions in absolute ethanol. The 2× guanidine extraction buffer is prepared at twice the concentrations above.

5. SDS-PAGE: Precast Tris–glycine SDS–polyacrylamide 4–20% gradient gels (Bio-Rad). Use the GSB and BPB solutions for loading the samples and the Laemmli buffer system (47) (see Subheading 2.3.2, items 3–5).

2.3.4. Immuno cytochemistry

1. Fixative: 2% paraformaldehyde in 0.1 M cacodylate buffer, pH 7.4. Dissolve 8 g of paraformaldehyde in 150 mL dH_2O in Ehrlenmeyer flask on hot plate in fume hood (do not exceed 65°C). Add ~2 mL of 1 N NaOH while stirring, and stir until solution is clear. Let solution cool for 15 min. Add 250 mL of 0.2 M cacodylate buffer, pH 7.4, and adjust pH if necessary.

2.3.5. Western Blotting

1. Reagents for SDS-PAGE (47) and autoradiography. Tris–glycine/SDS–polyacrylamide gels at polyacrylamide concentration appropriate for the size of matrix protein to be analyzed (*see* Subheadings 2.3.2 and 2.3.3).
2. Nylon-supported nitrocellulose membranes, Immobilon-P, 0.45-μm (Millipore, Billerica, MA).
3. Transfer buffer: 25 mM Tris–HCl, pH 7.6, 192 mM glycine, 10–20% (v/v) methanol.
4. Tris-buffered saline/Tween (TBST): 20 mM Tris–HCl, 137 mM NaCl, 0.1% (v/v) Tween-20, pH 7.6. Add 5% (w/v) nonfat dry milk (Carnation) as required.
5. Primary antibody: Dilute in TBST containing nonfat dry milk (BD Biosciences, San Jose, CA) or bovine serum albumin (Santa Cruz Biotechnology, Santa Cruz, CA) according to the supplier's instructions.
6. Horseradish peroxidase (HRP)-conjugated secondary antibody.
7. Enhanced chemiluminescence (ECL) substrate for HRP enzyme (GE Healthcare Biosciences, Pittsburgh, PA or Thermo Fisher Scientific, Rockford, IL among others).

2.3.6. Antibodies

Antibodies that detect human collagens and proteoglycans that are specific to cartilage are available commercially from Southern Biotechnology Associates, Inc. (Birmingham, AL; http://southernbiotech.com), IBEX Technologies, Inc. (Montreal, QC, Canada; http://www.ibex.ca/), and Chemicon International (Temecula, CA; http://www.chemicon.com). Some antibody preparations are useful for developing quantitative ELISA assays.

2.4. Analysis of mRNA

1. RNA extraction kit: The TRIzol® reagent (Invitrogen, Carlsbad, CA) or RNeasy® Plus Mini Kit (QIAGEN Inc., Valencia, CA) are suitable for extraction of total RNA from chondrocytes.
2. Sterile RNase-free solutions, polypropylene tubes, and other materials.

2.5. Analysis of Transcriptional Modulation

2.5.1. Luciferase Reporter Assays

1. EndoFree Plasmid Maxi Kit (QIAGEN Inc., Valencia, CA).
2. Lipid-based transfection reagent such as Lipofectamine™ PLUS™ Reagent (Invitrogen, Carlsbad, CA) or FuGENE 6 (Roche Applied Science, Indianapolis, IN).
3. Serum-free culture medium for transfections: Opti-MEM (Invitrogen, Carlsbad, CA) or DMEM/F-12 (test for optimal transfection efficiency).
4. Coomassie (Bradford) Protein Assay Reagent (Thermo Fisher Scientific, Rockford, IL).
5. Firefly and Renilla Luciferase Assay Systems (Promega, Madison, WI) or Dual-Luciferase Reporter Assay (Promega). When overexpressing Firefly luciferase with a second reporter gene, Passive (Promega) or Reporter (Promega) Lysis buffers are recommended.
6. Adenovirus producer cell line: 293 (ATTC CRL 1573; transformed primary human embryonic kidney).

2.5.2. siRNA-Mediated Knockdown

1. Lipid-based transfection reagent such as Lipofectamine™ PLUS™ Reagent (Invitrogen, Carlsbad, CA), Lipofectamine™ 2000 Reagent (Invitrogen, Carlsbad, CA) or DharmaFECT® transfection reagents (Thermo Fisher Scientific, Rockford, IL).
2. siRNA of interest and nontargeting siRNA (Thermo Fisher Scientific, Applied Biosystems/Ambion).
3. Antibiotic- and Serum-Free DMEM/F12.

2.5.3. Chromatin Immunoprecipitation (ChIP) Assays

1. ChIP-IT™ Express Chromatin Immunoprecipitation Kit (Active Motif, Carlsbad, CA #53008). The following reagents/materials are not provided in the Kit.
2. Formaldehyde cross-linking reagent: for preparing the formaldehyde Fixation Solution, add 0.54 mL of 37% formaldehyde (Sigma-Aldrich, St. Louis, MO) to 20 mL of cell culture medium per 150-mm × 25-mm plate (BD Biosciences, San Jose, CA).
3. Dounce Homogenizer (Kimble Chase Life Science, Vineland, NJ).
4. Enzymatic shearing: for preparing a working stock of the Enzymatic Shearing Cocktail, dilute the provided Enzymatic Shearing Cocktail 1:100 in a 50% solution of glycerol (Thermo Fisher Scientific, Rockford, IL) in nuclease-free water (QIAGEN Inc., Valencia, CA).
5. ChIP-grade antibodies.
6. QIAquick PCR Purification Kit (QIAGEN Inc., Valencia, CA) for DNA purification.

2.5.4. DNA Methylation Analysis

1. Methyl Detector™ Bisulfite Modification Kit (Active Motif, Carlsbad, CA) for bisulfite conversion of unmethylated cytosines to uracils.
2. Methylation-sensitive restriction enzyme/s (New England Biolabs, Ipswich, MA).
3. AllPrep DNA/RNA/Protein Mini Kit (QIAGEN Inc., Valencia, CA) for simultaneous purification of DNA and RNA.
4. CpGenome Universal Methylated DNA (Millipore, Billerica, MA).
5. Illustra GenomiPhi V2 DNA Amplification Kit (GE Healthcare Biosciences, Pittsburgh, PA).
6. 5-(Aza-deoxycytidine)aza-dC (Sigma-Aldrich, St. Louis, MO) and trichostatin A (Sigma-Aldrich, St. Louis, MO) for inhibition of DNA methyltransferase and histone deacetylase activities, respectively.

3. Methods

The methods described below outline (1) the isolation of human chondrocytes from cartilage and their primary culture in monolayer and after passage, (2) suspension culture systems for maintaining chondrocyte phenotype, analysis of the (3) synthesis and (4) mRNA expression of cartilage-specific matrix proteins, and (5) analysis of transcriptional modulation by transfection of regulatory DNA sequences, knockdown by siRNA transfection of *trans*-acting factors, chromatin immunoprecipitation assays, and DNA methylation analysis.

3.1. Isolation and Culture of Human Chondrocytes in Monolayer

1. Human adult articular cartilage is obtained, after Institutional Review Board approval, from the knee joints or hips after surgery for joint replacement or reconstruction or due to trauma, or at autopsy, and dissected free from underlying bone and any adherent connective tissue.
2. Place slices of cartilage in a 10-cm dish containing serum-free DMEM/F12. Cut cartilage into small pieces using a scalpel blade.
3. Add pronase solution (~25 mL per 10 g of tissue) and incubate at 37°C for 30 min with slow agitation.
4. Remove pronase solution and wash twice with PBS.
5. Add collagenase P solution (~25 mL per 10 g of tissue) and incubate at 37°C overnight (16–18 h) for articular cartilage and up to 48 h for costal cartilage, until the cartilage matrix is completely digested and the cells are free in suspension

Table 2
Culture vessel area vs. chondrocyte number required for plating density of ~2.5×10^4 cells/cm^2

Diameter	Area (cm^2)	No. of cells plated
16-mm well (24-well)	2	50,000
2.2-cm well (12-well)	3.8	100,000
3.5-cm well (6-well)	10	250,000
6-cm plate	28	750,000
10-cm plate	79	2×10^6

(see Note 2). Break up any clumps of cells by repeated aspiration of the suspension through a 10-mL pipette or a 12-cc syringe without a needle.

6. Transfer cell suspension to a sterile 50-mL conical polypropylene tube passing the suspension through a 40–70 μm filter (see Note 2). Centrifuge (5 min, 400 × *g*) to pellet cells and discard supernatant by aspiration.
7. Wash the cell pellet twice with PBS and once with DMEM/F12 containing 10% FCS, resuspending cells each time and centrifuging.
8. Resuspend the final pellet in DMEM/Ham's F-12 containing 10% FCS, perform cell count with a Coulter counter or hemocytometer (dead cells can be recognized by Trypan blue exclusion), and bring up to volume with culture medium to give 1×10^6 cells/mL. For monolayer culture, plate cells at 5×10^4/cm^2 (see Table 2), in dishes or wells containing culture medium, and agitate without swirling to distribute the cells evenly. Incubate at 37°C in an atmosphere of 5% CO_2 in air with medium changes after 2 days and every 3 or 4 days thereafter (see Note 3). Primary cultures of adult articular chondrocytes before and after passaging are shown in Fig. 2a, b, respectively.
9. Preparation of subcultured cells (see Note 4): Remove culture medium by aspiration with a sterile Pasteur pipette attached to a vacuum flask and wash with PBS. Add trypsin–EDTA (1 mL/10-cm dish) and incubate at room temperature for 10 min with periodic gentle shaking of dish and observation through microscope to assure that cells have come off the plate. If significant numbers of cells remain attached, continue the incubation for a longer time (≤20 min) or at a higher temperature (37°C) and/or scrape the cell layer with a sterile plastic scraper or syringe plunger. Repeatedly aspirate and expel the cell suspension into the plate using a 5- or 10-mL pipette

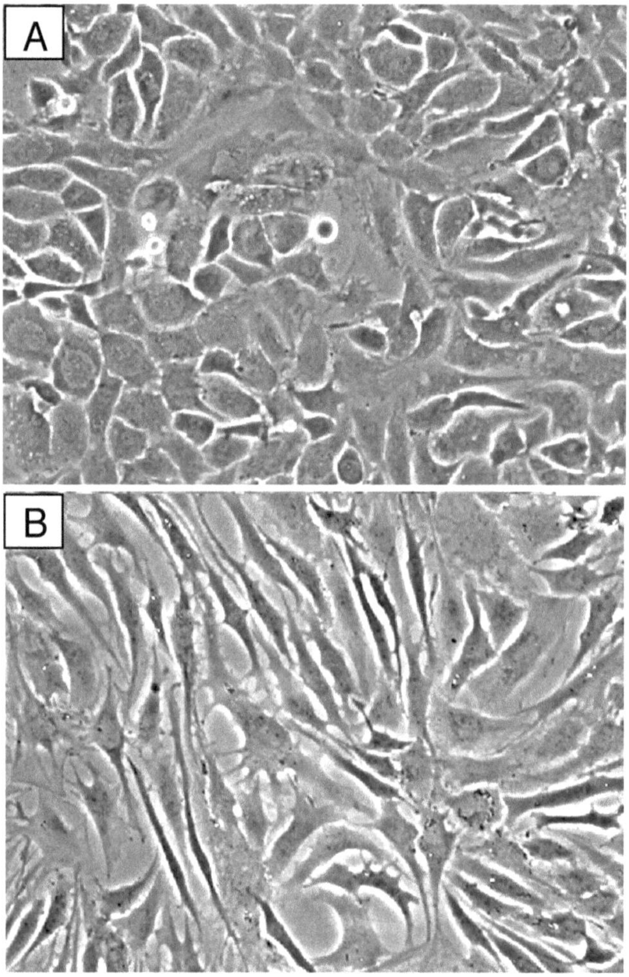

Fig. 2. Morphology of human articular chondrocytes grown in monolayer culture on plastic. Chondrocytes were isolated from articular cartilage and cultured in DMEM/F12 containing 10% FCS until confluent and then subcultured and grown again to confluence. (**a**) Primary chondrocytes display the characteristic cobblestone morphology. (**b**) Passaged chondrocytes display features of a dedifferentiated phenotype, including a portion of cells with fibroblast-like morphology.

containing culture medium, and then transfer to a sterile conical 15- or 50-mL polypropylene tube. Perform cell counts or determine the split ratio required (usually 1 dish into 2, or 1:2, for adult articular chondrocytes, or 1:5 for more rapidly growing, denser juvenile chondrocytes). Distribute equal volumes of the cell suspension in dishes or wells that already contain culture medium, rocking plates back-and-forth (not swirling) immediately after each addition to assure uniform plating density (see Note 5).

10. For experiments, plate cells in dishes or wells at a concentration of $1–2.5 \times 10^4$ cells/cm^2, depending upon the experimental approach, in DMEM/Ham's F-12 containing 10% FCS. When

the cultures are confluent, remove the growth medium and add fresh medium with additives appropriate for test conditions. If it is important that the cells be quiescent, as for analyzing serum-responsive genes, add test agent without changing medium or change to serum-free medium containing a serum substitute (see Note 6), such as Nutridoma-SP or ITS+, followed 1–24 h later by the test agent of interest. Continue incubation at 37°C for short-term time courses of 0.25–24 h or longer time courses of several days.

3.2. Three-Dimensional Culture Systems for Chondrocytes

Chondrocytes in monolayer culture are susceptible to loss of phenotype during prolonged culture and particularly after subculture. Thus, it is necessary to use culture conditions that maintain differentiated chondrocyte features or that permit redifferentiation (see Note 7). Freshly isolated chondrocytes may be cultured immediately in suspension, where they do not proliferate, or they may be expanded in monolayer culture and placed in suspension culture after several passages.

3.2.1. Suspension Cultures on Agarose- or PolyHEMA-Coated Dishes (See Note 8)

1. Trypsinize monolayer cultures, spin down cells, wash with PBS, centrifuge, and resuspend in culture medium containing 10% FCS at 1×10^6 cells/mL.
2. Transfer chondrocyte suspension to dishes that have been coated with 1% agarose or with 1% polyHEMA and culture for 2–4 weeks. The cells first form large clumps that begin to break up after 7–10 days and eventually form single-cell suspensions.
3. Change the medium weekly by carefully removing the medium above settled cells while tilting the dish, centrifuging the remaining suspended cells, and replacing them in the dish after resuspension in fresh culture medium.
4. To recover cells for direct experimental analysis, for redistribution in agarose- or polyHEMA-coated wells, or for culture in monolayer, transfer the cell suspension to 15- or 50-mL conical tubes, gently washing the agarose surface at least twice with culture medium to recover remaining cells, and spin down and resuspend cells in an appropriate volume of culture medium for plating or in extraction buffer for subsequent experimental analysis.

3.2.2. Suspension Culture Within Agarose

1. Precoat plastic tissue culture dishes with cell-free 1% agarose in culture medium (0.5 mL/3.5-cm, 1.5 mL/6-cm, or 4.5 mL/10-cm dish) and allow to gel at room temp. Add the same volume of 1% agarose in medium containing chondrocytes at a density of $1–4 \times 10^6$ cells/mL of gel, incubate at 37°C for 20–30 min to allow the cells to settle, and leave at room temp until the agarose forms a gel. Add culture medium containing 10% FCS and incubate at 37°C with medium changes every 3–4 days.

2. After incubations with test reagents and/or radioisotopes in minimal volumes of appropriate culture medium, the whole cultures may be stored frozen or medium and gel treated separately. For subsequent analysis, add appropriate guanidine extraction buffer directly to the gel (for proteoglycan or RNA extraction) or whole cultures may be adjusted to 0.5 M acetic acid, treated with pepsin, and neutralized, as described below for analysis of collagens. To remove agarose and debris, the samples are centrifuged in a high-speed centrifuge at >10,000 × *g* at 4°C.

3.2.3. Alginate Bead Cultures (See Note 9)

1. Trypsinize several 10-cm plates and wash the cells with PBS. Determine the cell count with a hemocytometer and pellet the cells.
2. Resuspend the pellet in a 1.2% solution of alginate in 0.15 M NaCl at a concentration of 1–4 × 10^6 cells/mL. Slowly express the alginate suspension in a dropwise manner through a 10-cc syringe equipped with a 22-gauge needle into a 50-mL polypropylene centrifuge tube containing 40 mL of 102 mM $CaCl_2$. Allow the beads to polymerize in the $CaCl_2$ solution for 10 min and wash twice with 25 mL of 0.15 M NaCl. The alginate beads should *not* be washed in PBS, as they will become cloudy.
3. Resuspend the beads at 7–15 beads per mL in growth medium supplemented with 25 μg/mL Na ascorbate (see Note 10) and decant to a culture dish or flask. Culture in DMEM/F12 with medium changes every 3 days, carefully pipetting the spent culture medium from the top of the settled beads.
4. At the end of the culture period (one to several weeks), add the appropriate guanidine extraction buffer or centrifuge at 500 × *g* for 10 min to recover the chondrocytes with pericellular matrix.
5. Alternatively, to recover cells from alginate, carefully aspirate the medium from the cultures and wash twice with PBS. Depolymerize the alginate by adding three volumes of a solution of 55 mM Na citrate/0.15 M NaCl and incubate at 37°C for 10 min. Aspirate the solution over the surface of the dish several times to dislodge adherent cells (the cells are sticky) and transfer the suspension to a 50-mL centrifuge tube. Because the solution is quite viscous, centrifuge the cells at 2,000 × *g* for a minimum of 10 min to completely pellet the cells. Wash the cells twice with PBS before using them for further analysis.

3.2.4. Culture on 3D Scaffolds (See Note 11)

1. Gelfoam®: Use sterile scalpel blade to cut into pieces of 1 × 1 × 0.5 cm^3 and place in wells of sterile 6-well plates. Inoculate by dropping 50 μL of growth medium containing 10^6 cells on each sponge. Place in incubator for 1.5–2 h, then add 100 μL medium and culture for an additional 1–3 h. Add medium to cover and continue incubation overnight or longer.

2. BD™ 3D Collagen Composite or OPLA® Scaffolds: Place scaffolds (0.5 cm^3) in the 48-well plates provided, in 96-well plates, or other plate as required. Seed scaffolds by dropping 100 μL of growth medium containing $1–5 \times 10^4$ cells. Incubate for 1 h, add 150 μL of medium to each scaffold, and incubate for 1.5–3 h. Add medium as required for further culture and experimental conditions.

3. Recovery of cells from scaffolds for analysis:

   (a) Prepare cell lysates for DNA analysis using 250–500 μL of cell lysis solution (0.2% v/v Triton X-100, 10 mM Tris–HCl, pH 7.0, 1 mM EDTA) per scaffold in 1.5-mL tube. Freeze samples at −70°C and subject to two freeze–thaw cycles, thawing at room temperature for 45–60 min. Break up scaffolds with pipette tip, centrifuge, and transfer lysates to fresh tubes. The cell lysates may be analyzed using the Picogreen Assay Kit according to the manufacture's protocol (Molecular Probes).

   (b) Recover cells for RNA extraction and other analyses by treatment of minced scaffolds with 0.03% (w/v) collagenase in HBSS for 10–15 min at 37°C. Collect cells by centrifugation, wash with PBS, and add appropriate extraction buffer to the final pellet.

3.2.5. Pellet Cultures

1. Recover the cells using trypsin/EDTA; determine the cell count and spin down to pellet the cells.

2. Resuspend the pellet to reach a density of 5×10^5 cells per 500 μL of DMEM/Ham's F-12 with10% FCS supplemented with 50 μg/mL ascorbic acid to distribute in centrifuge tubes.

3. Pellet by centrifugation at $1{,}000 \times g$ for 5 min at 4°C. Remove the medium, add fresh complete medium, and repeat the centrifugation step.

4. Remove the medium, add 1 mL of fresh complete medium and incubate at 37°C for 1–4 weeks, depending on the experimental purposes.

5. Change the medium every second day adding 500 μL of fresh complete medium.

6. If stimulation with test agents is required, the medium volume can be reduced to 100 μL and pellets can be incubated at various time points in the presence of the test agent or vehicle in a dose- and time-dependent manner.

7. Harvest the pellets for RNA and protein extraction (28, 48). For immunohistochemical analysis, the pellets are embedded in OCT, snap-frozen, and stored at −80°C till use (Fig. 3). Supernatants can be stored for analyses of secreted products.

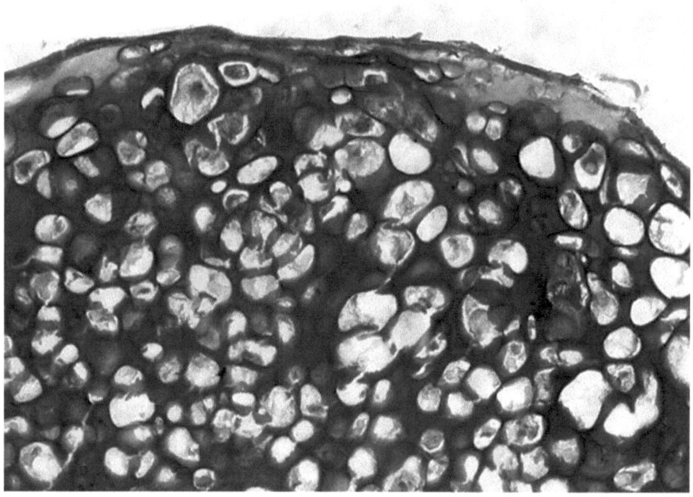

Fig. 3. Three-dimensional culture of articular chondrocytes. After 3 weeks in culture, pellets were embedded in OCT, snap frozen and stored at −80°C. Frozen sections were fixed in PBS/PFA for 5 min and stained with Toluidine blue to detect proteoglycan deposition.

3.3. Analysis of Matrix Protein Synthesis

Chondrocyte culture models are used to examine the effects of cytokines and growth/differentiation factors on the synthesis of chondrocyte phenotypic markers by staining the glycosaminoglycans with Alcian blue, characterizing the collagens and proteoglycans synthesized, and perform immunohistochemistry using specific antibodies against these proteins.

3.3.1. Alcian Blue Staining

1. Monolayer cultures: Remove culture medium from confluent cultures that have been incubated in the presence of 50 μg/mL ascorbic acid for at least 4 h and wash with PBS. Add Alcian blue/glutaraldehyde solution at room temperature for several hours, remove excess stain by washing with 3% acetic acid, and store cultures in 70% ethanol for subsequent examination by light photomicrography.
2. Alginate bead cultures:
   - (a) Using a 25-mL pipette, transfer five beads to a 12 cm × 15 cm tube, and wash twice with 2 mL PBS. Add 1 mL Alcian blue stain and 50 μL of a 50% glutaraldehyde solution. Store for 24 h at 4°C.
   - (b) Aspirate the stain from the beads, and wash twice with 2 mL of 3% (v/v) acetic acid. Destain the beads with rocking at room temp for 5 min sequentially with 2 mL of each of the following solutions: (1) 3% acetic acid, (2) 3% acetic acid–25% ethanol, and (3) 3% acetic acid–50% ethanol. Store the beads in 70% ethanol.

(c) Before photographing the stained beads, gently flatten beneath a glass coverslip, taking care not to disrupt the alginate matrix. Alternatively, the beads may be embedded and sectioned prior to photography.

3.3.2. Collagen Typing

1. Biosynthetic labeling of collagens (see Note 12): Remove serum-containing culture medium, wash with serum-free medium, and add (^{3}H)proline at 25 μCi/mL for a further 24 h in serum-free culture medium supplemented with 50 μg/mL ascorbate and 50 μg/mL β-APN (or without β-APN to retain collagen in the pericellular matrix). Remove culture medium and store at −20°C. Wash cell layer with PBS and solubilize by adding equal volumes of serum-free culture medium and 1 M ammonium hydroxide (an aliquot may be analyzed for DNA).
2. Collagen typing: To analyze pepsin-resistant collagens, add pepsin–acetic acid solution to equal volume of either labeled culture medium or solubilized cell solution for 16 h at 4°C, lyophilize, redissolve in 2× SDS sample buffer, and neutralize with 1 μL additions of 2 M NaOH to titrate the color of the from yellow–green to blue (but not to violet). To analyze procollagens and fibronectin, add 2× SDS sample buffer containing 0.2% β-ME to an equal volume of the culture medium. Heat samples to boiling for 10 min and load on SDS gels (5% acrylamide running gels or 7–15% gradient gels) that include a radiolabeled rat tail tendon collagen standard in one lane. Perform delayed reduction with 0.1% β-ME on pepsinized samples to distinguish α1(III) from α1 (I or II) collagens. Absence of the α2(I) collagen band generally indicate the absence of type I collagen synthesis. In cultures containing a mixture of type I and type II collagen, definitive identification of these collagens requires Western blotting using specific antibodies (see Subheading 3.3.5).

3.3.3. Proteoglycan Synthesis

1. Monolayer cultures: Aspirate the culture medium and wash the cell layer with PBS. Add DMEM/F12 containing 10% FCS supplemented with (^{35}S)sulfate at 50 μCi/mL and 25 μg/mL ascorbate. Incubate at 37°C for 18 h. Remove the conditioned medium to a 15-mL polypropylene tube and wash the cell layer three times with PBS.
2. Add an equal volume of 7 M urea to the medium. Mix and count 10 μL in liquid scintillation counter. Pass up to 4 mL through DEAE-Cellulose (DE52, Whatman) column, which has been preequilibrated with 7 M urea, to remove unincorporated cpm. Elute proteoglycans with 2 mL of 4 M guanidine extraction buffer. To 500 μL of column eluate, add two volumes of 100% ethanol and precipitate PGs for 2 h at −20°C. Spin at 10,000 × *g* for 20 min at 4°C. Wash final pellet with 70% ethanol.

3. Add 4 M guanidine extraction buffer to cell layer (e.g., 2 mL/25-cm² flask) and extract at 4°C for 24 h with rocking. Transfer extract to a 2-mL screw-cap microcentrifuge tube and spin at 10,000 × *g* for 20 min at 4°C to pellet particulate material from the sample. Remove supernatant to a fresh 2-mL tube and count 10 μL in a liquid scintillation counter. (Also, 10–50 μl aliquots may be taken at this point for DNA analysis.) Take 250 μL for precipitation of PGs by addition of three volumes of 100% ethanol at −20°C for 2 h (or overnight for alginate extracts), spin at 10,000 × *g* for 20 min at 4°C, and wash final pellet with 70% ethanol. Dry pellet at room temp for 30 min.
4. Alginate cultures: Aspirate the medium from a culture containing 50 beads in a 25-cm² flask cultured on end. Add 4 mL of growth medium supplemented with (^{35}S)sulfate at 50 μCi/mL and 25 μg/mL ascorbate. Incubate at 37°C for 18 h. Remove the conditioned medium to a 12-mL polypropylene tube and wash the beads three times with 5 mL of PBS (5 min/wash).
5. Medium extraction and purification: Perform as described for monolayer cultures in Subheading 3.3.3, step 2.
6. Extraction from alginate beads and purification: Transfer the beads (to a 15-mL polypropylene tube), extract radiolabeled PGs with 2 mL of 4 M guanidine extraction buffer at 4°C for 24 h with rocking. Transfer extract to a 2-mL screw-cap microcentrifuge tube and proceed as described for monolayers in Subheading 3.3.3, step 3.
7. Sulfate incorporation into proteoglycans: Dissolve ethanol-precipitated pellet in 100 μL of 1× GSB. Count 3 μL in a liquid scintillation counter and calculate the cpm incorporated, after accounting for dilutions, against the concentration of DNA in the cell extract. The (^{35}S)sulfate incorporation may also be determined after passing the guanidine extracts over Sephadex G-25M in PD 10 columns and eluting under dissociative conditions by scintillation counting.
8. SDS-PAGE analysis: Take a volume of cell or medium extract in GSB that corresponds to 0.25 μg of DNA in the cell extract and add DTT to a final concentration of 0.5 mM and 1% Bromophenol blue to a final concentration of 0.1%. Heat 5 min at 100°C and store remaining sample at −20°C. Electrophorese ~20,000 cpm on a 4–20% polyacrylamide gradient gel. Fix the gel in acetic acid/methanol for 1 h, dry, and expose to film at −80°C. Visualize radiolabeled proteoglycans by autoradiography, as shown in Fig. 4 (49–51) (see Note 13).

3.3.4. Immuno cytochemistry

1. Plate cells in plastic Lab-Tech 4-chamber slides (Nunc, Inc. Naperville, IL) at 6 × 10⁴ cells/chamber in culture medium containing 10% FCS. Add 25 μg/mL ascorbic acid with the first medium change and daily thereafter (see Note 10).

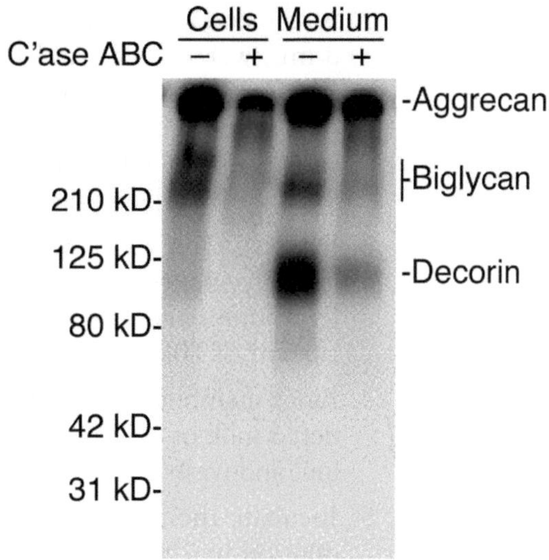

Fig. 4. SDS-PAGE analysis of proteoglycans synthesized by human articular chondrocytes. Primary cultures were grown to confluence in DMEM/F12 containing 10% FCS. The cultures were incubated with (^{35}S)sulfate in the presence of 25 μg/mL ascorbate during the final 18 h. Radiolabeled proteoglycans were purified by ethanol precipitation of 4 M guanidine extracts of cell monolayers or by DEAE ion-exchange chromatography and ethanol precipitation of 7 M urea extracts of the medium prior to SDS-PAGE on a 4–20% gradient gel. Equal volumes of cell and medium extracts representing equivalent proportions of the total culture were loaded on the gel. Molecular weight standards are indicated to the *left* of the *panel*. The migration of aggrecan, biglycan, and decorin is indicated to the *right*. Treatment with chondroitinase (C'ase) ABC resulted in the expected loss of sulfation.

2. When the cultures have reached confluence, add the desired test reagents. At the end of the incubation period, carefully wash the chambers three times with PBS and fix the cells with 2% paraformaldehyde in 0.1 M cacodylate buffer, pH 7.4, for 2 h at 4°C.
3. Rinse twice with 0.1 M cacodylate buffer. Add antibodies that recognize human type II collagen, aggrecan, etc. (see Subheading 2.3.6) to different chambers at concentrations recommended by the supplier. Incubate separate chamber slides with chondroitinase ABC for 30 min at 37°C prior to addition of monoclonal antibodies to expose epitopes.
4. Visualize the staining by incubation with a gold-conjugated secondary antibody (Auroprobe LM, Amersham) followed by silver enhancement (e.g., IntenSE Kit, Amersham).

3.3.5. Western Blotting

1. Plate cells in 6-well tissue culture plates at a density of 2.5×10^5 cells per well in culture medium containing 10% FCS. Add 25 μg/mL of ascorbic acid with the first medium change and daily thereafter (see Note 10). When the cultures have reached confluence, add the desired test reagents.

2. At the end of the incubation period, wash the cell layers with 3 mL/well of PBS. Add 1 mL of trypsin–EDTA to each well, and incubate at room temperature for ≤10 min to allow the cells to detach. Add serum-containing medium, transfer the suspension to a 15-mL centrifuge tube, and wash twice with 5 mL of PBS, pelleting cells by centrifugation at 1,000 × *g*.
3. Dissolve each pellet with 20 μL of GSB, electrophorese on a Tris–glycine/SDS–10% polyacrylamide precast gel, and transfer to Immobilon-P membrane by electroblotting with a transfer current of 100 V for 1 h.
4. Block membrane with 10 mL TBST containing 5% nonfat powdered milk or BSA (Blocking buffer) at 4°C for 1 h with rocking, and wash three times (10 min/wash) in 10 mL TBST.
5. Incubate the membrane with the antibody at the appropriate dilution in 5 mL of TBST in Blocking buffer for 1 h at room temperature with rocking.
6. Wash the membrane as in step 4, and incubate with an HRP-conjugated secondary antibody at the appropriate dilution in TBST or Blocking buffer.
7. Wash the membrane and detect bound antibody by enhanced chemiluminescence according to the manufacturer's protocol and expose to film for autoradiography.

3.4. Extraction and Analysis of RNA

1. Recovery of cells for extraction: For monolayer cultures, lyse cells directly on the culture dish by adding TRIzol® or RLT Plus lysis buffer reagents directly onto the 12 or 6-well plates. Depolymerize the alginate cultures as described in Subheading 3.2.3, step 3. Transfer cell suspension to sterile, RNase-free polypropylene tube of appropriate size, centrifuge at 1,000 × *g* at 4°C, washing two or three times with ice-cold PBS.
2. Extract total RNA using TRIzol® reagent, RNeasy® Plus Mini Kit, or other method as preferred, according to the manufacturer's instructions. If using the RNeasy® Plus Mini Kit, the use of QIAshredder columns is advisable to obtain a better homogenized lysate.
3. Resuspend RNA in nuclease-free water and determine the RNA concentration in a spectrophotometer. The final preparations should give yields of approximately 10 μg of RNA per 1×10^6 cells with the appropriate A_{260}/A_{280} ratio of approximately 2.0. It is advisable to determine the integrity of the isolated RNA by analyzing the apparent 28S:18S rRNA ratio (should be ~2:1) in a denaturing agarose gel electrophoresis, or by the use of microfluidics-based technology.
4. Store at −20°C in nonself-defrosting freezer or at −80°C.

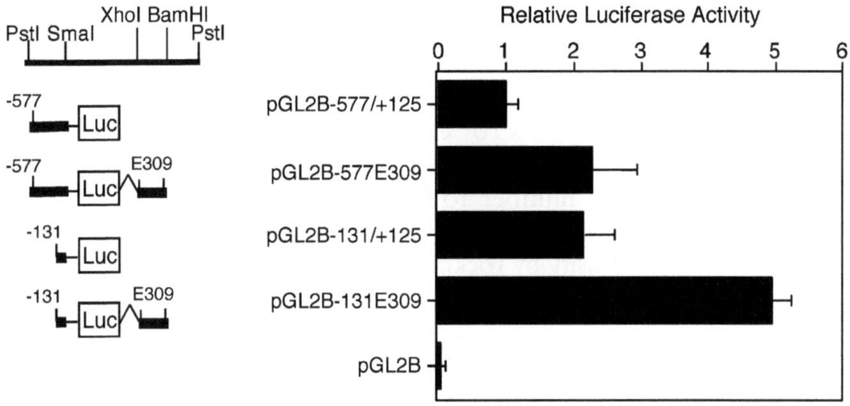

Fig. 5. Expression of the *COL2A1* promoter human articular chondrocytes. Luciferase reporter constructs containing *COL2A1* sequences were transfected in primary chondrocytes using LipofectAMINE+ and incubated in DMEM/F-12 containing 1% FCS for 24 h prior to harvest for luciferase assay. The promoter constructs, pGL2B–577/+125, and pGL2B–131/+125 without or with the E309 enhancer region (+2,388/+2,696 bp) from intron 1 were compared. Luciferase activity was normalized to the amount of protein and expressed as relative activity to that of untreated cells transfected with pGL2B–577/+125E309. Each value is calculated as the mean ± SD of the results from 3 to 6 wells and results are representative of at least three experiments. Note that the empty vector, pGL2-Basic, expressed at 1–2% of the levels of the *COL2A1* promoter constructs.

5. Analyze mRNAs by Northern blotting (15), semiquantitative RT-PCR (42, 52), or real-time PCR (53) by published methods. Type II collagen gene expression is relatively stable in primary cultures of human chondrocytes, at least through the first 12 days of culture (11, 54).

3.5. Analysis of Gene Transcription

Transfection studies are performed to analyze the DNA sequences and transcription factors involved in the regulation of gene expression using plasmid vectors, in which the expression of reporter genes such as CAT (55) or luciferase (42, 54, 56, 57) is driven by cis-acting regulatory sequences, such as those regulating type II collagen gene (*COL2A1*) transcription (Fig. 5). Coexpression of wild type or dominant-negative mutants of transcription factors, protein kinases, and other regulatory molecules, mediated by plasmid or adenoviral vectors may be performed to further dissect the mechanisms involved and to identify candidate *trans*-acting factors. The relative contribution of those factors in the transcriptional control of downstream target genes can be then addressed by (1) knockdown of the *trans*-acting factors, (2) chromatin immunoprecipitation assay of protein–DNA interactions in vivo within specific regulatory elements, and (3) epigenetic analysis of mechanisms such as the DNA methylation status of regulatory sequences (54, 58–60).

3.5.1. Transient Transfections Using Luciferase Reporter Plasmids

1. Prepare plasmids using the EndoFree Plasmid Maxi Kit, according to the manufacturer's instructions, to generate endotoxin-free DNA (see Note 14).
2. Seed 2.5×10^4 cells/cm^2 in 6-, 12- or 24-well tissue culture plates in DMEM/F-12 containing 10% FCS (determine optimal cell number to insure that cultures are greater than 50% confluent at the time of transfection, and assure uniform plating density by rocking plates back-and-forth (not swirling) immediately after seeding the cells (see Note 5)).
3. Change the medium to fresh growth medium 24 and 3 h before the transfection to insure that the cells are actively dividing. Hyaluronidase may be added during these medium changes to increase the transfection efficiency (see Note 15).
4. Prepare lipid/DNA complexes in serum-free DMEM/F-12 or Opti-MEM using LipofectAMINE$^+$or FuGENE 6, according to the manufacturer's protocol. Prepare in bulk for multiple transfections. Volumes indicated for transfections in 6-well plates. Do not vortex at any step:
   (a) .LipofectAMINE$^+$: For each well, add 92 μL of serum-free medium to a small sterile polypropylene tube, add 1 μL of plasmid DNA (maximum of 1 μg; see Note 16), and tap gently to mix. Add 6 μL of PLUS reagent, mix, and incubate for 15 min at room temperature. Dilute 4 μL of LipofectAMINE$^+$ reagent into 100 μL of serum-free medium, mix, and add to each reaction mixture. Mix and leave at room temperature for an additional 15–30 min at room temperature.
   (b) FuGENE 6: For each well, add 96 μL of serum-free medium to a small sterile polypropylene tube, add 3 μL of FuGENE 6 reagent, and tap gently to mix. Add 1 μL plasmid of DNA (maximum of 1 μg) to the prediluted FuGENE 6 reagent and incubate for 15 min at room temperature.
5. While the lipid–DNA complexes are forming, replace culture medium on cells with serum-free medium to give a final volume of 1 mL. Add the lipid–DNA complex mixture dropwise to the well and incubate for 4 h at 37°C.
6. Dilute the transfection medium by adding to the wells an equal volume of DMEM/F-12 containing 2% Nutridoma-SP (or 20% FCS) and incubate for 2 h to overnight. Add test agent without medium change and incubate further for 18 h (up to 48 h) (see Note 17).

3.5.2. Cotransfections Using Plasmid Vectors for Expression of Recombinant Proteins

1. Prepare plasmids as described above.
2. Titrate each expression vector and its corresponding empty vector, at amounts ranging from 10 to 200 ng per well, against a fixed amount of reporter vector. Equalize the total amount of

reporter plus expression plasmid in each well (<1 μg/well) by adding empty vector and maintain equal volumes (see Note 16).

3. After cotransfection, incubate the cells for 18–24 h to permit expression of recombinant protein prior to treatment with test reagents.

3.5.3. Adenovirus-Mediated Expression of Recombinant Proteins

1. Infect the 293 producer cell line with adenoviral vector containing the cDNA encoding the wild-type or mutant protein to be coexpressed and determine the titer (MOI) by standard techniques.
2. Incubate chondrocytes in DMEM/Ham's F-12 containing 10% FCS for 18 h following transfection of the reporter construct.
3. Remove medium and wash cells with PBS.
4. Add 1 mL of serum-free medium containing adenovirus at 1:125 MOI. Incubate at 37°C for 90 min.
5. Add 1 mL of DMEM/Ham's F-12 containing 20% FCS and continue incubation for 18 h.
6. Change medium to fresh DMEM/ Ham's F-12 containing 10% FCS or 1% Nutridoma-SP, incubate for 1 h, and treat with test agent for 18 h.

3.5.4. Luciferase Assay

1. Remove the medium from the cells, rinse with PBS and prepare cell lysates by extraction with 200 μL/well of a 6-well plate of Passive Lysis buffer, which will passively lyse cells without the requirement of a freeze–thaw cycle.
2. Rock culture plates for 5–10 min. Scrape adherent cells with a policeman and transfer solubilized cells to a 1.5-mL microcentrifuge tube.
3. Microcentrifuge 5 min at maximum speed at 4°C, transfer supernatant to a clean microcentrifuge tube, and store on ice.
4. Determine the protein content using the Coomassie (Bradford) Protein Assay Reagent.
5. Determine luciferase activities using the Luciferase Assay System or equivalent, according to manufacturer's protocol. Mix, manually or automatically, 20 μL of cell lysate with 100 μL of Luciferase Assay Reagent and read in a luminometer. Normalize to the amount of protein (or internal control such as β-galactosidase) and express as relative activity against the empty vector or untreated control. Perform each treatment in triplicate wells and each experiment at least three times to ensure reproducibility and significance.
6. Renilla luciferase control vectors may be used routinely or as necessary to check the purity of new plasmid preparations or relative activities of mutant and wild type constructs using the Dual-Luciferase Reporter Assay System.

3.5.5. siRNA-Mediated Knockdown

1. Seed 2.5×10^4 cells/cm^2 in 6-well plates (assuring uniform plating density) 24–36 h before transfection.
2. For each transfection sample, prepare tubes containing the siRNA:LipofectAMINE$^+$ mix as follows:
   (a) Dilute the appropriate amount of siRNA (see Note 18) in 125 μL of serum-free DMEM/Ham's F12 containing 7.5 μL of PLUS Reagent. Incubate for 15 min at room temperature.
   (b) Add 125 μL of serum-free media containing 5 μL of LipofectAMINE, mix by tapping, and incubate for 30 min at room temperature.
   (c) Add 250 μL of serum-free medium to give a final volume of 500 μL.
3. Remove complete medium from the 6-well plates and add 500 μL of serum-free medium per well.
4. Add 500 μL of the siRNA transfection mix on top of the serum-free medium added to the cells.
5. After overnight (16–18 h) incubation at 37°C, replace the transfection media with fresh serum-free or defined medium.
6. Incubate the cells at 37°C for the time required to achieve knockdown (KD) of the targeted gene (see Note 18).
7. When addressing involvement of the KD gene in downstream gene expression, add the testing agents at the optimal concentrations and times and proceed to RNA or protein isolation followed by real-time RT-PCR or Western blotting analysis.

3.5.6. Chromatin Immunoprecipitation Assay

The following is an adaptation with minor modifications of the Enzymatic Shearing ChIP Protocol suggested by ActiveMotif.

1. Plate 2.5×10^5 cells/plate in 150-cm^2 tissue culture plates (see Note 19) in DMEM/F12 containing 10% FCS. If test agents are to be used to compare between conditions, serum-deprivation may be required prior to add the test agents to the culture medium.
2. For cross-linking, remove media from the cells, add 20 mL of Fixation Solution per plate and incubate 10 min at room temperature with gentle agitation.
3. Remove Fixation Solution, rinse with PBS, add Glycine-Stop-Fix solution (10 mL per plate), and incubate for 5 min at room temperature with gentle agitation.
4. Remove Glycine-Stop-Fix solution and rinse with PBS.
5. Collect cells into a 15-mL Falcon tube by scraping with 2 mL of ice-cold PBS containing protease inhibitors.

6. Centrifuge the collected cells (10 min, 1,300 × *g* at 4°C) and discard the supernatant.
7. Resuspend pellets in 1 mL ice-cold Lysis buffer containing protease inhibitors and incubate on ice for 30 min.
8. Transfer cells to an ice-cold dounce homogenizer, homogenize on ice (10 strokes) and transfer lysates to clean prechilled 1.7-mL centrifuge tubes.
9. Centrifuge (10 min, 2,300 × *g* at 4°C), discard supernatant and resuspend pellets in 350 μL of Digestion buffer. Incubate at 37°C for 5 min.
10. Add 17 μL of the Enzymatic Shearing Cocktail to the reaction, vortex to mix, and incubate at 37°C for 8 min (see Note 20).
11. Stop reaction by adding 7 μL of ice-cold 0.5 M EDTA; chill on ice for 10 min and centrifuge at 16,100 × *g*, 10 min at 4°C.
12. Collect supernatant (~400 μL) containing the sheared chromatin. Use immediately for immunoprecipitation (see Note 20) or store in aliquots at −80°C (adding protease inhibitors) until use.
13. For preclearing: incubate from 2 to 5 μg of chromatin (see Note 20) with 25 μL of Protein G Magnetic Beads and 5 μg of normal IgG for 2 h at 4°C in an end-to-end rotator. After 2 h, spin the samples, discard bead pellets, collect supernatants and save 10% for input samples (store at −20°C).
14. Set up immunoprecipitation (IP) reactions in 1.7-mL siliconized tubes: make two preparations per condition, where the antibody of interest will be used in one IP reaction and a normal IgG in the other. Add ChIP buffer 1 (10–20 μL, depending on the volume of the sheared chromatin), 25 μL of Protein G Magnetic beads and nuclease-free water (up to 100 or 200 μL, depending on the volume of the sheared chromatin) to the precleared chromatin. Add 1–5 μg of antibody and incubate overnight (16–18 h) at 4°C in an end-to-end rotator.
15. Spin the samples, place the tubes in a magnetic stand and discard the supernatant. Wash twice with ChIP buffer 1 and twice with ChIP buffer 2 (3 min each, room temperature in an end-to-end rotator).
16. To elute the chromatin: resuspend the washed beads in 50 μL of elution buffer and incubate for 15 min in an end-to-end rotator.
17. For reverse cross-linking, spin the tubes and add 50 μL of Reverse Cross-linking buffer. Place the tubes on a magnetic stand to pellet the beads. Collect the supernatants and transfer them to clean 0.2-mL PCR tubes. Take input samples (stored at −20°C in step 13), add 88 μL of ChIP buffer 2 and 2 μL of NaCl. Incubate all the samples at 95°C for 15 min in a thermocycler.

18. Spin briefly and return tubes to room temperature; add 2 μL of Proteinase K, mix, and incubate for 1 h at 37°C in a thermocycler.
19. At room temperature, spin briefly and add 2 μL of Proteinase K Stop Solution. Purify the DNA using QIAquick columns or phenol/chloroform and proceed to the PCR analysis.

3.5.7. DNA Methylation Analysis

1. Immediately after isolation of articular chondrocytes, extract DNA and RNA from a portion of chondrocytes to assess the expression and the percentage of DNA methylation of the gene/s of interest (see Note 21).
2. Only chondrocytes isolated from deep-zone non-OA cartilage will be utilized for the cell culture experiments outlined below.
3. Seed 1×10^4 cells/cm^2 (see Note 22) in DMEM/Ham's F-12 containing 10% FCS.
4. Separate isolated deep-zone non-OA chondrocytes in five groups as follows (a) noncultured (for RNA and DNA isolation, as already stated in Subheading 3.5.7, step 2), (b) control (without treatment), (c) treated with 5-aza-dC (with a single addition of trichostatin A to facilitate access of 5-aza-dC), and (d) cultured with the test agent/s.
5. Twice per week, replace culture medium with fresh culture medium containing the different experimental treatments (i.e., 5-aza-dC or test agent/s).
6. When cells reach confluence (4–5 weeks in culture), harvest RNA and DNA from all the culture groups and proceed to analysis.
7. Perform real time RT-PCR analysis to address whether long-term exposure to the test agent/s leads to aberrant expression of the genes of interest as compared to the untreated control and the noncultured chondrocytes.
8. Address the DNA methylation status of the promoter/s of interest (from DNA isolated out of the different experimental groups) by bisulfite modification. Design specific CpG-free PCR primers (http://www.urogene.org/methprimer/) to PCR-amplify the target promoter sequence and clone the PCR products.
9. Send several colonies for sequencing and select candidate CpG site/s for quantification depending upon the average of the sequencing data.
10. Quantify the percentage of methylated candidate CpG site/s by digestion with methylation-sensitive restriction enzymes followed by real time PCR analysis (see Note 23). Use fully methylated and nonmethylated (obtained by demethylation of the CpGenome Universal Methylated DNA) DNA as internal

controls to adjust for the digestion efficiency and to generate standard curve/s for PCR calculations. Use nondigested DNA as a control for normalization of the respective DNA-digested samples.

4. Notes

1. Batches of serum should be tested and selected on the basis of the capacity to support expression of chondrocyte-specific matrix gene expression. High capacity to induce cell proliferation is not necessarily associated with the ability to maintain phenotype.
2. Chondrocytes are quite resilient and tolerate the prolonged incubation times required for complete dissociation of the matrix or the absence of serum for the digestion of other sources such as costal cartilage (15). If the digestion is not complete by the end of the allotted time, then more collagenase solution may be added, or the suspension may be recovered and the fragments left behind for further digestion. These conditions result in suspensions that are essentially single cell, and therefore, it is not necessary to resort to filtration through a nylon mesh, as has been done by others when shorter digestion times are used (12). However, we have introduced here the filtration of the final suspension through a filter that allows the cells to pass through and retains undigested material. These considerations are important for decreasing the loss of chondrocytes during their isolation from valuable human cartilage specimens.
3. After initial plating of the primary cultures, the chondrocytes require 2–3 days before they have settled down and spread out completely. Culture for approximately 4–7 days is required before reasonable amounts of total RNA may be extracted. Although the cultures may continue to express chondrocyte phenotype (e.g., type II collagen and aggrecan mRNAs) for several weeks, expression of nonspecific collagens I and III may begin as early as day 7 after isolation. Adult articular chondrocytes are strongly contact-inhibited and they may exhibit loss of phenotype within 1–2 weeks of monolayer culture. Juvenile costal chondrocytes continue to express chondrocyte phenotype (e.g., type II collagen mRNA) for several weeks and will form multilayer cultures. After they are subcultured, both types of chondrocytes cease the expression chondrocyte matrix proteins, but this loss of phenotype is reversible, and the cells may be redifferentiated in 3D or suspension culture.

4. Since chondrocytes adhere strongly to tissue culture plastic, possibly because of the presence of calcium ion-binding glycosaminoglycans in the pericellular matrix and cell membrane, a trypsin–EDTA solution rather than trypsin alone should be used for full recovery of chondrocytes from tissue culture plastic during passaging. It is preferable not to use any antibiotics so that any contamination that arises becomes apparent immediately. If necessary, particularly when antibiotics are used in other cultures in the same incubator or culture facility, standard concentrations of penicillin–streptomycin or other antibiotics that are suggested for fibroblast cultures are acceptable for use in chondrocyte cultures.
5. Primary chondrocyte cultures should be used for experimental analyses immediately before or just after confluence is reached to permit optimal matrix synthesis and cellular responsiveness. If the cells are not used or subcultured, they may be left at confluence for several weeks with weekly medium changes as long as the volume of the culture medium is maintained. If long-term culture results in the deposition of excessive matrix that is not easily digested with trypsin–EDTA, then a single cell suspension may be obtained by using a dilute solution of collagenase (0.25%) and trypsin (0.25%) in PBS.
6. The synthetic activities of chondrocytes in monolayer culture are inversely related to proliferative activities. Thus, the expression of genes encoding matrix proteins and their deposition into the extracellular matrix increase compared to cell growth-associated genes. For experiments, the growth medium should not be changed within 3 days before addition of the test agent. Alternatively, the cells should be made quiescent by changing to serum-free medium supplemented with an insulin-containing serum substitute such as Nutridoma-SP or ITS+, followed 18–24 h later by the addition (without medium change) of the test agent of interest. Confluent cultures may tolerate serum-free medium containing 0.3% bovine serum albumin for up to 48 h or longer.
7. While the growth and maintenance of chondrocytes in primary culture or after subculture requires the use of 10% FCS, the loss of phenotype that occurs under these conditions may be delayed if the cells are plated at four- to tenfold higher density. Since high cell yields are not usually attainable from human cartilage sources, the reversibility of the loss of phenotype may be exploited by expanding the chondrocyte populations in monolayer cultures, redifferentiating the cells in fluid suspension culture and replating them in monolayer immediately before performing the experimental procedure.
8. After several passages in monolayer, chondrocytes may be redifferentiated by 2 weeks or more of culture in alginate beads or in suspension over agarose or polyHEMA.

9. The method for culture of chondrocytes in alginate beads has been adapted from previously published methods (32, 33). For long-term alginate cultures, high viscosity alginate may provide more stable beads. Serum at concentrations as low as 0.5%, serum substitutes, or combinations of growth and differentiation factors or hormones have been used successfully, depending upon the experimental protocol, to permit chondrocyte phenotypic expression. Note that articular chondrocytes do not proliferate when cultured in fluid or gel suspension.
10. Ascorbate, which is required for synthesis and secretion of proteoglycans and collagens, is added daily to alginate or other 3D cultures to permit secretion and deposition of extracellular matrix, particularly when staining techniques are to be used. Add 25 μg/mL of ascorbate during the final 24–72 h of incubation when radiolabeling proteoglycans with ^{35}S-sulfate or collagens with ^{3}H-proline for characterization by SDS-PAGE.
11. Culture of immortalized chondrocytes in 3D scaffolds is a useful approach for tissue engineering applications. The commercially available methods are recommended because of their ease of use. Published methods are available for fabricating collagen sponges (34) and other 3D scaffolds, where the composition may be manipulated, for example, by using type II collagen and/or adding GAGs and other cartilage-specific matrix components. The biodegradable scaffolds are particularly useful if the cell-seeded scaffolds are to be implanted in animals. For studies entirely in vitro, where incubation periods of more than a few days are required, it is recommended that cultures be performed in wells that fit the size of the scaffolds. Otherwise, the culture surface of the well or dish should be coated with a nonadherent substrate or treated in such a way as to prevent attachment of cells that may migrate out from the sponges. Additional analytical methods have been described using Gelfoam® (61) and BD™ 3D scaffolds (see Web site at http://www.bdbiosciences.com/discovery_labware/Products/tissue_engineering/).
12. Biosynthetic labeling and immunocytochemistry procedures are readily performed on chondrocytes in a solid suspension system such as alginate, agarose, or collagen gels. Alginate culture may be the method of choice, since the chondrocytes are easily recovered by depolymerization of the alginate with a calcium chelator.
13. Various methods are available for analysis and characterization of proteoglycans. We have found the described methods to be a convenient and rapid approach for screening the relative amounts and molecular sizes of newly synthesized proteoglycans. Specific antibodies are available for more precise identification by either Western blotting or immunocytochemistry, as described in Subheading 2.3.6.

14. Although chondrocytes are generally less susceptible than monocyte/macrophages and other immune cells to endotoxin, it is possible that the transfection conditions, the proliferative state of the cells, or other factors may sensitize the cells to low concentrations of endotoxin (62). Endotoxin itself induces and activates transcription factors that are common to inflammatory responses and may thus upregulate or downregulate the promoter of interest, thereby masking the response to a cytokine or growth factor.
15. Hyaluronidase added before and/or during transfections has been shown to increase transfection efficiencies in chondrocytes (63–65).
16. The total amount of plasmid to be transfected, including reporter, expression and internal control plasmids, should not exceed 1 μg/well of 6-well plate and the optimal amount for the culture system should be tested empirically. If variable amounts of expression vector, for example, are included, the total amount of plasmid in each well should be equalized by the addition of the empty vector. Wells transfected with appropriate empty vector controls, without and with treatment with test agent, should also be included. Since primary chondrocytes are very difficult to transfect, we use them in definitive experiments in studies that mainly involve the use of immortalized chondrocyte cell lines (54, 56, 57).
17. The times of incubation following transfection before addition of the test agent may vary according to the cell density and culture condition and should be tested empirically. Nutridoma-SP, ITS+, or other serum substitute may be used for experiments that require quiescent cells for the reasons indicated in Note 6. Note also that test agents should be added without medium change to avoid induction of pathways of interest by serum growth factors or other constituents of the control medium.
18. The knockdown efficacy (degree of reduction of the targeted gene) depends upon transfection efficiency (successful delivery of the siRNA into the cells is usually cell dependent), efficacy of a given siRNA sequence/s, transcription rate of the gene of interest, and protein stability (time- and/or siRNA concentration-dependent). Therefore, those variables must be considered and empirically addressed when designing RNAi experiments. Transfection efficiencies can be determined with the use of fluorescent dye (e.g., rhodamine or FITC)-conjugated siRNA sequences, which allow for visualization of the positive siRNA-transfected cells as well as determination of siRNA subcellular localization and stability in time-course experiments. Knockdown efficacy can be determined in dose–response and time-course experiments, where different concentrations of the siRNA sequence/s (e.g., from 100 to 10 nM)

are delivered into the cells and the knockdown of the targeted gene is addressed at different time-points analyzing mRNA (usually, 24, 48, and 72 h posttransfection) and protein (usually, 48, 72, and 96 h posttransfection) levels as compared with those of mock-treated cells and cells transfected with non-targeting siRNA sequences (58, 59).

19. ChIP assays are generally limited by the requirement for large cell numbers, the different efficiencies with which different proteins cross-link with their interacting DNA – particularly relevant when addressing binding of transcription factors due to their lower abundance and somewhat dynamic and labile interaction with their target DNA sequences as compared to major chromatin constituents such as histones, the relative amount of a DNA-bound protein of interest within a cell type, and the availability of high-quality and high affinity antibodies against the protein of interest. The indicated number of cells will thus vary depending upon the protein–DNA interaction to be addressed. Therefore, several steps, including cross-linking and shearing, and controls for antibody specificity, must be carefully optimized for each specific assay. The optimization steps required for having a reproducible and specific ChIP assay may be not feasible using primary chondrocytes, given their usually limited availability. In that regard, the use of immortalized chondrocytes may be a useful tool to set the adequate conditions for the specific ChIP assay to be conducted, as it is possible to work with larger amounts of material in a more reproducible manner (54, 58, 59). The assays optimized using the cell lines may then be validated or assessed using primary cells.

20. Before conducting time- and cost-expensive experiments, several steps should be carefully optimized and empirically tested. Namely, it is advisable to perform small-scale experiments, where different cross-linking and shearing conditions are tested. Prior to performing immunoprecipitation, the sheared chromatin can be reverse cross-linked and analyzed in a 1–1.5% agarose gel (with the conditions described, enzymatic shearing will yield fragments between 250 and 1,000 bp) to ensure optimal shearing, and the yield of DNA can be quantified to use equal amounts of DNA for the immunoprecipitation and comparison of different conditions. Although not always necessary, preclearing or additional BSA-blocking steps may be required to increase the signal-to-noise ratio when using certain antibodies. For PCR analysis of the precipitated DNA, real time PCR analysis is recommended as it permits a more refined and quantitative analysis of the final result. When analyzing PCR products in agarose gels (Fig. 6), the PCR must be stopped in linear stage of amplification. Sequencing of the PCR products is suggested to ensure specific amplification.

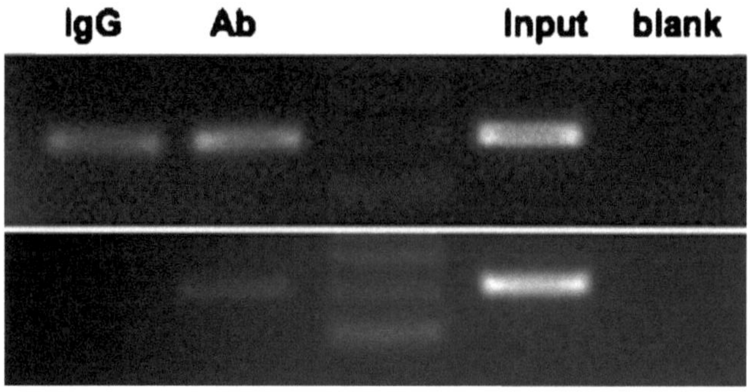

Fig. 6. ChIP analysis performed without or with a preclearing step. Chromatin was cross-linked and enzymatically sheared, and after reverse cross-linking of the DNA–protein complexes, the lysates were either (*upper panel*) used directly for immunoprecipitation without a preclearing step or (*lower panel*) precleared prior to overnight incubation with the antibody against a protein of interest (Ab1) or normal rabbit IgG (IgG). The precipitated DNA was purified and PCR-amplified, and the PCR products were resolved in a 2.5% agarose gel.

21. Previous studies have shown that there may be zonal variations in the methylation status of certain genes (60). For example, the chondrocytes from the superficial zone of normal cartilage may show increased/aberrant expression of proteases and cytokines and, therefore, chondrocytes from the middle to deep zones are used to assess physiological methylation status and compare with chondrocytes from osteoarthritis patients.
22. Experimental groups are designed with chondrocytes obtained from the same individual and, therefore, the number of cells may be limiting. More importantly, epigenetic events require long-term analysis and, thus, cells should be plated at low densities so they do not reach confluence quickly and allow for long-term treatment with the testing agent/s and inheritance of the epigenetic changes.
23. After screening of the methylation status of a given promoter by bisulfite modification and identification of candidate CpG site/s, platforms such as PyroMark Q96 Systems (QIAGEN Inc., Valencia, CA) allow for the quantitative analysis of multiple CpG sites simultaneously, whereas the enzyme-based method detailed here allows for the assessment of one CpG site at the time.

Acknowledgments

Dr. Goldring's research related to this project was supported in part by grants from the National Institutes of Health, AG022021, AR054887, and AR045378, and the Arthritis Foundation. The authors are grateful to Lujian Tan, Haibing Peng, Lii-Fang Suen, James Birkhead, Merrilee Flannery, James Robbins, Makoto

Osaki, and Bob Choy for supplying technical expertise and data, and to Dr. Thomas Sculco at The Hospital for Special Surgery and Dr. Benjamin Bierbaum at the New England Baptist Hospital for supplying cartilage samples. We also wish to acknowledge Dr. Helmtrud (Trudy) I. Roach, recently deceased, for her contributions to our understanding of epigenetics in chondrocytes and for transferring the technology to our laboratory.

References

1. Goldring, M. B. (2009) *Cartilage and Chondrocytes. In Kelley's Textbook of Rheumatology, 8th Edition* (Firestein, G. S., Budd, R. C., Harris, E. D. J., McInnes, I. B., Ruddy, S. and Sergent, J. S., eds.), Saunders an imprint of Elsevier, Inc., Philadelphia, Ch. 3, pp. 37–69.
2. Maroudas, A., Palla, G. and Gilav, E. (1992) Racemization of aspartic acid in human articular cartilage. *Connect Tissue Res* 28, 161–169.
3. Verzijl, N., DeGroot, J., Thorpe, S. R., Bank, R. A., Shaw, J. N., Lyons, T. J., Bijlsma, J. W., Lafeber, F. P., Baynes, J. W. and TeKoppele, J. M. (2000) Effect of collagen turnover on the accumulation of advanced glycation end products. *J Biol Chem* 275, 39027–39031.
4. Maroudas, A., Bayliss, M. T., Uchitel-Kaushansky, N., Schneiderman, R. and Gilav, E. (1998) Aggrecan turnover in human articular cartilage: use of aspartic acid racemization as a marker of molecular age. *Arch Biochem Biophys* 350, 61–71.
5. Otero, M. and Goldring, M. B. (2007) Cells of the synovium in rheumatoid arthritis. Chondrocytes. *Arthritis Res Ther* 9, 220.
6. Goldring, M. B., Otero, M., Tsuchimochi, K., Ijiri, K. and Li, Y. (2008) Defining the roles of inflammatory and anabolic cytokines in cartilage metabolism. *Ann Rheum Dis* 67 Suppl 3, iii75–82.
7. Goldring, M. B. and Marcu, K. B. (2009) Cartilage homeostasis in health and rheumatic diseases. *Arthritis Res Ther* 11, 224.
8. Holtzer, J., Abbott, J., Lash, J. and Holtzer, A. (1960) The loss of phenotypic traits by differentiated cells in vitro. I. Dedifferentiation of cartilage cells. *Proc. Natl. Acad. Sci. USA* 46, 1533–1542.
9. Ham, R. G. and Sattler, G. L. (1968) Clonal growth of differentiated rabbit cartilage cells. *J. Cell. Physiol.* 72, 109–114.
10. Green, W. T., Jr. (1971) Behavior of articular chondrocytes in cell culture. *Clin. Orthopaed. Rel. Res.* 75, 248–260.
11. Goldring, M. B., Sandell, L. J., Stephenson, M. L. and Krane, S. M. (1986) Immune interferon suppresses levels of procollagen mRNA and type II collagen synthesis in cultured human articular and costal chondrocytes. *J. Biol. Chem.* 261, 9049–9056..
12. Aulthouse, A. L., Beck, M., Friffey, E., Sanford, J., Arden, K., Machado, M. A. and Horton, W. A. (1989) Expression of the human chondrocyte phenotype in vitro. *In Vitro Cell. Devel. Biol.* 25, 659–668.
13. Kolettas, E., Buluwela, L., Bayliss, M. T. and Muir, H. I. (1995) Expression of cartilage-specific molecules is retained on long-term culture of human articular chondrocytes. *J Cell Sci* 108 (Pt 5), 1991–1999.
14. Goldring, M. B. (1987) Control of collagen synthesis in human chondrocyte cultures by immune interferon and interleukin-1. *J Rheumatol* 14 Spec No, 64–66.
15. Goldring, M. B., Birkhead, J., Sandell, L. J., Kimura, T. and Krane, S. M. (1988) Interleukin 1 suppresses expression of cartilage-specific types II and IX collagens and increases types I and III collagens in human chondrocytes. *J Clin Invest* 82, 2026–2037.
16. Schnabel, M., Marlovits, S., Eckhoff, G., Fichtel, I., Gotzen, L., Vecsei, V. and Schlegel, J. (2002) Dedifferentiation-associated changes in morphology and gene expression in primary human articular chondrocytes in cell culture. *Osteoarthritis Cartilage* 10, 62–70.
17. Barbero, A., Grogan, S. P., Mainil-Varlet, P. and Martin, I. (2006) Expansion on specific substrates regulates the phenotype and differentiation capacity of human articular chondrocytes. *J Cell Biochem* 98, 1140–1149.
18. von der Mark, K., Gauss, V., von der Mark, H. and Muller, P. (1977) Relationship between cell shape and type of collagen synthesised as chondrocytes lose their cartilage phenotype in culture. *Nature* 267, 531–532.
19. Watt, F. M. (1988) Effect of seeding density on stability of the differentiated phenotype of pig articular chondrocytes in culture. *J Cell Sci* 89 (Pt 3), 373–378.
20. Kuettner, K. E., Pauli, B. U., Gall, G., Memoli, V. A. and Schenk, R. K. (1982) Synthesis of cartilage matrix by mammalian chondrocytes

in vitro. I. Isolation, culture characteristics, and morphology. *J. Cell Biol.* 93, 743–750.

21. Bassleer, C., Gysen, P., Foidart, J. M., Bassleer, R. and Franchimont, P. (1986) Human chondrocytes in tridimensional culture. *In Vitro Cell. Devel. Biol.* 22, 113–119.
22. Thonar, E. J., Buckwalter, J. A. and Kuettner, K. E. (1986) Maturation-related differences in the structure and composition of proteoglycans synthesized by chondrocytes from bovine articular cartilage. *J Biol Chem* 261, 2467–2474.
23. Norby, D. P., Malemud, C. J. and Sokoloff, L. (1977) Differences in the collagen types synthesized by lapine articular chondrocytes in spinner and monolayer culture. *Arthritis Rheum* 20, 709–716.
24. Glowacki, J., Trepman, E. and Folkman, J. (1983) Cell shape and phenotypic expression in chondrocytes. *Proc. Soc. Exp. Biol. Med.* 172, 93–98.
25. Castagnola, P., Moro, G., Descalzi-Cancedda, F. and Cancedda, R. (1986) Type X collagen synthesis during in vitro development of chick embryo tibial chondrocytes. *J Cell Biol* 102, 2310–2317.
26. Reginato, A. M., Iozzo, R. V. and Jimenez, S. A. (1994) Formation of nodular structures resembling mature articular cartilage in long-term primary cultures of human fetal epiphyseal chondrocytes on hydrogel substrate. *Arthritis Rheum.* 37, 1338–1349.
27. Croucher, L. J., Crawford, A., Hatton, P. V., Russell, R. G. and Buttle, D. J. (2000) Extracellular ATP and UTP stimulate cartilage proteoglycan and collagen accumulation in bovine articular chondrocyte pellet cultures. *Biochim Biophys Acta* 1502, 297–306.
28. Olivotto, E., Vitellozzi, R., Fernandez, P., Falcieri, E., Battistelli, M., Burattini, S., Facchini, A., Flamigni, F., Santi, S., Facchini, A. and Borzi, R. M. (2007) Chondrocyte hypertrophy and apoptosis induced by GROalpha require three-dimensional interaction with the extracellular matrix and a co-receptor role of chondroitin sulfate and are associated with the mitochondrial splicing variant of cathepsin B. *J Cell Physiol.* 210, 417–427.
29. Gibson, G. J., Schor, S. L. and Grant, M. E. (1982) Effects of matrix macromolecules on chondrocyte gene expression: Synthesis of a low molecular weight collagen species by cells cultured within collagen gels. *J. Cell Biol.* 93, 767–774.
30. Benya, P. D. and Shaffer, J. D. (1982) Dedifferentiated chondrocytes reexpress the differentiated collagen phenotype when cultured in agarose gels. *Cell* 30, 215–224.
31. Aydelotte, M. B. and Kuettner, K. E. (1988) Differences between sub-populations of cultured bovine articular chondrocytes. I. Morphology and cartilage matrix production. *Conn. Tiss. Res.* 18, 205–222.
32. Guo, J., Jourdian, G. W. and MacCallum, D. K. (1989) Culture and growth characteristics of chondrocytes encapsulated in alginate beads. *Conn. Tiss. Res.* 19, 277–297.
33. Hauselmann, H. J., Aydelotte, M. B., Schumacher, B. L., Kuettner, K. E., Gitelis, S. H. and Thonar, E. J.-M. A. (1992) Synthesis and turnover of proteoglycans by human and bovine adult articular chondrocytes cultured in alginate beads. *Matrix* 12, 116–129.
34. Mizuno, S., Allemann, F. and Glowacki, J. (2001) Effects of medium perfusion on matrix production by bovine chondrocytes in three-dimensional collagen sponges. *J Biomed Mater Res* 56, 368–375.
35. Xu, C., Oyajobi, B. O., Frazer, A., Kozaci, L. D., Russell, R. G. and Hollander, A. P. (1996) Effects of growth factors and interleukin-1 alpha on proteoglycan and type II collagen turnover in bovine nasal and articular chondrocyte pellet cultures. *Endocrinology* 137, 3557–3565.
36. Adolphe, M., Froger, B., Ronot, X., Corvol, M. T. and Forest, N. (1984) Cell multiplication and type II collagen production by rabbit articular chondrocytes cultivated in a defined medium. *Exp. Cell Res.* 155, 527–536.
37. Gerstenfeld, L. C., Kelly, C. M., Von Deck, M. and Lian, J. B. (1990) Comparative morphological and biochemical analysis of hypertrophic, non-hypertrophic and $1,25(OH)_2D_3$ treated non-hypertrophic chondrocytes. *Connective Tissue Research.* 24, 29–39.
38. Adams, S. L., Pallante, K. M., Niu, Z., Leboy, P. S., Golden, E. B. and Pacifici, M. (1991) Rapid induction of type X collagen gene expression in cultured chick vertebral chondrocytes. *Exp. Cell Res.* 193, 190–197.
39. Poole, A. R. (1989) Honor Bridgett Fell, Ph.D., D.Sc. F.R.S., D.B.E., 1900–1986. The scientist and her contributions. *In Vitro Cell Dev Biol* 25, 450–453.
40. Benoit, B., Thenet-Gauci, S., Hoffschir, F., Penformis, P., Demignot, S. and Adolphe, M. (1995) SV40 large T antigen immortalization of human articular chondrocytes. *In Vitro Cell. Dev. Biol.* 31, 174–177.
41. Steimberg, N., Viengchareun, S., Biehlmann, F., Guenal, I., Mignotte, B., Adolphe, M. and Thenet, S. (1999) SV40 large T antigen expression driven by col2a1 regulatory sequences immortalizes articular chondrocytes but does not allow stabilization of type II collagen expression. *Exp Cell Res* 249, 248–259.
42. Robbins, J. R., Thomas, B., Tan, L., Choy, B., Arbiser, J. L., Berenbaum, F. and Goldring, M. B. (2000) Immortalized human adult articular

chondrocytes maintain cartilage-specific phenotype and responses to interleukin-1β. *Arthritis Rheum* 43, 2189–2201.

43. Grigolo, B., Roseti, L., Neri, S., Gobbi, P., Jensen, P., Major, E. O. and Facchini, A. (2002) Human articular chondrocytes immortalized by HPV-16 E6 and E7 genes: Maintenance of differentiated phenotype under defined culture conditions. *Osteoarthritis Cartilage* 10, 879–889.
44. Piera-Velazquez, S., Jimenez, S. A. and Stokes, D. (2002) Increased life span of human osteoarthritic chondrocytes by exogenous expression of telomerase. *Arthritis Rheum* 46, 683–693.
45. Goldring, M. B. (2004) Immortalization of human articular chondrocytes for generation of stable, differentiated cell lines. *Methods Mol Med* 100, 23–36.
46. Goldring, M. B., Birkhead, J. R., Suen, L. F., Yamin, R., Mizuno, S., Glowacki, J., Arbiser, J. L. and Apperley, J. F. (1994) Interleukin-1 beta-modulated gene expression in immortalized human chondrocytes. *J Clin Invest* 94, 2307–2316.
47. Laemmli, U. K. (1970) Cleavage of structural proteins during the assembly of the head of bacteriophage T4. *Nature* 227, 680–685.
48. Olivotto, E., Borzi, R. M., Vitellozzi, R., Pagani, S., Facchini, A., Battistelli, M., Penzo, M., Li, X., Flamigni, F., Li, J., Falcieri, E., Facchini, A. and Marcu, K. B. (2008) Differential requirements for IKKalpha and IKKbeta in the differentiation of primary human osteoarthritic chondrocytes. *Arthritis Rheum* 58, 227–239.
49. Vogel, K. G., Sandy, J. D., Pogany, G. and Robbins, J. R. (1994) Aggrecan in bovine tendon. *Matrix Biology* 14, 171–179.
50. Robbins, J. R. and Vogel, K. G. (1997) Mechanical loading and TGF-β regulate proteoglycan synthesis in tendon. *Arch. Biochem. Biophys.* 342, 203–211.
51. Kokenyesi, R., Tan, L., Robbins, J. R. and Goldring, M. B. (2000) Proteoglycan production by immortalized human chondrocyte cell lines cultured under conditions that promote expression of the differentiated phenotype. *Arch. Biochem. Biophys.* 383, 79–90..
52. Dell'Accio, F., De Bari, C. and Luyten, F. P. (2001) Molecular markers predictive of the capacity of expanded human articular chondrocytes to form stable cartilage in vivo. *Arthritis Rheum* 44, 1608–1619.
53. Bau, B., Gebhard, P. M., Haag, J., Knorr, T., Bartnik, E. and Aigner, T. (2002) Relative messenger RNA expression profiling of collagenases and aggrecanases in human articular chondrocytes in vivo and in vitro. *Arthritis Rheum* 46, 2648–2657.
54. Peng, H., Tan, L., Osaki, M., Zhan, Y., Ijiri, K., Tsuchimochi, K., Otero, M., Wang, H., Choy, B. K., Grall, F. T., Gu, X., Libermann, T. A., Oettgen, P. and Goldring, M. B. (2008) ESE-1 is a potent repressor of type II collagen gene (COL2A1) transcription in human chondrocytes. *J Cell Physiol* 215, 562–573.
55. Goldring, M. B., Fukuo, K., Birkhead, J. R., Dudek, E. and Sandell, L. J. (1994) Transcriptional suppression by interleukin-1 and interferon-γ of type II collagen gene expression in human chondrocytes. *J. Cell. Biochem.* 54, 85–99.
56. Osaki, M., Tan, L., Choy, B. K., Yoshida, Y., Auron, P. E., Cheah, K. S. E. and Goldring, M. B. (2003) The TATA-containing core promoter of the type II collagen gene (COL2A1) is the target of interferon-γ-mediated inhibition in human chondrocytes: requirement for Stat1α, Jak1, and Jak2. *Biochem. J.* 369, 103–115.
57. Tan, L., Peng, H., Osaki, M., Choy, B. K., Auron, P. E., Sandell, L. J. and Goldring, M. B. (2003) Egr-1 Mediates Transcriptional Repression of COL2A1 Promoter Activity by Interleukin-1beta. *J Biol Chem* 278, 17688–17700.
58. Ijiri, K., Zerbini, L. F., Peng, H., Otu, H. H., Tsuchimochi, K., Otero, M., Dragomir, C., Walsh, N., Bierbaum, B. E., Mattingly, D., van Flandern, G., Komiya, S., Aigner, T., Libermann, T. A. and Goldring, M. B. (2008) Differential expression of GADD45beta in normal and osteoarthritic cartilage: potential role in homeostasis of articular chondrocytes. *Arthritis Rheum* 58, 2075–2087.
59. Tsuchimochi, K., Otero, M., Dragomir, C. L., Plumb, D. A., Zerbini, L. F., Libermann, T. A., Marcu, K. B., Komiya, S., Ijiri, K. and Goldring, M. B. (2010) GADD45beta enhances Col10a1 transcription via the MTK1/MKK3/6/p38 axis and activation of C/EBPbeta-TAD4 in terminally differentiating chondrocytes. *J Biol Chem* 285, 8395–8407.
60. Hashimoto, K., Oreffo, R. O., Gibson, M. B., Goldring, M. B. and Roach, H. I. (2009) DNA demethylation at specific CpG sites in the IL1B promoter in response to inflammatory cytokines in human articular chondrocytes. *Arthritis Rheum* 60, 3303–3313.
61. Wu, Q. Q. and Chen, Q. (2000) Mechanoregulation of chondrocyte proliferation, maturation, and hypertrophy: ion-channel dependent transduction of matrix deformation signals. *Exp Cell Res* 256, 383–391..
62. Cotten, M., Baker, A., Saltik, M., Wagner, E. and Buschle, M. (1994) Lipopolysaccharide is a frequent contaminant of plasmid DNA

preparations and can be toxic to primary human cells in the presence of adenovirus. *Gene Ther* 1, 239–246.

63. Lu Valle, P., Iwamoto, M., Fanning, P., Pacifici, M. and Olsen, B. R. (1993) Multiple negative elements in a gene that codes for an extracellular matrix protein, collagen X, restrict expression to hypertrophic chondrocytes. *J Cell Biol* 121, 1173–1179.

64. Viengchareun, S., Thenet-Gauci, S., Steimberg, N., Blancher, C., Crisanti, P. and Adolphe, M. (1997) The transfection of rabbit articular chondrocytes is independent of their differentiation state. *In Vitro Cell Dev Biol Anim* 33, 15–17.

65. Madry, H. and Trippel, S. B. (2000) Efficient lipid-mediated gene transfer to articular chondrocytes. *Gene Ther* 7, 286–291.

Chapter 22

Isolation and Culture of Human Osteoblasts

Alison Gartland, Robin M.H. Rumney, Jane P. Dillon, and James A. Gallagher

Abstract

The skeleton is a dynamic organ that is constantly active throughout life. The highly coordinated actions of bone cells early in life determine the body's shape and form, whilst the constant remodelling (bone resorption followed by an equal amount of bone formation) during adulthood helps to maintain skeletal mass and repair microdamage. When the balance of bone resorption and bone formation becomes unequal, bone diseases, such as osteoporosis, occur. In order to develop drugs to combat bone disease, it is important to know the regulatory systems involved in normal bone formation and resorption.

In this chapter, we concentrate on bone formation, providing a detailed guide to isolating and culturing primary human osteoblasts in bone explant cultures, as well as the methodology used to characterise and monitor the function of osteoblasts. In combination, these methods provide a powerful tool in bone cell biology and in the development of new novel treatments for bone disease.

Key words: Osteoblasts, Bone derived, Alkaline phosphatase, Primary cell culture, Explant, Mineralisation, SaOS-2

1. Introduction

Bone is a complex tissue that contains at least four different cell types of the osteoblastic lineage. (1) Active osteoblast – a plump, polarised, cuboidal cell rich in organelles involved in the synthesis and secretion of matrix proteins. (2) Osteocyte – an osteoblast with low metabolic activity that has been engulfed in matrix during bone formation and entombed in lacunae. (3) Bone-lining cell – osteoblasts that have avoided entombment in lacunae and lost their prominent synthetic function; these cells cover most of the bone surfaces in mature bone. (4) Pre-osteoblast – a fibroblastic proliferative cell with osteogenic capacity. In addition, bone contains cells of a distinct lineage, the osteoclast (reviewed in Chapter 23).

Ragai R. Mitry and Robin D. Hughes (eds.), *Human Cell Culture Protocols*, Methods in Molecular Biology, vol. 806, DOI 10.1007/978-1-61779-367-7_22,

The complex structure of bone tissue, the heterogeneity of cell types, as well as the cross-linked extracellular matrix and the mineral phase all combine to make bone a difficult tissue from which to extract cells and to study at the cellular and molecular level (1). Consequently, early attempts to culture osteoblasts relied on enzymic digestion of poorly mineralised, highly cellular fetal or neonatal tissue from experimental animals, and avoided mature, mineralised human bone. Although these studies undoubtedly furthered our knowledge of bone cell biology, they had obvious drawbacks due to the differences in cell physiology between the species, and also between adults and neonates within a species. In order to understand fully the pathological mechanisms that underlie bone diseases, including age-related bone loss, the ability to culture human bone cells is essential.

Early attempts to isolate cells from adult human bone were reported by Bard et al. (2), using demineralisation and collagenase digestion, and by Mills et al. (3), using the alternative approach of explant culture. The first successful attempts to isolate large numbers of cells that expressed an osteoblastic phenotype from human bone were undertaken in the early 1980s in Graham Russell's laboratory at the University of Sheffield. The defining characteristics of these studies were the use of explant culture, which avoided the need for digestion of the tissue, and the availability of an appropriate phenotypic maker. Successful culture of any cell type can be achieved only if there is a specific marker of the phenotype that can be used to confirm the identity of the cells in vitro. In the case of bone, the marker was the then recently discovered bone-gla protein as measured by a radioimmunoassay developed by Poser and colleagues (4, 5).

Twenty-five years later, bone-gla protein, now known as osteocalcin, undoubtedly remains the most specific marker of the osteoblastic phenotype. Although formation of bone is the most conspicuous function of the osteoblasts, research over the past decade has revealed that cells of the osteoblastic lineage play a major role in regulating bone resorption. Researchers have increasingly used tissue culture techniques to investigate the roles of osteoblasts in bone formation and bone resorption, and the culture of human bone cells is now the predominant technique used in the investigation of bone biology.

The techniques for culturing primary human osteoblasts have been reviewed widely over the past few years (6–8); however, in this rapidly moving field, there have been many new developments. The aim of this chapter is to describe in detail the basic techniques of human bone cell culture (both primary and clonal) and the methodology used to characterise and measure the function of these cells. We hope that this comprehensive guide will aid researchers in their investigations of bone cell biology and the development of new novel treatments for bone disease.

2. Materials

2.1. Tissue Culture Media and Supplements

1. Phosphate-buffered saline (PBS) without calcium and magnesium, pH 7.4 (Invitrogen, Paisley, UK).
2. Serum-free medium (SFM) – Dulbecco's modification of minimum essential medium (D-MEM)–Glutamax™ containing 100 U/mL penicillin and 100 μg/mL streptomycin.
3. Complete medium – D-MEM–Glutamax™ containing 100 U/mL penicillin, 100 μg/mL streptomycin, 10% FBS (previously tested for ability to support human bone-derived cell (HBDC) growth), and 50 μg/mL freshly prepared L-ascorbic acid (if needed – see Subheading 3.1.4).
4. D-MEM with Glutamax™ (Invitrogen) (see Note 1).
5. Penicillin–Streptomycin (Invitrogen).
6. Fetal bovine serum (FBS) (see Note 2).
7. In cultures in which matrix synthesis or mineralisation is investigated 50 μg/mL freshly prepared L-ascorbic acid (Sigma–Aldrich, Dorset, UK) (see Note 3).
8. Vented tissue culture flasks (typically 75 cm^2) or Petri dishes (typically 90- or 100-mm diameter) (see Note 4).

2.2. Instruments for Preparation of Explants

1. Bone rongeurs from any surgical instrument supplier.
2. Solid stainless-steel scalpels with integral handles (VWR, Leicestershire, UK).

2.3. Passaging and Secondary Culture

1. Trypsin–EDTA solution: 0.05% trypsin and 0.02% EDTA in Ca^{2+}- and Mg^{2+}-free Hank's BSS (HBSS), pH 7.4 (Invitrogen).
2. 0.4% Trypan blue in 0.85% NaCl (Sigma–Aldrich).
3. 70 μm "Cell Strainer" (BD, Oxford).
4. Neubauer Hemocytometer (VWR).
5. SFM containing 25 U/mL purified collagenase and 2 mM additional calcium (1:500 dilution of a filter-sterilised stock solution of 1 M $CaCl_2$). Collagenase type VII from Clostridium histolyticum (Sigma–Aldrich).
6. Osteogenic medium (complete medium supplemented with 10 nM dexamethasone and 50 μg/mL ascorbic acid).
7. 1 M $CaCl_2$ stock solution (filter sterilised).
8. DNAse I (Sigma–Aldrich).

2.4. Phenotypic Characterisation

1. *p*-Nitrophenyl phosphate (*p*-NPP) substrate (Sigma–Aldrich). Prepare at 1 mg/mL in 0.1 M glycine, pH 10.4, with 1 mM $MgCl_2$ and 1 mM $ZnCl_2$. Stock solutions can be kept frozen at −20°C for several months.

2. NBT/BCIP Ready-To-Use tablets (Roche), reconstituted in PBS. Tablets stable at room temperature until the expiry date, and solution should be made up immediately prior to use.

2.5. In Vitro Mineralisation

1. L-Ascorbic acid (Sigma–Aldrich) (see Note 3).
2. 500 mM phosphate solution, pH 7.4. Prepared by mixing 500 mM solutions of Na_2HPO_4 and NaH_2PO_4 in a 4:1 (v/v) ratio. The filter-sterilised stock solution can be stored at 4°C indefinitely.
3. Dexamethasone (Sigma–Aldrich).
4. Hematoxylin (VWR).
5. Alizarin Red S (ARS) (Sigma–Aldrich).
6. Ethanol.
7. DPX (WVR).

3. Methods

3.1. Bone Explant Culture System

In this chapter, we use the term HBDCs to describe cells expressing osteoblastic characteristics derived from human bone. Since the description of the original methods, many investigators have turned to the use of HBDCs (extensively reviewed in ref. 8). Whereas some have developed techniques for the isolation and culture of HBDCs that differ significantly from the original method (see Note 5), most have continued to use the explant technique with only a few minor modifications. A scheme outlining the basic explant culture technique is shown in Fig. 1.

3.1.1. Establishment of Primary Explant Cultures of Human Osteoblastic Cells

1. All scalpels, bone rongeurs, scissors, blades, and forceps should be sterilised prior to use. The work should be performed in the laminar flow cabinet.
2. Transfer tissue, removed at surgery or biopsy, into a sterile container with PBS or SFM for transport to the laboratory with minimal delay, preferably on the same day (see Note 6). An excellent source is the upper femur of patients undergoing total hip replacement surgery for osteoarthritis. Cancellous bone is removed from this site prior to the insertion of the femoral prosthesis and would otherwise be discarded (consult local Ethical Committee for guidelines on seeking ethical approval and patient permission). The tissue obtained is remote from the hip joint itself, and thus from the site of pathology, and is free of contaminating soft tissue (see Note 7).
3. Remove extraneous soft connective tissue from the outer surfaces of the bone by scraping with a sterile scalpel blade.

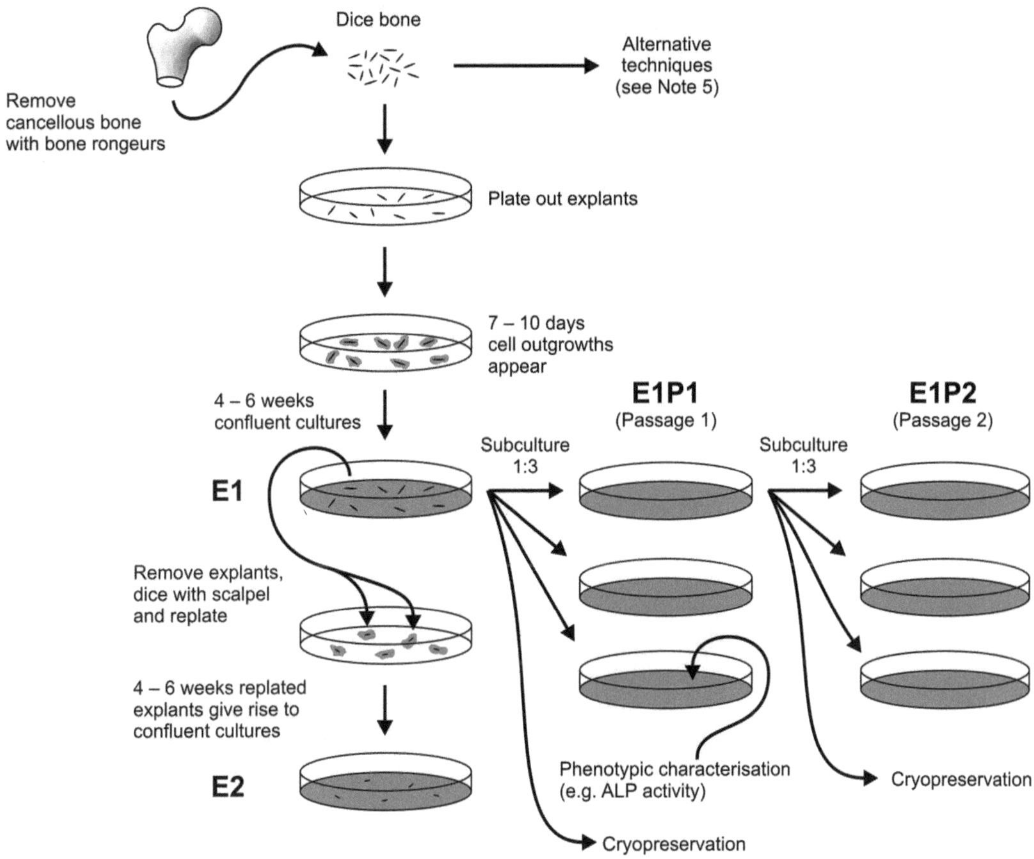

Fig. 1. Schematic representation of the technique used to isolate HBDCs from explanted cancellous bone.

Rinse the tissue in sterile PBS and transfer to a sterile Petri dish containing a small volume of PBS (5–20 mL, depending on the size of the specimen). If the bone sample is a femoral head, remove cancellous bone directly from the open end using sterile bone rongeurs or a solid stainless steel blade with integral handle. Disposable scalpel blades may shatter during this process. With some bone samples (e.g. rib), it may be necessary to gain access to the cancellous bone by breaking through the cortex with the aid of the sterile surgical bone rongeurs.

4. Transfer the cancellous bone fragments to a clean Petri dish containing 2–3 mL of PBS and dice into pieces 3–5 mm in diameter. This can be achieved in two stages using a scalpel blade first, and then fine scissors. Decant the PBS and transfer the bone chips to a sterile 30-mL "Universal container" with 15–20 mL of PBS. Vortex the tube vigorously three times for 10 s and then leave to stand for 30 s to allow the bone fragments to settle. Carefully decant off the supernatant containing haematopoietic tissue and dislodged cells, add an additional 15–20 mL of PBS, and vortex the bone fragments as before.

Repeat this process a minimum of three times or until no remaining haematopoietic marrow is visible and the bone fragments have assumed a white, ivory-like appearance.

5. Culture the washed bone fragments as explants at a density of 0.2–0.6 g of tissue/100-mm-diameter Petri dish or 75-cm^2 flask (see Note 4) in 10 mL of complete medium at 37°C in humidified atmosphere of 93% air, 7% CO_2.
6. Leave the cultures undisturbed for 7 days after which time replace the medium with an equal volume of fresh complete medium taking care not to dislodge the explants. Feed again at 14 days and twice weekly thereafter.

   With the exception of small numbers of isolated cells, which probably become detached from the bone surface during the dissection, the first evidence of cellular proliferation is observed on the surface of the explants, and this normally occurs within 5–7 days of plating. After 7–10 days, cells can be observed migrating from the explants (denoted by * in Fig.2) onto the surface of the culture dish (see Fig. 2). If care is taken not to dislodge the explants when feeding and they are left undisturbed between media changes, they rapidly become anchored to the substratum by the cellular outgrowths. Typical morphology of the cells is shown in Fig. 3, but cell shape varies between donors, from fibroblastic to cobblestone-like (see Note 8). Cultures generally attain confluence 4–6 weeks post plating, and typically achieve a saturation density of 29,000 ± 9,000 cells/cm^2 (mean ± SD, n = 11 donors).

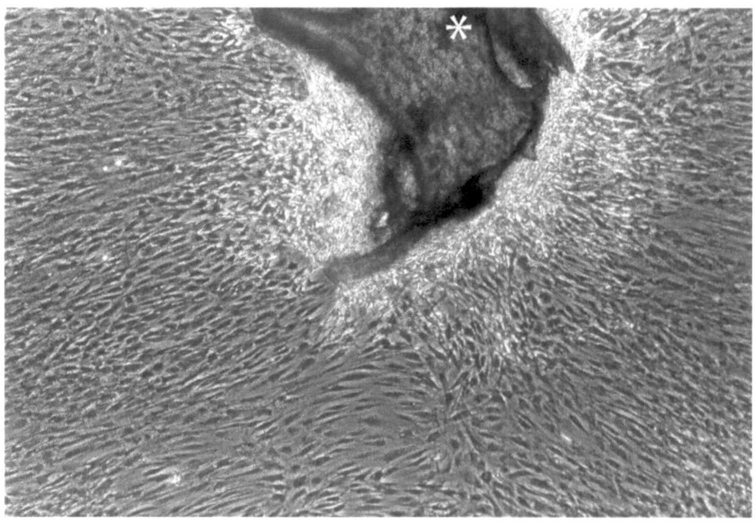

Fig. 2. Migration of HBDCs from an explanted cancellous bone chip.

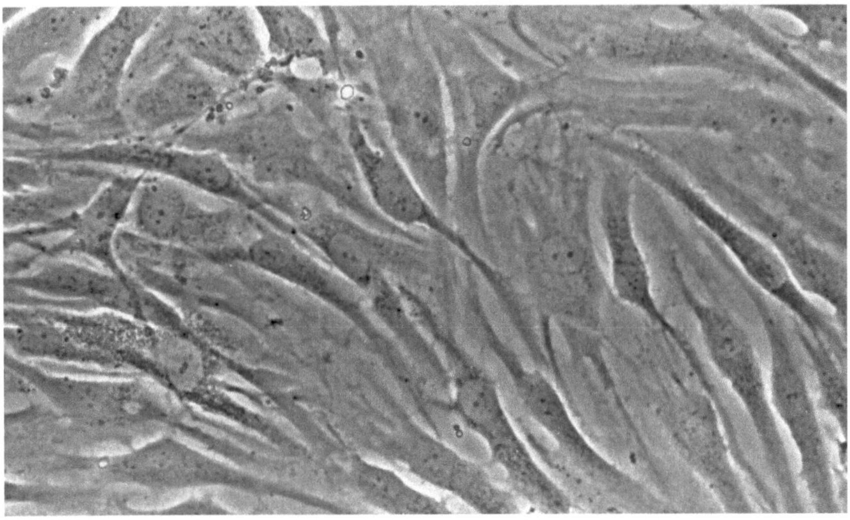

Fig. 3. Typical morphology of HBDCs from explanted cancellous bone.

3.1.2. Passaging Primary Cultures (E1–E1P1)

1. Remove the bone chips using a sterile needle or forceps to a new dish if re-plating is required (see Subheading 3.1.3) or discard. Remove and discard the spent medium.
2. Gently wash the cell layers three times with 10 mL of PBS without Ca^{2+} and Mg^{2+}.
3. To each flask, add 5 mL of freshly thawed trypsin–EDTA solution at room temperature (20°C) and incubate for 5 min at 37°C with gentle rocking of the flask every 30 s to ensure that the entire surface area of the flask and explants is exposed to the trypsin–EDTA solution.
4. Remove the flasks from the incubator and examine under the microscope. Look for the presence of rounded, highly refractile cell bodies floating in the trypsin–EDTA solution. If none, or only a few, are visible, tap the base of the flask sharply on the bench top in an effort to dislodge the cells. If this is without effect, incubate the cells for a further 5 min at 37°C.
5. When the bulk of the cells have become detached from the tissue culture plastic, transfer the cells to a "Universal container" with 10 mL of complete medium to inhibit tryptic activity. Wash the flask 2–3 times with 10 mL of SFM and pool the washings with the original cell isolate. To recover the cells, centrifuge at $250 \times g$ for 5 min at room temp.
6. Remove and discard the supernatant, invert the tube, and allow it to drain briefly. Re-suspend the cells in 2 mL of SFM. If the cells are clumping, see Note 9. If required, the cell suspension can be filtered through a 70-μm "Cell Strainer" to remove any bone spicules or remaining cell aggregates. For convenience and ease of handling, the filters have been designed to fit into

the neck of a 50-mL polypropylene tube. Wash the filter with 2–3 mL of SFM and add the filtrate to the cells.

7. Take 20 μL of the mixed cell suspension and dilute with 80 μL with SFM. Add 100 μL of trypan blue solution, mix, and leave for 1 min before counting viable (round and refractile) and nonviable (blue) cells in a Neubauer haemocytometer. Using this procedure, typically $1–1.5 \times 10^6$ cells are harvested per 75-cm^2 flask of which ≥75% are viable.

8. Plate the harvested cells at a cell density suitable for the intended analysis and culture until required at 37°C in humidified atmosphere of 93% air, 7% CO_2. We routinely subculture at $5 \times 10^3–10^4$ cells/cm^2 and achieve plating efficiencies measured after 24 h of ≥70% (see Note 10).

   If dishes have reached confluence but the cells are not required immediately, the cells can be stored by cryopreservation (see Subheading 3.3.1).

3.1.3. Re-Plating Explant Cultures (E1–E2)

As a matter of routine, we perform our studies on P1 cells. In practical terms, this presents real difficulties because it is often desirable to obtain large numbers of HBDCs from a single donor. Other investigators have studied the effects of repeated subculture on the phenotypic stability of HBDCs and found that they lose their osteoblast-like characteristics. We would encourage readers who choose to use cells at P2 cells (and any further passage) to verify that the cells do still maintain an osteoblast phenotype (see Subheading 3.4). As an alternative to repeated subculture, trabecular explants can be re-plated at the end of primary culture into a new flask (see Fig. 1). Using this technique, it is possible to obtain additional cell populations that continue to express osteoblast-like characteristics, including the ability to mineralise their extracellular matrix and maintain their cytokine expression profile (7). Presumably, these cultures are seeded by cells that are situated close to the bone surfaces, and that retain the capacity for extensive proliferation and differentiation. The continued survival of these cells may be related to the gradual release over time in culture of the cytokines and growth factors that are known to be present in the extracellular bone matrix, many of which are known to be produced by mature cells of the osteoblast lineage (9–12).

1. Remove the bone chips from the confluent, initial, explant culture using a sterile needle or forceps to a new dish containing 10 mL of complete medium.

2. Using a sterile disposable scalpel, carefully cut the bone chips into smaller pieces to re-stimulate proliferation from the bone surfaces.

3. Leave the cultures undisturbed at 37°C in humidified atmosphere of 93% air, 7% CO_2 for 7 days after which time replace the medium with an equal volume of fresh complete medium taking care not to dislodge the explants. Feed again at 14 days and twice weekly thereafter.

3.1.4. Modifications to Basic Culture System: Culture and Passage of HBDC in the Continuous Presence of Ascorbate

L-ascorbic acid (vitamin C) functions as a cofactor in the hydroxylation of lysine and proline residues in collagen and is essential for its normal synthesis and secretion. In addition, it increases pro-collagen mRNA gene transcription and mRNA stability (13). Addition of L-ascorbic acid (50 μg/mL to 25 μM) (see Note 3) to HBDCs in secondary culture (E1P1) increases proliferation, produces a sustained increase in the steady-state levels of α1 (I)-pro-collagen mRNA, and dramatically increases the secretion of type I collagen. There is also an increase in non-collagenous protein synthesis, most importantly bone sialoprotein and osteocalcin, and increased deposition of matrix. Because of their synthesis and secretion of an extensive collagen-rich extracellular matrix, HBDCs cultured in the continuous presence of ascorbate cannot be subcultured using trypsin–EDTA alone. They can, however, be subcultured if first treated with purified collagenase.

The basic procedure is as follows:

1. Add L-ascorbic acid (50 μg/mL to 25 μM) to HBDCs with every medium change.
2. When confluent/needed for experimentation, rinse the cell layers twice with SFM (10 mL/75-cm^2 flask).
3. Incubate the cells for 2 h at 37°C in 10 mL of SFM containing 25 U/mL collagenase. Gently agitate the flask for 10–15 s every 30 min.
4. Terminate the collagenase digestion by discarding the medium (check that there is no evidence of cell detachment at this stage). Gently rinse the cell layer twice with 10 mL of Ca^{2+}- and Mg^{2+}-free PBS.
5. Proceed from Subheading 3.1.2, step 3.

Typically, this procedure yields ~3.5–4 × 10^6 cells/75-cm^2 flask after 28 days in culture. Cell viability is generally ≥90%.

3.1.5. Modifications to Basic Culture Sys tem: In Vitro Mineralisation

The major function of the mature osteoblast is to form bone. Despite the overwhelming evidence that cultures of HBDC contain cells of the osteoblast lineage, initial attempts to demonstrate the presence of osteogenic (i.e. bone forming) cells proved unsuccessful. Subsequently, several authors reported that culture of HBDCs in the presence of ascorbate and millimolar concentrations of the organic phosphate ester β-glycerophosphate (β-GP) led to the formation of mineralised structures resembling the nodules that form in cultures of fetal or embryonic animal bone-derived cells (reviewed in ref. 14). These have been extensively characterised and shown by a variety of morphological, biochemical, and immunochemical criteria to resemble embryonic/woven bone formed in vivo. An alternative to the use of β-GP is to provide levels of inorganic phosphate sufficient for supporting the process of cell-mediated mineralisation in vitro, and the preferred method when studying HBDCs is supplementation of the culture medium with 5 mM inorganic phosphate.

The protocol for inducing matrix mineralisation in cultures of HBDCs is as follows.

1. Fragments of human trabecular bone are prepared as described in Subheading 3.1.1 and cultured in medium additionally supplemented with 100 μM L-ascorbic acid 2-phosphate (see Note 11) and either 200 nM hydrocortisone or 10 nM dexamethasone, which approximates to a physiological dose of glucocorticoid (see Note 12). For studies of in vitro mineralisation, it is preferable to obtain cancellous bone from sites containing haematopoietic marrow. In practice, this is usually from the upper femur or iliac crest.
2. When the cells have attained confluence and synthesised a dense extracellular matrix, typically after 28–35 days, subculture the cells using the sequential collagenase/trypsin–EDTA protocol and plate the cells at a density of 10^4 viable cells/cm^2. Change the medium twice weekly.
3. After 14 days, supplement the medium with 5 mM inorganic phosphate. This is achieved by adding 0.01% (v/v) of a 500 mM phosphate solution, pH 7.4.
4. After 48–72 h, the cell layers are washed 2–3 times with 10 mL of PBS prior to fixation with 100% ethanol at 4°C.

3.2. Human Clonal Osteoblast Cell Lines

Osteosarcomas are malignant tumours of bone derived from cells of the osteoblast lineage. An osteosarcoma is a relatively uncommon tumour, although it is the most frequently occurring primary malignant tumour of bone, predominantly occurring in children and adolescents. It is also reported to occasionally occur in elderly patients with long-standing Paget's disease of bone. Histologically, the tumour is heterogeneous, the cells being associated with immature bone matrix or osteoid. The tumour cells are generally poorly differentiated and pleomorphic with high mitotic activity. The tumours are highly vascularised and the cells have a tendency to metastasise. As osteosarcoma cells express osteoblastic genes, synthesise bone matrix proteins, and, in many cases, respond to calcium-regulating hormones, including parathyroid hormone, they are extremely useful as an osteoblast cell model. Their widespread use has meant that researchers do not always take into account that some of the characteristics of these cells are related to their oncogenic potential rather than osteoblastic characteristics.

Nevertheless, many significant advances in our knowledge of bone biology have been made via experiments with these cells. The most commonly used human clonal osteoblast cell lines are SaOS-2, MG63, Te85 (also referred to as HOS), and U2-OS (HTB96). These cell lines were all derived from osteosarcomas; however, they differ in their responsiveness to certain hormones, i.e. oestrogen and progesterone. This is most likely because of their stage of differentiation along the osteoblastic phenotype. Therefore, the choice of which cell line to use is solely dependent on the hypothesis being tested and preference of the investigator. An advantage of transformed human osteoblast cell lines is that they are easily manipulated and, as such, can be utilised to make stable reporter cell lines, which are powerful research tools (15). Alternatives to the "classic" osteosarcoma-derived cell lines are the cell line hFOB 1.19 (16) and, more recently, the osteoprecursor cell line OCP1 (17), both of which were derived from human fetal bone tissue (see Note 13).

3.2.1. Routine Culture of the Human Clonal Osteoblast Cell Line SaOS-2

1. Resurrect a vial of frozen SaOS-2 cells according to Subheading 3.3.2.
2. Culture cells in complete medium at 37°C in humidified atmosphere of 95% air, 5% CO_2.
3. When cells reach approximately 70–80% confluence, passage as per Subheading 3.1.2, **steps 2–8**. We routinely seed SaOS-2 cells at between 1 and 2.5×10^6 cells/T75 flask depending upon how frequently they are needed for use.

   It is inadvisable to let the stock SaOS-2 cell cultures grow beyond 100% confluence (see Fig. 4). If this happens, it is advisable to remove a fresh vial from liquid nitrogen stock.

3.2.2. Culture of SaOS-2 Cells for In Vitro Mineralisation

1. Culture SaOS-2 cells in osteogenic medium until confluent.
2. Once cells have become confluent, culture for a further 7–10 days (it is advisable to reduce the FBS level in the osteogenic medium to prevent the cell layers from lifting off the tissue culture plastic – between 0.5 and 2% FBS are suitable levels), with 5 mM inorganic phosphate (or 2 mM β-GP) being added to the culture medium for the final 3 days of culture.
3. Wash twice with 10 mL of PBS prior to fixation with 100% ethanol at 4°C.

3.3. Long-Term Storage of Cells

Clonal osteoblasts and, if required, HBDCs can be stored frozen at −80°C for short periods (2–3 months maximum) or for extended periods in liquid nitrogen or in ultralow-temperature (−135°C) cell-freezer banks. For best results, the cells should be re-suspended in 90% serum with 10% cryopreservant (DMSO) and then frozen gradually at a rate of −1°C/min in a −70°C freezer using one of the many devices available that allow the rate of cooling to be controlled precisely (e.g. Mr. Frosty, Nalgene).

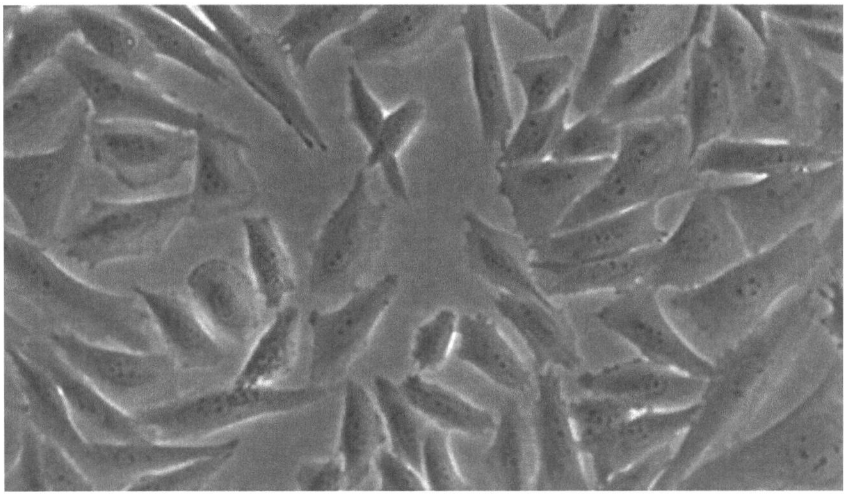

Fig. 4. Typical morphology of SaOS-2 cells.

3.3.1. Cryopreservation

1. When cells are in the log phase of growth/about 80–90% confluent, harvest the cells in the same manner used for routine subculture (Subheading 3.1.2, steps 1–7). After obtaining a cell count, re-pellet the cells by centrifuging at 250 × *g* for 5 min.
2. Re-suspend the cell pellet in an appropriate amount of freezing medium to achieve $2–4 \times 10^6$ cells/mL.
3. Close cryovials (Nalgene) tightly and freeze using a Mr. Frosty or the following protocol: 5°C/min to 4°C, 1°C/min to −80°C (cells can be "held" at −80°C for several months without loss of viability on recovery).
4. Transfer cells directly to liquid nitrogen or ultralow-temperature (−135°C) cell-freezer bank.

3.3.2. Resurrection of Cells

1. Remove a frozen vial of cells from liquid nitrogen storage and thaw quickly in a water bath at 37°C.
2. Add the thawed cell suspension dropwise to a T75 flask containing 15 mL complete medium.
3. Incubate at 37°C in humidified atmosphere of 95% air, 5% CO_2 (93% air, 7% CO_2 for HBDCs).
4. After 24 h, replace the medium with the normal volume of fresh complete medium and continue to culture cells at 37°C in humidified atmosphere of 95% air, 5% CO_2 (93% air, 7% CO_2 for HBDCs). The efficiency of plating obtained after 24 h using this method is typically ≥70%.

3.4. Phenotypic Characteristics of Osteoblasts

The phenotypic characterisation of HBDCs is described in detail in ref. 6. The most specific phenotypic marker is osteocalcin. This is a protein of M_r 5800 containing residues of the vitamin

K-dependent amino acid γ-carboxyglutamic acid. In humans, synthesis is restricted to mature cells of the osteoblast lineage. It is an excellent late-stage marker for cells of this series despite the fact that its precise function in bone has still to be established. $1,25(OH)_2D_3$ increases the production of osteocalcin in cultures of HBDCs, but not fibroblasts obtained from the same donors. There are many commercially available kits for osteocalcin and the reader should refer to the specific manufacturer for detailed protocols. Another method of characterising osteoblasts is to perform reverse transcription–polymerase chain reaction (RT-PCR) to examine the expression of osteoblastic-specific genes using either SYBR-Green (18) or TaqMan® technologies.

The simplest and most inexpensive phenotypic marker to investigate is the glyco-protein enzyme alkaline phosphatase (ALP), a widely accepted marker of early osteogenic differentiation. ALP activity can be measured both on whole cell lysates as it is primarily bound to the cell surface and in the culture medium as the phosphoinositol linkage which binds the enzyme to the cell can be cleaved. Surface ALP expression can also be measured on intact cells by histochemical staining. Basal activity is initially low, but increases with increasing cell density. Treatment of HBDC with $1,25(OH)_2D_3$ increases ALP activity. SaOS-2 cells express higher levels of ALP than Te85 cells (19); however, it should be noted that the ALP activity of SaOS-2 cells is not classically affected by $1,25(OH)_2D_3$ (20). The basic principle of the ALP enzyme activity assay is that the colourless substrate *p*-NPP when dephosphorylated by ALP forms the yellow-coloured product *p*-nitrophenol (*p*-NP) in alkaline solution, the absorbance of which can be measured at 405 nm.

3.4.1. Alkaline Phosphatase Enzymatic Activity Assay

1. At experimental end point, remove the supernatant from the cells – retain and transfer 100 μL to a 96-well plate if required for assay (store at −80°C if not assayed immediately).
2. Wash the cells twice in ice-cold PBS and lyse in 0.1% Triton-X (see Note 14) (store at −80°C if not assayed immediately).
3. If not already cultured in a 96-well plate, transfer 100 μL of the lysate to a 96-well plate.
4. Add 100 μL of *p*-NPP solution to the sample (lysate or supernatant), immediately read the absorbance at 405 nM, and then take readings every 5 min for 30 min (to make sure that the reaction is in the linear range).
5. The ALP activity in the sample is calculated from Beer–Lambert's law as follows:

$$U = \frac{(OD_{t1} - OD_{t0}) \times V}{t \times \varepsilon \times l},$$

where:

U = ALP activity (nmol Pi per min).

ODt_1 = absorbance at 405 nM of sample at the end time point.

ODt_0 = absorbance at 405 nM of sample at start (background).

V = volume in microlitre of sample and reagent in well that is measured.

t = reaction time in min.

$\varepsilon = 17.8$ μL/nM/cm for *p*NPP.

l = path length of light in cm (specific to the plate reader).

6. Alternatively, allow the reaction to take place at 37°C for 40 min, add 50 μL of 3 M NaOH to terminate the reaction, and read the absorbance at 405 nM within 15 min of terminating the reaction. All the values should then be normalised to the control measurements and expressed as percentage of change from control.

   It is important when performing these assays to correct the ALP activity values to the cell density of each individual sample. This is because different cell cultures or treatments may result in changes in cell density which would affect ALP activity. This can be done by a variety of methods, including setting parallel plates up to perform cell counts on or basic protein (Bradfords/BCA) or dsDNA (PicoGreen) assays on the same cell lysates. The ALP activity can then be expressed as U/μg protein or U/ng DNA, etc.

3.4.2. Alkaline Phosphatase Histochemical Staining

Cell surface ALP can be quantified by staining with the substrates Nitro blue tetrazolium chloride (NBT) and 5-Bromo-4-chloro-3-indolyl phosphate, toluidine salt (BCIP). In this assay, BCIP is dephosphorylated by ALP to give a dark-blue indigo dye as an oxidation product and then NBT serves as the oxidant and gives a dark-blue dye.

1. At experimental end point, remove the cell culture medium from cells, and wash cell layers carefully in ice-cold PBS.
2. Add an appropriate amount of NBT/BCIP solution to cover the cells and incubate at room temperature in the dark for 5 minutes or until a blue product is observed (HBDC may take longer, SaOS-2 cells react very quickly).
3. Decant the solution, wash briefly with PBS, and then air dry. If assays performed in well plates scan on a high resolution scanner for quantification using an image analysis package, such as ImageJ (NIH), if cells cultured onto glass slides, then mount in a non-xylene-based mounting media and record images with light microscope.

   As with the ALP activity assay, it is important to correct for any variation in cell number – this can be done again by either setting up parallel cultures to perform cell counts on or by counterstaining either with a nuclear-specific stain (i.e. haematoxylin or DAPI) or a general total cell stain (i.e. 0.1% toluidine blue).

3.4.3. Alizarin Red-S Staining of Mineralisation

The effect of cytokines and/or novel drugs on the extent of mineralisation observed in bone cultures is an excellent indication of their pro- or anti-osteogenic nature (21, 22). A relatively simple and fast method for quantifying mineralisation in both HBDC and SaOS-2 cells once the cells have reached the end of the culture is to stain with ARS, which stains calcium deposits on the surface of the cell monolayer.

1. At the end of the experiment, fix cells overnight at 4°C in 100% ethanol.
2. Rinse in PBS, and incubate in 2% alizarin red (pH 4.2) for 1 h at room temperature with gentle agitation.
3. Wash extensively in 50% ethanol (until no more unbound dye is released) and air dry.
4. Quantification of the area of ARS staining per well (see Note 15) can be achieved by scanning the plates with a high-resolution flat bed scanner equipped with a transparency adaptor. From the whole plate image, the percentage of ARS staining for each well can be determined using an image analysis package, such as Image J (NIH). Plates can then be counterstained with 0.1% toluidine blue, re-scanned, and the total area of the well occupied by cells determined.

4. Notes

1. The authors routinely use D-MEM with Glutamax™ (Invitrogen™) as this avoids the need to add fresh L-glutamine to D-MEM (L-glutamine is an essential nutrient in cell cultures for energy production as well as protein and nucleic acid synthesis). However, L-glutamine spontaneously degrades in cell culture media generating ammonia as a toxic by-product. Ammonia can affect cell viability, lower protein production, and change glycosylation patterns.
2. Batches of serum vary in their ability to support the growth of HBDCs. It is advisable to screen batches and reserve a large quantity of serum once a suitable batch has been identified. HBDCs grow in autologous and heterologous human serum, but as yet no comprehensive studies have been performed to identify the effects on growth and differentiation.

3. Beresford et al. introduced the more stable analog L-ascorbic acid 2-phosphate (Wako Pure Chemical Industries, Ltd.) which does not have to be added daily (see ref. 23 for details).
4. The authors have obtained consistent results with plasticware from Sarstedt and Becton Dickinson. Smaller flasks or dishes can be used if the amount of bone available is <0.2 g.
5. Robey and Termine used prior digestion of minced bone with *Clostridial* collagenase and subsequent culture of explants in medium with reduced calcium concentrations. In contrast, Wergedal and Baylink have used collagenase digestion to directly liberate cells. Marie et al. have used a method in which explants are first cultured on a nylon mesh. These alternative methods are described in greater detail in ref. 6.
6. Bone can be stored for periods of at least 24 h, possibly longer, at 4°C in PBS or SFM prior to culture without any deleterious effect on the ability of the tissue to give rise to populations of osteoblastic cells.
7. Bone cells have been successfully cultured from many anatomical sites, including tibia, femur, rib, vertebra, patella, and digits.
8. The available evidence indicates that cultures of HBDCs contain cells of the osteogenic lineage at all stages of differentiation and maturation. This conclusion is consistent with the expression of both early (ALP)- and late (osteocalcin, bone sialoprotein)-stage markers of osteoblast differentiation. In addition, in ascorbate-treated cultures, there is a small subpopulation (≤5%) of cells that express the epitope recognised by the monoclonal antibody (MAb) STRO-1, which is a cell surface marker for clonogenic, multipotential marrow stromal precursors capable of giving rise to cells of the osteogenic lineage in vitro. The presence of other cell types, including endothelial cells and those derived from the haematopoietic stem cell, has been investigated using a large panel of MAbs and flow cytometry and/or immunocytochemistry. The results of these studies reveal that at first passage there are no detectable endothelial, lymphoid, or erythroid cells present. A consistent finding, however, is the presence of small numbers of cells (≤5%) expressing antigens present on cells of the monocyte/macrophage series.
9. If the cells are clumping, re-suspend in 2 mL of SFM containing 1 μg/mL DNase I for each dish or flask treated with trypsin–EDTA, and using a narrow bore 2-mL pipette, repeatedly aspirate and expel the medium to generate a cell suspension.
10. In our experience, the minimum plating density for successful subculture is 3,500 cells/cm^2. Below this, the cells exhibit extended doubling times and often fail to grow to confluence.

11. The ability of the cells to mineralise their extracellular matrix is dependent on ascorbate being present continuously in primary culture. The addition of ascorbate in secondary culture, even for extended periods, cannot compensate for its omission in primary culture and cells cultured without ascorbate, irrespective of the presence or absence of glucocorticoid, secrete little extracellular matrix and do not mineralise. This finding provides further evidence to support the hypothesis that maintenance of adequate levels of ascorbate during the early stages of explant culture is of critical importance for the survival of cells that retain the ability to proliferate extensively and give rise to precursors capable of undergoing osteogenic differentiation. The presence of mineralised matrix is indicative of the true osteogenic nature of the cultured cell population. This is supported by the fact that HBDCs cultured continuously in the presence of ascorbate and glucocorticoid retain the ability to form bone when implanted in vivo within diffusion chambers in athymic mice (24).
12. HBDCs cultured in the continuous presence of the long-acting ascorbate analogue and glucocorticoid produce a dense extracellular matrix that mineralises extensively following the addition of inorganic phosphate. This is the case for the original cell population (E1P1) and that obtained following re-plating of the trabecular explants (E2P1), which further attests to the phenotypic stability of the cultured cells. When cells are cultured in the continuous presence of ascorbate and then treated with glucocorticoid only after first passage although they possess similar amounts of extracellular matrix and ALP activity, they show a more localised and patchy pattern of mineralisation in comparison to cells cultured in the continued presence of both ascorbate and glucocorticoid.
13. Clonal human osteoblasts can be obtained from either the American Type Culture Collection (ATCC) or the European Collection of Cell Culture (ECCAC) (http://www.lgcstandards-atcc.org/ or http://www.hpacultures.org.uk).
14. For accurate quantification of ALP activity, it is essential to achieve full lysis of the cell layer. Other lysis buffers or methods can be used – such as simple lysis in water with repeated freeze–thaw cycles. Ultimately, the lysis solution needs to be compatible with the subsequent protein or DNA quantification method employed.
15. Other researchers use ARS extraction as an alternative method of quantification. In our experience, the amount of extracted dye that is recovered from the wells is inconsistent and extremely variable even after extended extraction periods and with various extraction reagents. For those readers who wish to attempt this protocol, see ref. 25.

Acknowledgements

This work was supported in part by Arthritis Research UK (17538) and the European Commission under the 7th Framework Programme (proposal #202231) performed as a collaborative project among the members of the ATPBone Consortium (Copenhagen University, University College London, University of Maastricht, University of Ferrara, University of Liverpool, University of Sheffield, and Université Libre de Bruxelles), and is a sub-study under the main study "Fighting osteoporosis by blocking nucleotides: purinergic signalling in bone formation and homeostasis".

References

1. 1. Gallagher, J. A., Beresford, J. N., Caswell, A., and Russell, R. G. G. (1987) Subcellular investigations of skeletal tissue., in *Subcellular Pathology of Systemic Disease* (Peters, T. J., Ed.), pp 377–397, Chapman and Hall, London.
2. Bard, D. R., Dickens, M. J., Smith, A. U., and Zarek, J. M. (1972) Isolation of living cells from mature mammalian bone, *Nature 236*, 314–315.
3. Mills, B. G., Singer, F. R., Weiner, L. P., and Holst, P. A. (1979) Long-term culture of cells from bone affected by Paget's disease, *Calcif Tissue Int 29*, 79–87.
4. Beresford, J. N., Gallagher, J. A., Poser, J. W., and Russell, R. G. (1984) Production of osteocalcin by human bone cells in vitro. Effects of 1,25(OH)2D3, 24,25(OH)2D3, parathyroid hormone, and glucocorticoids, *Metab Bone Dis Relat Res 5*, 229–234.
5. Gallagher, J. A., Beresford, J. N., McGuire, M. K., Ebsworth, N. M., Meats, J. E., Gowen, M., Elford, P., Wright, D., Poser, J., Coulton, L. A., and et al. (1984) Effects of glucocorticoids and anabolic steroids on cells derived from human skeletal and articular tissues in vitro, *Adv Exp Med Biol 171*, 279–291.
6. Gallagher, J. A., Gundle, R., and Beresford, J. N. (1996) Isolation and Culture of Bone-Forming Cells (Osteoblasts) from Human Bone, Methods Mol Med. 2:233–262.
7. Gundle, R., Stewart, K., Screen, J., and Beresford, J. N.(1998) Isolation and culture of human bone derived cells, in *Marrow Stromal Cell Culture* (Beresfod, J. N., and Owen, M., Eds.), Cambridge University Press, Cambridge.
8. Gallagher, J. A. (2003) Human Osteoblast Culture, pp 3–18.
9. Birch, M. A., Ginty, A. F., Walsh, C. A., Fraser, W. D., Gallagher, J. A., and Bilbe, G. (1993) PCR detection of cytokines in normal human and pagetic osteoblast-like cells, *J Bone Miner Res 8*, 1155–1162.
10. Canalis, E., Pash, J., and Varghese, S. (1993) Skeletal growth factors, *Crit Rev Eukaryot Gene Expr 3*, 155–166.
11. Chaudhary, L. R., Spelsberg, T. C., and Riggs, B. L. (1992) Production of various cytokines by normal human osteoblast-like cells in response to interleukin-1 beta and tumor necrosis factor-alpha: lack of regulation by 17 beta-estradiol, *Endocrinology 130*, 2528–2534.
12. Mohan, S., and Baylink, D. J. (1991) Bone growth factors, *Clin Orthop Relat Res*, 263: 30–48.
13. Prockop, D. J., and Kivirikko, K. I. (1984) Heritable diseases of collagen, *N Engl J Med 311*, 376–386.
14. Beresford, J. N., Graves, S. E., and Smoothy, C. A. (1993) Formation of mineralized nodules by bone derived cells in vitro: a model of bone formation? *Am J Med Genet 45*, 163–178.
15. Bowler, W. B., Dixon, C. J., Halleux, C., Maier, R., Bilbe, G., Fraser, W. D., Gallagher, J. A., and Hipskind, R. A. (1999) Signaling in human osteoblasts by extracellular nucleotides. Their weak induction of the c-fos proto-oncogene via Ca2+ mobilization is strongly potentiated by a parathyroid hormone/cAMP-dependent protein kinase pathway independently of mitogen-activated protein kinase, *J Biol Chem 274*, 14315–14324.
16. Harris, S. A., Tau, K. R., Enger, R. J., Toft, D. O., Riggs, B. L., and Spelsberg, T. C. (1995) Estrogen response in the hFOB 1.19 human fetal osteoblastic cell line stably transfected with the human estrogen receptor gene, *J Cell Biochem 59*, 193–201.

17. Winn, S. R., Randolph, G., Uludag, H., Wong, S. C., Hair, G. A., and Hollinger, J. O. (1999) Establishing an immortalized human osteoprecursor cell line: OPC1, *J Bone Miner Res 14*, 1721–1733.
18. Gartland, A., Buckley, K. A., Dillon, J. P., Curran, J. M., Hunt, J. A., and Gallagher, J. A. (2005) Isolation and culture of human osteoblasts, *Methods Mol Med 107*, 29–54.
19. Farley, J. R., Kyeyune-Nyombi, E., Tarbaux, N. M., Hall, S. L., and Strong, D. D. (1989) Alkaline phosphatase activity from human osteosarcoma cell line SaOS-2: an isoenzyme standard for quantifying skeletal alkaline phosphatase activity in serum, *Clin Chem 35*, 223–229.
20. Murray, E., Provvedini, D., Curran, D., Catherwood, B., Sussman, H., and Manolagas, S. (1987) Characterization of a human osteoblastic osteosarcoma cell line (SAOS-2) with high bone alkaline phosphatase activity, *J Bone Miner Res 2*, 231–238.
21. Jones, S. J., Gray, C., Boyde, A., and Burnstock, G. (1997) Purinergic transmitters inhibit bone formation by cultured osteoblasts, *Bone 21*, 393–399.
22. Rao, L. G., Liu, L. J., Murray, T. M., McDermott, E., and Zhang, X. (2003) Estrogen added intermittently, but not continuously, stimulates differentiation and bone formation in SaOS-2 cells, *Biol Pharm Bull 26*, 936–945.
23. Gundle, R., Stewart, K., Screen, J., and Beresford, J. N.(1998) Isolation and culture of human bone derived cells, in *Marrow Stromal Cell Culture* (Beresfod, J. N., and Owen, M., Eds.), Cambridge University Press, Cambridge.
24. Gundle, R., Joyner, C. J., and Triffitt, J. T. (1995) Human bone tissue formation in diffusion chamber culture in vivo by bone-derived cells and marrow stromal fibroblastic cells, *Bone 16*, 597–601.
25. Gregory, C. A., Gunn, W. G., Peister, A., and Prockop, D. J. (2004) An Alizarin red-based assay of mineralization by adherent cells in culture: comparison with cetylpyridinium chloride extraction, *Anal Biochem 329*, 77–84.

Chapter 23

Human Osteoclast Culture and Phenotypic Characterization

Ankita Agrawal, James A. Gallagher, and Alison Gartland

Abstract

Bone remodelling occurs throughout life via the coupled actions of bone resorption and bone formation. When the balance of bone resorption and formation becomes unequal, bone diseases, such as osteoporosis occur, while the absence of *functional* osteoclasts leads to diseases such as osteopetrosis and pycnodysostosis. In order to develop effective treatments for bone disease the normal regulatory systems involved in bone resorption need to be fully elucidated. The only cell in the body capable of resorbing bone is the osteoclast – a highly specialized cell of haematopoietic origin. Until relatively recently, the ability to study the formation and function of human osteoclasts in vitro has been limited.

In this chapter, we provide an up-to-date detailed guide to isolating and culturing primary osteoclasts from human peripheral blood. In addition, we detail the methodology used to characterize osteoclasts and how to quantify resorption in vitro. In combination these methods provide a powerful tool in the study of human osteoclasts and the development of new novel treatments for bone disease.

Key words: Osteoclast, Resorption, TRAP, RANKL, Multinucleated, Monocyte

1. Introduction

Research involving human osteoclasts has historically been difficult to undertake due to the lack of supply and the fact that osteoclasts are terminally differentiated cells that cannot simply divide and be maintained in culture; therefore, any osteoclast supply must constantly be replenished. Osteoclasts are large, multinucleated cells formed by the fusion of osteoclast precursors found in the monocyte fraction of blood. The osteoclast is thought to be the only cell capable of excavating authentic resorption lacunae in calcified substrates in vivo and in vitro.

Until relatively recently, the study of human osteoclasts in vitro relied on giant cell tumours removed at surgery becoming available and the subsequent isolation of osteoclasts from these tissues (1).

Ragai R. Mitry and Robin D. Hughes (eds.), *Human Cell Culture Protocols*, Methods in Molecular Biology, vol. 806,
DOI 10.1007/978-1-61779-367-7_23, © Springer Science+Business Media, LLC 2012

The discovery of receptor activator for nuclear factor kappaB (NF-κB) ligand (RANKL) (2) provided significant advances in understanding the events leading to osteoclast formation as well as the development of techniques to routinely produce functional human osteoclasts (3). These techniques are based on the finding that co-activation of receptors for RANKL and macrophage colony-stimulating factor (M-CSF) expressed by osteoclast precursors results in the fusion of these cells to form multinucleated osteoclasts.

Osteoblasts secrete M-CSF and following stimulation by factors, such as 1,25-dihydroxy vitaminD_3 and dexamethasone, express elevated levels of RANKL; therefore, co-cultures of these cells and human peripheral blood mononuclear cells (PBMC), which contain osteoclast precursors, can be used to generate human osteoclasts. This co-culture method has generally been superseded due to the commercial availability of a recombinant RANKL protein which negates the need for osteoblasts to be cultured with PBMC to generate osteoclasts (4–8). These techniques for human osteoclast generation allow the study of both osteoclast formation and resorption. Whereas these procedures are highly valuable in that they provide a constant supply of human osteoclasts, their one disadvantage is that they are time consuming as osteoclasts only form after ~7–21 days in culture. This chapter describes the generation of human osteoclasts from isolated PBMC in both co-cultures and cultures supplemented with recombinant RANKL. Osteoclast characterization and the subsequent assay of osteoclast activity in vitro are also provided.

2. Materials

2.1. Blood Isolation and Separation

1. Lithium-Heparin Vacutainer® tubes (BD, Oxford, UK) or some other heparinized, sterile container.
2. Histopaque®-1077 for blood separation (Sigma–Aldrich, Poole, UK).
3. Human CD14 Microbeads and MACS separator (*only required for optional* $CD14^+$ *enrichment*) (Miltenyi Biotech, Germany).
4. Ethylenediaminetetraacetic acid (EDTA) and Bovine serum albumin (BSA) (*only required for optional* $CD14^+$ *enrichment*) (Invitrogen, Paisley, UK).

2.2. Tissue Culture

1. Phosphate-buffered saline (PBS) without calcium and magnesium, pH 7.4 (Invitrogen, Paisley, UK).
2. α-Modification of minimum essential medium (α-MEM) with Glutamax™ (Invitrogen) (see Note 1).
3. Penicillin–Streptomycin (Invitrogen).
4. Fetal bovine serum (FBS) (see Note 2) (Invitrogen).

5. Human recombinant M-CSF (Cambridge, MA, USA).
6. Human recombinant RANKL (Insight Biotechnology, Wembley, UK).
7. 1,25-dihydroxy vitaminD_3 (Sigma–Aldrich).
8. Dexamethasone (Sigma-Aldrich).
9. 6-mm coverslips (see Note 3, Richardson's of Leicester, Leicester, UK).
10. Dentine disks (http://www.dentinedisks.com, UK).
11. 96-Well tissue culture plates (see Note 3).
12. UMR-106 cells (for co-culture).

2.3. Fixation

Choice of fixative is dependent upon which procedure you subsequently intend to perform on the cells/dentine. Traditionally, a cacodylate–glutaraldehyde solution was used for fixation of osteoclasts on dentine as they would routinely be examined by scanning electron microscopy; should you require this, please refer to Walsh et al., (1). We now routinely fix dentine discs in 10% buffered formalin.

1. Sodium dihydrogen phosphate dehydrate (VWR).
2. Sdi-Sodium hydrogen phosphate (VWR).
3. Formaldehyde 39% w/v (VWR).

Using analytical-grade reagents, mix to dissolve 8 g sodium dihydrogen phosphate dehydrate and 13 g di-sodium hydrogen phosphate in 200 mL formaldehyde and 200 mL warm distilled water (this helps to dissolve the phosphate buffers). Once dissolved, make up to 2 L with water. The solutions do not need to be sterile, but should be prepared in clean containers. Wear gloves and mask and prepare in a containment hood. The solution can be stored at room temperature for at least 6 months (after which time, the solution starts to become acidic – if this is the case, discard and make a fresh batch).

2.4. Staining

2.4.1. Toluidine Blue Solution

Use analytical-grade reagents and distilled water for the solutions. The solutions do not need to be sterile, but should be prepared in clean containers and can be stored at room temperature for several years.

1. di-sodium tetraborate (Sigma-Aldrich).
2. Toluidine blue (Sigma-Aldrich).

Dissolve 19.07 g of disodium tetraborate in 100 mL of water to make 0.5 M sodium tetraborate solution. Then, dissolve 1 g of toluidine blue in 100 mL of 0.5 M sodium tetraborate solution to give a 1% toluidine blue solution.

2.4.2. Tartrate-Resistant Acid Phosphatase Histochemistry

Use analytical-grade reagents and distilled water for the solutions.

1. Acetic acid (VWR).
2. Dimethylformamide (Fisher Scientific, Loughborough, UK).
3. Hydrochloric acid (VWR).
4. Naphthol AS-BI phosphate (sodium salt) (Sigma-Aldrich) (store at -20°C).
5. Pararosaniline (Sigma-Aldrich).
6. Sodium acetate trihydrate (Sigma-Aldrich).
7. Sodium nitrite (Sigma-Aldrich).
8. Sodium tartrate (dihydrate) (Sigma-Aldrich).
9. Gill's haematoxylin (Merck, Germany).
10. DPX (VWR).

Stock pararosaniline: Wearing a mask, weigh out 1 g pararosaniline and add to 20 mL distilled water in a glass conical flask. Carefully add 5 mL concentrated HCl in a fume hood (the solution turns brown), mix gently and cover lightly with foil/lid. Heat carefully while stirring on hotplate at a low setting for 15 min, then increase temperature and continue heating until solution is simmering (up to 45 min). Allow to simmer for 5–10 min, cool to room temperature, filter into bottle and protect from light with foil wrapping. The solution is stable for a few months at room temperature.

The following steps are to prepare enough staining solutions to theoretically stain ~500 wells of a 96-well plate (or enough slides that fit in a coplin jar that takes 50 mL; it can be scaled up by adjusting all ratios accordingly). Most solutions need to be prepared fresh on the day of staining – and in some instances only min prior to use; therefore, it is advisable to read the complete staining protocol prior to making up the solutions.

Acetate–tartrate buffer: Dissolve 5.44 g sodium acetate trihydrate in 200 mL distilled water in a large beaker. Adjust pH to 5.2 with 50 mL 1.2% acetic acid (this acetate buffer can be stored at 4°C if not used immediately). Add 4.6 g sodium tartrate to the acetate buffer and warm to 37°C for 1–2 h.

Solution A: To prepare the naphthol/dimethylformamide solution, weigh 20 mg naphthol AS-BI phosphate into a 50 mL glass, screw-cap bottle. Add 1 mL dimethylformamide (N.B. this dissolves certain plastics). The solution turns a straw yellow colour. Add 50 mL acetate–tartrate buffer to the naphthol/dimethylformamide solution in a glass, screw-cap bottle. This is done in a fume hood.

Solution B: To prepare the sodium nitrite solution, dissolve 80 mg sodium nitrite in 2 mL distilled water. Hexazotise the pararosaniline by mixing 2 mL pararosaniline stock with 2 mL of 4% sodium nitrite solution. Allow to stand for 5–15 min (as this solution is effervescent, especially in large quantities). Immediately prior to use, add 2.5 mL of this hexazotised solution to 50 mL acetate–tartrate buffer in a pre-warmed bottle and mix well.

3. Methods

3.1. Preparation of Dentine Discs and Coverslips

Sterile, pre-cut, 5-mm Ø dentine discs can be obtained from http://www.dentinedisks.com or alternatively prepared in the following way.

1. Cut thin wafers of dentine (180–200 μm) using a water-cooled, low-speed diamond saw (Labcut 1010).
2. Soak the wafers for 1 h in distilled water and then punch discs from these wafers using a circular punch (5-mm diameter) (see Fig. 1). Dentine discs may be marked with a pencil on one side for later identification of the surface not containing cells.

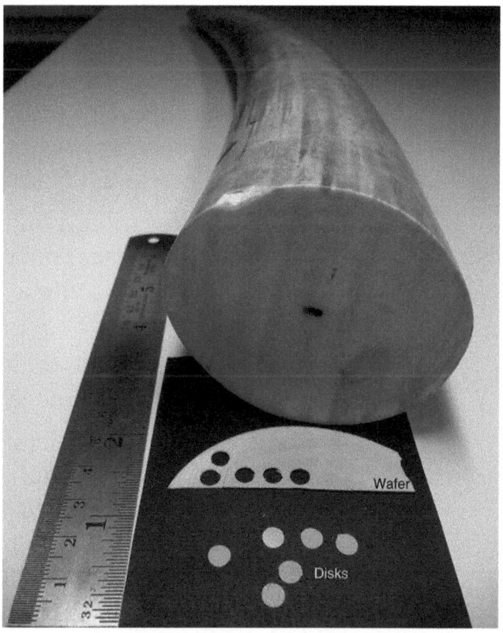

Fig. 1. Dentine wafers are cut from blocks of dentine using a low-speed diamond saw. Uniform 5-mm Ø discs are then punched from the wafer.

3. Sonicate dentine discs in double-distilled water for two 15-min washes. Wash the discs in 70% ethanol, allow to air dry and then arrange the discs in a Petri dish, pencil-marked-side face down. Place the Petri dish over a mirror and sterilize under ultraviolet (UV) light in a flow cabinet for 2 h. If a UV light is not available, then keeping the dentine in 70% ethanol overnight should be sufficient sterilization.
4. Coverslips can be similarly sterilized by UV irradiation for 2 h or by baking in an oven at 180°C for at least 2 h.

3.2. Isolation of Human Peripheral Blood Mononuclear Cells

Human osteoclasts can be generated from their precursors found in the monocyte fraction of blood or bone marrow. Obtaining a regular source of human blood for osteoclast cultures may be problematic and all studies certainly require local Ethical Committee approval. One alternative is to apply to the local blood transfusion service to use the buffy coat that is isolated from donated blood – but again, this usually requires ethical approval and the cells are normally a few days old and may not arrive in the laboratory at a convenient time. Commercial sources of isolated PBMC are available (i.e. from Lonza), but this is an expensive option as they require the purchase of their own brand culture reagents for successful osteoclast generation.

3.2.1. Isolation of PBMC by Density Gradient

1. Collect 10–100 mL human peripheral blood into Lithium-Heparin Vacutainers® or some other heparinized, sterile container (see Note 4). Mix the blood with an equal volume of α-MEM and then layer 9 mL of this mixture over 5 mL of Histopaque®-1077 in 15-mL Falcon tubes and centrifuge at 400×*g* for 30 min at 20°C. It is important that the brakes of the centrifuge are set to the off position to prevent the layers from collapsing under harsh braking. Therefore, this step takes ~45 min.
2. After centrifugation, remove the clear upper layer and collect the opaque interface containing the monocyte and lymphocyte fraction of blood using a fine pipette and dilute into 20 mL α-MEM.
3. Centrifuge these cells at 300×*g* for 20 min and then wash the cell pellet a second time in α-MEM.
4. Resuspend the cell pellet in an appropriate volume of α-MEM containing 10% FBS and 100 U/mL penicillin, 100 μg/mL streptomycin (referred to as complete medium) (see Note 5), take an aliquot of cell suspension, mix 1:1 with 5% acetic acid to lyse contaminating red cells and determine the cell number by performing a cell count.
5. Proceed to culture as described in Subheading 3.3.1 or 3.3.2 as required.

3.2.2. Isolation of CD14+ Enriched PBMC

It is now known that osteoclast precursors derive from CD14 positive ($CD14^+$) monocytes (9, 10) and with the recent availability of anti-CD14-coated magnetic beads (11), isolation of $CD14^+$ enriched monocytes from human peripheral blood is now possible. The advantage of this method is that it removes the need for the time-consuming wash step once the cells have been settled onto the coverslips or dentine and it yields reproducibly even numbers of cells per well; the disadvantage is the need for specialist equipment and increased costs associated with this procedure (see Note 6). It is important that the cells are kept cold at all incubation stages and pre-cooled reagents are used. This will prevent capping of antibodies on the cell surface, therefore prevent non-specific cell labelling.

1. Collect blood and isolate the monocyte fraction as per Subheading 3.2.1, steps 1–2, with the exception of mixing the blood with equal amounts of PBS containing 2mM EDTA (Buffer I) and diluting the cells in 40–50mL of Buffer I, instead of α-MEM.
2. Centrifuge these cells at 300 × *g* for 20 min and then wash the cell pellet a second time using 40-50mL of Buffer I. Dilute 1:10 with 5% acetic acid to lyse contaminating red cells and determine the cell number by performing a cell count.
3. Pellet the cells by centrifugation at 300 × *g* for 20 min. (Reserve $1.5–2.5 \times 10^5$ cells to check for purity of separation by flow cytometry, if required).
4. Resuspend the pellet in 80 μL cold Buffer I containing 0.5% BSA (Buffer II) per 1×10^7 cells. Mix well, clumps of cells will reduce the efficiency at which the microbeads attach the cells hence reducing purity.
5. Add 20 μL Microbeads per 1×10^7 cells and leave for 15 min at 4°C. Make sure to mix the Microbeads properly with the cell suspension to avoid clumping of cells. Also, it is a good idea to gently flick the mixture at 5-min intervals.
6. Add 2 mL buffer II per 1×10^7 cells to wash excessive/unattached microbeads.
7. Centrifuge at 300 × *g* for 10 min.
8. Apply the column (various sized columns are available, see Note 7) to the magnet firmly and prepare column by adding 500 μL cold Buffer II to wet it (see Note 8).
9. Resuspend the cell–bead pellet to have up to 1×10^8 in 500 μL cold Buffer II.

   In the next two steps, it is essential that extra care should be taken not to remove the column from the magnetic field as the magnetically labelled cells need a high magnetic field to be retained in the column matrix.

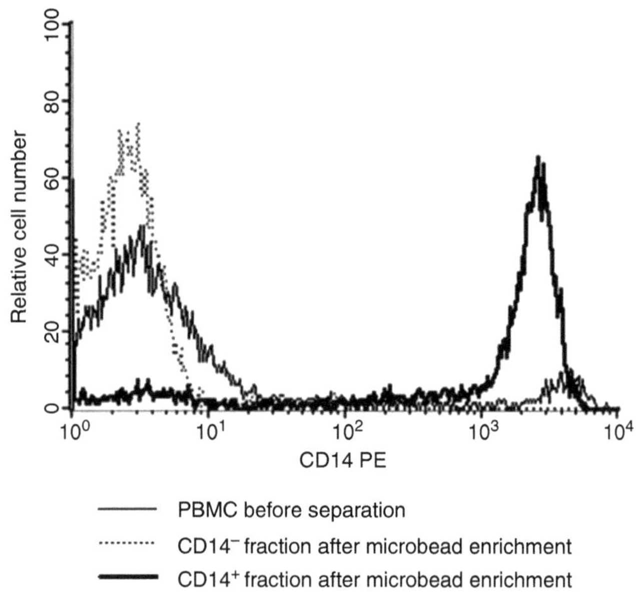

Fig. 2. CD14 flow cytometry profile of cells from the PBMC fraction of blood before and after enrichment using anti-CD14 beads.

10. Pass the suspension through the column and collect the non-labelled cells in a universal held in a rack below the column outlet.
11. Wash the column three times with 500 μL cold Buffer II to completely flush through the non-labelled cells.
12. Remove the column from the magnet and use 1 mL cold Buffer II to flush the magnetically labelled cells by firmly pushing the plunger into the column and collect in a separate universal (volume of Buffer II depends on the size of column).
13. Perform a cell count (reserve $1.5–2.5 \times 10^5$ cells to check for purity of separation by flow cytometry if required, see Fig. 2).
14. Proceed to culture as described in Subheading 3.3.1 or 3.3.2 as required.

3.3. Human Osteoclast Cultures

3.3.1. Co-Culture

The basic principle of co-cultures for human osteoclast formation is that osteoblasts in these cultures express and supply the RANKL, which induces the fusion of monocytic osteoclast precursors also present in the culture. However, it could be possible that these osteoblasts may also be expressing factors in addition to RANKL we are as yet unaware of that also influence the process of osteoclastogenesis. The co-culture described here utilizes the rat osteosarcoma cell line UMR-106, but it cannot be assumed that all osteoblast cell lines are able to induce osteoclast formation in these cultures (see Note 9).

1. The dentine discs or coverslips to be used should be pre-wetted by soaking in 100 μL complete α-MEM for no less than 1 h at 37°C and preferably overnight.
2. UMR-106 cells should be grown to confluence, then washed twice with warm PBS and incubated for 5 min (see Note 10) with 1 mL of warm trypsin–EDTA. Wash the detached cells and resuspend them in complete α-MEM. Perform a cell count and add the appropriate volume of cell suspension containing 2×10^4 UMR-106 cells to wells of a 96-well plate containing the pre-wetted dentine discs or coverslips and then incubate overnight at 37°C in a humidified atmosphere of 95% air and 5% CO_2.
3. Isolate the PBMC or $CD14^+$ enriched PBMC as per Subheading 3.2.1 or 3.2.2, respectively.
4. Add an appropriate volume of cell suspension containing 1×10^5 cells to wells of the 96-well plate containing UMR-106 cells and dentine discs or coverslips.
5. Incubate the cells at 37°C for a minimum of 1 h to allow the blood mononuclear cells to adhere and then remove the non-adherent cells by removing the dentine discs or coverslips from the 96-well plate (see Note 11) and washing in a Petri dish containing α-MEM. This should remove the lymphocytes because these cells do not adhere. This step is not required if using $CD14^+$ enriched PBMCs.
6. Transfer the dentine discs or coverslips to a 24-well plate, taking care to ensure that the surface containing the cells is facing upwards. Three coverslips or dentine discs can be placed in the same well. Incubate in 1 mL complete α-MEM containing 25 ng/mL M-CSF, 10^{-7} M 1,25-dihydroxy vitaminD_3 (see Note 12) and 10^{-8} M dexamethasone at 37°C in a humidified atmosphere of 93% air and 7% CO_2 for ~3 weeks.
7. Replace the complete α-MEM, containing all of the above added factors, every 2–3 days.
8. After 2 weeks in culture, it is advisable to transfer coverslips or dentine discs to a fresh 24-well plate because by this stage the UMR-106 cells will have grown to cover the surface of the well. By the end of the culture period, the UMR-106 cell layer should have detached from the dentine disc or coverslip surface to reveal large multinucleated osteoclasts beneath.

3.3.2. Recombinant RANKL Cultures

This procedure negates the need for osteoblasts in the culture by utilizing commercially available recombinant RANKL. In our experience, it is an easier and more reliable method of osteoclast production than the co-culture system. The choice of which system to use, however, ultimately lies with the researcher and the hypothesis to be tested.

1. The dentine discs or coverslips to be used should be pre-wetted by soaking in 100 μL complete α-MEM for no less than 1 h at 37°C and preferably overnight.
2. Isolate the PBMC or CD14+ enriched PBMC as per Subheading 3.2.1 or 3.2.2, respectively.
3. In this culture, add an appropriate volume of cell suspension containing 5×10^5 cells to the pre-wetted coverslips or dentine discs in the 96-well plate (see Note 13) or 4.5×10^4 cells if using CD14+ enriched PBMC.
4. Incubate the cells at 37°C for a minimum of 90 min to allow the blood monocytes to adhere and then remove the non-adherent cells (i.e. the lymphocytes) by washing the dentine discs or coverslips by repeated pipetting (see Note 14), while they are still in the wells, in α-MEM (without FBS as this would create bubbles). This wash step is not required if using CD14+ enriched PBMC.
5. Incubate the remaining adherent cells in 100 μL complete medium containing 25 ng/mL M-CSF and 30 ng/mL recombinant RANKL (see Note 15) at 37°C in a humidified atmosphere of 93% air and 7% CO_2 for ~2–3 weeks. In our experience, osteoclasts first appear between day 7 and 14, but larger multinucleated osteoclasts are more commonly seen from day 14 onwards (see Fig. 3 and Note 16).
6. Replace the complete medium containing 25 ng/mL M-CSF and 30 ng/mL RANKL every 2–3 days.

3.4. Characterization of Cells Produced in Human Osteoclast Cultures

The cells produced in the above cultures may be identified as osteoclasts by the detection of certain osteoclast markers expressed by the cells. In addition, these markers can allow the differentiation of monocytes into osteoclasts to be monitored over time. Some typical markers used are the following.

1. Cathepsin K: Thought to be the predominant lysosomal cysteine protease enzyme expressed by osteoclasts and is responsible for the degradation of type 1 collagen (12).
2. Calcitonin receptor (CTR): The polypeptide hormone calcitonin is a potent inhibitor of osteoclastic bone resorption. Cells produced in this manner have displayed desensitization to calcitonin following exposure, owing to downregulation of receptor expression (13), therefore confirming that the behaviour of these osteoclasts is identical to mature osteoclasts isolated from tissue samples.
3. Tartrate-resistant acid phosphatase (TRAP): Enzyme cytochemical studies have indicated that osteoclasts in situ and giant cell tumours express high levels of TRAP (14, 15). TRAP expression, however, extends to other cells of the monocyte/macrophage

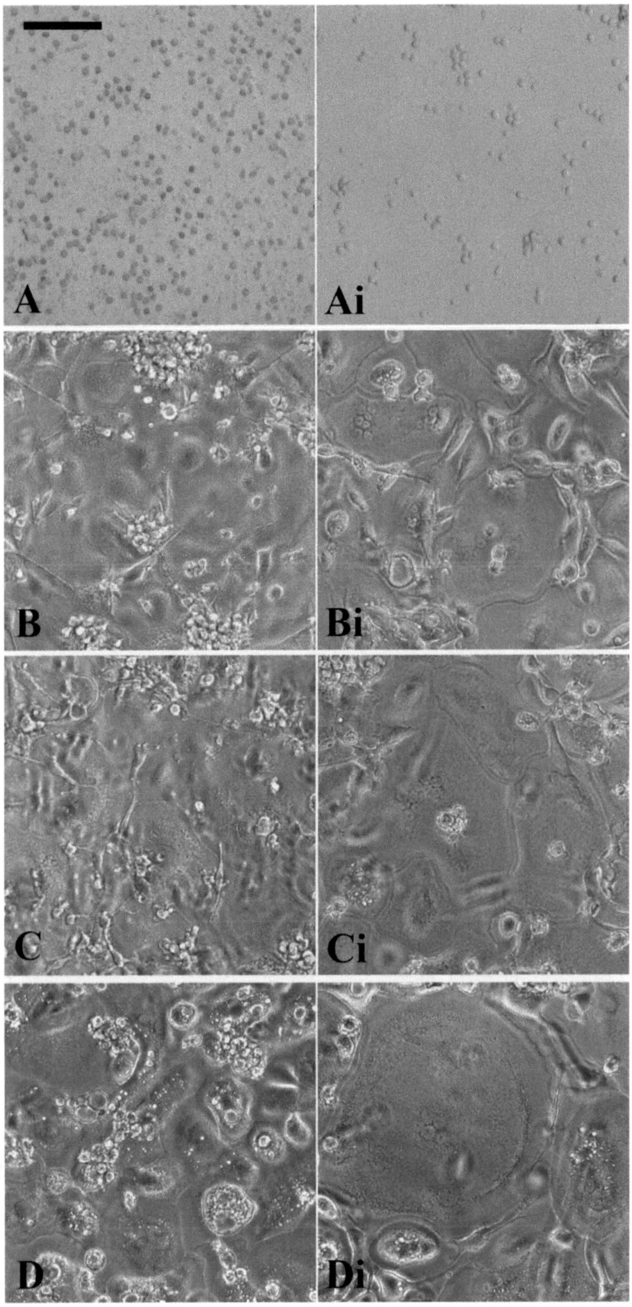

Fig. 3. Human osteoclasts formed in vitro from PBMC. Phase-contrast microscopy images of PBMC or (i) $CD14^{+}$ enriched PBMC cultured with M-CSF and RANKL at (**a**) day 0, (**b**) day 7, (**c**) day 14 and (**d**) day 21. At day 0, cells are small, spherical and mononuclear; by day 7, cells are more spread out with a larger cytoplasm, tend to cluster together and are beginning to fuse. At day 14, cells are larger, multinucleated, and display an osteoclast-like appearance, with the appearance of vacuoles; by day 21, cells are larger again, with larger numbers of nuclei and a distinct osteoclast-like appearance. Some may begin to apoptose after this time period. Scale bar = 100 micron, all images to same scale.

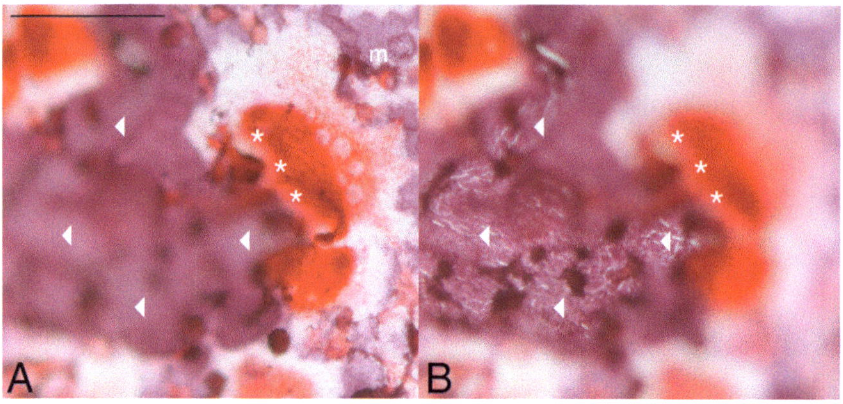

Fig. 4. TRAP-stained osteoclasts generated from human PBMC. Human PBMC were cultured with M-CSF and RANKL for 21 days on a dentine disc after which time the disk was fixed and TRAP stained. A large multinucleated TRAP positive osteoclast (white asterisks) can be seen at the leading edge of a large pit resorption (white arrowheads). TRAP positive mononuclear cells can also be distinguished (m); these cells are also capable of resorption and are often referred to as mono-nuclear osteoclasts. Images were taken on a transmitted light microscope and are of the same field of view, A is focused on the surface of the dentine disk and the osteoclast, whilst B is focused *within* the dentine disk at the bottom of the resorption pit. Scale bar = 100 micron.

lineage; therefore, this marker does not effectively distinguish between osteoclasts and their precursors. We have found that both mono- and multinucleated cells in RANKL-supplemented cultures express this enzyme (see Fig. 4). TRAP staining, however, remains the most widely used method to describe the formation of osteoclasts in these cultures (osteoclasts cultured in vitro often being defined as TRAP positive cells with three or more nuclei).

4. Other markers indicative of an osteoclast phenotype include type II carbonic anhydrase and the vitronectin receptor (integrin $\alpha v\beta 3$). Type II carbonic anhydrase is abundantly expressed by osteoclasts (16), where it catalyzes the production of protons. The vitronectin receptor is functionally involved in adhesion and possibly resorption (17), but is also expressed by many other cells of the monocyte/macrophage lineage.

We have previously shown that the mRNAs for Cathepsin K, CTR and TRAP are expressed in a time-dependent manner in osteoclast cultures derived from human blood and cultured as above (5). Antibodies for the above markers are also commercially available – their utilization requires optimisation by the end user as for any commercially available antibody.

3.4.1. TRAP Staining

We routinely stain both coverslips and dentine discs with TRAP to differentiate the osteoclasts from any stromal cells that may still be present in the culture. Previously, we have used the commercially available staining kits; however, we found that these kits gave very high levels of staining to all cells in these cultures. Therefore, we prefer to use the protocol below.

1. Following the culture of osteoclasts on dentine discs, remove the culture medium, wash the dentine disks in PBS and then transfer the discs to clean wells containing enough fixative solution (see Subheading 2.3) to cover the discs. Fix for at least 5 min (fixative times of up to 1 week are also acceptable).
2. Add 4.6 g sodium tartrate to acetate buffer (see Subhead ing 2.4.2) in a large beaker and warm to 37°C for 1–2 h with foil/lid over it. At the same time, incubate at 37°C coplin jars, conical flasks labelled A and B, small and large measuring cylinders.
3. Remove the fixative from the wells containing the coverslips and/or dentine discs and wash twice with tap water just before starting the staining procedure.
4. Add 100 μL acetate–tartrate buffer (see Subheading 2.4.2) to the wells and incubate at 37°C for 5 min.
5. Remove the acetate–tartrate buffer and add 100 μL solution A (see Subheading 2.4.2) to the wells and incubate for 30 min at 37°C.
6. Remove solution A and add solution B (see Subheading 2.4.2) to the wells for 15 min at 37°C. Check control for staining, but do not incubate for more than an additional 5 min as the solution begins to precipitate.
7. Counterstain slides for 40 s in Gill's haematoxylin and rinse twice thoroughly in tap water (see Note 17).
8. Completely air dry coverslips and mount using DPX. Dentine discs can be stored dry after the staining (see Fig. 4).

Ultimately, these phenotypic markers can still only be regarded as *suggestive* of osteoclastic differentiation as the only definitive marker for osteoclasts remains the ability of these cells to form resorption lacunae (often referred to as pits) on calcified substrates.

3.4.2. Detection and Quantification of Resorption

Osteoclast activity in vitro can be assessed visually by scanning electron microscopy (SEM), reflected light microscopy (RLM) (18) and transmitted light microscopy (TLM). Although SEM produces images of excellent quality and stereophotogrammetry of scanning electron micrographs enables the area and volumes of resorption lacunae to be determined, this procedure requires expertise and expensive equipment, and is impractical for the assessment of a large number of discs because it is highly time consuming. RLM, using an adapted standard microscope, avoids many of the problems associated with SEM providing accurate information on the area, volume and depth of resorption lacunae, especially if there is dense layer of cells. TLM can be routinely used, however, especially when a good counter stain has been achieved or if the cells have been removed from the cell surface (see Fig. 5).

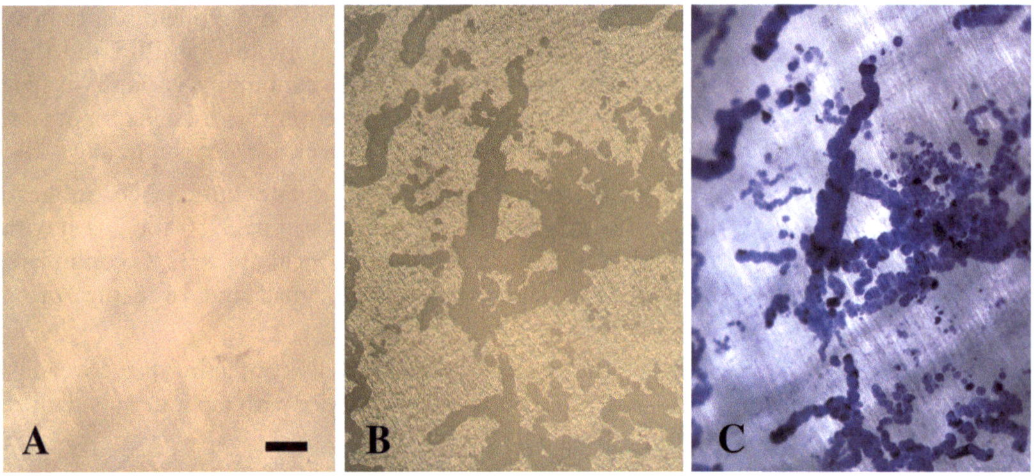

Fig. 5. Identification of resorption pits produced by human osteoclasts generated from PBMC. Human PBMCs were cultured with M-CSF and RANKL for 21 days on a dentine disc after which time the cells were removed from the disc and viewed (**a**) by TLM and (**b**) by RLM. The disc was subsequently stained with 1% toluidine blue and then (**c**) viewed by TLM (all images are from the same field of view and were taken at the same magnification). Authentic resorption pits excavated by osteoclasts are clearly visible without any staining using RLM (darker structures) but require toluidine blue staining to be viewed by TLM at a level required for quantification (compare A and C; faint outlines of the resorption pits are visible on the dentine disc in A, but there is not enough contrast for accurate quantification). Scale bar = 100 micron.

In addition to visually detecting resorption pits, osteoclastic resorption can be monitored by measuring the presence of the type I collagen breakdown product C-terminal telopeptide (CTX) released into the cell culture medium during active resorption (19). CTX is an aspartic acid 1211 residue of the 1209AHDGGR1214 sequence (20), which can undergo spontaneous post-translational modifications as bone ages, resulting in the α-aspartic acid form converting to the β-form (21). Because dentine discs used in our experiments are obtained from mature animals, β-CTX is likely to be the predominant bone-degradation product. Various immunoassays are commercially available to quantify β-CTX (including Elecsys® β-CrossLaps Serum assay from Roche and Crosslaps® for Culture ELISA by IDS). We would advise collection of the cell culture medium once resorption has started at each medium change until the end of the experiment. We would also recommend that any assay is fully validated prior to use.

3.4.2.1. Toluidine Blue Staining

As with fixatives, the choice of staining is dependent upon the procedure you wish to use to examine the osteoclasts and the discs. Toluidine blue solution is a quick and simple method to detect both cells and resorption pits and can be used in combination with the cacodylate–glutaraldehyde fixative for SEM studies. The solutions do not need to be sterile, but should be prepared in clean containers and can be stored at room temperature for several years.

1. Following the culture of osteoclasts on dentine discs, remove the culture medium, wash the dentine discs in PBS and then transfer the discs to clean wells containing enough fixative solution (see Subheading 2.3) to cover the discs and fix for 5 min. Alternatively, if the retention of the cells is not required, remove them from the surface of the disc by rubbing with a cotton bud soaked in 70% ethanol.
2. Wash with distilled water and then stain for 2 min with a 1% solution of toluidine blue buffered with sodium tetraborate (see Subheading 2.4.1). Remove excess stain by washing in 70% ethanol.
3. Rinse in tap water and air dry before examining the wafers under the microscope. Wafers can then be stored indefinitely at room temperature.

3.4.2.2. Area Measurements

Using reflected light resorption, lacunae can be clearly identified even when the dentine disc is covered by a layer of cells. While reflected light offers the best discrimination, resorption pits can also be easily seen using TLM when the cells have been removed and the disc stained with toluidine blue (see Fig. 5). In our laboratory, we routinely stain dentine discs with TRAP (see subheading 3.4.1) and use TLM to count both TRAP+ve cells and resorption pits (see Fig. 4).

1. Using a microscope fitted with a CCTV camera, a drawing-arm, or an eye-piece grid, quantification can be done manually by point counting or can be automated using custom-designed counting software (22). The magnification should be adjusted to ensure that the grid covers an area of 1 mm^2 on the surface of the dentine disc.
2. Area of resorption can be measured over the whole disc or alternatively random fields can be selected and resorption expressed as a percentage of the area. When selecting random fields, repeat until the measurements between fields become consistent to ensure an accurate representation of resorption on the whole disc (22).
3. To determine resorption area on the dentine disc, record the number of points of the grid falling onto resorption lacunae.
4. Resorption can be expressed as a relative measurement compared to resorption produced in vehicle wells, where no compounds in addition to RANKL and M-CSF are introduced.

3.4.2.3. Volume Measurements

The quantification of mean plan area of resorption in most cases provides a good indicator of osteoclast activity; however, this method is not without its pitfalls! (23) If a more accurate measurement of osteoclast activity is required, however, the volume of resorption lacunae may be determined. The following method is based on using an Olympus BH2 microscope; slight modifications are required for other instruments and therefore the manufacturer should be consulted.

1. Bring the surface of the dentine disc into focus using the ×50 objective lens (numerical aperture 0.75). The position of the fine-focus knob of the microscope at the surface of the dentine disc should be noted.
2. Move the fine-focus knob one division at a time, representing a change in depth of 2 μm. At each depth, record the area of the lacuna by counting the number of points of the grid falling within the part of the lacuna that is in focus or has yet to come into focus.
3. Repeat this procedure for ten resorption lacunae chosen at random.
4. Calculate the volume of resorption lacunae by application of the Cavalieri estimator of volume (24). This states that the volume of a sectioned object is "the sum of the products of the sectioned areas × the section separation".

4. Notes

1. The authors routinely use α-MEM with Glutamax™ (Invitrogen™) as this avoids the need to add fresh L-glutamine to α-MEM (L-glutamine is an essential nutrient in cell cultures for energy production as well as protein and nucleic acid synthesis). However, L-glutamine spontaneously degrades in cell culture media generating ammonia as a toxic by-product. Ammonia can affect cell viability, lower protein production and change glycosylation patterns.
2. Batches of serum vary in their ability to support the growth of cells. It is advisable to screen batches and reserve a large quantity of serum once a suitable batch has been identified.
3. We have found that osteoclasts do not grow very well on tissue culture plastic surfaces, instead preferring glass.
4. We have found that it is better to only half fill the lithium-heparin containers with blood to fully prevent coagulation. This, then, requires a considerable number of tubes for 100 mL and so a butterfly needle is preferred. Consult your phlebotomist if in doubt.
5. Typically, for each 20 mL of blood used, the isolated monocyte fraction is finally resuspended in between 3 and 4 mL of complete α-MEM to obtain a cell concentration that allows the addition of an appropriate volume of this suspension that is held by wells of a 96-well plate.

6. One of the major disadvantages with the CD14 magnetic microbead isolation is the initial set-up cost as you need to buy the MACS magnet and stand for efficient separation, and then also the constant added cost of the columns as these cannot be reused. Access to a flow cytometer is also needed in order to assess the efficiency of your procedure.
7. To increase the purity of magnetic labelled cells, cells need to be eluted using the appropriate column. MS columns (130-042-201) are for 1×10^7 magnetically labelled cells from up to 2×10^8 total cells and LS columns (130-042-401) are for 1×10^8 magnetically labelled cells from up to 2×10^9 total cells.
8. This step is essential as the column contains a hydrophilic coating which allows rapid filling and so Buffer II washes this coating before separation.
9. It has become apparent that this osteogenic potential of osteoblasts is not universal to all osteoblast cell lines, some supporting and others not supporting osteoclast formation. This difference is due to the ratio of expression of RANKL and osteoprotegerin (OPG) (a soluble decoy receptor to RANKL) by the osteoblasts following stimulation with osteotropic factors, such as 1,25 dihydroxy vitaminD3 and dexamethasone (25). Binding of OPG to RANKL prevents osteoclastogenesis from occurring (26). In our experience, the human osteosarcoma cell line SaOS-2 is unable to support osteoclast formation.
10. UMR-106 cells should be incubated with trypsin–EDTA for the minimum length of time possible, as prolonged exposure impairs the osteogenic capacity of this cell line.
11. To remove dentine discs or coverslips from wells of 96-well plates, use a fine needle to lever the discs or coverslips into an upright position and then pick up with fine forceps.
12. Although osteoblasts produce M-CSF, the rat M-CSF produced by UMR-106 cells is not recognized by the human M-CSF receptor on the osteoclast precursors; therefore, human recombinant M-CSF must be added. 1,25 dihdroxy vitaminD_3 should be stored under nitrogen at −70°C and in the absence of light to maintain its activity.
13. 20 mL of blood provides approximately enough cells for 50 wells, depending on the donor.
14. In our experience, this washing step is very important in the resulting number and quality of osteoclasts produced in the culture. Ensure that the whole surface of the dentine disc or coverslip is washed evenly and take care not to flip the dentine discs over. A rough guide to washing would be to use two changes of medium and with each medium change pipette up and down four times, firmly but not vigorously.

15. This is the standard concentration of M-CSF and RANKL used; however, it is advisable to test a range of concentrations with new supplies of these factors and if possible to screen batches and reserve a large quantity once a suitable batch has been identified.
16. If cultures are being used to examine the effect of compounds/ treatment on osteoclast resorption, it is advisable to set up extra dentine discs to remove from the cultures and stain with toluidine blue (Subheading "Toluidine Blue Staining") to monitor the onset of resorption – this can occur as early as 10 days and as late as 18 days and is donor dependent.
17. A stronger nuclear stain can be achieved by an additional step of incubating in 0.1% Triton X/PBS for 10 min at room temperature followed by a rinse in tap water prior to addition on Gill's Haematoxylin.

Acknowledgements

This work was supported in part by the European Commission under the 7th Framework Programme (proposal #202231) performed as a collaborative project among the members of the ATPBone Consortium (Copenhagen University, University College London, University of Maastricht, University of Ferrara, University of Liverpool, University of Sheffield and Université Libre de Bruxelles): it is a sub-study under the main study "Fighting osteoporosis by blocking nucleotides: purinergic signalling in bone formation and homeostasis".

References

1. Walsh CA, Carron JA, Gallagher JA. The Isolation of Osteoclasts from Human Giant Cell Tumors and Long-Term Marrow Cultures. 1996: 263–76.
2. Yasuda H, Shima N, Nakagawa N, et al. Osteoclast differentiation factor is a ligand for osteoprotegerin/osteoclastogenesis-inhibitory factor and is identical to TRANCE/RANKL. *Proc Natl Acad Sci USA*. 1998; 95: 3597–602.
3. Quinn JM, Fujikawa Y, McGee JO, Athanasou NA. Rodent osteoblast-like cells support osteoclastic differentiation of human cord blood monocytes in the presence of M-CSF and 1,25 dihydroxyvitamin D3. *Int J Biochem Cell Biol.* 1997; 29: 173–9.
4. Agrawal A, Buckley KA, Bowers K, Furber M, Gallagher JA, Gartland A. The Effects of P2X7 Receptor Antagonists on the Formation and Function of Human Osteoclasts in vitro. *Purinergic Signalling.* 2010;6: 307–15.
5. Buckley KA, Hipskind RA, Gartland A, Bowler WB, Gallagher JA. Adenosine triphosphate stimulates human osteoclast activity via upregulation of osteoblast-expressed receptor activator of nuclear factor-kappa B ligand. *Bone.* 2002; 31: 582–90.
6. Chan BY, Gartland A, Wilson PJ, et al. PPAR agonists modulate human osteoclast formation and activity in vitro. *Bone.* 2007; 40: 149–59.
7. Gartland A, Buckley KA, Bowler WB, Gallagher JA. Blockade of the pore-forming P2X7 receptor inhibits formation of multinucleated human osteoclasts in vitro. *Calcif Tissue Int.* 2003; 73: 361–9.

8. Neale SD, Smith R, Wass JA, Athanasou NA. Osteoclast differentiation from circulating mononuclear precursors in Paget's disease is hypersensitive to 1,25-dihydroxyvitamin D(3) and RANKL. *Bone.* 2000; 27: 409–16.
9. Massey HM, Flanagan AM. Human osteoclasts derive from CD14-positive monocytes. *Br J Haematol.* 1999; 106: 167–70.
10. Nicholson GC, Malakellis M, Collier FM, et al. Induction of osteoclasts from CD14-positive human peripheral blood mononuclear cells by receptor activator of nuclear factor kappaB ligand (RANKL). *Clin Sci (Lond).* 2000; 99: 133–40.
11. Sorensen MG, Henriksen K, Schaller S, et al. Characterization of osteoclasts derived from CD14+ monocytes isolated from peripheral blood. *J Bone Miner Metab.* 2007; 25: 36–45.
12. Kamiya T, Kobayashi Y, Kanaoka K, et al. Fluorescence microscopic demonstration of cathepsin K activity as the major lysosomal cysteine proteinase in osteoclasts. *J Biochem.* 1998; 123: 752–9.
13. Samura A, Wada S, Suda S, Iitaka M, Katayama S. Calcitonin receptor regulation and responsiveness to calcitonin in human osteoclast-like cells prepared in vitro using receptor activator of nuclear factor-kappaB ligand and macrophage colony-stimulating factor. *Endocrinology.* 2000; 141: 3774–82.
14. Minkin C. Bone acid phosphatase: tartrate-resistant acid phosphatase as a marker of osteoclast function. *Calcif Tissue Int.* 1982; 34: 285–90.
15. Scheven BA, Kawilarang-De Haas EW, Wassenaar AM, Nijweide PJ. Differentiation kinetics of osteoclasts in the periosteum of embryonic bones in vivo and in vitro. *Anat Rec.* 1986; 214: 418–23.
16. Hall GE, Kenny AD. Role of carbonic anhydrase in bone resorption induced by 1,25 dihydroxyvitamin D3 in vitro. *Calcif Tissue Int.* 1985; 37: 134–42.
17. Ross FP, Teitelbaum SL. alphavbeta3 and macrophage colony-stimulating factor: partners in osteoclast biology. *Immunol Rev.* 2005; 208: 88–105.
18. Walsh CA, Beresford JN, Birch MA, Boothroyd B, Gallagher JA. Application of reflected light microscopy to identify and quantitate resorption by isolated osteoclasts. *J Bone Miner Res.* 1991; 6: 661–71.
19. Buckley KA, Chan BY, Fraser WD, Gallagher JA. Human osteoclast culture from peripheral blood monocytes: phenotypic characterization and quantitation of resorption. *Methods Mol Med.* 2005; 107: 55–68.
20. Okabe R, Nakatsuka K, Inaba M, et al. Clinical evaluation of the Elecsys beta-CrossLaps serum assay, a new assay for degradation products of type I collagen C-tlopeptides. *Clin Chem.* 2001; 47: 1410–4.
21. de la Piedra C, Calero JA, Traba ML, Asensio MD, Argente J, Munoz MT. Urinary alpha and beta C-telopeptides of collagen I: clinical implications in bone remodeling in patients with anorexia nervosa. *Osteoporos Int.* 1999; 10: 480–6.
22. Armour KJ, van 't Hof RJ, Armour KE, Torbergsen AC, Del Soldato P, Ralston SH. Inhibition of bone resorption in vitro and prevention of ovariectomy-induced bone loss in vivo by flurbiprofen nitroxybutylester (HCT1026). *Arthritis Rheum.* 2001; 44: 2185–92.
23. Boyde A, Jones SJ. Pitfalls in pit measurement. *Calcif Tissue Int.* 1991; 49: 65–70.
24. Sterio DC. The unbiased estimation of number and sizes of arbitrary particles using the dissector. *Journal of Microscopy.* 1983; 134: 127–36.
25. Nagai M, Sato N. Reciprocal gene expression of osteoclastogenesis inhibitory factor and osteoclast differentiation factor regulates osteoclast formation. *Biochem Biophys Res Commun.* 1999; 257: 719–23.
26. Tsuda E, Goto M, Mochizuki S, et al. Isolation of a novel cytokine from human fibroblasts that specifically inhibits osteoclastogenesis. *Biochem Biophys Res Commun.* 1997; 234: 137–42.

Chapter 24

Effects of Temperature Generated from the Holmium: YAG Laser on Human Osteoblasts in Monolayer Tissue Culture

Moustafa I. Hafez, Anne Sandison, Richard R.H. Coombs, Ian D. McCarthy, and Al-Shymaa M. Hafez

Abstract

With the use of lasers for ablation purposes in spinal surgery, the tissue temperature increases above the boiling point of water, leading to tissue ablation by vaporisation. Due to the thermal environment engendered by the use of lasers, there is concern about the safety of the surrounding important structures, such as dura mater, dorsal root ganglia, and nerve roots.

Key words: Laser, Holmium:YAG, Osteoblast, Dura mater

1. Introduction

The results obtained from cadaveric bone ablation experiments using lasers with and without saline irrigation suggest that irrigation is important not only to cool the lens of the endoscope, but also to keep the surrounding tissues at a safe temperature. Using saline irrigation at 27 ml/min keeps the temperature at specific points in the surrounding structures to around 37°C. To date, however, there has been very limited work on the safety of the cellular environment and the consequences on cells arising from the use of lasers.

Heat-shock proteins (HSPs), as the name suggests, are proteins produced by cells in response not only to thermal stress, but also to other physiological stress, such as extreme pH changes. HSPs can, thus, be used as a non-specific marker for physiological stress. If two identical populations of cells are stressed and their levels of HSPs are measured subsequent to this treatment, those that are more stressed can be expected to produce more heat-shock protein, up to the physiological limit (1, 2).

Ragai R. Mitry and Robin D. Hughes (eds.), *Human Cell Culture Protocols*, Methods in Molecular Biology, vol. 806, DOI 10.1007/978-1-61779-367-7_24,

Cells produce HSPs in response to stress caused by heat, poisons, or signals from nerves or hormones. HSPs are sometimes called molecular chaperones because their role is to protect and use other protein molecules around the cells (3).

Osteoblasts are fully differentiated cells responsible for the production of bone matrix. An osteoblast is a typical protein-producing cell with a prominent Golgi apparatus and a well-developed rough endoplasmic reticulum (4). It secretes Type I collagen and the non-collagenous proteins of the bone matrix. An osteocyte is a mature osteoblast within the bone matrix and is responsible for its maintenance. These cells have the capacity not only to synthesise, but also to resorb matrix to a limited extent (5).

In vitro studies are of increasing importance for evaluating new devices by allowing precise control of the experimental environment. A single-cell type may be reliably studied and in vivo animal experimentation is reduced or avoided. An osteoblast cell culture model has been used to study a wide range of issues in bone biology, including the regulation of differentiation and the metabolic activity of normal bone cells. Such studies permit investigation of the human osteoblast response to artificial materials or devices, such as lasers, and promote an understanding of cell reactions providing morphological and biochemical information on osteoblast development.

In the past, many cell lines obtained from different species have been used to evaluate cell proliferation and differentiation for osteoblast-like cells (6–8). The proliferation, differentiation, and osteo-activity of primary human osteoblasts have been widely used in assessing the biocompatibility of different materials (9). Moreover, any stressful phenomenon, such as heat or laser energy, can affect the function of osteoblast cells and may result in cell death, according to the threshold of the stressful factor.

Numerous in vitro models using human and animal primary, immortalised, and transformed osteoblastic cells have been described. Primary cultures of osteoblast-like cells are advantageous for studies of bone differentiation and metabolism because they retain a normal cell genotype (10, 11). The degree of stress produced by temperature increase was then studied together with the level of HSP produced and short-term cell viability using the Trypan blue exclusion method (12).

This chapter shows how to study the in vitro effects of heat generation equivalent to laser exposure on human osteoblasts in monolayer cultures. Osteoblast cultures could be exposed to different treatment regimens with variations in temperature equivalent (45, 55, and 65°C) to nearby lasing with the Holmium:YAG (Ho:YAG) probe, with an average exposure time for lasing of 15 min.

2. Materials

2.1. General Media, Supplements, Chemicals, and Antibodies

1. Dulbecco's modified Eagle's medium (DMEM).
2. Fetal bovine serum (FBS).
3. L-glutamine.
4. Penicillin G/streptomycin B.
5. Amphotericin B.
6. Trypsin–EDTA (1×) in HBSS without calcium and magnesium.
7. Phosphate-buffered saline (PBS).
8. Trypan blue (0.4%).
9. Xylene.
10. Hydrogen peroxide.
11. Methanol.
12. Fluoromount mountant.
13. Primary anti-HSP70 antibody.

2.2. Maintenance/ Growth Medium

Cells were maintained in DMEM, supplemented with 10% FBS (v/v), 2 mm L-glutamine, 50 U/ml penicillin G, 50 μg/ml streptomycin B, and 0.3 μg/ml amphotericin B.

2.3. Equipment

1. Haemocytometer: Improved Neubauer.
2. Standard humidified incubator set to 37°C, 5% CO_2.
3. Inverted light microscopy fitted with IX-phase-contrast option and digital camera.
4. Holmium:YAG laser machine – Coherent Versa Pulse (Coherent Inc., CA, USA).

2.4. Other Items

Disposable culture vessels, glass microscope slides, glass coverslips, sterile graduated pipettes, electric rechargeable pipettor.

3. Methods

3.1. Osteoblast Isolation and Plating

Appropriate ethical approvals and patient consent must be obtained prior to starting the study.

1. Human osteoblasts could be isolated from human femoral heads which are excised from the hips during routine total hip replacement.
2. Fresh femoral head should be immediately placed in sterile DMEM containing penicillin, streptomycin, and amphotericin B.

3. Osteoblasts can be isolated using the protocol given in Chapter 22.
4. The cell number and viability are estimated using a haemocytometer and standard Trypan blue exclusion test.
5. Cells are plated onto 6-well plates at a density of 9×10^4 cells/well or on a sterile glass microscope slide placed in a sterile Petri dish. Maintain cultures in DMEM/supplements for 48 h to achieve confluency.

3.2. Osteoblasts Exposure to Heat Stress Using Pre-heated Medium

Heat stress is delivered to the cells in 6-well plates by replacing the culture medium with pre-heated fresh medium.

1. Pre-heat medium to the desired temperature (control: 37°C; treatment: 45, 55, or 65°C).
2. Maintain the temperature of each osteoblast culture by placing the plate in a water bath set to the target temperature (see Note 1).
3. The effects of varying the exposure time period (immediate to 3 h) could be studied.
4. After exposure to heat, culture medium is replaced with fresh medium pre-warmed to 37°C (representing normal body temperature).
5. Incubate plates for 15 min in an incubator at the same temperature as the medium while the fourth plate is maintained at 37°C as an untreated control (see Note 2).
6. Wash all wells with PBS.
7. For each plate (including controls), three out of six wells are trypsinised and cell viability is estimated using Trypan blue exclusion test.
8. The other three wells are washed with PBS and attached cells can be directly stained with Trypan blue for 2 min and then washed with PBS. Using an inverted light microscope, cells in the wells are examined for morphological changes, and for an estimation of cell viability using the Trypan blue exclusion test (Figs. 1 and 2).

3.3. Exposure of Cells to Ho:YAG Laser Source

1. Prior to the use of the laser Ho:YAG on osteoblasts cultured on glass slides, raise each side of the slide on a glass coverslip. The culture medium underneath each slide absorbs the laser light preventing it from reaching the Petri dish (see Note 3).
2. Expose cells to a Ho:YAG laser source at 2.1 μm (Fig. 3). Exposure to the laser source should be fixed at total laser energy of 0.01 kJ (set at 1 J/pulse and 10 pulses/s).
3. Remove culture medium, and wash cells with PBS.
4. Check cell viability by direct staining with Trypan blue, then after 2 min wash cells with PBS, and the number of viable cells

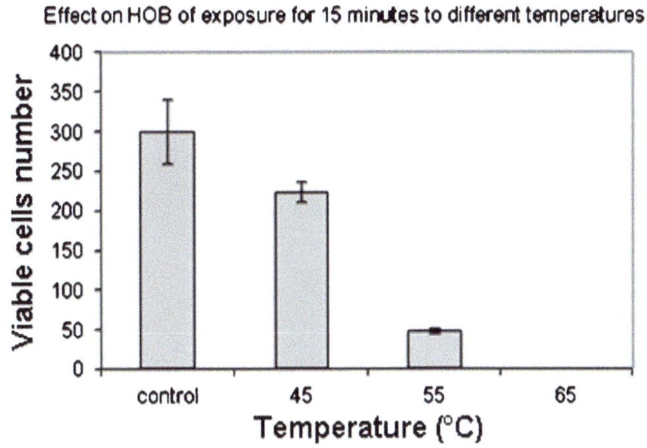

Fig. 1. Effects of heat stress on human osteoblasts in culture following exposure to various temperatures for 15 min.

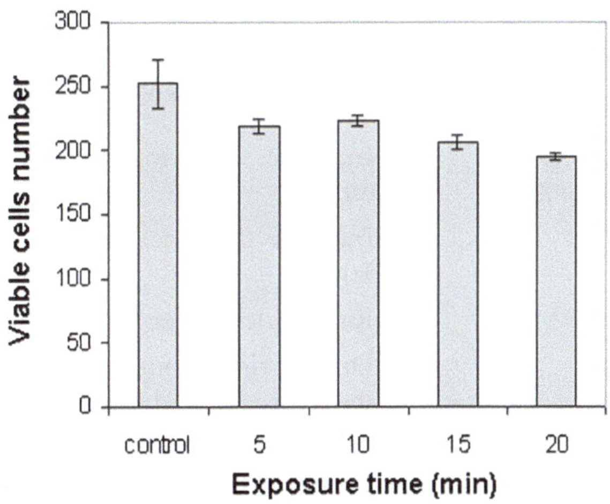

Fig. 2. Human osteoblasts under light microscope following exposure to heat stress (45°C) for various time periods.

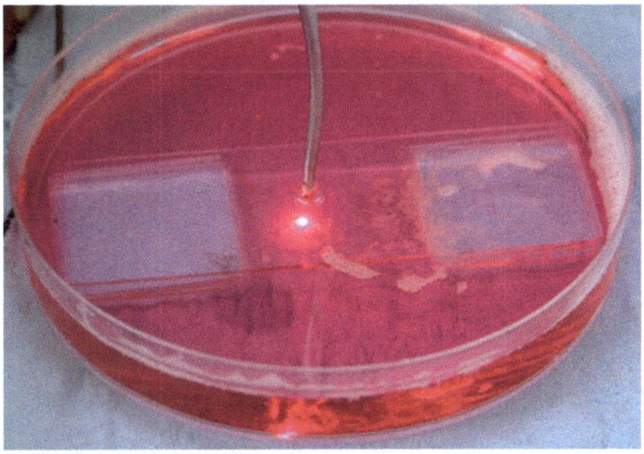

Fig. 3. Cells on a glass microscope slide mounted with supplementary portions of glass, exposed to Ho:YAG laser light source within culture medium.

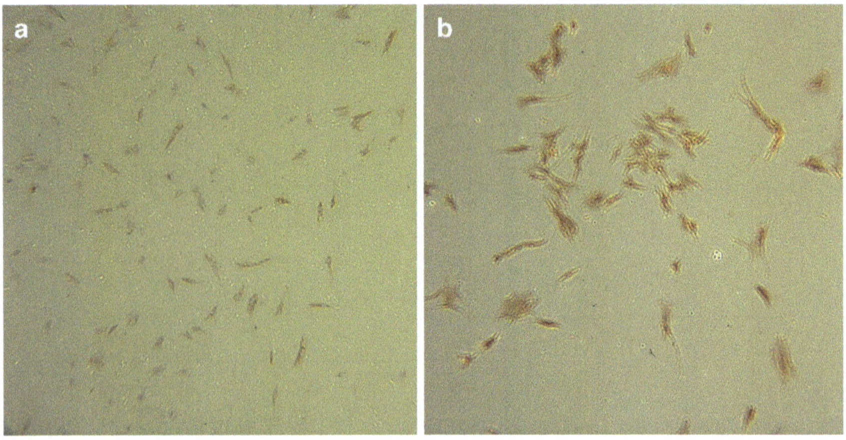

Fig. 4. HSP70 staining of osteoblasts exposed for 15 min to (**a**) 45 and 65°C (**b**).

could be estimated under light microscope (count both the total number and viable cells in several fields). Estimate the number of viable cells at the border of ablated areas.

3.4. Immunocytochemistry: HSP70 Staining

1. Immerse slides in Xylene for 1 min, followed by 20 s in ethanol.
2. Incubate the slides in 2% hydrogen peroxide in methanol for 10 min.
3. Rinse slides under running tap water.
4. Draw a ring around each cell areas using a PAP pen, place slides in a humid chamber, and cover cells within each ring with PBS.
5. Drain off PBS, then cover cells with FCS, and incubate slides for 15 min.
6. Drain off FCS, then place diluted primary HSP70 antibody (1:3,000) on cells, and incubate slides for 60 min.
7. Rinse off the primary antibody, then place biotinylated secondary antibody, and incubate slides for 40 min.
8. Rinse off secondary antibody, apply AB complex, and then counterstain with haematoxylin. The antigenic sites stain dark brown and nuclei stain blue (Fig. 4).

4. Notes

1. In order to accurately maintain the desired temperature of the culture medium (around ±0.5°C), thermocouples could be used. This helps in maintaining the cells at the right temperature

for the whole period of exposure, mimicking the length of time for which the laser would be active during surgery.

2. Another permutation to this experiment could be having a set of five 6-well plates incubated at 45°C. Each plate is maintained for different time periods (e.g. 5, 10, 15, or 20 min), i.e. fixing temperature and varying exposure time.
3. Direct application of laser beam to cells in culture can lead to perforation in the underlying plastic, resulting in a leakage of medium. To prevent that from happening, place a glass microscope slide supported on both sides with 3-mm-thick pieces of glass cut from another microscope slide, and mounted together with a Fluoromount mountant.

References

1. Craig E A, Weissman JS and Horwich A L (1994). Heat Shock Proteins and Molecular Chaperones: Mediators of Protein Conformation and Turnover in the Cell. Cell; 78: 365–372.
2. Garrido C, Gurbuxani S, Ravagnan L and Kroemer G (2001). Heat shock proteins: endogenous modulators of apoptotic cell death. Biochem Biophys Res Commun; 24: 433–442.
3. Henderick J P and Hartl F U (1993). Molecular Chaperone Functions of Heat Shock Proteins. Annu Rev Biochem; 62: 349–384.
4. Marks SCJ and Hermey D C (1996) The structure and development of bone. In: Bilezikian J P, Raisz LG, Rodan G A(eds) Principles of bone biology Academic Press, San Diego, CA ;3–14.
5. Buckwalter J A, Glimcher MJ, Cooper RR and Recker R (1995) Bone Biology. Part I. Structure, blood supply, cells, matrix and mineralisation. Jour Bone Joint Surg 77-A: 1256–1275.
6. Andrianarivo A G, Robinson JA, Mann KG and Tracy R P (1992). Growth of type I collagen promotes expression of the osteoblastic phenotype in human osteosarcoma MG-63 cells. Jour Cell Physiol; 153: 256–265.
7. Price N, Bendall S P, Frondoza C, Jinnah R H and Hungerford D S (1997). Human osteoblast-like cells (MG63) proliferate on a bioactive glass surface. Jour Biomed-Mater Res; 37: 394–400.
8. Hing, K A, Best SM, Tanner K E, Revell P A and Bonfield W (1998). Histomorphological and biomechanical characterization of calcium phosphates in the osseous environment. Proc Inst Mech Eng H; 212: 437–451.
9. Malik M A, Puleo D A, Bizios R and Doremus R H (1992). Osteoblasts on hydroxyapatite, alumina and bone surfaces in vitro: morphology during the first two hours of attachment. Biomaterials; 13: 123–128.
10. Harris S A, Enger R J, Riggs B L and Spelsberg TC (1995). Development and characterization of a conditionally immortalized human fetal osteoblastic cell line. Jour Bone Miner Res; 10: 178–86.
11. Sudo H, Kodama H A, Amagai Y, Yamamoto S and Kasai S (1983). In vitro differentiation and calcification in a new clonal osteogenic cell line derived from newborn mouse calvaria. Jour Cell Biol; 96: 191–198.
12. Wieder R (1999). Selection of methods for measuring proliferation. In : Studzinski GP (ed) Cell growth, differentiation and senescence. Oxford University Press, Oxford, UK.

Chapter 25

Laser Microdissection Microscopy: Application to Cell Culture

Ahlam Mustafa, Cathy Cenayko, Ragai R. Mitry, and Alberto Quaglia

Abstract

Laser microdissection (LMD) microscopy allows isolation of specific cell populations to target their molecular profile. There are several different types of LMD microscopes, but they are all based on the same principle. A laser beam is used to cut out cells or tissues of interest from a histological section, cytology preparations, or live cells from tissue cultures. Live cells can be isolated using LMD and processed for downstream molecular work. RNA, DNA, and protein isolation is possible from a small number of cells and the material is suitable for further real-time PCR, ELISA, Western Blotting, and protein microarray analysis.

Key words: Laser microdissection, HepG2, RNA, DNA, Protein and live cell microdissection

1. Introduction

Laser microdissection (LMD) is a well-established technique that allows the isolation of specific cell populations from tissue samples (1). In this chapter, we summarise the protocols we have developed for the isolation of live cells from tissue cultures using LMD and applications in molecular biology.

2. Materials

2.1. HepG2 Cell Line

Human hepatocellular carcinoma cell line (HepG2, Cat. No. 85011430; Sigma-Aldrich Ltd., Dorset, UK).

Ragai R. Mitry and Robin D. Hughes (eds.), *Human Cell Culture Protocols*, Methods in Molecular Biology, vol. 806, DOI 10.1007/978-1-61779-367-7_25, © Springer Science+Business Media, LLC 2012

2.2. Chemicals and Solutions

The following is a list of chemicals and solutions used in tissue culture of HepG2 cells and further LMD, RNA extraction, and protein lysis.

1. RPMI 1640 with L-glutamine (Cat No. 21875; Invitrogen Ltd., Paisley, UK).
2. Foetal calf serum (FCS), heat inactivated (Cat. No. 10108-165; Invitrogen Ltd.).
3. Penicillin–streptomycin (Cat. No. 15140163; Invitrogen Ltd.).
4. Rnase away (Cat. No. 732-2351; VWR International, Leicestershire, UK).
5. TRIZOL® reagent (Cat. No. 15596-0026; Invitrogen Ltd.).
6. Chloroform (Cat. No. 2432-25 ml; Sigma–Aldrich Ltd.).
7. 2-Propanol (isopropyl alcohol, Cat. No. 19516 – 25 ml; Sigma–Aldrich Ltd.).
8. Ethanol (Cat. No. E7023 – 500 ml; Sigma–Aldrich Ltd.).
9. DEPC-treated Water (DEPC-H_2O, Cat. No. AM9915G; Invitrogen Ltd.).
10. 1× Phosphate-buffered saline (PBS) solution (Cat. No. 10010-05; Invitrogen Ltd.).
11. Aprotonin (Cat. No. A6279-10 ml; Sigma–Aldrich Ltd.).
12. Phenylmethylsulfonyl fluoride (PMSF, Cat. No. P7626; Sigma–Aldrich Ltd.).
13. Sodium orthovanadate (Cat. No. S-6508; Sigma–Aldrich Ltd.).
14. Igepal CA-630 (NP-40, Cat. No. 56741; Sigma–Aldrich Ltd.).
15. Deoxycholic acid "sodium salt" (Cat. No. D-5670; Sigma–Aldrich Ltd.).
16. Sodium dodecyl sulphate (SDS, Cat. No. 444464T; VWR International).
17. Sodium carbonate anhydrous (Na_2CO_3, Cat. No. 102404H; VWR International).
18. L(+)Tartaric acid "dipotassium salt" (Cat. No. T6897; Sigma–Aldrich Ltd.).
19. Sodium hydroxide pellets (NaOH, Cat. No. 28244.262; VWR International).
20. Cupric sulfate pentahydrate ($CuSO_4$, Cat. No. C-7631; Sigma–Aldrich Ltd.).
21. Folin and Ciocalteu's solution (Cat. No. F9252 – 100 ml; Sigma–Aldrich Ltd.).
22. Bovine serum albumin (BSA) 10 mg/ml, (Cat. No. A2153; Sigma–Aldrich Ltd.).

2.3. Preparation of Solutions

1. The culture medium should be stored at 4–8°C and prepared by adding the following supplements to 500 ml of RPMI.
   (a) 50 ml FCS.
   (b) 5 ml penicillin–streptomycin.
2. Prepare RIPA buffer by adding the following to PBS, and store at room temperature (RT).
   (a) NP-40 (1% final concentration).
   (b) Sodium deoxycholate (0.5%).
   (c) SDS (0.1%).
3. Before using RIPA in preparing tissue lysate, freshly add the following:
   (a) 10 μl/ml PMSF.
   (b) 30 μl/ml aprotinin.
   (c) 10 μl/ml sodium orthovanadate.
4. Prepare alkaline copper reagent by dissolving the following into 80 ml of distilled water, and top the solution up to 100 ml with distilled water.
   (a) 10 g Na_2CO_3.
   (b) 100 mg tartaric acid.
   (c) 2 g NaOH.

   Dissolve 50 mg $CuSO_4$ in a few drops of distilled water and then add to the above mixture.
5. Dilute Folin and Ciocalteu's solution by adding 8 ml of distilled water to 0.5 ml of the stock (this has to be prepared freshly prior to immediate use).

2.4. Laser Microdissection Microscope and Accessories

1. Microscope Leica DM6000 B (Leica Microsystems Ltd., Knowlhill, Milton Keynes, UK).
2. Laser FTSS355-50 (Cat. No. 11501466; Leica Microsystems).
3. Camera DFC300 FX (Cat. No. 12730046; Leica Microsystems).
4. Operating System Leica Application Suite: Version 6.5.0.3104 (Leica Microsystems).
5. Scanning stage collector: 0.5 ml PCR tube (Cat. No. 11505228; Leica Microsystems).
6. Scanning stage collector: 8-well strip (Cat. No.11 505 230; Leica Microsystems).
7. Scanning stage Holder: Petri dish, ∅ 5 cm (Cat. No.11 505 227; Leica Microsystems).
8. 50 mm Petri dish with PEN membrane (Cat. No.11505172; Leica Microsystems).

9. 0.5 ml PCR tubes (Cat. No. 30124502; Leica Microsystems).
10. 8-Well Strips (Cat. No.11505240; Leica Microsystems).

2.5. Others

1. Refrigerated bench-top centrifuge for tubes and plates.
2. Water bath.
3. Suction pump.
4. Laminar flow cabinet.
5. Humidified incubator (37°C, 5% CO_2).
6. Maxwell®16 instrument (Cat. No. AS2000-SX; Promega UK, Southampton, UK).
7. Maxwell®16 FFPE Tissue LEV DNA Purification Kit (Cat. No. AS1130, Promega, UK).
8. Flat-bottom, 96-well plates with lids (Cat. No. 734-2097; VWR International).
9. Nanodrop® spectrophotometer (Labtech international, UK).
10. Plate reader to read 96-well plate at 650nm.

3. Methods

3.1. Cell Culture

1. Quickly thaw cryopreserved HepG2 cells stored in −140°C by gentle swirling of cryotube in a water bath at 37°C, taking care not to overheat the cells, i.e. cell suspension should remain cold after thawing.
2. In a laminar flow cabinet, transfer cells into a fresh 50-ml falcon® tube by passing them through a 70-μm sieve to remove clumps and clots. Slowly add medium (drop by drop) in a ratio of 10× the volume of the frozen Cells. (For 1 ml cells, add 10 ml medium).
3. Centrifuge solution to pellet cells at 450×*g*, 4 min, at RT. Discard supernatant. Re-suspend cells by tapping on the bottom of the tube, and then add 5 to 7 ml medium per 1 ml of cells available.
4. Estimate the cell count and viability using the standard Trypan blue exclusion technique (2).
5. Place 0.5×10^6 cells per Petri dish and top up with medium to total of 1.5 ml. Incubate cells at 37°C for 24–48 h.

3.2. Laser Microdissection

1. Switch on the Leica microscope electronics box, computer hard drive, and the laser.
2. Select PCR tube as collection device mode for samples to be used for further molecular work and eightfold holder with 8-well strip for re-cultivating LMD cells.

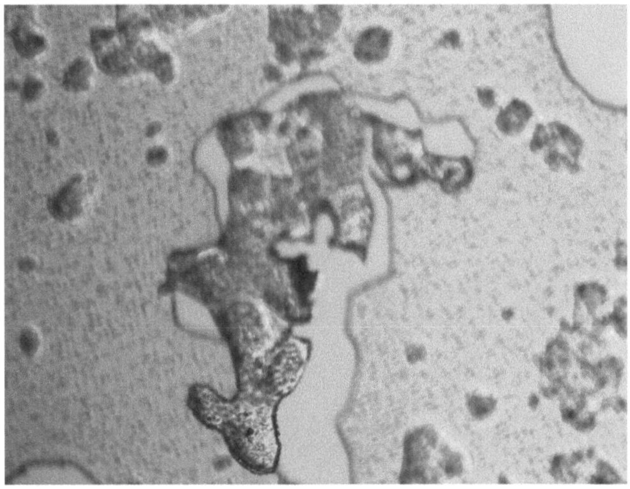

Fig. 1. HepG2 cell area falling into the PCR cap following LMD.

3. Insert the 50 mm Petri dish with PEN membrane and cultured cells into the sample holder.
4. Calibrate laser to the following parameters:
   (a) Power 42.
   (b) Speed 06.
   (c) Specimen Balance 13.
5. Select cutting area: Identify cells of interest. Select "Draw shape" options and trace around cells and cut using the laser (Fig. 1). Differential interference contrast may be used to clearly identify cells. The software is able to identify the area size being dissecting in μm^2.
6. Cells fall into the cap of a collection PCR tube or into one of the 8-well strip compartments for re-cultivating. Always check cells in the cap or the well to ensure proper localisation (Fig. 2).
7. Remove sample: It is advised to remove each PCR tube after each allocation of cells is reached; the 8-well strip can remain in the holder for longer (see Note 1).

3.3. Total RNA Extraction

Total RNA was extracted using TRIZOL® according to manufacturer's protocols with some modifications as below.

1. Clean all surfaces with RNase away.
2. Place 25 μl of TRIZOL® in the cap of a sterile 0.5 ml PCR tube prior to LMD. Following the collection of LMD cells, add a further 25 μl of TRIZOL® to the cap, and carefully close it.

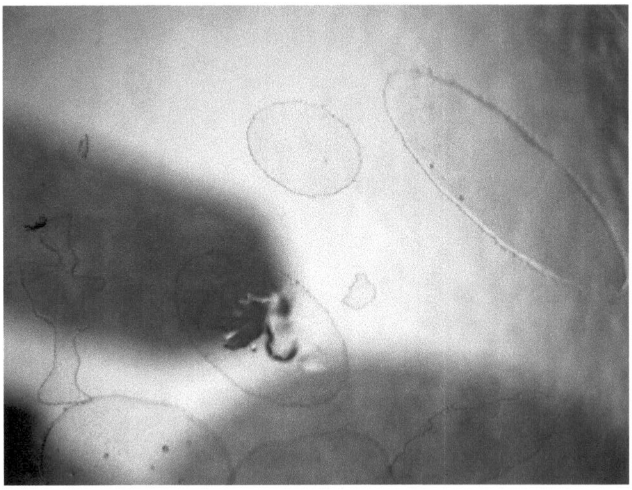

Fig. 2. Laser-microdissected HepG2 areas seen as oval shapes in the collecting PCR tube cap.

3. Allow LMD samples to lyse at RT for 5 min, followed by brief vortexing (see Note 2).
4. Add 10 μl of chloroform to each sample and shake vigorously for 15 s followed by incubation at RT for 3 min. Then, centrifuge at 12,000 × *g*, 4°C, for 10 min.
5. Transfer the aqueous layer carefully to a fresh sterile tube. Add 25 μl isopropyl alcohol per sample and mix well.
6. Incubate at RT for 10 min followed by centrifugation at 12,000 × *g*, 4°C, for 10 min.
7. Discard the supernatant, wash RNA pellet carefully with 60 μl of 75% ethanol, and centrifuge at 7,500 × *g*, 4°C, for 5 min.
8. Discard supernatant and leave pellet to air dry for 7 min in laminar flow cabinet. Dissolve each RNA pellet in 6.2 μl sterile DEPC–H_2O at 55°C for 7 min.
9. Read RNA concentration at 260 nm and assess quality (260/280 ratio) using the Nanodrop® spectrophotometer. It is possible to obtain an average of 79.7 ng/μl obtain from an area of 2.5×10^6 μm^2 of cells with an average 260/280 ratio of 1.7.
10. Extracted RNA is suitable for real-time PCR.

3.4. DNA Extraction

1. Place 25 μl of PBS in the cap of a sterile 0.5 ml PCR tube prior to LMD. Following sample collection, add a further 75 μl of PBS, close cap carefully, and spin tubes to ensure mixing (see Note 3).
2. Add two volumes of lysis buffer (provided in the DNA extraction kit – Subheading 2.5) to each sample and vortex briefly. Extract DNA using the Maxwell® instrument, DNA

purification Kit, according to manufacturer's protocol. It is possible to obtain an average of 57.3 ng/μl is possible to obtain from an area of 2.5×10^6 μm² of cells.

3. Extracted DNA is suitable for gene sequencing (see Note 4).

3.5. Tissue Lysate and Protein Concentration

Measuring protein concentration using Lowry technique (3) with some modification.

1. Place 25 μl of ice-cold RIPA buffer in the cap of sterile 0.5-ml PCR tube prior to dissection. After LMD is complete, add a further 25 μl of the lysis buffer to the cap and carefully close tubes.
2. Leave the LMD samples to lyse on ice for 30 min and then centrifuge at 15,000 × *g*, 4°C, for 20 min. Carefully collect supernatant (tissue lysate (see Note 5)).
3. Use tissue lysate to estimate protein concentration using Lowry technique with modifications as follows.
4. Prepare BSA standards using dilutions in Table 1.
5. Place 50 μl of the test sample or BSA standard per well. (Samples should be diluted using PBS to 1 in 3 volumes due to high concentration).
6. Add 50 μl of alkaline copper reagent to each well and mix content; incubate plate for 10 min at RT.
7. Add 200 μl of freshly prepared diluted Folin and Ciocalteu's reagent per well and mix content; incubate plate in RT for 15 min.
8. Read absorbance at 650 nm against blank. Plot the BSA standard curve and determine the protein concentration in each test sample. It is possible to obtain an average of 1,562 μg/ml from an area of 2.5×10^6 μm² of cells.
9. Tissue lysate is suitable for ELISA, Western Blotting, and protein microarray (see Note 6).

3.6. Isolating Cells for Re-Culturing

1. Place 50 μl of RPMI medium per well of the 8-well strip. After LMD of cells is complete, carefully collect all medium by pipetting and place into an allocated Petri dish.

Table 1
BSA standard curve

	Volume (μl)					
PBS	500	495	490	485	480	475
BSA	0	5	10	15	20	25
Final concentration (μg/ml)	0	100	200	300	400	500

2. Add an extra 1.5 ml of RPMI medium and incubate at 37°C for 24 h. Specific cells could be isolated using the same technique from a heterogeneous sample.

4. Notes

1. LMD cell areas are minute in size; to ensure full lysis in all protocols, keep microfuge tube inverted during the incubation period. Then, turn right side up and centrifuge to collect lysate at the bottom of the tube.
2. Samples may be frozen at −80°C for later analysis at two points (following the completion of steps 2 and 4) in the RNA extraction protocol.
3. In the DNA extraction protocol, volume of PBS used should not exceed 100 μl.
4. Methods of RNA, DNA, and protein extraction used in this chapter are not exclusive; other techniques may be applied with careful optimisation to render the procedure applicable to the small number of cells being analysed.
5. In the tissue lysis protocol, if the tissue lysate is not being processed immediately, add an extra 3 μl of PMSF per sample should be added for every 30 min of delay following the completion of step 2.
6. To use the protein lysate for protein microarray, exclude NP-40 and SDS from RIPA buffer and increase the cell areas being dissected by two- and threefold of minimum recommended in the protocol.

References

1. Espina V, Wulfkuhle JD, Calvert VS, VanMeter A, Zhou W, Coukos G, et al. Laser-capture microdissection. Nat Protoc 2006;1(2):586–603.
2. Freshney R. Culture of Animal Cells. New York, NY: Wiley-Liss; 2000.
3. Lowry OH, Rosebrough NJ, Farr AL, Randall RJ. Protein measurement with the Folin phenol reagent. J Biol Chem 1951 Nov;193(1):265–75.

Chapter 26

Automated Adherent Human Cell Culture (Mesenchymal Stem Cells)

Robert Thomas and Elizabeth Ratcliffe

Abstract

Human cell culture processes developed at research laboratory scale need to be translated to large-scale production processes to achieve commercial application to a large market. To allow this transition of scale with consistent process performance and control of costs, it will be necessary to reduce manual processing and increase automation. There are a number of commercially available platforms that will reduce manual process intervention and improve process control for different culture formats. However, in many human cell-based applications, there is currently a need to remain close to the development format, usually adherent culture on cell culture plastic or matrix-coated wells or flasks due to deterioration of cell quality in other environments, such as suspension. This chapter presents an example method for adherent automated human stem cell culture using a specific automated flask handling platform, the CompacT SelecT.

Key words: Alkaline phosphatase, Automation, Automated, Culture, Flow cytometry, Human, Mesenchymal, Osteogenic, Osteogenesis, Process, Stem cell

1. Introduction

Automation plays an increasing role in human cell culture and dependent scientific disciplines. The fundamental purpose of an automated system is to remove human operator dependency and variation from a process (1). Less process noise and reduced uncontrolled process events allow higher experimental resolution and more efficient characterization of process parameters (2). Automation also allows relatively economic scale out of processing without deterioration of process performance (3, 4). Automation is therefore attractive for fundamental cell research through regulated cell manufacture. These advantages are amplified when specific processes, such as human stem cell culture, exhibit poor reproducibility due to sensitivity to the culture environment.

Ragai R. Mitry and Robin D. Hughes (eds.), *Human Cell Culture Protocols*, Methods in Molecular Biology, vol. 806, DOI 10.1007/978-1-61779-367-7_26, © Springer Science+Business Media, LLC 2012

As stem cell culture for screening or therapeutic use has received increasing attention, a number of automated solutions have begun to emerge. Some of these are incremental or partial solutions such as systems for colony picking, dissection, or cell counting (5, 6). Others are grounded in conventional bioprocess technology, such as stirred tanks, that reduce manual intervention, increase process control, and enable simple scalability (7, 8). The design of automated technology is determined by the production scale requirements for the cell, the intended use (and therefore regulatory considerations), the unit of production (single patient or single use, multipatient or multiuse), and the environmental demands of the cells. Research exploring the limits of stem cell tolerance to processing environments with regard to transfer to existing industrial platforms is currently highly relevant. In this chapter, we detail methods based on the CompacT SelecT robotic T-flask processing platform that have been validated for scaled automated production of human mesenchymal stem cells (hMSCs). Availability of a good manufacturing practice compliant version of this equipment ensures that these protocols are potentially clinically relevant. Similar methods have been validated for other adherent noncommercial cell types and proprietary cell types. Reference to the notes section should allow adaptation of most adherent cell lines to automation.

Fig. 1. The CompacT SelecT automated cell culture platform. The *inset* shows an enlarged image of the manipulation chamber and pipette head. Major processing components are labeled: (**a**) Robot arm; (**b**) flask incubator; (**c**) plate incubator; (**d**) flask de-cappers; (**e**) flask holders; (**f**) media pumps; (**g**) pipette head; (**h**) Cedex-automated cell counter.

In brief, the CompacT SelecT (The Automation Partnership, Cambridge, UK; Fig. 1) is a fully automated cell culture platform consisting of a robot arm that can access 90 × T175 flask or multiwell plate incubators. Flasks are bar-coded for identification and cell process tracking. Two flask de-cappers and flask holders, automated media pumping (or pipetting for volumes of <10 ml), medium warmers, and a Cedex-automated cell counter are also integrated within a Class II culture cabinet. This system allows most cell culture activities, such as passage or media changes, to be conducted and controlled to a schedule in a sterile environment with minimal human interference.

2. Materials

2.1. Mononuclear Cell Isolation

1. Fresh human adult bone marrow (25 ml) (observed a range of cells on the receipt of $16–23 \times 10^6$ cells/ml) (Lonza, Walkersville, MD, USA).
2. hMSC Expansion media: Dulbecco's modified Eagle's medium 1.0 g/L glucose (DMEM-LG) (Lonza) supplemented with 10% fetal calf serum MSC qualified (Invitrogen-Gibco), 1% 100× antibiotic/antimycotic (Invitrogen), 1% 100× Glutamax-I (Invitrogen-Gibco), and 1% 100× non-essential amino acid mixture (Lonza).
3. 50 ml Accuspin tubes (Sigma-Aldrich).

2.2. Automated Adherent Mononuclear Cell Culture

1. CompacT SelecT Robotic cell processing platform (The Automation Partnership).
2. BD Falcon T175 bar-coded cell culture flasks.
3. hMSC expansion media (as discussed in Subheading 2.1).
4. Trypsin/EDTA 0.25% porcine (Sigma-Aldrich).

2.3. Differentiation Cell Culture

1. Dexamethasone stock [4.7 mg cell culture tested Dexamethasone (Sigma-Aldrich) dissolved in 12 ml 100% ethanol]. Sterile filter into 120 ml HBSS (Lonza). Stored in aliquots at −20°C.
2. Ascorbic acid stock [36.6 mg L-Ascorbic acid phosphate (WAKO) dissolved in 10 ml HBSS]. Use fresh aliquots.
3. Osteogenic induction media: As hMSC expansion media but with 1% supplement of Dexamethasone stock and Ascorbic acid stock (as described above).

2.4. Flow Cytometry and Alkaline Phosphatase Activity Analysis

1. Pharmingen stain buffer (BD Biosciences).
2. Anti-human CD105-PE-conjugated antibody, anti-human CD166-PE-conjugated antibody, anti-human CD90-PE-conjugated antibody (all Beckman Coulter).

3. Alkaline phosphatase substrate working solution (1.0 mg/ml *p*-nitrophenyl phosphate in 0.2 M Tris buffer). This is made up from Sigmafast™ tablets: *p*-nitrophenyl phosphate tablets (Sigma) and Tris buffer tablets (Sigma). Dissolve one of each in 20 ml de-ionized water.

3. Methods

3.1. Isolation of Mononuclear Cells

1. The 25 ml fresh bone marrow is diluted with an equal volume (25 ml) of hMSC expansion media in a 50 ml centrifuge tube, and then gently mixed with a 25 ml pipette.
2. Take two Accuspin tubes (containing 15 ml Histopaque and frit). Check that the Histopaque liquid is beneath the frit. This may need precentrifuging if it has been disrupted in storage. Split the diluted bone marrow equally between the two Accuspin™ tubes. Do not mix.
3. Centrifuge the Accuspin tubes on a swing-arm rotor at 1,000 × *g* for 10 min.
4. After centrifugation, the tubes will have four distinct layers. Remove from centrifuge taking care not to disturb the separated layers. The top translucent layer is plasma, the milky thin second layer is the monocyte cell population of interest that will contain a small proportion of mesenchymal stem cells among other cell types, the transparent third layer containing the frit is the Histopaque, and the dark pellet at the base are the red blood cells.
5. Carefully remove the plasma with a pipette taking care not to disturb the monocyte layer. Remove the monocyte layers from both tubes and place into a single fresh centrifuge tube. Dispose of the plasma and remaining contents of the Accuspin tube according to local bio-safety procedures.
6. Add 10 ml hMSC expansion media to the recovered monocytes, gently mix, and then centrifuge at 300 × *g* for 10 min. Aspirate and dispose of the supernatant. Re-suspend the cell pellet in 30 ml of hMSC expansion media. Count using an automated cell counting system to ensure accuracy [i.e., Cedex (Innovatis) or Countess (Invitrogen)].
7. Based on the count, calculate the volume of monocyte suspension containing 10.5×10^6 cells and transfer to a bar-coded T175 flasks (gives a seeding density of 60,000/cm^2) and make up to 25 ml with hMSC expansion media (we have found a linear relationship between seeding density and adherent cells recovered at first passage up to approximately 80,000/cm^2, however this is subject to natural variation in the sample).

3.2. Automated Cell Culture

NB these are the protocol timings for the automated process steps. The example programs for the procedures themselves are given as an Appendix and can be used to extract all detailed handling parameters. Reference to the notes section is required to understand the automated processing.

1. Register the newly seeded flasks and import into the automated platform.
2. On day 2 and day 5, implement an automated media top-up (see Appendix and Note 5). All processing is conducted with liquids at *ambient* temperature 17–23°C (liquids have been used after up to 48 h at ambient temperature for delayed out of hours processing).
3. On day 8, implement an automated media change (see Appendix).
4. On day 12, implement an automated passage (see explanatory Notes 1–7 and Appendix). In the primo culture period, cells will form colonies, as shown in Fig. 2 (NB the cell growth in this period will be subject to the quality of serum used and the characteristics of the donor patient's cells, and this passage time may need to be altered). Cells should be passaged before colonies become too dense (see Fig. 2). In our laboratory, this protocol will produce more than 2×10^6 adherent cells/flask. Seeding density is only 5,000 cells/cm^2 from this point on as most nonadherent cells (that do not contribute to culture) are lost at first passage.
5. Run the feed protocol and passage protocol alternately at 3 day intervals. During this expansion period, cells will grow in monolayer as shown in Fig. 2. Again this is subject to some donor and culture component quality variation, so time intervals may need to be altered when accustomed to process.

3.3. Cell Output Assays

hMSCs are definitively identified by their ability to differentiate into cells of mesenchymal lineage, i.e., undergo adipogenesis, myogenesis, chondrogenesis, or osteogenesis. Standard differentiation protocols and assays are available for directing and detecting all these lineages and the population assay(s) chosen will depend on the use for which the cells need to be validated. There are also a range of cell surface markers that are recognized as characteristic of, although not specific to, hMSCs (9). In our laboratory, we run preliminary checks on hMSCs from the automated platform for expression of surface markers CD105, CD90, and CD166 using flow cytometry (Fig. 5), as well as validating for early osteogenic potential through incubation in osteogenic media and assay for subsequent increase in alkaline phosphates activity.

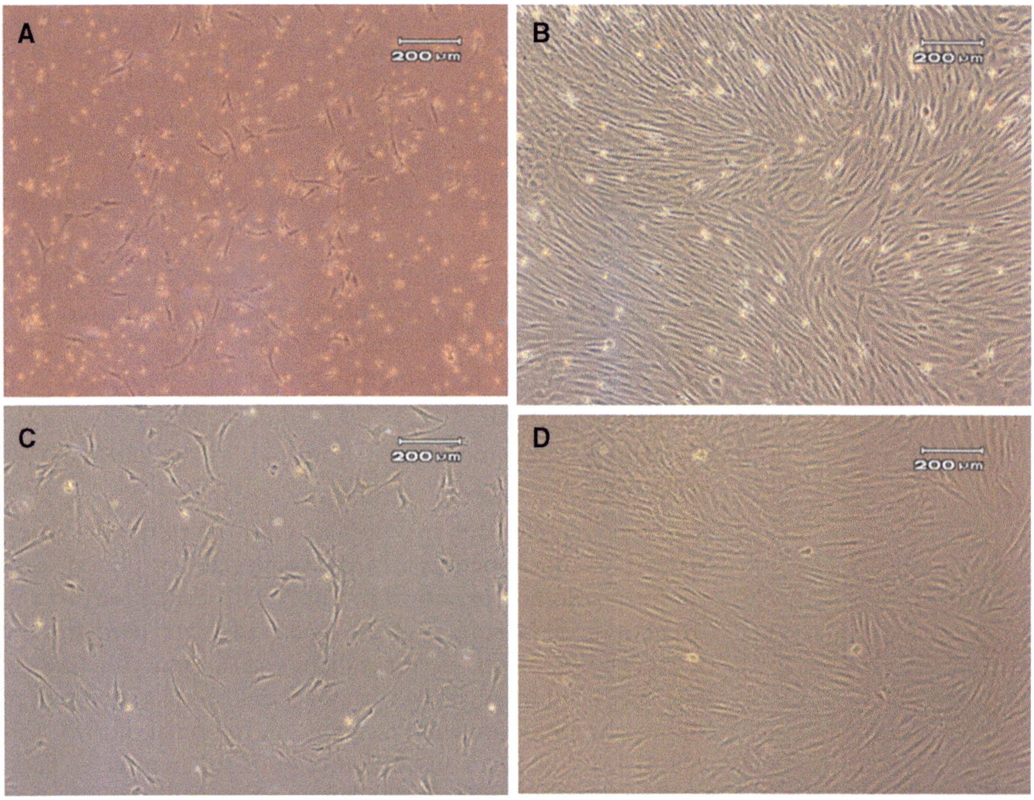

Fig. 2. Cell morphologies are shown at different stages of culture. In primo culture (**a**) early and (**b**) advanced adherent cell colonies are both seen after ~12 days. The cell colonies are not synchronized and cell cultures should be passaged before any colony centers get denser than shown in (**b**) leading to loss of proliferative capacity. After the first passage, cells grow in a monolayer format (**c**), and should be passaged at ~80% confluent: (**d**) shows a cell culture a little past this point.

3.3.1. Osteogenic Induction and Alkaline Phosphatase Activity Assay

1. Take a pooled cell suspension of known concentration from the automated process, either unseeded excess for an in process check or at the end of the expansion. Take two separate samples containing 10,000 cells and centrifuge at 300×*g* for 5 min.
2. Aspirate and discard the supernatant. Re-suspend one sample of cells at a concentration of 5,000 cells/ml in osteogenic induction media (2 ml), and the other at the same density in hMSC expansion media as a no induction control (2 ml). Pipette each sample into separate sets of replicate wells of a 96-well plate at a density of 1,000 cells (200 μl)/well.
3. Incubate the cells for 4 days.
4. Remove media from the wells and add 100 μl/well of Alkaline phosphatase substrate working solution. Add substrate to additional wells that do not contain cells as a blank reading. Incubate at room temperature for 1 h.

5. After incubation read sample absorbance at 405 nm wavelength using a microplate reader. The wells without cells provide a baseline control for zero enzyme activity; the noninduced hMSCs will have a small amount of alkaline phosphatase activity. The induced cells should show significantly more. Standard curves should be used to check the sensitivity of the assay using both cell concentration and incubation time.

3.3.2. Flow Cytometry

1. Take a pooled cell suspension of known concentration from the automated process. Surplus cells from the passage process can be used to check in process quality, or end of process cells for final quality.
2. Transfer a well mixed volume of this suspension containing approximately 1×10^5 cells to the required number of 1.5 ml Eppendorf tubes. Spin in a microcentrifuge at $300 \times g$ and discard supernatant.
3. Use one Eppendorf for each antibody. Add 10 μl of each antibody directly to the pellet and re-suspend. Add a further 200 μl Pharmingen stain buffer and incubate out of direct light at room temperature for 20 min (appropriate negative controls, nonspecific isotypes, and no antibody should be run in parallel in different tubes).
4. After incubation, centrifuge samples at $300 \times g$, aspirate and discard supernatant, and re-suspend in 300 μl Pharmingen stain buffer twice to wash. Analyze re-suspended cells on a flow cytometer (Fig. 3 shows typical profiles).

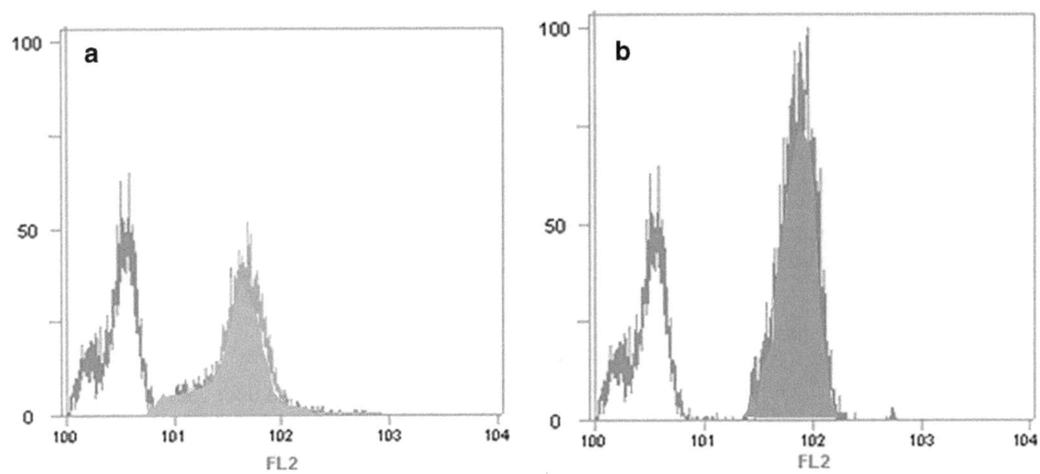

Fig. 3. Typical flow cytometry profiles for two characteristic hMSC cell surface markers (**a**) CD166 and (**b**) CD105.

4. Notes

4.1. Notes for Automated hMSC Culture

1. The basic automated elements of the passage process (Fig. 4) involve: (a) fetching a flask from the cell incubator, (b) disposing of spent cell culture media, (c) addition of digestion enzyme, (d) disposal of digestion enzyme, (e) incubation, and (f) addition of a small volume of cell culture media to re-suspend the cells. At this point, there is an option: cells can be poured to a pool flask, and then further flasks are processed as described above to build a single pool for further processing. Alternatively, the single flask can be processed independently. Either way, (g) the pool flask or the individual flask is automatically sampled and counted through the integrated cell count facility. (h) The liquid handling will dilute the flask or pool flask based on this count to a user-programmed cell concentration. (i) A user-programmed volume of cell culture media will then be dispensed into new flasks followed by (j) a defined volume of cell suspension. This delivers a defined seeding density for further expansion. These key automated steps are shown in Fig. 4 reference to which will be required to understand the following points.

   The appended programming steps can be used to process flasks in the CompacT SelecT for the protocol described in

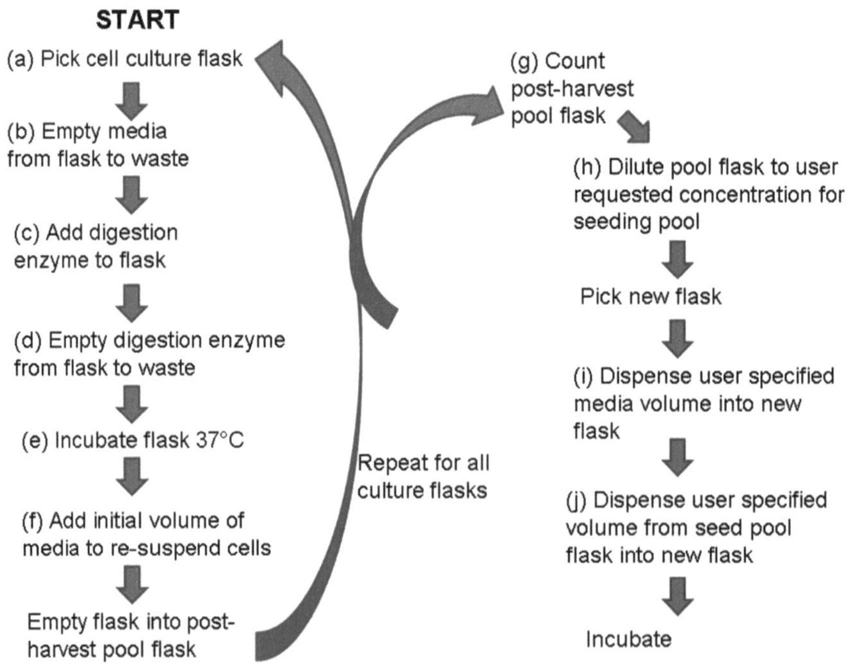

Fig. 4. The key steps in the automated passage process as a reference to assist understanding of the notes section.

Subheading 3.2 through to primo culture passage. The automation has a user-interface that allows simple programming of culture steps to produce programs such as those shown in Appendix. However, there are some important points of variation noted below with regard to passage steps when culturing primary cells or any cells that alter their growth characteristics as they are cultured.

2. The cell yield per flask, and hence the concentration after the initial media addition to re-suspend the cells, must be known reasonably accurately to program a protocol to prevent either excessive dilution volumes or insufficient cell densities for subsequent re-seeding. For example, if a request for a seeding pool cell concentration is programmed that is tenfold lower than the postharvest pool, then a 100 ml postharvest pool would require a 1 L dilution which would exceed flask volume. Alternatively, a programmed request to create a higher concentration seeding pool than is achieved in the postharvest pool will result in a program error (the automation can only dilute, not concentrate, cell suspensions). As these values are preprogrammed, the requested concentrations in the protocol need to be changed to account for different cell number yields as passage increase or if using cells with variable growth characteristics. In the example given (Appendix), based on experience we anticipate 2×10^6 cells/flask yield and request an initial media addition for cell re-suspension of 8 ml/flask. This will result in a cell concentration of approximately 0.25×10^6/ml in the postharvest pool. We need to be sure that the cell concentration requested for the seeding pool is lower than the lowest likely value of the postharvest pool to avoid an error. In this example, the concentration requested for the seeding pool is 0.125×10^6/ml allowing a twofold safety margin. The seeding volume into new flasks given this seed pool concentration is therefore 7 ml to achieve 0.875×10^6 cells/flask (5,000 cells/cm^2).

   In a typical later passage, we may expect to achieve eight million cells/flask from 12 individual flasks. If the initial media addition for re-suspension remained at 8 ml/flask, it would be reasonable to expect a postharvest pool concentration of 1×10^6 cells/ml. We could therefore program a seed pool concentration of 437,500/ml (to keep a similar margin of error), and new flask seed volume would be reduced to 2 ml to retain seeding density.

   The protocol shown would also need to be adjusted for the new flask numbers depending on anticipated yield at each passage (12 here is based on a primo culture of six flasks seeded at 60,000 cells/cm^2 and anticipated to yield in excess of two million cells each. This should comfortably seed 12 flasks at 0.875×10^6 cells/flask.

3. It is also worth noting that the postharvest pool dilution volume can have a significant impact on the proportion of cells used in the counting process (1 ml) if a single flask or low numbers of flasks are processed as such early in a primary culture. For example, if a single flask has a postharvest and precount addition of 10 ml media, 10% of that volume and cell density will be used for the counting process. If ten flasks are similarly processed, but are pooled prior to a single count, only 1% of cell number is lost. A simple method of reducing this loss with low flask numbers is to increase the volume of the initial media addition for re-suspension. However, the dilution must not be so great that it risks delivering a cell suspension diluter than required for further seeding, or beneath the measurement capability of the integrated counter (varies, but, e.g., 1×10^5/ml).

4.2. Additional Notes for Developing Variant Automated Protocols

4. *Cell harvest pooling strategy*: The program described here (see Appendix) will pool a number of flasks to create a homogenous suspension before counting and seeding daughter flasks at each passage. If possible this is more time efficient as flask processing can be interleaved, allowing up to four flasks to be trypsinized and pooled simultaneously (or more or less depending on length of trypsinization – Fig. 5 explains this process).

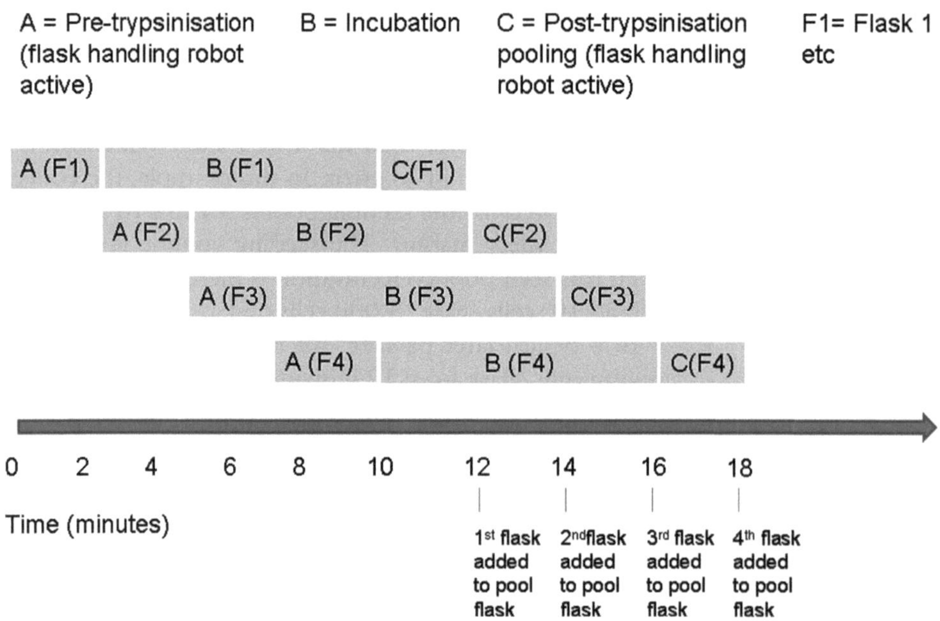

Fig. 5. The chart explains how the robotic processing can be most efficiently used during cell culture passage. The number of flasks that can be interleaved (i.e., liquid handling occurs on one flask while others are incubated with dissociation enzyme) is calculated by the number of times the preprocess time (robot active) for a single flask can be fitted into the incubation time (robot inactive and therefore available to preprocess further flasks). This only works accurately if the post-process handling time is *shorter* than the preprocess time, or flasks will become backlogged in the incubator. Clearly from the *chart*, processing four flasks sequentially would take 48 min, but interleaving four in this manner takes only 18 min. If eight flasks were to be processed, the total interleaved process would therefore take 36 min.

However, there is the option to process flasks independently; for example, if separate patient cells or treatment conditions are being handled. If using the efficient route of pooling cells, it is necessary to validate cells for their response to pooling time, and design the protocol so that process batches do not exceed tolerated pooling time. The number of flasks pooled will also affect the volume of the pooled cell suspension. If this increases past several hundred milliliters, it is preferable to program the system to pick up and swirl the flask prior to seeding rather than mix with the integrated pipette.

5. *In process liquid storage*: Cell culture solutions are used at room temperature (17–23°C) in all our protocols covering multiple human primary cells and cell lines. This has not caused us problems and enables standardization: it is important to allow liquids to equilibrate to RT and not to heat in a water bath prerun to maintain consistent results. Small volumes of cell culture fluid and additives can be stored in the processing chamber in a centrifuge tube. However, due to a significantly higher than room temperature operating environment, this method should be validated before using for delayed additions of growth factors or enzymes that degrade rapidly above room temperature.
6. *Enzymatic digestion*: The removal of digestion enzyme before incubation to harvest the cell sheet is important because this cannot be removed by centrifugation after cells are detached. The residual enzyme remaining after emptying the flask is adequate to remove cells but may take slightly longer to release cells than in manual culture. In manual or automated culture, there will always be some enzyme contamination after passage; however, in automated culture the level of contaminating enzyme depends on dilution levels after harvest. There is a linear correlation between enzyme contamination (trypsin) and reduced hMSC growth and if a relatively low number of cells are harvested this can lead to a detectable effect. This effect can be particularly problematic in serum-free culture. We have addressed this when necessary by using gentler proprietary digestion enzymes such as Accutase or TryplE and/or through addition of an enzyme neutralization step (such as Soy bean trypsin inhibitor) after passage. Digestion times with these enzymes can be reduced through washing the cell sheet with a neutral buffer rinse step (i.e., PBS) prior to digestion, and this may be a necessity for well attached cells.
7. *Cell aggregates*: Some cell types (i.e., neural progenitors) tend to clump during growth and passage; others may require passage as aggregates rather than single cell suspensions. Achieving the same level of aggregates in the automated process as the manual process is necessary to achieve a comparable cell output. Aggregates can occur in the automation with certain cell types,

presumably due to differences in pipette nozzle size and aspiration speeds relative to a manual process. We have overcome these problems when they occurred through trialing different digestion enzymes, optimizing pipette mixing speeds and volumes and, in one case, introducing a DNase treatment step to reduce stickiness from leaked DNA.

Appendix (Automated Protocols)

Cell culture media top-up (Subheading 3.2; step 2)

```
<steps>
<fetch>
<dispense liquid=“hMSC Expansion media” volume=“25 ml”/>
<store passage=“no”/>
</fetch>
</steps>
```

Cell culture media change (Subheading 3.2; step 3)

```
<steps>
<fetch>
<dump pause=“3 s”/>
<dispense liquid=“hMSC Expansion media” volume=“40 ml”/>
<store passage=“no”/>
</fetch>
</steps>
```

Cell passage (Subheading 3.2; step 4)

```
<steps>
<new flasktypegroup=“Single”>
<putdown name=“pool”/>
</new>
<fetch maxrepeat=“9” staggertime=“0 s” interleave=“3”>
<dump pause=“4 s”/>
<dispense liquid=“Trypsin/EDTA” volume=“5 ml”/>
<swirl repeat=“1” speed=“100%” pause=“0 s” capped=“no”/>
<dump pause=“4 s”/>
<incubate period=“10 m”/>
<shake repeat=“30” speed=“100%” pause=“0 s” capped=“yes”/>
<dispense liquid=“hMSC Expansion media” volume=“8 ml”/>
<swirl repeat=“1” speed=“100%” pause=“1 s” capped=“no”/>
<pour name=“pool” pause=“4 s” robotspeed=“100%”/>
<dispose robotspeed=“100%”/>
</fetch>
<mix name=“pool”
    volume=“10 ml”
    repeat=“5”
    fromheight=“2 mm”
```

```
      toheight = “20 mm”
      mixspeed = “10 ml/s”
      finaldispensespeed = “10 ml/s”
      newtip = “yes”/>
<count name = “pool”
        fromheight = “5 mm”
        aspiratespeed = “5 ml/s”
        dispensespeed = “1 ml/s”
        pause = “2 s”/>
<pickup name = “pool”/>
<dispense liquid = “hMSC Expansion media”
       volume = “10 ml”
       cellconc = “125,000”
       minvolume = “0 ml”
       maxvolume = “300 ml”/>
<putdown name = “pool”/>
<new repeat = “12” flasktypegroup = “Single”>
<dispense liquid = “hMSC Expansion media” volume = “13 ml”/>
<putdown name = “output”/>
<mix name = “pool”
      volume = “10 ml”
      repeat = “3”
      fromheight = “2 mm”
      toheight = “10 mm”
      mixspeed = “10 ml/s”
      finaldispensespeed = “10 ml/s”/>
<pipette fromname = “pool”
    toname = “output”
    volume = “7 ml”
    fromheight = “2 mm”
    toheight = “20 mm”
    aspiratespeed = “4 ml/s”
    dispensespeed = “4 ml/s”
    pause = “2 s”/>
<pickup name = “output”/>
<swirl repeat = “1” speed = “75%” pause = “1 s” capped = “yes”/>
<store passage = “yes”/>
</new>
<pickup name = “pool”/>
<dispose/>
</steps>
```

References

1. Liu Y, Hourd P, Chandra A, Williams DJ. 2010. Human cell culture process capability: a comparison of manual and automated production. J Tissue Eng Regen Med. 4(1): 45–54.
2. RJ Thomas, PC Hourd, DJ Williams. 2008. Application of process quality engineering techniques to improve the understanding of the in vitro processing of stem cells for therapeutic use. J Biotech. 136:148–155.
3. RJ Thomas, AD Hope, P Hourd, M Baradez, EA Miljan, JD Sinden, DJ Williams. 2009. Automated, serum-free production of CTX0E03: a therapeutic clinical grade human neural stem cell line. Biotech letters. 31:1167–72.
4. RJ Thomas, PC Hourd, DJ Williams. 2009. Auto-mated, scalable culture of human embryonic stem cells in feeder-free conditions. Biotech Bioeng. 102: 1636–1644.
5. Terstegge S, Laufenberg I, Pochert J. 2007. Automated maintenance of embryonic stem cell cultures. Biotechnol Bioeng. 96:195–201.
6. Joannides A, Fiore-Hériché C, Westmore K. 2006. Automated mechanical passaging: a novel and efficient method for human embryonic stem cell expansion. Stem Cells. 24:230–5.
7. SKW Oh, AK Chen, Y Mok, X Chen, U-M Lim, A Chin, ABH Choo, S Reuveny. 2009. Long-term microcarrier suspension cultures of human embryonic stem cells. Stem Cell Research 2:219–230.
8. Serra M, Brito C, Sousa MF, Jensen J, Tostões R, Clemente J, Strehl R, Hyllner J, Carrondo MJ, Alves PM. 2010. Improving expansion of pluripotent human embryonic stem cells in perfused bioreactors through oxygen control. J Biotechnol. 148(4):208–15.
9. Haasters F, Prall WC, Anz D, Bourquin C, Pautke C, Endres S, Mutschler W, Docheva D, Schieker M. 2009. Morphological and immunocytochemical characteristics indicate the yield of early progenitors and represent a quality control for human mesenchymal stem cell culturing. J Anat. 214(5):759–67.

Chapter 27

Culturing and Differentiating Human Mesenchymal Stem Cells for Biocompatible Scaffolds in Regenerative Medicine

William R. Otto and Catherine E. Sarraf

Abstract

Mesenchymal stem cells from a variety of sites are a natural resource that using appropriate skills can be cultured in the laboratory, in scaffolds, to provide differentiated-cell replacement tissues, for clinical application. To perform such work with human cells, strict ethical integrity must be observed at all stages. Adipocytes, osteocytes and chrondrocytes are amongst the most desirable end-point cells. Hydrolytic degradable scaffolds allow implanted cells to synthesise their own extracellular matrix in situ after implantation, degeneration of the foreign scaffold to temporally match creation of the new innate one. For preliminary in vitro stem cell differentiation protocols, initial investigation is commonly performed with stem cells in commercially available porous collagen sponges or cell-free small intestinal submucosa. Differentiation of stem cells to a specific phenotype is achieved by culturing them in apposite culture media under precise conditions. Once the cells have differentiated, they are checked and characterised in a wide variety of systems. This chapter describes differentiation media for adipocytes, osteocytes, chondrocytes, myocytes and neural precursors and methods of observing their characteristics by microscopy using phase contrast microscopy, standard light microscopy and electron microscopy with tinctorial, immunocytochemical and electron dense stains, respectively. Cell sorting techniques are not dealt with here. Immunocytochemistry/microscopy staining for specific differentiated-cell antigens is an invaluable procedure, and the range of commercially available antibodies is wide. Precautions need to be considered for using actively proliferating cells in vivo, so that implanted cells remain controlled by the body's molecular signals and avoid development of malignancy.

Key words: Mesenchymal stem cells, Scaffolds, Differentiation

1. Introduction

In tissue engineering, cells are cultured and grown in appropriate matrix scaffolds; such constructs can be fashioned into three-dimensional apposite anatomical volumes, shapes and proportions, usable in regenerative medicine as substitutes for diseased or damaged native tissue (1). Stem cells can be cultured and

Ragai R. Mitry and Robin D. Hughes (eds.), *Human Cell Culture Protocols*, Methods in Molecular Biology, vol. 806,
DOI 10.1007/978-1-61779-367-7_27,

differentiated into any number of adult phenotypes, commonly adipocytes (2), osteocytes (3), chondrocytes (4) and many others ("totipotent" stem cells – into all types of differentiated cell, "pluripotent" ones into many types of differentiated cell). There are hosts of difficulties to be overcome, but the aim is clear – to be able to synthesise perfect human replacement tissues and organs.

These days, the aim of a construct is that seeded cells should take over all functions of the tissues they are replacing, and that in time, the foreign scaffold will naturally deteriorate, being replaced as it does so by cells of the construct synthesising their own extracellular matrix. Biocompatibility of the initial construct-scaffold is vital, as this is implanted into the host. The prime property of a scaffold is that its synthesis results in it being riven with interconnecting cell-size passages and sinuses. When applied, populating cells must be able to infiltrate the body of the scaffold fully through these, and once there must be able to proliferate – then stop proliferating when appropriate. Cell growth factors and other stimulatory proteins can be bound to structural surfaces of the scaffold to enhance its overall population. This chapter concerns itself with examples of bioengineered scaffolds and wholly biological surrogates only; there are other types, for example, non-biological prostheses, not dealt with here.

Hydrolytic degradable scaffolds are popular. They provoke little or no immune response in a host and generally cause only minor local inflammatory reactions. Variations of polyglycolic acid (PGLA) polymer substrates presented in mesh formations are very serviceable, initially pioneered as heart valve models by Hoerstrup et al. (5), living autologous constructs provide a stem cells update to that protocol (6). PGLA has also been used with fibres of silk in knitted hybrid nano-microscaffolds (7) and polyethersulphone nanofibre lattices improved epidermal cell infiltration and adherence when used as dressings during wound healing (8). Alternatively, biologically modified tissues such as de-epithelialised connective tissues have enjoyed success. Experimentally, such matrix has been shown to resorb, leaving functionally competent cells and de novo matrix behind (9). In the clinic, autologous stem cells have been used successfully on human-donated natural connective tissue scaffolds for tracheal implantation (10). For initial in vitro work, alginate scaffolds are highly compatible with bone marrow derived stem cells for their successful differentiation into functional hepatocytes (11). Electrospun collagen nanofibres have been developed as interlacing matrices, on which neural stem cells have proliferated and been characterised (12) and also with neural prostheses, electrospun nanocomposites based on poly(epsilon-caprolactone) have shown cytocompatibility (13). Collagen sponges crosslinked or non-crosslinked with diphenylphosphoryl azide are indispensable as "starter" matrices (14), while deriving new protocols for stem cell differentiation, but these tend to be friable, thus of little use for transplantation. Also in this category fall three dimensional microcarriers used to support and up-scale human bone marrow-derived mesenchymal stem cells during their differentiation into adipogenic and osteogenic lineages (15).

In our laboratory, we use a commercially available porous matrix made from animal small intestinal submucosa (Cook Biotech Inc., IN, USA) and porous non-crosslinked collagen porcine sponges (Perous SARL, Lyon, France). There are very many types of scaffold available commercially. We apply appropriate cells to the chosen construct and incubate for as long as is necessary for the cells to infiltrate the scaffold and proliferate successfully there. It is stressed that experimental components are not necessarily suitable for transplantation into humans; there are immunological and safety prohibitions of performing cross-species transplantation. Some extrinsically synthesized tissue surrogates have (at least partially) animal-derived components; collagen from animal sources is a major part of many tissue-engineered constructs. Improvements in synthesis of recombinant collagen mean that a pure and safe source of such human material can be produced from molecular first principles.

To obtain human stem cells for scientific research, precise ethical requirements must be adhered to; then the cells need to be differentiated in the laboratory in vitro. There are various sources of human stem cells, embryonic ones having been the first to be described and challenged (16). However, at present sampling the inner cell mass of a human blastocyst results in death of the embryo, this is a serious ethical impediment. Many laboratories work with human stem cells (17), fulfilling all regional and ethical requirements (18, 19), but laws governing obtaining the cells are different in different countries. The most usual source of human embryonic stem cells is that obtained after in vitro fertilisation therapy, when there may be embryos in excess of the number necessary for a successful implantation.

The next most common source of human stem cells for use in the laboratory is from cell populations of the bone marrow; this was first shown by Friedenstein et al. (20). Again, all ethical approval must be obtained and conditions complied with, which in essence in this case means that volunteers accept to undergo bone marrow aspiration to provide the cells. They sign appropriate waiver forms indicating that the precise purpose of the cells has been explained to them, they understand, and that they freely desire to provide the donation. There are two types of bone marrow stem cell: haematopoietic stem cells (HSCs) (from which all blood cells originate) and mesenchymal stem cells (MSCs), which have the property of being able to differentiate into a wide variety of mature non-haematopoietic cells. Further valuable sources of human stem cells are umbilical cord/blood and placental stem cells (21, 22) and residual tissue from elective breast-reduction surgery (23) and liposuction. There are others, but in our laboratory we have used bone marrow- and peripheral blood-derived MSCs originally obtained from Thalmeier et al. (24) or freshly derived MSC from abdominal adipose-tissue taken during liposuction operations, as described by Quirici et al. (25).

Once obtained, bone marrow-derived stem cells need to be selected (in our case, and as described below, separating the MSC component from the more numerous HSCs of the bone marrow). The next step is to expand their number by repeated culturing, splitting cultures and re-culture in the laboratory. In tissue engineering, substantial cell numbers need to be available to successfully populate any matrix to be a sustainable replacement proto-organ in regenerative medicine.

Differentiation of MSCs to a desired end depends on the conditions in which they are cultured, most commonly meaning the components of their various specific differentiation media. Once the desired cell cultures have been appropriately grown for the correct times, resultant cells need to be characterised – to demonstrate, and be certain of their final phenotypes. Some of the methods to do this are by microscopy (phase contrast, light microscopy and electron microscopy), first to observe the cells' distinctive features, then to test their veracity by tinctorial (histochemical) methods and specific immunocytochemical staining, with appropriate antibodies.

2. Materials

Scaffolds need to chosen carefully. If protocols for differentiation of the stem cells are still being worked-up, porous non-crosslinked or crosslinked collagen sponges are advised. Subsequently, more adventurous matrices can be explored. Chemicals, media and additives need to be of the highest analytical purity and quality available.

2.1. Isolation of MSCs

(i) Bone Marrow

1. Ice-chilled phosphate buffered saline (PBS; NaCl 9 g/l, Na_2HPO_4 1.15 g/l, KH_2PO_4 0.2 g/l; pH 7.4).
2. Syringes (BD Biosciences, Oxford UK).
3. 21-, 23- and 25-Gauge needles (BD Biosciences, Oxford, UK).
4. 30 ml Universal containers (Sterilin, Caerphilly, Wales).
5. 40 μm Mesh filters (BD Falcon, Oxford, UK).
6. Swing-out bench top centrifuge for 30 ml universals (MSE, London, UK or equivalent).

7. Culture medium (see below).

(ii) Adipose Tissue

8. Collagenase, 3-times crystallized (Worthington Biochemicals Corp, Lakewood, NJ, USA).

2.2. Culture of MSCs

1. Base medium: 60% Low glucose (1 g/l) Dulbecco's modified Eagle's medium (DMEM, Gibco, Invitrogen, Paisley, UK), 40% MCDB201 (Sigma, 1.77 g/100 ml). αMEM (Invitrogen) also cultures MSC well. Special MSC media are available, such as Mesencult with its MSC Stimulatory Supplement (Stem Cell Technologies, Grenoble, France), which works well but is more expensive.
2. Foetal calf serum (FCS) (Seralab, Haywood's Heath, UK; use at 20% in medium).
3. Penicillin and streptomycin, 100× stocks stored frozen. Renew every week (Sigma; final concentrations 100 U/ml, 0.1 mg/ml respectively).
4. pH meter (Mettler Toledo, Leicester, UK).
5. 8 M NaOH (BDH Chemicals, Poole, UK; Analar grade or equivalent).
6. 0.2 μm sterilising filters (Nalge-Nunc, Roskilde, Denmark or Millipore, Watford, UK).
7. Leukaemia inhibitory factor (LIF) (Millipore, Watford, UK; final concentration 1,000 U/ml) (26).
8. Platelet derived growth factor – b chain homodimer (PDGF-BB) (R&D Systems, Minneapolis, MN, USA; final concentration 10 ng/ml).
9. Epidermal growth factor (EGF) (Sigma; final concentration 10 ng/ml).
10. Plasma fibronectin (Sigma)-coated surfaces (20 μg/ml for 30 min).
11. Collagenase type 4 (lowest trypsin variety: Worthington Biochemicals, Lakewood, NJ, USA).
12. Tissue culture flasks (BD Falcon).
13. Incubator to provide humidified CO_2 atmosphere (Heraeus; Thermo Scientific, Basingstoke, UK).
14. 0.25% Trypsin (3× crystallised; Worthington Biochemicals).
15. 1 mM Ethylenediaminetetraacetic acid diSodium salt (EDTA; Sigma).
16. Accutase (Millipore).

2.3. MSC Differentiation Cocktails

Table 1 shows components of typical cocktails to stimulate differentiation into adipocytes, osteocytes, chondrocytes, myocytes and neural cells.

Table 1
Components of typical cocktails to stimulate differentiation into adipocytes, osteocytes, chondrocytes, myocytes and neural cells

Cell type	Reagent	Supplier	Final concentration
Adipocytic	Hydrocortisone	EMD Chemicals, San Diego, USA	0.5 μM
	Isobutyl methyl xanthine	Sigma, Poole, UK	0.2 mM
	Indomethacin	Sigma	60 μM
	Insulin	Sigma	5 μg/ml
Osteocytic	Dexamethasone	Sigma	10nM
	Vitamin C phosphate	Sigma	0.2 mM
	Na β-glycerophosphate	Sigma	10 mM
Chondrocytic	FGF2	Sigma/R&D Systems, Minneapolis, MN, USA	1 ng/ml
	TGFβ1	Oncogene, Siemens Healthcare, Cambridge, MA, USA	5 ng/ml
Myocytic	5′-Azacytidine	Sigma	3 μM
	Vit C phosphate	Sigma	0.1 mM
Neurogenic	Butylated hydroxyanisole	Sigma	200 μM
Medium A	KCl	BDH Chemicals, Poole, UK	5 mM
	Valproic acid	Sigma	2 μM
	Forskolin	Sigma	10 μM
	Hydrocortisone	EMD Chemicals	1 μM
	Insulin	Sigma	5 μg/ml
Neurogenic	EGF	Sigma	10 ng/ml
Medium B	FGF2	Sigma/R&D	20 ng/ml

2.4. Phenotyping Differentiated Adherent MSC, Including Tinctorial Stains

1. Glass 8-chambered microscope slides or 24-well plates (BD Falcon).
2. Plastic 8-chambered microscope slides (Permanox Labtek, Nalge-Nunc, Thermo Fisher, Roskilde, Denmark).
3. 4-Well plates (Nalge-Nunc) or 24-well plates (Falcon; BD Biosciences).
4. 40% Formaldehyde solution (BDH): To use as neutral buffered formalin (NBF) fixative, dilute 1:10 in sodium phosphate buffer pH 7.4 (NaH_2PO_4 4.0 g/l, Na_2HPO_4 (anhydrous) 6.5 g/l]
5. Paraformaldehyde (Sigma): 4% in PBS, use fresh, or freshly thawed from −20°C.

2.4.1. Adipocytic Differentiation

1. 60% Isopropanol (BDH Chemicals, Poole, UK).
2. 1% Oil red O (ORO) in 60% isopropanol.
3. Harris' haematoxylin (RA Lamb, part of Thermo-Fisher).

2.4.2. Osteocytic Differentiation

1. Alizarin red S (Sigma; 2% in distilled water; adjust pH to 4.2 with 10% ammonium hydroxide; stable for 1 month).
2. Alternative protocol: von Kossa stain for calcium salts. 1% Aqueous silver nitrate (Sigma).
3. 5% Aqueous sodium thiosulphate (Sigma).
4. van Gieson counterstain (RA Lamb).

2.4.3. Chondrocytic Differentiation

1. To embed pellet cultures: 1% Agarose in PBS (Sigma; to dissolve, heat to boiling in microwave with loose cap to vessel. Beware! Super-heated liquids can froth suddenly. Use just before gelling, at around 40°C).
2. Alcian blue 8GX stain (Sigma; 1 in 3% acetic acid, pH 2.5).
3. 1% Aqueous neutral red nuclear counterstain. Filter to use.
4. Alternative stain: Safranin O (RA Lamb; 0.1% in distilled water).
5. 0.02% Aqueous Fast Green (RA Lamb).
6. Wiegert's iron haematoxylin (RA Lamb).
7. 1% Glacial acetic acid (BDH, Poole, UK) in distilled water.

2.5. Immunohistochemical Analysis

1. Antibodies for some specific antigens likely to be expressed by differentiated MSC are shown in Table 2. Those shown for HSC should be negative in MSC cultures.
2. 4% Paraformaldehyde (see Subheading 2.4.5).
3. 30% H_2O_2 (BDH).
4. Foetal calf serum (FCS; Seralab, Haywards Heath, UK; use as antibody block and diluent at 5% in PBS).
5. Biotinylated rabbit anti-mouse IgG antibody (Dako, Glostrup, Denmark; 1:300).
6. Biotinylated swine anti-rabbit IgG antibody (Dako; 1:500).
7. Streptavidin-peroxidase (Dako; 1:500).
8. Diaminobenzidine (DAB) Chromogen kit (Dako).
9. Harris' haematoxylin counterstain (RA Lamb, Thermo Scientific).
10. Coverslips (RA Lamb).
11. Glycergel mounting medium (Dako).
12. DePeX mounting medium (RA Lamb).

Table 2
Antibodies used to immunostain differentiated cells

Cell type	Antigen	Supplier	Dilution
MSC	CD71	Dako	1:100
	CD105	Dako	1:100
	NGF-R (p75)	Chemicon/Millipore, Watford, UK	1:100
	STRO1	Developmental Studies Hybridoma Bank, Iowa, USA	1:100
HSC	CD34	Dako	1:25
	CD45	Dako	1:100
	CD133	BD Pharmingen, Oxford, UK	1:100
Muscle	MyoD	Abcam, Cambridge, UK	1:200
	Myosin Heavy Chain	Dako	1:10
	Troponin C	Abcam	1:5,000
Neuron	Gap43	Sigma	1:1,000
	NeuN	Chemicon/Millipore	1:100
	Nestin	Chemicon/Millipore	1:50
	Neurofilament 200	EuroPath, Bude, UK	1:50
Cytoplasmic filaments	α-Smooth muscle actin	Sigma	1:4,000
	Vimentin	Dako	1:200
	Desmin	Dako	1:100
	Glial fibrillar acidic protein (rabbit)	Dako	1:100

Antibodies are mouse monoclonals except where indicated

2.6. Electron Microscopy

1. 2% Glutaraldehyde (Sigma-Aldrich, St Louis, MO, USA).
2. Phosphate buffered saline solution buffer (PBS).
3. Range of graded alcohols.
4. Osmium tetroxide (Huntsman Advanced Materials, Basel, Switzerland).
5. Glass microscope slides and coverslips.
6. Toluidine blue O stain (RA Lamb; 1 in 2% aqueous sodium borate; 0.2-μm filter before use).
7. Araldite (Huntsman Advanced Materials, Basel, Switzerland).
8. Copper electron microscope grids (Ted Pella Inc., Redding, CA, USA).
9. Uranyl acetate (Structure Probe Inc. Chem™, West Chester, PA, USA).
10. Lead citrate (Structure Probe Inc. Chem™, West Chester, PA, USA).
11. Ultramicrotome [such as Reichert-Jung (Leica) Ultracut, ISS Group, Manchester, UK].
12. Electron microscope, such as Philips 100 (Philips, Eindhoven, Netherlands).

3. Methods

Isolation of MSCs was first shown in vitro by Friedenstein et al. (20). This group derived fibroblast-like cells after adherence of bone marrow-derived cells to plastic tissue culture plates. The cells were subsequently found to differentiate into adipocytes and osteoblasts. In our laboratory we use a modification of that method, as described by Lennon et al. (27). We apply the stem cells to the chosen scaffold, submerged in the correct differentiation culture medium. The construct is then incubated at 37°C in a closed incubator for days or weeks (under observation and fresh medium supplementation) for as long as is necessary for the cells to fully populate the scaffold and differentiate appropriately.

To characterise cells' differentiation, samples from cultures must be taken for microscopy and the various types of staining. In tissue engineering of proto-tissue, live cells are applied to the chosen scaffold and are allowed to infiltrate and proliferate. Collagen sponges are a good starting point for stem cell population; cells/sponge matrix can be treated in the same way as natural tissue. Culture medium is washed off and the construct is fixed and embedded in paraffin wax (for light microscopy, LM) or resin (for electron microscopy, EM). Sections are cut and stained with haematoxylin and eosin (H&E) for LM and uranyl acetate and lead citrate for EM. After these stages, it is ascertained that the cells have behaved as expected; that is to say, by the use of markers of proliferation, tinctorial staining and immunocytochemistry on the tissue sections.

We use tinctorial stains first to characterise differentiation lineages of some of the resultant cells (much cheaper than immunocytochemistry – which then follows). The ability of adipocytes to take up the lipophilic dye ORO is definitive, and for osteocytes, extracellular calcium deposits stain with alizarin red or the von Kossa stain; Safranin O or Alcian blue staining can be used for the extracellular matrix produced by chondrocytes.

By immunocytochemistry, we stain undifferentiated cells for the presence of subsets of specific antigens, appropriate markers of MSCs (and also for HSC lineages, for original sorting), using commercial antibodies as shown in Table 2, using the well known three-layer method. A fuller list of MSC markers is given in Otto and Rao (28). After appropriate blocking of tissue endogenous peroxidase (common in many cell types), the first layer is the selected specific primary antibody; this is applied to the cells at appropriate dilution, and incubated. The antibody binds to its specific antigen on the cell, but at this stage is not visible. After washing, the biotinylated second layer antibody is applied and incubated; this amplifies the signal from primary antibody staining. Next comes horseradish peroxidase-conjugated streptavidin, the third and final layer; when diaminobenzidine is added with peroxide catalyst a brown precipitate is deposited, which visualises the location of the secondary antibody, bound as it is at the location of the primary antibody/antigen of interest.

There are very many more antigens of interest used to characterise differentiated stem cells. For example, human embryonic stem (ES) cells express stage-specific embryonic antigens SSEA-3 and SSEA-4; CD45 or CD34 are typically expressed by HSCs, and ES-derived endothelial cells express endothelial markers including CD31, Flk-1, VEGF, VE-cadherin, PECAM, Tie-1, and Tie-2. Human bone marrow-derived mesenchymal cells are $CD34^{-ve}$ and $CD45^{-ve}$, yet $CD73^{+ve}$, $CD105^{+ve}$ and express STRO-1. Other cell types that stain best with immunocytochemistry include myocytes, using specific antibodies, such as anti-MyoD or myosin heavy chain, or, for neuronal cells, antibodies against Gap43 or NeuN. The list is considerable – antibodies to the antigens are widely commercially available and alternative antibodies to each may be compared online (http://www.biocompare.com).

3.1. Isolation of Mesenchymal Stem Cells

MSC may be obtained from consenting donors undergoing procedures to obtain bone marrow for rescue of leukaemia patients, from abdominal adipose tissue of patients electing for liposuction or from elective breast reduction. All procedures should be authorised well in advance by the institute's or hospital's local Research Ethics Committees as appropriate.

3.1.1. Bone Marrow-Derived MSC

1. Bone marrow is aspirated from donors under anaesthesia by surgery, using 20- to 50-ml sterile heparinised syringe and a 10-G to 15-G wide bore needle. Several sites may be sampled to obtain sufficient cells for transplantation, and any surplus is available for in vitro study. Samples are kept on ice until processed, as outlined below, beneath a sterile flow hood in a cell culture suite.
2. Pooled BMCs are then triturated by pipette in an excess of cold PBS and further dispersed from a syringe by gently passing through 21- and 23-gauge needles, which breaks up any clumps. Frothing and aerosol production are to be avoided.
3. The cell suspension is sieved through a 40-μm mesh filter (BD Falcon, Oxford, UK). It is useful to perform this by holding the filter with artery clamps over a small beaker, since the diameter of the filters is a tight fit in most commercial 50-ml poylpropylene conical tubes (BD Falcon), which can make flow of fluid into the tube difficult.
4. Cell suspensions are then centrifuged at $400\times g$ for 5 min, and supernatant is discarded. The preparation is further spin-washed in 10–20 ml PBS.
5. Pellets are resuspended in sterile culture medium and a small aliquot is counted using a haemacytometer, before plating into culture dishes as described below.

3.1.2. Adipose Tissue MSC

1. Lipoaspirates obtained from the operating theatre should be processed to cell suspensions as soon as possible, in the culture laboratory in a sterile culture flow hood.
2. The adipose tissue is washed three times in PBS, minced with sterile scissors and then incubated in freshly made 1 mg/ml sterile collagenase in serum-free medium. Incubations are for 30–60 min at 37°C in a shaking water bath at around 40–60 cycles/min.
3. Cells released by collagenase are collected after sieving through a 40-μm mesh, as above, and spin-washed twice in PBS to remove collagenase.
4. Pellets of cells are then resuspended in full culture medium of choice and plated at around 5×10^4 to $10^5/\mathrm{cm}^2$ in vessels of choice depending on yield. At this point it may be suitable to send cells for FACS immunophenotyping or sorting specific sub-populations; MSC can be enriched considerably by separation of NGF-R, Stro-1, CD105 positive cells for example.

3.2. Culture of MSCs

1. All cell populations must be tested for mycoplasmas and found to be negative (see Note 1).
2. MCDB-201 is dissolved at 1.77 g/100 ml and brought to pH 7.4 with 8 M NaOH, and must be sterilized using a 0.2-μm filter. For initial cell expansion, LIF, platelet derived growth factor – b chain homodimer (PDGF-BB, 10 ng/ml) and epidermal growth factor (EGF, 10 ng/ml) can be added (27). We have found that the main active agent here is PDGF-BB. Others recommend FGF2 and TGF-β in combination (29).
3. BMC isolates can be plated on plasma fibronectin-coated surfaces: a solution at 20 μg/ml is incubated for 30 min in the culture vessels and then air-dried in the flow hood. Cells are then plated at a density of 5×10^4 to 10^5 cells/cm^2 and incubated at 37°C in a humidified atmosphere containing 8–10% CO_2. Other basement membrane components are suitable as pro-proliferative substrates, with the exception of laminin-5 for chondrogenesis (30).
4. Non-adherent cells (largely HSCs in BM preparations) are removed after 3 days by washing with PBS, and remaining cells are refed every 3 days!.
5. At 70–80% confluence, adherent mesenchymal cells are passaged and sub-cultured at a split-ratio of 1:3 using 2.5 mg/ml trypsin in PBS containing 0.5 mM ethylenediaminetetraacetic acid (EDTA; whole buffer also known as "trypsin-versene").
6. At this stage, portions of cell preparations should be characterized for their ability to undergo chondrogenic, osteogenic and adipogenic differentiation, and 8-chambered culture slides can be inoculated with cells for immunophenotyping (see Note 2).

7. Any promising parent cell line can then be single-cell cloned using the same growth conditions. Cell suspensions are diluted to three cells/ml and 100 μl/well placed into a 96-well plate (0.3 cells/well). Individual wells are inspected for presence of single cells and the lid marked accordingly (see Note 3).
8. Successful clones are grown up by serial sub-culturing in bulk to larger vessels, until enough cells exist to test for lineage differentiation or be chosen for further study. It is useful to choose several 10 to 12 clones from each parent line as their differentiation potentials and characteristics can vary.

3.3. Differentiation of Adherent BMCs

1. Cells are usually grown to subconfluence in glass or plastic 8-chambered culture-treated microscope slides or 4- or 24-well plates, then switched to the appropriate cocktails for each differentiation assay (Table 1; see Note 4).
2. Differentiation additives are conveniently dispensed from sterile 1,000× stock solutions, but an exception is β-glycerophosphate used for osteogenic cultures. Final concentration used (10 mM) is too high for a 1,000× stock and we make a 1 M (100×) stock solution. If this constitutes too great a change in final volume, an alternative is to prepare a small aliquot of medium with freshly dissolved salt at the final concentration (3.06 mg/ml for sodium β-glycerophosphate, $5H_2O$ salt), which is sterilised by 0.2-μm filtration.
3. In our method, cells for chondrocytic differentiation are also cultured as pellets (100×*g* for 5 min) of 50,000 cells in 1 ml medium plus additives, for up to 1 month. Such cultures need to be harvested by rinsing in PBS, fixation in fresh PFA for 20 min, rinsing in PBS again and then embedding in 1% agarose/PBS to create a larger specimen, which will not be lost during paraffin embedding. Phenotyping is then performed on 5–6-μm sections cut through the pellets on to microscope slides (see Note 5).

3.4. Adipocyte Differentiation Assay

1. Fix cells in freshly thawed 4% PFA in PBS for 20 min.
2. Wash them in PBS and equilibrate in 60% isopropanol.
3. Incubate cultures in ORO (1 in 60% isopropanol) for 15 min.
4. Rinse quickly in 60% isopropanol for 5 s.
5. Wash cells in distilled water.
6. Counterstain in haematoxylin for 30 s to visualise nuclei.
7. Rinse specimens gently in tap water to blue the nuclei.
8. Mount preparations in glycergel. Lipid-containing droplets stain bright red, with nuclei being blue.

3.5. Chondrocytic Differentiation Assay

1. Fix pellet culture samples in 4% PFA in PBS for 30 min. For immediate staining of monolayer cultures, rinse gently in PBS and go to step 6.
2. Embed in 1% agarose in PBS, just above gelling temperature, about 40°C.
3. Dehydrate in ten volumes of 70% ethanol for 2 h. Samples can be stored in 70% ethanol.
4. Embed in paraffin wax for sectioning.
5. Dewax sections in xylene and bring to water via a series of graded ethanol solutions (100, 90, 70, and 50%) 1 min each.
6. Incubate sections in 0.1% aqueous Safranin O for 5 min.
7. Dehydrate through a series of graded alcohols (reverse of above step 5 sequence).
8. Air-dry.
9. Mount in glycergel. Cartilaginous extracellular matrix turns red.
10. Stain further sections by the diastase-periodic acid-Schiff (PAS) reaction, for carbohydrates other than glycogen. The matrix will stain pink here.
11. Stain further sections in Alcian blue stain at pH 2.5 for acid mucopolysaccharides.

3.6. Osteocytic Differentiation

1. Fix cells in 4% PFA as above
2. Stain in 2% aqueous alizarin red S, for extracellular calcium, pH 4.2, for 3 min. pH is important here.
3. Remove excess dye by rinsing in distilled water (see Note 6).
4. Cultures on plastic slides may be mounted in Glycergel. Glass slide cultures can be dehydrated in ethanol, rinsed 3 times in xylene and mounted in DPX.
5. Alternative stain: von Kossa, for extracellular calcium.
6. Cells are fixed as step 1 above.
7. Rinse three times in distilled water.
8. Incubate in fresh aqueous 1% silver nitrate (Sigma) for 10–60 min under strong light – sunlight or quartz-halogen illumination is good for this. It is useful to monitor the reaction under the microscope and stop at an appropriate time.
9. Wash in distilled water three times.
10. Fix in 2.5% sodium thiosulphate (Sigma; "hypo") for 5 min.
11. Wash in tap water three times.
12. Counterstain for 30–60 s in 0.1% nuclear fast red/5% aluminium sulphate, rinse in tap water, and mount in Glycergel (if using cell culture plastic), or dehydrate in 100% ethanol, air-dry and mount in DPX (glass slides). The extracellular calcium deposits are black, nuclei red.

3.7. Myocytic Differentiation

1. Incubate cultured MSC monolayers for 2 weeks up to 3 months to ascertain muscle cell differentiation potentiality, either in the demethylation agent 5′-azacytidine (ref. 31; 3 μM) or in vitamin C phosphate-containing (ref. 32; 100 μM) media (see Note 7).
2. Perform immunohistochemical analysis as below to determine whether stimulated cells express troponin C, a cardiac-specific protein; MyoD, a developmental transcription factor regulator of skeletal muscle differentiation; or myosin heavy chain, a marker of mature skeletal myocytes.

3.8. Neurogenic Differentiation

1. Grow MSC in neurogenic medium for up to 2 weeks.
2. Rinse cultures with PBS and fix cells in 4% paraformaldehyde (PFA) for 20 min, rinse again in PBS.
3. Perform 3-layer immunocytochemistry (IHC) as described below, with antibodies to appropriate neuronal markers, as described in Table 2 (see Note 8).

3.9. Antibody Staining for Cell Type Characterisation

Note: All steps are carried out at room temperature (21–25°C) in a humid environment to prevent drying of the samples, which would cause non-specific binding of the reagents and raise background signals. High specificity antibodies can be incubated for less than 60 min, and it is good practice to test the minimum time needed for specific signals without background.

1. The well known three-layer method is used starting with cultures being fixed in 4% PFA in PBS for 20 min.
2. Wash cells in PBS.
3. Block against activity of endogenous peroxidase by using 0.3% H_2O_2 in PBS for 30 min.
4. Wash again in PBS.
5. Normal serum blocking step with 5% FCS, or pre-immune serum of the second layer host animal, in PBS for 30 min.
6. All antibodies are diluted in the step 5 blocking solution and applied for 60 min.
7. Three PBS washes of 5 min.
8. Detection of mouse monoclonal antibodies is by using biotinylated rabbit anti-mouse antibodies (Dako; 1:300). Rabbit polyclonal antibodies are detected using biotinylated swine anti-rabbit antibodies (Dako; 1:500). Both are incubated for 60 min.
9. Three PBS washes of 5 min.
10. Application of streptavidin-peroxidase (Dako; 1:500) for 60 min.

11. Sites of antibody binding are detected using diaminobenzidine, 0.5 mg/ml in PBS in the presence of 0.03% H_2O_2 (see Note 9).
12. Counterstain in Harris' haematoxylin (RA Lamb) for 30 s, blue nuclei in tap water for 10 s.
13. Coverslip slide preparations under Glycergel (Dako).

3.10. Electron Microscopy (see Note 10)

1. Collect cells and constructs of all components of the study from culture environments by trypsinising cultures in the normal way. Wash in buffer, spin cells at 100–200 × *g*, and aspirate supernatant.
2. Fix in 2% glutaraldehyde in PBS for 1 h. Wash 2× in PBS, spin cells at 100–200 × *g*, aspirate supernatant.
3. Osmicate 1% osmium tetroxide in PBS, 1 h. Wash 2× in PBS, spin cells at 100–200 × *g*, aspirate supernatant.
4. Dehydrate through a series of graded alcohols, always spinning cells at 100–200 × *g*, and removing supernatant.
5. Embed preparations in liquid Araldite (Huntsman Advanced Materials, Basel, Switzerland).
6. Polymerize at 60°C overnight.
7. When the blocks have cooled, on the ultramicrotome cut 1 μm semi-thin sections for sampling at the light microscope level (see Note 11).
8. Stain these with 1% aqueous toluidine blue in 2% sodium borate for 1–2 min to select areas for ultrathin sections. The main block is trimmed down to the area of interest.
9. Then cut 100-nm sections and collect them on copper electron microscope grids.
10. Sections are stained with 5% aqueous uranyl acetate for 15 min and Reynolds lead citrate for 5 min. See Note 12 for formula for lead citrate solution.
11. View on an electron microscope, for example, a Philips 100 apparatus (Eindhoven, Netherlands).

4. Conclusions

Serious consideration needs to be taken on whether grafts populated with the differentiated stem cells would be safe for transplantation, and there are justified concerns with reference to the extent and quality of proliferation of seeded cells. In developing such grafts, stem cells should not be allowed to reach the "Hayflick" limit to their proliferative potential when they begin to senesce and may pass on deleterious genetic mutations to their progeny. It may be feared that stem cells could have phenotypic, molecular and/or mechanistic traits similar to those of malignant cells. Whether the undifferentiated state of the stem cell might render it vulnerable to neoplastic changes or whether deregulation of normally tightly controlled molecular signals that govern cell proliferation may initiate neoplasia, remains to be seen.

5. Notes

1. Mycoplasma detection: It is useful to set up a separate culture flask for this purpose as soon as possible after cell harvest. Reliable kits are available, which identify most common strains quickly by PCR, such as MycoSensor (Agilent Technologies, Santa Clara, CA, USA), or by 16S RNA oligonucleotide plate assay (Mycoprobe, R&D Systems, Minneapolis, MN, USA).
2. At the stage of expanding numbers of cells originally obtained, it is advised to always freeze down aliquots of cells. Even in the best laboratories, accidents happen, and one would not want to lose all cultures after months of work at selecting, expanding and differentiating very precious cells. Cells surplus to immediate requirements can be cryopreserved in screw-cap or heal-sealed (safer) 1.5-ml vials growth medium containing 10% FCS and 10% DMSO. It is usual to place 10^6 to 10^7 cells/vial in a well-insulated box filled with cotton wool at –80°C overnight before transfer to liquid nitrogen for long-term storage.
3. Some authors recommend cloning in the presence of similar (the parent line or equivalent) feeder cells, which have been lethally irradiated (10 Gy), or treated with the anti-proliferative agent mitomycin C (100 μg/ml) for 60 min. Both methods have disadvantages. Irradiation necessitates that the wells with single cells are filled with other cells that could contaminate the clone and certainly mask the identity of growing cloned cells until transferred to other vessels. Mitomycin C could leach out of dead or dying treated feeder cells and poison the cloned cell(s).
4. Culture vessels for differentiation assays: Users are advised that plastic 8-chamber slides are somewhat prone to distortion, which can cause the gasket to leak between wells, which subverts future immunohistochemistry or other assays dependent on the integrity of each single well. Glass types do not suffer from this problem, but cell biology on glass can be quite different from on plastic, notably adherence and cell behaviour, so that each type has its drawbacks. 4-Well plates are convenient since a control well can accompany one each of adipogenic, osteogenic and chondrogenic differentiation in one vessel, which is useful when time-course assays are in progress. These provide

a two- to threefold larger surface area for analysis than the 8-chambered slides, at the price of needing more cells to inoculate them and more medium to feed them.

5. When using monolayer chondrocytic cultures, the cell population grows to confluence and then increasingly begins to secrete extracellular matrix. The cells often then detach from the culture base, specially if on a glass substrate, and preferentially adhere to the matrix and remodel it by contraction. It thus becomes both harder to visualise the cells by phase-contrast microscopy, and easier to lose them when removing medium at refeeding. It is also difficult to stain them by IHC or histochemically without losing them during the many steps each procedure entails. Such cultures resemble pellet cultures above and can be treated accordingly if needed. A simple toluidine blue stain after fixation may indicate degree of differentiation the cells are undergoing, though this is not particularly specific.
6. Alizarin red may be quantified by destaining cultures from step 3 in 5% perchloric acid for 15 min and reading absorbance of the solutions in a 96-well plate spectrophotometer at 490 nm. A standard curve of alizarin red, double-diluted from 0.1%, may be used to estimate amounts of dye that has bound (33).
7. 5′-Azacytidine treatment results in wholesale and random genome demethylation and is non-specific in nature. As such it is not a true indicator of differentiation potential of putative MSC as the reagent may be applied to other non-stem cell populations such as fibroblasts, with similar results. Clones of muscle cells so derived may be useful for experimental purposes, but are not suitable for consideration in transplantation for clinical use. The use of vitamin C phosphate would be more useful in this regard. In our experience, degree of muscle differentiation is usually low: 1 in 10^4 cells or less, which may reflect the prior degree of lineage restriction that has occurred in bone-marrow cell populations.
8. Note that neuronal *functional* proteins such as GAP43 or TRPV1 are more persuasive of differentiation into neurons than morphology or expression of GFAP, an intermediate filament often present in MSC cells before stimulus. More persuasive still is a functional stimulus, such as that provided by capsaicin on intracellular calcium signalling, which can be measured using real-time fluorescence imaging in preloaded cells (34).
9. Pre-made DAB kits (Dako) are useful to avoid handling this potential carcinogen. Peroxidase reacts quickly, so 5 min is a useful maximum incubation time, but this may be complete in 1 min if a highly expressed antigen is being detected (say cytokeratin or other intermediate filament). See Lunn and Lawler (35) for special considerations on disposal of spent toxic DAB solution.

10. Electron microscopy (EM) is not for the uninitiated, and notes below are at a level of help for the already skilled electron microscopist/technician. Each cell group must be cultured in as many duplicate dishes as necessary to provide 10^6 cells per cohort for further processing. There are many treatments and washings along the way for EM, and although the greatest care must be taken to get the cells through, it is inevitable that some will be lost. The final cell pellet must be big enough to handle and have sections cut from it.
11. At the 1 μm section stage, sample sections collected on glass slides are stained with toluidine blue. For staining paraffin wax-embedded material, haematoxylin and eosin would always have been of choice; however, these stains only very weakly react with resin-embedded sections. Toluidine blue stains them perfectly well, but of course all the components of the cells are thus stained in various shades of blue, rather than the typical blue nuclei/pink cytoplasm of H&E. The operator just gets used to it.
12. To prepare 50 ml Reynolds lead citrate for EM staining, add these chemicals to distilled water in following order: Lead nitrate 1.33 g, distilled water 30 ml. Add sodium citrate, dihydrate 1.76 g. The solution becomes cloudy when sodium citrate is added. 1 N NaOH 5 ml: the solution becomes clear when NaOH is added. Stir for 10 min to dissolve and make up to 50 ml with an additional 15 ml of distilled water. Store solution for 3–6 months at 4°C. Note: the mount of NaOH is very important. Too much will make solution cloudy (modified from IHCWorld; See Additional Reading).

Acknowledgements

The authors would like to thank George Elia (CRUK, London) for exemplary histochemical expertise, Dr Uma Anand (Imperial College, London) for advice on neuronal cell culture, Professor Ketan Patel and Dr Anthony Otto (Reading University) for muscle cell growth conditions.

References

1. Sarraf, C.E., Harris, A.B., McCulloch, A.D. and Eastwood, M. (2002) Tissue engineering of biological cardiovascular system surrogates. *Heart Lung Circ.* **11**, 142–150.
2. Du, M., Yin, J. and, Zhu, M.J. (2010) Cellular signaling pathways regulating the initial stage of adipogenesis and marbling of skeletal muscle. *Meat Sci.* **86**, 103–109.
3. Owen, M.E., Cavé, J. and Joiner, C.J. (1987) Clonal analysis in vitro of osteogenic differentiation of marrow CFU-F. *J Cell Sci.* **87**, 731–738.

4. Gazit, D., Karmish, M., Holzman, L. and Bab, I. (1990) Regenerating marrow induces systemic increase in osteo- and chondrogenesis. *Endocrinology* **126**, 2607–2613.
5. Hoerstrup, S.P., Sodian, R., Daebritz, S., Wang, J., Bacha, E.A., Martin, D.P. et al. (2000) Functional living tri-leaflet heart valves grown in vitro. *Circulation* **102** (3), 11144–11149.
6. Schmidt, D., Dijkman, P.E., Driessen-Mol, A., Stenger, R., Mariani, C., Puolakka, A. et al (2010) Minimally-invasive implantation of living tissue engineered heart valves: a comprehensive approach from autologous vascular cells to stem cells. *J. Am. Coll. Cardiol.* **56**, 510–520.
7. Sahoo, S., Lok Toh, S. and Hong Goh, J.C. (2010) PLGA nanofiber-coated silk microfibrous scaffold for connective tissue engineering. *J. Biomed. Mater. Res. B Appl. Biomater.* **95**(1), 19–28.
8. Babaeijandaghi, F., Shabani, I., Seyedjafari, E., Safaei Naraghi, Z., Vasei, M. and Haddadi-Asi, V. et al. (2010) Accelerated epidermial regerneration and improved dermal reconstruction achieved by polyethersulfone nanofibres. *Tissue Engineering Part A.* **16**(11), 3527–3536.
9. Matheny, R.G., Hutchison, M.L., Dryden, P.E., Hiles, M.D., Shaar, C.J. (2000) Porcine small intestine submucosa as a pulmonary valve leaflet substitute. *J. Heart Valve Dis.* **9**, 769–774.
10. Bader, A. and Macchiarini, P. (2010) Moving towards in situ tracheal regeneration: the bionic tissue engineered transplantation approach. *J. Cell Mol. Med.* **14**(7), 1877–1889.
11. Lin, N., Lin, J., Liu, B., Weidong, P., Chen, S., Xu, R. (2010) Differentiation of bone marrow-derived mesenchymal stem cells into hepatocye-like cells in an alginate scaffold. *Cell Proliferation* **43**(5), 427–434.
12. Liu, T., Teng, W.K., Chan, B.P., Chew, S.Y. (2010)Photochemical crosslinked electrospun collagen nanofibers: synthesis, characterisation and neural stem cell interactions. *J. Biomed. Mater. Res A.* **95**(1), 276–282.
13. Bianco, A., Del Gaudio, C., Baiguera, S., Armentano, I, Bertarelli C., and Dottori, M. (2010) Microstructure and cytocompatability of electrospun nanocomposites based on poly (epsilon-caprolactone) and carbon nanostructures. *Int. J. Artif. Organs* **33**, 271–282.
14. Sarraf, C.E., Harris, A.B., McCulloch, A.D. and Eastwood M (2002) Tissue engineering of cardiovascular system surrogates. *Heart Lung Circ.* **11**, 142–150.
15. Sun, L-Y., Hsieh, D-K., Syu, W-S., Li, Y-S., Chiu, H-T., Chiou, T-W. (2010) Cell proliferation of human bone marrow mesenchymal stem cells on biodegradable microcarriers enhances *in vitro* differentiation potential. *Cell Proliferation* **43**(5), 445–456.
16. Smith, A.G., Heath, J.K., Donaldson, D.D., Wong, G.G. Moreau, J. et al. (1988) Inhibition of pluripotential embryonic stem cell differentiation by purified polypeptides. *Nature* **15**, 688–690.
17. Dick, E., Rajamohan, D., Ronksley, J. and Denning, C. (2010) Evaluating the utility of cardiomyocytes from human pluripotent stem cells for drug screening. *Biochem. Soc. Trans.* **38**, 1037–1045.
18. Carrasco de Paula, I. (2008) General principles regarding the use of adult stem cells. *Cell Prolif.* **41**(1), 78–84.
19. Gluckman, E. (2000) Ethical and legal aspects of placental/cord blood banking and transplant. Eurocord Netcord Organisation. *Hematol. J.* **1**, 67–69.
20. Friedenstein, A. J., Gorskaja, J.F. and Kulagina, N.N. (1976) Fibroblast precursors in normal and irradiated mouse hematopoietic organs. *Exp. Hematol.* **4**, 267–274.
21. von Drygalski, A. and Adamson, J.W. (2000) Placental/umbilical cord blood (PCB) stem cells for transplantation: early clinical outcomes and the status of ex vivo expansion strategies. *Keio J. Med.* **49**, 141–151.
22. Sasayama, N., Kashiwakura, I., Tokushima, Y., Wada, S., Murakami, M. Hayase, Y. et al. (2001) Expansion of megakaryocyte progenitors from cryopreserved leukocyte concentrates of human placental and umbilical cord blood in short-term liquid culture. *Cytotherapy* **3**, 117–126.
23. Mizuno, H. and Hyakusoku, H. (2001). Fat grafting to the breast and adipose-derived stem cells: recent scientific consensus and controversy. *Aesthet. Surg. J.* **30**, 387–389.
24. Thalmeier, K., Meissner, P., Moosmann, S., Sagebiel, S., Wiest, I., Huss, R. (2001) Mesenchymal differentiation and organ distribution of established human stromal cell lines in NOD/SCID mice. *Acta Haematol.***105**(3), 159–165.
25. Quirici, N., Scavullo, C., de Girolamo, L., Lopa, S., Arrigoni, E., Deliliers, G.L., Brini, A.T. (2010) Anti-L-NGFR and -CD34 monoclonal antibodies identify multipotent mesenchymal stem cells in human adipose tissue. *Stem Cells Dev.* **19**, 915–925.
26. Williams, R.L., Hilton, D.J., Pease, S., Willson, T.A., Stewart, C.L., Gearing, D.P., Wagner,

E.F., Metcalf, D., Nicola, N.A., Gough, N.M. (1988) Myeloid leukaemia inhibitory factor maintains the developmental potential of embryonic stem cells. *Nature.* **336**(6200), 684–687.

27. Lennon, D.P., Haynesworth, S.E., Young, R.G., Dennis, J.E., Caplan, A.I. (1995) A chemically defined medium supports in vitro proliferation and maintains the osteochondral potential of rat marrow-derived mesenchymal stem cells. *Exp. Cell Res.* **219**, 211–222.
28. Otto, W.R., Rao, J. (2004)Tomorrow's skeleton staff: mesenchymal stem cells and the repair of bone and cartilage. *Cell Prolif.* **37**, 97–110.
29. Jung, S., Sen, A., Rosenberg, L., Behiel, L.A. (2010) Identification of growth and attachment factors for the serum-free isolation and expansion of human mesenchymal stromal cells. *Cytotherapy* **12**, 637–657.
30. Lindner, U., Kramer, J., Behrends, J., Driller, B., Wendler, N.-O., Boehrnsen, F., Rohwedel, J., Schlenke, P. (2010) Improved proliferation and differentiation capacity of human mesenchymal stromal cells cultured with basement-membrane extracellular matrix proteins. *Cytotherapy* **12**(8), 992–1005.
31. Xu, W., Zhang, X., Qian, H., Zhu, W., Sun, X., Hu, J. et al. (2004) Mesenchymal stem cells from adult human bone marrow differentiate into a cardiomyocyte phenotype in vitro. *Exp. Biol. Med.* **229**, 623–631.
32. Takahashi, T., Lord, B., Schulze, P.C., Fryer, R.M., Sarang, S.S., Gullans, S.R., Lee, R.T. (2003) Ascorbic acid enhances differentiation of embryonic stem cells into cardiac myocytes. *Circulation* **107**, 1912–1916.
33. Sittichokechaiwut, A., Edwards, J.H., Scutt, A.M., Reilly, G.C. (2010) Short bouts of mechanical loading are as effective as dexamethasone at inducing matrix production by human bone marrow mesenchymal stem cell. *Eur. Cell Mater.* **20**, 45–57.
34. Anand, U., Otto, W.R., Casula, M.A., Day, N.C., Davis, J.B., Bountra, C., Birch, R., Anand, P. (2006) The effect of neurotrophic factors on morphology, TRPV1 expression and capsaicin responses of cultured human DRG sensory neurons. *Neurosci. Lett.* **399**, 51–56.
35. Lunn, G., Lawler, G. (2003) Unit 2A: Laboratory Safety, in Current Protocols in Protein Science, John Wiley & Sons, Chichester, UK. DOI: 10.1002/0471140864.psa02as28.

Additional Reading

IHCWorld: http://www.ihcworld.com/_protocols/em/em_routine_tissue.htm Routine transmission electron microscopy staining protocol for tissues: Reynold's Lead Citrate Solution. Visited 20.9.10.

Page KM (1982) Bone and the preparation of bone sections, in Bancroft, JD and Stevens A (Eds) Theory and practice of histological techniques. Churchill Livingstone, Edinburgh, pp 297–331.

INDEX

Ragai R. Mitry and Robin D. Hughes (eds.), *Human Cell Culture Protocols*, Methods in Molecular Biology, vol. 806, DOI 10.1007/978-1-61779-367-7, © Springer Science+Business Media, LLC 2012

C

D

H

I

R

S

T

U

V

MIX
Papier aus verantwortungsvollen Quellen
Paper from responsible sources
FSC® C105338

If you have any concerns about our products,
you can contact us on
ProductSafety@springernature.com

In case Publisher is established outside the EU,
the EU authorized representative is:
Springer Nature Customer Service Center GmbH
Europaplatz 3, 69115 Heidelberg, Germany

Printed by Libri Plureos GmbH
in Hamburg, Germany